# Betaine

## Chemistry, Analysis, Function and Effects

**Food and Nutritional Components in Focus**

*Series Editor:*
Professor Victor R Preedy, *School of Medicine, King's College London, UK*

*Titles in the Series:*
1: Vitamin A and Carotenoids: Chemistry, Analysis, Function and Effects
2: Caffeine: Chemistry, Analysis, Function and Effects
3: Dietary Sugars: Chemistry, Analysis, Function and Effects
4: B Vitamins and Folate: Chemistry, Analysis, Function and Effects
5: Isoflavones: Chemistry, Analysis, Function and Effects
6: Fluorine: Chemistry, Analysis, Function and Effects
7: Betaine: Chemistry, Analysis, Function and Effects

*How to obtain future titles on publication:*
A standing order plan is available for this series. A standing order will bring delivery of each new volume immediately on publication.

*For further information please contact:*
Book Sales Department, Royal Society of Chemistry, Thomas Graham House, Science Park, Milton Road, Cambridge, CB4 0WF, UK
Telephone: +44 (0)1223 420066, Fax: +44 (0)1223 420247
Email: booksales@rsc.org
Visit our website at www.rsc.org/books

# *Betaine*

## *Chemistry, Analysis, Function and Effects*

Edited by

**Victor R Preedy**
*School of Medicine, King's College London, UK*
*Email: editor@publicationeditor.org.uk*

THE QUEEN'S AWARDS
FOR ENTERPRISE:
INTERNATIONAL TRADE
2013

Food and Nutritional Components in Focus No. 7

Print ISBN: 978-1-84973-886-6
PDF eISBN: 978-1-78262-844-6
ISSN: 2045-1695

A catalogue record for this book is available from the British Library



Published by The Royal Society of Chemistry,
Thomas Graham House, Science Park, Milton Road,
Cambridge CB4 0WF, UK

Registered Charity Number 207890

For further information see our web site at www.rsc.org

Printed by CPI Group (UK) Ltd, Croydon, CR0 4Y

# *Preface*

Recently, there have been major advances in our understanding of the chemistry and function of nutritional components. This has been enhanced by rapid developments in analytical techniques and instrumentation. Chemists, food scientists and nutritionists are, however, separated by divergent skills, and professional disciplines. Hitherto, this transdisciplinary divide has been difficult to bridge.

The series *Food and Nutritional Components in Focus* aims to cover in a single volume the chemistry, analysis, function and effects of single components in the diet or its food matrix. Its aim is to link scientific disciplines so that information becomes more meaningful and applicable to health in general.

The series *Food and Nutritional Components in Focus* covers the latest knowledge base and has a structured format with major subsections covering.

- Compounds in context
- Chemistry
- Analysis
- Function and Effects

In some books the section on Chemistry is also linked with Biochemistry. Each chapter has a novel cohort of features namely by containing:

- *Summary Points*
- *Key Facts* (areas of focus explained for the lay person)
- *Definitions of Words and Terms*

The series covers numerous classes of dietary components including, for example, minerals, vitamins, food additives, and so on. The chapters are written by national or international experts, specialists and leaders in the field.

*Betaine* has the following coverage. Section 1 Betaine in Context has chapters on betaine in general, sugar beet, food plants, synthesis and content.

Food and Nutritional Components in Focus No. 7
Betaine: Chemistry, Analysis, Function and Effects
Edited by Victor R Preedy

Published by the Royal Society of Chemistry, www.rsc.org

In Section 2 Chemistry and Biochemistry there is coverage of the chemistry of betaine, chemical interactions, folate, biomolecules, uptake, the betaine/GABA transporter, and betaine-homocysteine s-methyltransferase. Section 3 Analysis has chapters on NMR spectroscopy, CE Tandem Mass Spectrometry, hydrophilic interaction liquid chromatography, characterization analysis of seeds and fruits. The final Section 4 Function and Effects has contributions covering metabolic stress, restraint stress, memory impairment, serum betaine, bariatric surgery, folate, pregnancy, alcohol- and nonalcohol-induced liver injury, DNA methylation, age-related inflammation, athletic performance, body composition, osmoprotection, ocular surfaces, choline and cancer.

*Betaine* is specifically designed for chemists, analytical scientists, forensic scientists, food scientists, dieticians, nutritionists, food scientists and research academics. The series is suitable for lecturers and teachers in food and nutritional sciences. Additionally, and importantly, the series will be a valuable resource for college or university libraries as a reference guide.

Professor Victor R Preedy
*King's College London*

# Contents

## Betaine in Context

**Chapter 1 Betaine in Context** 3
*Vinood B. Patel and Kosha Mehta*

Summary Points 5
Key Facts of Alcoholic Liver Disease 5
Definitions and Explanations of Key Terms 5
List of Abbreviations 6
References 6

**Chapter 2 Betaine in Sugar Beet** 9
*Teruhiro Takabe, Yoshito Tanaka, and Tetsuko Takabe*

2.1 Introduction 9
2.2 Physiologic Function of GB in Plant Cells 10
2.2.1 GB as Compatible Solute 10
2.2.2 GB as an Osmoprotectant 10
2.3 GB Accumulation in *Beta vulgaris* 11
2.3.1 Subcellular Localization of GB in Beet Root Cells 11
2.3.2 Distribution in Whole Plant 11
2.3.3 GB in Storage Root 12
2.3.4 GB in Leaves 13
2.4 Pathway and Sites of GB Biosynthesis in Higher Plants 14
2.4.1 Choline to GB 14
2.4.2 Ethanolamine to Choline 14

Food and Nutritional Components in Focus No. 7
Betaine: Chemistry, Analysis, Function and Effects
Edited by Victor R Preedy

Published by the Royal Society of Chemistry, www.rsc.org

2.4.3 Serine to Ethanolamine 16
2.4.4 Enzymes Involved in Regeneration Cycle of *S*-Adenosylmethionine 16
2.4.5 Regulation of Gene Expression 16
2.5 Translocation of GB in *Beta vulgaris* 17
2.6 Utilization of GB from Sugar Beet 18
2.6.1 Exogenous Application for Crop Production 18
2.6.2 Protective Effects and Risks of GB Application 20
2.7 Genetic Engineering of GB Biosynthesis in Plants 21
Summary Points 24
Key Facts for Betaine in Sugar Beet 24
Definitions of Words and Terms 25
List of Abbreviations 26
References 26

**Chapter 3 Betaine in Food Plants: A Focus on Synthesis and Content 29**
*Rosario A. Muñoz-Clares*

3.1 Introduction 29
3.2 Glycine Betaine is an Important Osmoprotectant in Plants 30
3.3 Biosynthesis of Glycine Betaine in Plants 31
3.3.1 Betaine Aldehyde Dehydrogenase (BADH) 33
3.3.2 Choline Mono-Oxygenase (CMO) 36
3.3.3 Synthesis and Transport of Choline 37
3.4 Concluding Remarks 38
Summary Points 39
Key Facts 39
Key Facts of Enzyme Catalysis 39
Definition of Words and Terms 40
Abbreviations 41
Acknowledgements 41
References 41

## Chemistry and Biochemistry

**Chapter 4 The Chemistry of Betaine 47**
*Panteleimon G. Takis, Konstantinos D. Papavasileiou, Anastassios N. Troganis, and Vasilios S. Melissas*

4.1 Introduction 47
4.2 Nomenclature and Generic Properties 48
4.3 Physical and Chemical Properties – Chemistry of Important Functions 49

4.4 Chemistry of Important Functions 49
4.5 Spectral Data 54
4.6 Crystal Structure 55
4.7 Handling and Toxicity 56
Summary Points 56
Key Facts for Betaines and Some of Its Derivatives Function and Utility 56
Definitions of Words and Terms 57
List of Abbreviations 58
References 59

**Chapter 5 Chemical Interactions of Betaine with Folate and Other Biomolecules 62**
*Michael R. Duff Jr, and Elizabeth E. Howell*

5.1 Introduction to Osmolytes and Models of Osmolyte Action 62
5.2 Osmolytes Function as Chemical Chaperones 63
5.3 Preferential Interactions of Betaine with Macromolecules 64
5.4 Betaine Association with Folate and Its Derivatives 65
5.4.1 Do These *In Vitro* Observations Translate into *In Vivo* Effects? 69
5.5 Cation–π Interactions Provide a Handle to Bind Betaine 70
5.6 Conclusions 71
Summary Points 71
Definitions and Explanation of Key Terms 72
Abbreviations 73
References 73

**Chapter 6 Betaine Uptake with a Focus on the Betaine/GABA Transporter 76**
*Y. Zhou, S.A. Kempson, and N.C. Danbolt*

6.1 Introduction 76
6.2 Betaine Acts as a Compatible Osmolyte and as a Methyl Donor 77
6.3 Identification of the Betaine-GABA Transporter (BGT1; slc6a12) 77
6.4 BGT1 is Expressed in the Brain, Kidney and Liver 78
6.4.1 The Roles of BGT1 in the Kidney 78
6.4.2 The Role of BGT1 in the Liver 80
6.4.3 The Role of BGT1 in the Brain 82
6.4.4 The Role of Betaine in the Testis 84

Summary Points 84
Definitions of Words and Terms 84
List of Abbreviations 85
Acknowledgements 85
References 85

**Chapter 7 Betaine-Homocysteine S-Methyltransferase 90**
*Timothy A. Garrow*

7.1 Introduction 90
7.2 Early Studies of Bet-Dependent Methyltransferase Activity 92
7.3 Studies Using Recombinant BHMT Enzymes 93
7.4 BHMT Structure and Characterization of Ligand Binding 94
Summary Points 98
Key Facts 99
Definitions of Words and Terms 99
List of Abbreviations 99
References 100

## Analysis

**Chapter 8 Assays of Betaines using $^1$H NMR Spectroscopy 105**
*Panteleimon G. Takis, Konstantinos D. Papavasileiou, Anastassios N. Troganis, and Vasilios S. Melissas*

8.1 Introduction 105
8.2 $^1$H-NMR Detection and Quantification of Betaines in Biofluids 107
8.2.1 Selected Glycine Betaine Assays by $^1$H-NMR 109
8.2.2 Selected Other Betaines' and *N*-Methylated Compounds Assays by $^1$H-NMR 111
Summary Points 113
Key Facts for NMR Contribution to Metabolites Function Revelation 113
Definitions of Words and Terms 114
List of Abbreviations 115
References 116

**Chapter 9 Determination of Betaines in Vegetable Oils by Capillary Electrophoresis-Ion Trap Mass Spectrometry 120**
*Laura Sánchez-Hernández, Maria Luisa Marina Alegre, and Antonio L. Crego Navazo*

9.1 Introduction 120

9.2 Capillary Electrophoresis Coupled to Mass Spectrometry for the Determination of Betaines 122
9.2.1 Derivatization of Betaines 123
9.2.2 Separation of Betaines by CE/MS 124
9.2.3 Validation of the CE/MS$^2$ Method 126
9.3 Application to the Analysis of Vegetable Oils 129
9.3.1 Sample Preparation 130
9.3.2 Determination of Betaines in Vegetable Oils 130
Summary Points 133
Key Facts 133
Key Facts for Betaines 133
Key Facts for the Extraction Procedure 134
Key Facts for the Separation Technique 134
Definitions of Words and Terms 134
List of Abbreviations 135
Acknowledgements 136
References 136

**Chapter 10 Hydrophilic Interaction Liquid Chromatography for Determination of Betaine 139**
*Jonathan M. Curtis and Si Mi*

10.1 Introduction 139
10.1.1 Basics of HILIC 139
10.1.2 Summary and Comparison to Other Analytical Techniques Used for Analysis of Betaine 143
10.2 The Use of HILIC in the Determination of Betaine Concentrations 144
10.2.1 Betaine in Standard Mixtures 144
10.2.2 Betaine in Samples of Plant Origin 147
10.2.3 Betaine in Samples of Animal Origin 148
10.2.4 Betaine in Food Samples 150
10.3 Conclusions and Future Perspectives 152
Summary Points 152
Key Facts of Betaine Analysis by HILIC 153
Definitions of Words and Terms 153
List of Abbreviations 154
References 155

**Chapter 11 Characterization of Betaine from Seeds 159**
*Priti Patil and Pratima Tatke*

11.1 Introduction 159

11.2 Isolation of Betaine from *Achyranthes aspera* seeds 162
11.2.1 Procurement of Plant Material 162
11.2.2 Extraction of Alkaloids from *Achyranthes aspera* 163
11.2.3 Isolation of Alkaloids from Seeds of *Achyranthes aspera* 163
11.2.4 Thin-Layer Chromatographic Studies 164
11.2.5 Purification of Alkaloid 164
11.2.6 Characterization of Alkaloid as Betaine 165
11.2.7 Analysis of Betaine 165
11.3 Isolation of Betaine from Kancolla Seeds 168
11.3.1 Extraction and Isolation 168
11.3.2 Characterization of Betaine Analogs 168
11.3.3 Betaine Analysis 168
11.4 Isolation of L-(+)-Isoleucine Betaine from *Cannabis* Seeds 170
11.4.1 Extraction and Purification of Isoleucine Betaine 170
11.4.2 Characterization of Isoleucine Betaine 170
11.4.3 Analysis of Isoleucine Betaine 171
11.5 Conclusion 171
Summary Points 172
Key Facts 172
Key Facts of $^1$H Correlation Spectroscopy (COSY) 172
Key Facts of Distortionless Enhancement by Polarization Transfer (DEPT NMR) 173
Key Facts of Electrospray Ionization-Mass Spectroscopy (ESI-MS) 173
Definition of Words and Terms 173
List of Abbreviations 175
References 175

**Chapter 12 Occurrence and Analysis of Betaines in Fruits** 178
*Luigi Servillo, Alfonso Giovane, Rosario Casale, Nunzia D'Onofrio, Giovanna Ferrari, Domenico Cautela, Domenico Castaldo, and Maria Luisa Balestrieri*

12.1 Introduction 178
12.2 Biosynthesis of Betaines and QACs 180
12.3 Analytical Methods for Determination of Betaines and QACs 184
12.4 Identification and Quantification of Betaines and QACs by ESI-Tandem Mass Spectrometry 187

12.5 Betaines and QACs Distribution in Fruits 190
12.5.1 Glycine Betaine 191
12.5.2 Betaines and QAC Deriving from Proline 192
12.5.3 γ-Butyrobetaine 192
12.5.4 Pipecolic Acid Betaine and β-Alanine Betaine 192
12.5.5 Trigonelline and Choline 192
12.5.6 $N^{\varepsilon}$-trimethyllysine 193
12.5.7 Hercynine and Ergothioneine 193
12.6 Conclusions 194
Summary points 194
Key Facts of Betaines 195
Definitions and Explanations of Key Terms 195
List of Abbreviations 196
References 197

## Function and Effects

**Chapter 13 Betaine and Metabolic Stress in Different Tissues 203**
*Kaisa Olli and Kirsti Tiihonen*

13.1 Introduction 203
13.1.1 Physicochemical Properties of Betaine 203
13.1.2 Digestion, Absorption and Distribution 204
13.2 Betaine and Methylation in the Liver 206
13.2.1 Methylation Functions 206
13.2.2 Betaine and Liver Fat Metabolism 207
13.2.3 Liver Macrophages and the Immunological Effects of Betaine 208
13.3 Betaine and Osmoregulation in the Kidney 209
13.3.1 Osmoregulation 209
13.3.2 The Role of Betaine in the Kidney 209
13.4 Betaine in Epithelial Cells 211
13.4.1 Betaine in the Gut Epithelium 211
13.4.2 Betaine in Skin and Oral Mucosa 211
13.5 Betaine in Muscle Cells 212
13.5.1 Betaine and Lean Weight Gain 212
13.5.2 Betaine and Carnitine Metabolism in Muscle 212
13.6 Betaine and Fat Cells 212
13.6.1 Betaine and Hypoxic Stress in Adipose Tissue 212
13.6.2 Betaine and a High-Fat Diet 214
13.6.3 Betaine and Fat Metabolism 214

13.7 Concluding Remarks 214
Summary Points 215
Key Facts 215
Definitions of Words and Terms 216
List of Abbreviations 217
Acknowledgements 217
References 218

**Chapter 14 Beneficial Effects of Betaine on Water-Immersion Restraint Stress-Induced Memory Impairment in Mice** **222**
*Masayuki Hiramatsu, Yukari Takahashi, Anna Amano, Kiwamu Kido, Takuya Matsukura, Toru Yamanashi, Yusuke Kakehi, Toshitaka Nabeshima, and Masaya Miwa*

14.1 Introduction 222
14.2 Methods 224
14.2.1 Animals 224
14.2.2 Drugs 224
14.2.3 Experimental Schedules 225
14.2.4 Water-Immersion Restraint Stress (WIRS) 225
14.2.5 Spontaneous Alternation Behavior in Y-Maze Test 225
14.2.6 Step-Down-Type Passive Avoidance Test 226
14.2.7 Responses to Electric Shock 226
14.2.8 Corticosterone Assay 226
14.2.9 Real-Time RT-PCR 227
14.2.10 Determination of Lipid Peroxidation Level 228
14.2.11 Protein Assay 228
14.2.12 Data Analysis 228
14.3 Results 229
14.3.1 Onset of Memory Impairment Induced by Water-Immersion Restraint Stress 229
14.3.2 Effects of Repeated Administration of Betaine on Learning and/or Memory Impairment after Water-Immersion Restraint Stress 230
14.3.3 Effects of Subacute Administration of Betaine or Catechin on Learning and/or Memory Impairment after Water-Immersion Restraint Stress 231
14.3.4 Effects of Acute Administration of Betaine on Learning and/or Memory Impairment after Water-Immersion Restraint Stress 231

14.3.5 Effects of Acute Administration of Betaine on Plasma Corticosterone Level after Water-Immersion Restraint Stress 231
14.3.6 Effects of Acute Administration of Betaine on Interleukin-1β mRNA Level in the Hippocampus and Frontal Cortex after Water-Immersion Restraint Stress 232
14.3.7 Effects of Acute Administration of Betaine on MDA Level in the Hippocampus after Water-Immersion Restraint Stress 234
14.3.8 Effects of Acute Administration of Betaine on the Expression of GABA Transporter mRNA Levels in the Hippocampus after Water-Immersion Restraint Stress 234
14.4 Discussion 235
14.5 Conclusions 238
Competing Interests 238
Authors' Contributions 238
Summary Points 238
Key Facts 239
Key Facts of Delayed Neurodegeneration 239
Definitions of Words and Terms 239
Abbreviations 240
Acknowledgements 241
References 241

**Chapter 15 Serum Betaine Concentrations and the Effects of Bariatric Surgery** **245**
*Julian Swierczynski, Areta Hebanowska, and Tomasz Sledzinski*

15.1 Introduction 245
15.2 The Role of Betaine in Pathological Conditions 249
15.3 Serum Betaine and Dimethylglycine Concentrations in Health and Disease 250
15.4 Human Plasma Betaine Concentration after Betaine Supplementation 254
15.5 Betaine and Body Composition 255
15.6 Serum Betaine Concentrations in Obese Patients after Bariatric Surgery 257
15.7 Directions of Future Research 259
Summary Points 260
Definitions of Words and Terms 260
Key Facts of Bariatric Surgery 261

List of Abbreviations 262
Acknowledgements 262
References 262

**Chapter 16 Folate and Betaine Interactions during Pregnancy 267**
*Michelle M. Murphy*

16.1 One-Carbon Metabolism 267
16.1.1 Folate and Betaine and One-Carbon Metabolism 267
16.2 Early Development 270
16.2.1 Betaine, BHMT, Embryo and Fetal Development 270
16.3 Pregnancy and Fetal Development 271
16.3.1 One-Carbon Metabolism, Pregnancy and Fetal Development 271
Summary Points 278
Key Facts 279
Definitions of Words and Terms 279
Abbreviations 280
References 280

**Chapter 17 Multiple Roles of Betaine in Protecting Against Alcohol-Induced Liver Injury 285**
*Paul G. Thomes, Sarah M. Bligh, and Kusum K. Kharbanda*

17.1 Introduction 285
17.2 Mechanisms of Alcohol-Induced Liver Injury 286
17.3 Methionine Metabolism 286
17.4 Effect of Ethanol Consumption on Methionine Metabolism 288
17.5 Functional Consequences of Ethanol-Induced Alteration in Methionine Metabolism 288
17.5.1 Defects in Crucial Methylation Reactions 289
17.5.2 Homocysteine Toxicity 289
17.6 Protective Role of Betaine in Alcoholic Liver Injury 290
17.6.1 Betaine Attenuates Alcohol-Induced Steatosis 291
17.6.2 Betaine Attenuates Alcohol-Induced Apoptosis 295
17.6.3 Betaine Prevents Accumulation of Ethanol-Induced Damaged Proteins 295
17.6.4 Betaine Prevents Impairments in PRMT-Catalyzed Reactions 297
17.6.5 Betaine Affects Histone Methylation and Mallory–Denk Body Formation 297

17.6.6 Betaine Prevents Blood-Alcohol Cycling 298
17.6.7 Betaine Treatment Attenuates Alcohol-Induced Alterations to the Mitochondrial Respiratory Chain Proteome 298
17.6.8 Betaine Treatment Attenuates Oxidative Stress and Inflammation 298
17.7 Future Directions 299
Summary Points 300
Key Facts of Alcohol and Liver Injury 300
Definition and Explanations of Key Terms Used 301
List of Abbreviations 301
Acknowledgements 302
References 302

**Chapter 18 Betaine, DNA Methylation and Nonalcoholic Fatty Liver Disease 311**
*Zhu Huilian and Wang Lijun*

18.1 Epidemiologic Character and Nutritional Risk Factors of NAFLD 311
18.2 Experimental and Clinical Evidence for Betaine Treatment of NAFLD 313
18.3 The Underlying Genetic Role of Betaine in the Treatment of NAFLD 318
18.4 Regulation of DNA Methylation by Betaine in the Treatment of NAFLD 319
18.4.1 Epigenetic Profile of NAFLD 319
18.4.2 DNA Methylation Connects Methyl Donors and NAFLD 319
Summary Points 322
Key Facts of Nutriepigenomics 322
Definitions of Words and Terms 323
List of Abbreviations 324
References 325

**Chapter 19 The Beneficial Role of Dietary Betaine in the Suppression of Age-Related Inflammatory NF-κB 329**
*Hae Young Chung, Eun Kyeong Lee, and Byung Pal Yu*

19.1 Introduction 329
19.2 Modulation of NF-κB by Betaine in the Aging Process 330
19.2.1 Betaine as a Redox Mediator 330
19.2.2 Sensitivity of NF-κB to Redox Changes 332
19.2.3 Changes in NF-κB by Thiol Status 333
19.2.4 Activation of NF-κB during Aging 335

19.3 Protective Roles of Betaine Against Age-Related Changes 335
19.3.1 Attenuation of Oxidative Stress-Induced NF-κB Pathway in the Aged Kidney 335
19.3.2 Suppression of Adhesion Molecule Expression 336
19.3.3 Enhancement of Thiol Status in Attenuating Age-Related NF-κB 337
19.4 Summary 337
Summary Points 338
Key Facts in the Development of Betaine as an Antiaging Agent 339
Definition of Words and Terms 339
List of Abbreviations 340
Acknowledgements 341
References 341

**Chapter 20 Betaine in Athletic Performance and Body Composition 344**
*Stuart A. S. Craig*

20.1 Introduction 344
20.2 Human Performance Studies 345
20.2.1 Acute Dose Studies 345
20.2.2 Chronic Dose Studies 347
20.3 Mechanism Studies 355
20.3.1 In Humans 355
20.3.2 In Animals 357
20.3.3 *In Vitro* 359
20.4 Conclusions 361
Summary Points 362
Key Facts 362
Key Facts on Betaine as a Nutrient 362
Definitions of Words and Terms 363
List of Abbreviations 364
References 364

**Chapter 21 Betaine, Osmoprotection, and Health of the Ocular Surface in Dry-Eye Disease 368**
*Qian Garrett and Mark D. P. Willcox*

21.1 The Ocular Surface and Tear Film 368
21.2 Dry-Eye Disease and Tear-Film Hyperosmolarity 369
21.2.1 Hyperosmolarity, Inflammation and Apoptosis 370
21.2.2 Osmoprotection – Organic Osmolytes 371

21.3 Betaine and Ocular Osmoregulation 371
21.3.1 Betaine Presence in Human Tears 372
21.3.2 Ocular Betaine Transport 372
21.3.3 Betaine and Hyperosmotic Stress-Induced Cell Shrinkage 374
21.3.4 Betaine and Hyperosmotic Stress-Induced Cell Apoptosis 375
21.3.5 Betaine in Prevention and Treatment of Animal Dry Eye 380
21.4 Summary 380
Summary Points 381
Key Facts 381
Definitions of Words and Terms 383
List of Abbreviations 384
Acknowledgements 384
References 384

**Chapter 22 Betaine and Choline and Cancer** **390**
*Jung Eun Lee and Eunyoung Cho*

22.1 Introduction 390
22.2 Assessment of Betaine and Choline Intake by Food Frequency Questionnaire (FFQ) in Epidemiological Studies 391
22.3 Assessment of Circulating Levels of Betaine and Choline in Epidemiological Studies 392
22.4 Intake Distribution and Food Sources of Betaine and Choline 392
22.5 Case-Control Studies of Dietary Betaine and Choline Intake and Cancer Risk 393
22.6 Cohort Studies of Dietary Betaine and Choline Intake and Cancer Risk 398
22.7 Circulating Levels of Betaine and Choline and Cancer Risk 405
22.8 Mortality Studies of Dietary Betaine and Choline Intake among Cancer Patients 405
22.9 Summary and Conclusion 411
Summary Points 411
Key Facts 412
Key Facts of Betaine and Choline 412
Key Facts of Measures in Epidemiologic Studies 412
Definitions of Words and Terms 412
Abbreviations 413
Acknowledgements 413
References 413

**Subject Index** **418**

# Section I
# Betaine in Context

CHAPTER 1

# *Betaine in Context*

VINOOD B. PATEL*[a] AND KOSHA MEHTA[a]

[a]Department of Biomedical Science, Faculty of Science & Technology, University of Westminster, 115 New Cavendish Street, London, W1W 6UW, UK
*E-mail: v.b.patel@westminster.ac.uk

Betaine, also referred to as trimethyglycine, oxyneurine and glycine-betaine is a naturally occurring human nutrient that was first discovered in sugar beets and was later found in several micro-organisms, marine invertebrates, plants and animals. Chemically, it is a neutral methyl derivative of glycine with a positively charged tri-methylammonium group and a negatively charged carboxyl group (Figure 1.1), a specific type of zwitterion that performs methylation in addition to osmoregulation (Craig, 2004).

Humans can obtain their daily intake of betaine (1–2.5 g) from exogenous sources such as wheat bread (201 mg per 100 g), beets (114–297 mg per 100 g), spinach (600–645 mg per 100 g) and wheat bran (1339 mg per 100 g) or it can be synthesized endogenously from its metabolic precursor choline (Craig, 2004). Dietary betaine is absorbed from the duodenal enterocytes into circulation and maintained between 20 and 70 $\mu$mol $L^{-1}$ with a median plasma concentration of 27.8 $\mu$mol $L^{-1}$, although being slightly higher in men than in women (Lever *et al.*, 1994; Awwad *et al.*, 2014). It is carried to the liver and kidneys where it is catabolized by a series of enzyme-catalyzed reactions, predominantly in the mitochondria, and participates in the methionine cycle. Here, betaine donates a methyl group to homocysteine to form *N,N*-dimethylglycine and L-methionine *via* the enzyme betaine-homocysteine methyltransferase (BHMT). This is a zinc thiol-enzyme expressed mainly in the liver and kidneys that also

Food and Nutritional Components in Focus No. 7
Betaine: Chemistry, Analysis, Function and Effects
Edited by Victor R Preedy

Published by the Royal Society of Chemistry, www.rsc.org

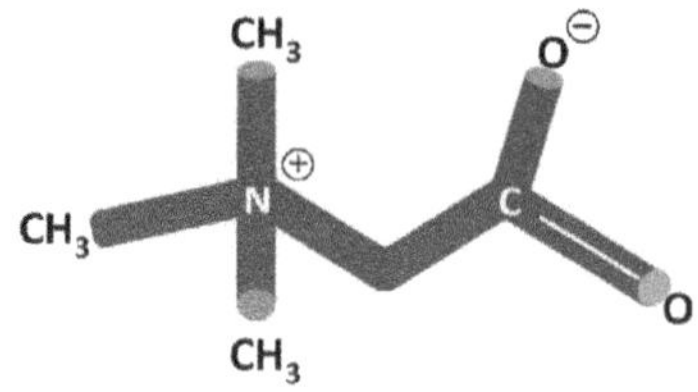


**Figure 1.1** The structure of betaine. Structure of glycine-betaine, also known as trimethyglycine and oxyneurine.

functions in glycine, serine and threonine metabolisms (Pajares and Perez-*Sala*, 2006). The transmethylation reaction mediated by BHMT not only helps to detoxify homocysteine but also increases serum methionine levels (Storch *et al.*, 1991) and *S*-adenosylmethionine (SAM) (Barak *et al.*, 1993), which itself serves as a methyl donor in DNA methylation and many anabolic pathways of phospholipids, hormones and proteins. Removal of betaine from the body is primarily by metabolism with minimal urinary excretion, even after high amounts of betaine consumption (Lever and Slow, 2010).

Uncatabolized betaine acts as an osmolyte and confers protection to the cells against environmental stresses like osmotic irregularity, adverse temperatures and dehydration. By regulating the surface tension of water, it aids in water retention and thereby maintains cellular volume. Some examples of its osmoregulatory performances are the hydration of albumin, where it forms a single layer of water around albumin (Courtenay *et al.*, 2000), hydration and transport of molecules across the intestine (Kettunen *et al.*, 2001) and sustenance of haemoglobin solvation (Hundahl *et al.*, 2003). Additionally, while betaine can protect renal cells from high concentrations of electrolytes (Horio *et al.*, 2001) it also protects myosin in the skeletal muscle from urea-induced structural changes (Ortiz-Costa *et al.*, 2002) Thus, betaine helps to preserve the optimal functions of several organs and cell types. Interestingly, hepatic and renal concentrations of betaine and its availability as an osmoprotectant are controlled by the regulation of BHMT, which in turn, is dependent on tonicity (Pajares and Perez-*Sala*, 2006).

Since betaine provides methyl groups, lack of betaine in the diet can cause hypomethylation and elevated homocysteine and reduced SAM concentrations resulting in perturbed methionine metabolism. In addition, a betaine-deficient diet can lead to liver steatosis (Craig, 2004) and increase the predisposition to stroke, cardiovascular disorders, Alzheimer's (Kittner *et al.*, 1999; Finkelstein, 2000; Seshadri *et al.*, 2002) and atypical DNA methylation leading to carcinogenesis (Cooney, 1993). A betaine- or choline-rich diet can increase the plasma concentrations of methionine, and thereby the rate of methylation and consequently lower the risk of various neoplasms (Nitter *et al.*, 2014; Zeng *et al.*, 2014) and also decrease homocysteine levels in patients with inherited disorders homocysteinemia and homocystinuria (Dudman *et al.*, 1993; Tangerman *et al.*, 2000). Alongside, several animal studies have been conducted that present betaine as a promising therapeutic agent.

Based on these studies, betaine-mediated restoration of normal DNA methylation of specific genes to ameliorate the pathogenesis of nonalcoholic fatty liver disease and alcoholic liver disease has been envisaged (Dou *et al.*, 2014; Wang *et al.*, 2014). In alcohol-fed animals betaine prevented and partially reversed alcoholic liver damage by increasing SAM levels and decreasing liver injury caused by endoplasmic reticulum stress (Barak *et al.*, 1997).

## Summary Points

- Betaine helps to maintain the normal physiological functions of vital organs like the heart, liver, brain and kidneys.
- It acts as an osmolyte and as a methyl donor, whereby it plays an important role in the methionine cycle.
- Lack of betaine in the diet can result in hypomethylation, elevated homocysteine and reduced SAM concentrations.
- This could cause disturbances in numerous biochemical pathways and lead to diverse physiological implications.
- Animal studies and preliminary studies with human subjects suggest that betaine supplementation can attenuate the pathogenesis of several diseases.
- Further investigation is required to elucidate the mechanism of action of betaine, along with clinical trials to determine suitable pharmaceutical dosage and the consequences of prolonged supplementation.

## Key Facts of Alcoholic Liver Disease

- It is caused due to excessive alcohol consumption.
- It is manifested as liver damage and can show a wide variety of symptoms.
- The disease encompasses several stages of liver damage beginning from increased deposition of fat, followed by inflammation and excessive collagen production, subsequently leading to severe liver scarring.
- Alcohol abstinence during the early stages can prevent the progression of the disease.

## Definitions and Explanations of Key Terms

- DNA methylation – Addition of methyl group(s) to adenine or cytosine residues on the DNA. It is one of the ways in which gene expression is regulated.
- Haemoglobin solvation – Refers to the solubility of haemoglobin in an aqueous environment; here in the plasma.
- Homocystinuria and homocysteinemia – Genetic disorders of methionine metabolism that elevate homocysteine levels in blood.
- Hypomethylation – Reduction in methylation.
- Liver steatosis – Deposition of high amounts of fats in the liver.

- Methylation – Addition of methyl group(s).
- Osmoprotectant – A molecule that protects the cells/organisms from osmotic pressures.
- Thiol-enzyme – An enzyme that contains sulfhydryl group, *i.e.* the –SH group
- Transmethylation – Transfer of methyl group from one molecule to another.
- Zwitterion – A molecule that has both positive and negative charge(s) and is therefore electrically neutral.

## List of Abbreviations

BHMT Betaine-homocysteine methyltransferase
SAM *S*-Adenosylmethionine

## References

Awwad, H. M., Kirsch, S. H., Geisel, J. and Obeid, R., 2014. Measurement of concentrations of whole blood levels of choline, betaine, and dimethylglycine and their relations to plasma levels. *Journal of Chromatography B: Analytical Technologies in the Biomedical and Life Sciences*. 957: 41–45.

Barak, A. J., Beckenhauer, H. C., Badakhsh, S. and Tuma, D. J., 1997. The effect of betaine in reversing alcoholic steatosis. *Alcoholism: Clinical and Experimental Research*. 21: 1100–1102.

Barak, A. J., Beckenhauer, H. C., Junnila, M. and Tuma, D. J., 1993. Dietary betaine promotes generation of hepatic S-adenosylmethionine and protects the liver from ethanol-induced fatty infiltration. *Alcoholism: Clinical and Experimental Research*. 17: 552–555.

Cooney, C. A., 1993. Are somatic cells inherently deficient in methylation metabolism? A proposed mechanism for DNA methylation loss, senescence and aging. *Growth, Development and Aging*. 57: 261–273.

Courtenay, E. S., Capp, M. W., Anderson, C. F. and Record, M. T., Jr., 2000. Vapor pressure osmometry studies of osmolyte-protein interactions: implications for the action of osmoprotectants *in vivo* and for the interpretation of "osmotic stress" experiments *in vitro*. *Biochemistry*. 39: 4455–4471.

Craig, S. A., 2004. Betaine in human nutrition. *American Journal of Clinical Nutrition*. 80: 539–549.

Dou, X., Xia, Y., Chen, J., Qian, Y., Li, S., Zhang, X. and Song, Z., 2014. Rectification of impaired adipose tissue methylation status and lipolytic response contributes to hepatoprotective effect of betaine supplementation in a mouse model of alcoholic liver disease. *British Journal of Pharmacology*. 171: 4073–4086.

Dudman, N. P., Wilcken, D. E., Wang, J., Lynch, J. F., Macey, D. and Lundberg, P., 1993. Disordered methionine/homocysteine metabolism in premature vascular disease. Its occurrence, cofactor therapy, and enzymology. *Arteriosclerosis and Thrombosis*. 13: 1253–1260.

Finkelstein, J. D., 2000. Homocysteine: a history in progress. *Nutrition Reviews*. 58: 193–204.

Horio, M., Ito, A., Matsuoka, Y., Moriyama, T., Orita, Y., Takenaka, M. and Imai, E., 2001. Apoptosis induced by hypertonicity in Madin Darley canine kidney cells: protective effect of betaine. *Nephrology, Dialysis, Transplantation*. 16: 483–490.

Hundahl, C., Fago, A., Malte, H. and Weber, R. E., 2003. Allosteric effect of water in fish and human hemoglobins. *Journal of Biological Chemistry*. 278: 42769–42773.

Kettunen, H., Peuranen, S. and Tiihonen, K., 2001. Betaine aids in the osmoregulation of duodenal epithelium of broiler chicks, and affects the movement of water across the small intestinal epithelium in vitro. *Comparative Biochemistry and Physiology, Part A: Molecular and Integrative Physiology*. 129: 595–603.

Kittner, S. J., Giles, W. H., Macko, R. F., Hebel, J. R., Wozniak, M. A., Wityk, R. J., Stolley, P. D., Stern, B. J., Sloan, M. A., Sherwin, R., Price, T. R., Mccarter, R. J., Johnson, C. J., Earley, C. J., Buchholz, D. W. and Malinow, M. R., 1999. Homocyst(e)ine and risk of cerebral infarction in a biracial population: the stroke prevention in young women study. *Stroke*. 30: 1554–1560.

Lever, M., Sizeland, P. C., Bason, L. M., Hayman, C. M. and Chambers, S. T., 1994. Glycine betaine and proline betaine in human blood and urine. *Biochimica et Biophysica Acta*. 1200: 259–264.

Lever, M. and Slow, S., 2010. The clinical significance of betaine, an osmolyte with a key role in methyl group metabolism. *Clinical Biochemistry*. 43: 732–744.

Nitter, M., Norgard, B., De Vogel, S., Eussen, S. J., Meyer, K., Ulvik, A., Ueland, P. M., Nygard, O., Vollset, S. E., Bjorge, T., Tjonneland, A., Hansen, L., Boutron-Ruault, M., Racine, A., Cottet, V., Kaaks, R., Kuhn, T., Trichopoulou, A., Bamia, C., Naska, A., Grioni, S., Palli, D., Panico, S., Tumino, R., Vineis, P., Bueno-De-Mesquita, H. B., Van Kranen, H., Peeters, P. H., Weiderpass, E., Dorronsoro, M., Jakszyn, P., Sanchez, M., Arguelles, M., Huerta, J. M., Barricarte, A., Johansson, M., Ljuslinder, I., Khaw, K., Wareham, N., Freisling, H., Duarte-Salles, T., Stepien, M., Gunter, M. J. and Riboli, E., 2014. Plasma methionine, choline, betaine, and dimethylglycine, in relation to colorectal cancer risk in the European prospective investigation into cancer and nutrition (Epic). *Annals of Oncology*. 25: 1609–1615.

Ortiz-Costa, S., Sorenson, M. M. and Sola-Penna, M., 2002. Counteracting effects of urea and methylamines in function and structure of skeletal muscle myosin. *Archives of Biochemistry and Biophysics*. 408: 272–278.

Pajares, M. A. and Perez-Sala, D., 2006. Betaine homocysteine S-methyltransferase: just a regulator of homocysteine metabolism? *Cellular and Molecular Life Sciences*. 63: 2792–2803.

Seshadri, S., Beiser, A., Selhub, J., Jacques, P. F., Rosenberg, I. H., D'agostino, R. B., Wilson, P. W. and Wolf, P. A., 2002. Plasma homocysteine as a risk factor for dementia and Alzheimer's disease. *New England Journal of Medicine*. 346: 476–483.

Storch, K. J., Wagner, D. A. and Young, V. R., 1991. Methionine kinetics in adult men: effects of dietary betaine on l-[$^{2}H_3$-methyl-1-$^{13}C$]methionine. *American Journal of Clinical Nutrition*. 54: 386–394.

Tangerman, A., Wilcken, B., Levy, H. L., Boers, G. H. and Mudd, S. H., 2000. Methionine transamination in patients with homocystinuria due to cystathionine beta-synthase deficiency. *Metabolism*. 49: 1071–1077.

Wang, L. J., Zhang, H. W., Zhou, J. Y., Liu, Y., Yang, Y., Chen, X. L., Zhu, C. H., Zheng, R. D., Ling, W. H. and Zhu, H. L., 2014. Betaine attenuates hepatic steatosis by reducing methylation of the MTTP promoter and elevating genomic methylation in mice fed a high-fat diet. *Journal of Nutritional Biochemistry*. 25: 329–336.

Zeng, F. F., Xu, C. H., Liu, Y. T., Fan, Y. Y., Lin, X. L., Lu, Y. K., Zhang, C. X. and Chen, Y. M., 2014. Choline and betaine intakes are associated with reduced risk of nasopharyngeal carcinoma in adults: a case-control study. *British Journal of Cancer*. 110: 808–816.

CHAPTER 2

# *Betaine in Sugar Beet*

TERUHIRO TAKABE*[a], YOSHITO TANAKA[b], AND TETSUKO TAKABE[a]

[a]Research Institute, Meijo University, 1-501 Shiogamaguchi, Tenpaku-ku, Nagoya, Aichi 468-8502, Japan; [b]Graduate School of Environmental and Human Sciences, Meijo University, 1-501 Shiogamaguchi, Tenpaku-ku, Nagoya, Aichi 468-8502, Japan
*E-mail: takabe@meijo-u.ac.jp

## 2.1 Introduction

In response to water stresses, many plants, animals and bacteria accumulate highly soluble, low molecular weight compounds called osmolytes that allow the organism to take up and retain their cellular water. Glycine betaine (GB) is a typical osmolyte in the organisms living in sever environmental conditions. GB, trimethylglycine, is a zwitterionic quaternary ammonium compound with a molecular weight of 117.2. In 1866, GB was isolated by Scheibler from sugar beet (*Beta vulgaris*) and named "betaine". Plants in taxonomically distant species can synthesize GB and accumulate larger amounts when they are exposed to abiotic stress conditions, such as drought, salt and cold stress (Rhodes and Hanson, 1993). However, many plants do not accumulate GB and accumulate other osmolytes. Sugar beet (*Beta vulgaris*), spinach (*Spinacia oleracea*), barley (*Hordeum vulgare*), wheat (*Triticum aestivum*), and sorghum (*Sorghum bicolor*), are known as GB accumulators.

*Beta vulgaris* is a GB-accumulating dicotyledonous plant of the Chenopodiaceae family, including many varieties of spinach beet, Swiss chard, garden

Food and Nutritional Components in Focus No. 7
Betaine: Chemistry, Analysis, Function and Effects
Edited by Victor R Preedy

Published by the Royal Society of Chemistry, www.rsc.org

beet, fodder beet and sugar beet. The ancestor of cultivated beets is probably wild sea beet (*Beta vulgaris* ssp. *maritima*), a halophytic plant that grows around the coasts of the Europe, Mediterranean and North Africa. *Beta vulgaris* was first grown on a field as fodder beets. As fodder beets, beet plants have been selected for large amounts of dry matter in their roots. Sugar beet was selected from fodder beet with high sugar concentration early in the eighteenth century and most of sugar beet is grown at latitudes between 30 and 60′N, in semiarid and arid conditions. Sugar beet has high economic value because it is one of the two main sources of sucrose. The total production of sugar now approaches 140 Mt in 2000 (Draycott, 2006). In the beet sugar industry, GB has been thought to be "impurity" of sugar beet root sap that prevents crystallization of sucrose. GB content in root sap is usually about 0.2–0.3%, GB is concentrated in molasses up to 3–6%. Nowadays, however, GB and several other compounds (*i.e.* raffinose) contained in molasses became economically worth, so these compounds and remaining sucrose are purified by chromatographic separation.

## 2.2 Physiologic Function of GB in Plant Cells

### 2.2.1 GB as Compatible Solute

The accumulation of GB at high concentration in halotolerant plants is explained in terms of the compatibility of the osmolyte with macromolecular structure and function (Rhodes and Hanson, 1993). Osmolyte compatibility is the chemical characteristic with the absence of perturbing effects on the interaction of macromolecule and surrounding water molecules. Inorganic ions readily perturb the hydration sphere of proteins, forcing unfolding of macromolecules. Compatible solute such as GB, however, is excluded from the surface of macromolecules with their hydration sphere in cells. This results in the stabilization of correctly folded state of macromolecular structure in relatively low water potential (Bolen and Baskakov, 2001). Compatible solutes in plants would have to be localized in the cytoplasm, because they are generally present at modestly high concentration to be effective.

### 2.2.2 GB as an Osmoprotectant

GB is not only "compatible", it can protect proteins against thermodynamic perturbation caused by dehydration and heat denaturation. GB also protects inhibition of several enzyme activity induced by NaCl. GB protects *Beta vulgaris* root membranes against heat destabilization and spinach thylakoids against freezing stress. The concentrations of GB required to produce these protective effects are often high (>0.5 M). The physicochemical effect of osmoprotectant is called "osmophobic" (Bolen and Baskakov, 2001).

# 2.3 GB Accumulation in *Beta vulgaris*

## 2.3.1 Subcellular Localization of GB in Beet Root Cells

GB is thought to be distributed exclusively in cytoplasm and chloroplasts in plant cells (Figure 2.1). However, clear experimental data have not been reported. The amount of GB in the chloroplasts of spinach leaves was estimated to be close to 50% of the total GB in plant cells (Rhodes and Hanson, 1993). Concentrations of GB of up to 0.3 M (20 times as high as the concentration calculated if GB distributes uniformly in leaf tissue) have been reported for chloroplasts isolated from salinized spinach plants. Such concentrations could contribute to chloroplast osmotic adjustment, facilitating maintenance of chloroplast volume and photosynthetic activity at low leaf-water potentials. The calculated concentrations of GB in vacuole were always lower than that in the cytoplasm. It was reported that the GB concentration was below detection limit in vacuoles of *Suaeda maritima* leaf cells. Major solutes in vacuole are considered to be inorganic ions in leaves and sucrose in roots (Bell *et al.*, 1996).

## 2.3.2 Distribution in Whole Plant

The GB content of the whole leaf system of young plants increased throughout the vegetative period and reached a maximum at the time of flowering. Moreover, young seedling leaves of *Beta vulgaris* contained much less GB than young actively growing leaves sprouting from the crown of a mature root. The concentration of GB in the storage root of sugar beet, fodder beet and Swiss chard was reported 86, 113 and 75 μmol $g^{-1}$ dry weight (DM), respectively (Hoffman, 2013). Under salt stress, all types of beet plants increased

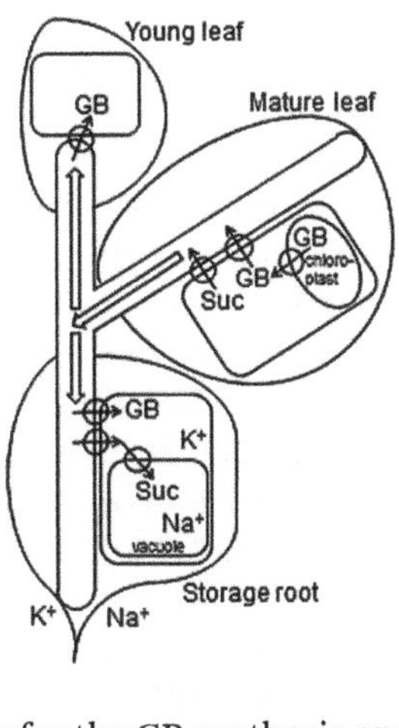


**Figure 2.1** Schematic figure for the GB synthesis and transport in plants.

GB concentrations 2- to 3-fold, as averages for all plants, 108 to 263 μmol $g^{-1}$ in shoots and 99 to 212 μmol $g^{-1}$ in roots. The shoot GB concentrations were closely correlated with leaf water potential. Salt-induced GB accumulation in the shoot could also contribute to cytoplasmic osmoregulation in beet leaves, and so play an important role in their salt tolerance. However, the contribution of GB to achieve osmotic balance in roots varied among genotypes.

The response of sugar beet, fodder beet and Swiss chard differed with regard to the accumulation of solutes in the storage root after drought stress (Hoffman, 2013). Sugar beet had by far the lowest concentration of alkali ions ($K^+$ and $Na^+$) in the root (197 mmol $kg^{-1}$ DM) compared with fodder beet (366 mmol $kg^{-1}$ DM) and Swiss chard (367 mmol $kg^{-1}$ DM). After drought stress treatment, GB content was significantly increased in sugar beet storage root (from 86 to 169 mmol $kg^{-1}$ DM), whereas the increase in GB contents in fodder beet and Swiss chard were not significant. The concentration of potassium and sodium did not change in sugar beet and fodder beet but increased in Swiss chard. The sucrose concentration was slightly reduced in fodder beet and Swiss chard but the decrease in sucrose content was not significant in sugar beet. There was a negative relationship between the sucrose concentration and the concentration of nonsucrose solutes (the sum of $K^+$, $Na^+$, amino acids and GB) in sugar beet and fodder beet, but no relation was seen in Swiss chard. The negative correlation between nonsucrose solutes and sucrose suggests the limitation of the storage root as a major sink organ. It has been reported that nonsucrose solutes are replaced by sucrose in the storage root of sugar beet during growth (Bell *et al.*, 1996). The data suggested that drought stress might force elevation of the level of amino acids and GB in root cells, and as a result, sucrose could not be accumulated to the same extent as under normal conditions.

### 2.3.3 GB in Storage Root

In general, GB concentration in sugar beet tap root increases in parallel to the increase in sucrose content during growth. It was reported that GB concentration in sugar beet root is more highly related to the concentration of $K^+$. When sugar beet was grown in the field with a supply of $K^+$, potassium and sodium are present at concentrations of 50–70 mM and 20–50 mM, respectively, at an earlier stage of plant development, but their concentrations fall during growth and are 10–20 mM or less at maturity. In leaves and petioles significant increases of the GB content were found with the increase in $K^+$ supply from 0 to 30 mg $K^+$ 100 $g^{-1}$ soil followed by decreases with the highest $K^+$ (40 mg $K^+$ 100 $g^{-1}$ soil). In the storage root, on the contrary, the GB content increased from 0 to 40 mg $K^+$ 100 $g^{-1}$ soil. Apparently, more GB was translocated into the storage root with a higher supply of $K^+$ in soil. The data from the individual storage roots indicated a close positive correlation between GB and $K^+$ concentration. The correlations with sucrose for both GB and $K^+$ were also positive and highly significant, although there is considerable variation for sucrose/GB and sucrose/$K^+$ compared with the correlation between GB and $K^+$ (Bell *et al.*, 1996). The relationships between sucrose concentrations and root sap osmotic pressure indicate that there is a basal osmotic potential of 200–400 mOsm $kg^{-1}$ contributed by other solutes. The higher levels of $Na^+$ and $K^+$ in young roots probably indicate that these are major vacuolar osmotica at this stage but then are replaced by sucrose as the root grows. Potentially, all salts in vacuole could be replaced by sucrose as there might be no physiologic function in vacuole other than to contribute to osmotic pressure. Concentrations of sucrose and $K^+$ are inversely related across different zones of the root, supporting the view that sucrose replaces $K^+$ in the vacuole. However, some $K^+$ must be present in the cytosol because some metabolic processes in cytoplasm require $K^+$ selectively. GB functions as a "compatible" solute in plants and is accumulated in the cytosol to compensate for increases in vacuolar osmotic pressure, thus maintaining water-potential equilibrium between the two compartments. If the cytosolic volume is 10% of the vacuolar volume, then the maximum tissue GB concentrations (37 mM) suggests a cytosolic GB concentration of just over 400 mM and this would osmotically balance about 50% of the sucrose in the vacuole. Strong correlation of GB levels with $K^+$ concentrations in storage root suggested that cytosolic osmotic regulation may also involve this cation. The effects of nitrogen (N) fertilization on the accumulation level of amino N and GB have been examined (Hoffmann, 2005). N application treatments were varied 0 to 200 kg N $ha^{-1}$ for four varieties of sugar beet on a field. Varieties significantly differed in amino N concentration, GB concentration and the (amino N)/GB ratio. However, for all varieties, amino N concentration nearly doubled with the increase of N supply (8–10 to 14–20 mmol $kg^{-1}$ beet). In contrast to amino N, GB concentration increased only slightly with N supply (12–14 to 13–16 mmol $kg^{-1}$ beet).

### 2.3.4 GB in Leaves

It was shown that nearly 95% of the $K^+$ in red-beet can be replaced by $Na^+$, with $Na^+$ levels reaching close to 4000 μmol $g^{-1}$ DM, which is similar to the $K^+$ levels of normally grown hydroponic plants (Subbarao *et al.*, 2001). When red-beet plants were grown with medium replacements of 95% of $K^+$ to $Na^+$, lamina $K^+$ levels ranged from 65 to 95 μg $g^{-1}$ DM. Lamina $K^+$ concentrations were about 3–4% of the $K^+$ levels under nonlimiting $K^+$ supply. The fact that $Na^+$ comprised 70–90% of the total cations and the relationship between lamina $Na^+$ levels and solute potential of the leaf sap suggest that $Na^+$ replaced $K^+$ as the inorganic osmoticum for the maintenance of cell turgor. GB levels increased parallel with the increase in $Na^+$ levels during growth period under nonsaline condition (<5 mM NaCl). This indicates that internal Na alone, not osmotic stress, could accelerate synthesis of GB in red-beet. Further increases in GB under salinity (50 mM NaCl) could be the result of an independent trigger from osmotic stress. If GB functions as a compatible solute in cytoplasm, its synthesis should be carefully regulated to balance the osmotic pressure between intracellular compartments. The linear relationship between lamina GB levels to the solute potential of the leaf sap in red-beet supports this notion. The lamina GB levels observed in red-beet could only contribute to 1–2% to the solute potential of the leaf sap if distributed uniformly in leaf cells. However, if it is assumed that GB can accumulate preferentially in the cytoplasm, the contribution of GB to the cytoplasmic solute potential could be 50–60% in the lamina of red-beet. GB synthesis could be accelerated at elevated $Na^+$ levels in the lamina to protect cytosolic enzymes from the adverse effect of $Na^+$. Tissue $K^+$ levels ranged from 60 to 90 μmol $g^{-1}$ DM correspond to about 6.0 to 9.0 mmol $L^{-1}$, if uniformly distributed in leaf tissue. This is much lower than the 100 to 200 mmol $L^{-1}$ $K^+$ assumed to be required for normal cytoplasmic functioning. However, if this $K^+$ was localized almost exclusively in the cytoplasm, the concentration of $K^+$ would have been 120 to 180 mmol $L^{-1}$, well within a prefered range. Thus, GB and most of the $K^+$ were likely localized in the cytoplasm, providing the osmotic balance with $Na^+$ in vacuoles. In contrast, there was no apparent relationship between $Na^+$ and GB levels in storage roots. Sodium levels in storage roots were only about 30% of the shoot $Na^+$ levels, but GB levels were similar to the shoot. GB synthesized in chloroplasts of the shoots may have been translocated to the storage roots contributing to cytoplasmic solute potential similar extent as in leaves irrespective of the $Na^+$ levels in the roots. It was also reported that higher GB levels were maintained in a red-beet cultivar that showed salt-tolerant characteristics in growth and photosynthetic activity (Subbarao *et al.*, 1999). In contrast, the GB levels at the higher $Na^+$ levels in salt-sensitive cultivar returned to the lower levels.

## 2.4 Pathway and Sites of GB Biosynthesis in Higher Plants

### 2.4.1 Choline to GB

In Chenopods including sugar beet, it has been known that GB is synthesized by a two-step oxidation of choline in chloroplasts (Rhodes and Hanson, 1993). The first step is catalyzed by choline mono-oxygenase (CMO) that converts choline to betaine aldehyde (Figure 2.2). CMO is a Rieske-type iron–sulfur enzyme that is detected in Amaranthaceae including Chenopodiaceae and consists of homo-dimer or trimer of subunit Mr 42 kDa (Rathinasabapathi *et al.*, 1997; Hibino *et al.*, 2002). The second step is the conversion of betaine aldehyde to betaine that is catalyzed by $NAD^+$-dependent betaine aldehyde dehydrogenase (BADH).

### 2.4.2 Ethanolamine to Choline

Carbon-14-tracer experiments showed the synthesis of choline and GB *via* phosphoryl base intermediates in salinized sugar beet leaves (Rhodes and Hanson, 1993) (Figure 2.3). In all plants, choline consists of charged moiety

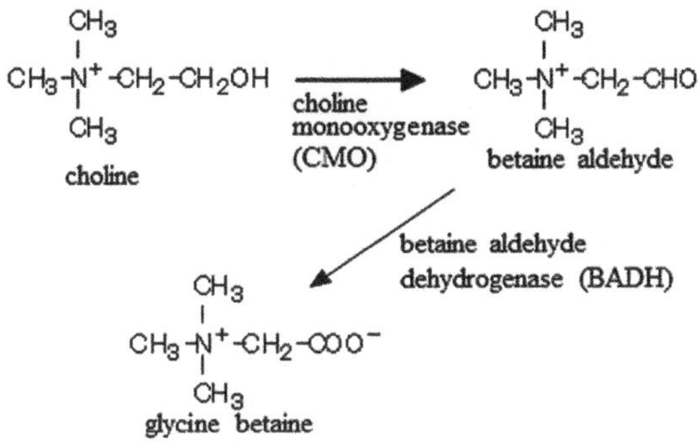


**Figure 2.2** Biosynthetic pathway of GB in plants.

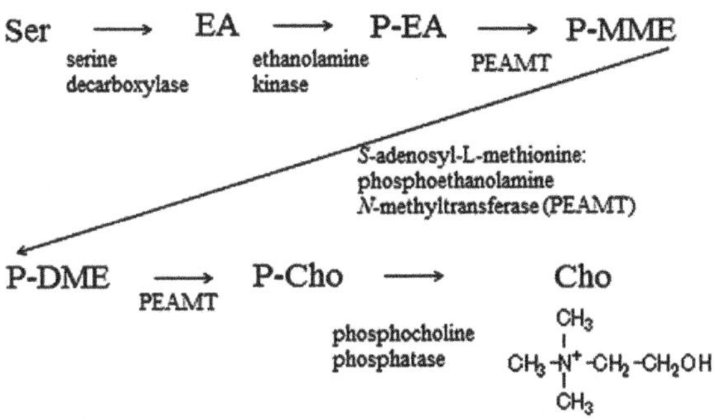


**Figure 2.3** Biosynthetic pathway of choline from serine in plants.

of phosphatidylcholine, which is a major component of phospholipids in plant cell membrane. Phosphatidylcholine plays essential roles not only as a structural component of the membrane but also as a signaling molecule. Biosynthetic routes to choline through three parallel, interconnected series of *N*-methylation reactions at the free base, phospho-base or phosphatidyl-base levels are possible among higher plants. Different routes predominating in different species are used. Phospho-base methylation is the major route of choline synthesis in sugar beet. In plants, the biosynthesis of choline occurs exclusively in the cytosol (McNeil *et al.*, 2000). The *N*-methylation of phosphoethanolamine is catalyzed by *S*-adenosyl-L-methionine: phosphoethanolamine *N*-methyltransferase (PEAMT). A gene-encoding PEAMT was isolated from spinach (Nuccio *et al.*, 2000). Polypeptide of PEAMT comprises two similar, tandem methyltransferase domains, implying that PEAMT arose by gene duplication and fusion. PEAMT catalyzes not only the first *N*-methylation of phosphoethanolamine but also the two subsequent *N*-methylations, yielding phosphocholine. Enzyme activities were inhibited by the reaction product phosphocholine at a concentration of the physiological level. This suggests that PEAMT is the committing step in the synthesis of choline and exerts major control over the metabolic flux to GB in spinach. It has been reported that the activity of PEAMT was 30 to 100 times lower in tobacco, GB nonaccumulator, compared with GB-accumulator spinach (Nuccio *et al.*, 2000). *In vivo* tracer experiments indicate that in the Chenopodiaceae, phosphocholine is hydrolyzed to choline. A member of the HAD (Haloacid Dehalogenase) superfamily was reported to act as phosphoethanolamine/phosphocholine phosphatase in *Arabidopsis*. Ethanolamine kinase was partially purified from spinach leaf. Spinach ethanolamine kinase exhibited specificity towards ethanolamine exclusively. Based on sequence homology, ethanolamine kinase can be distinguished from choline kinases. The activity of the gene product has not yet been examined.

### 2.4.3 Serine to Ethanolamine

Pyridoxal 5′-phosphate-dependent L-serine decarboxylase activity can be detected in soluble extracts from spinach leaves (Rontein *et al.*, 2001). The activity was higher in spinach compared with GB nonaccumulator *Arabidopsis*, and was further increased by salinization. A gene-encoding serine decarboxylase was isolated from *Arabidopsis thaliana.* The deduced polypeptide lacked obvious targeting signals and was shown to exist as a tetramer and to contain pyridoxal 5′-phosphate. It did not catalyze decarboxylation of other L-amino acids, L-phosphoserine or phosphatidylserine.

### 2.4.4 Enzymes Involved in Regeneration Cycle of *S*-Adenosylmethionine

During methylation reactions by PEAMT, *S*-adenosylmethionine (AdoMet) is converted to *S*-adenosylhomocysteine (AdoHcy), which is a potent inhibitor of methylation enzyme which uses AdoMet. AdoHcy is hydrolyzed to homocysteine and adenosine by AdoHcy hydrolase. Spinach and sugar beet, GB accumulators, showed salt-responsive increases in AdoHcy hydrolase and adenosine kinase activities, whereas tobacco and canola (*Brassica napus*), which do not accumulate GB, did not show comparable changes in these enzymes. The importance of adenosine kinase for the salvage synthesis of adenine monophosphate from adenosine and ATP to sustain transmethylation reaction was demonstrated.

### 2.4.5 Regulation of Gene Expression

It has been known that the gene expression of CMO and BADH is induced by salt and drought stress. Production of GB involves expression of not only the above-mentioned genes but also many genes that encode enzymes such as AdoMet synthetase, AdoHcy hydrolase and methionine synthase. Tabuchi *et al.* examined transcriptional regulation of the GB-related genes in leaf beet (Tabuchi *et al.*, 2006). The expression of GB-related genes in leaf tissue showed induction under salt stress. Their induction showed diurnal rhythms were reduced in dark conditions. Salinization of spinach plants increases PEAMT mRNA abundance and enzyme activity in leaves by about 10-fold, consistent with the high demand in stressed plants for choline to support GB synthesis (Nuccio *et al.*, 2000).

## 2.5 Translocation of GB in *Beta vulgaris*

Halotolerant plants often accumulate GB under nonstress condition. GB of 420 μmol $g^{-1}$ FW accumulated in young leaves of *Beta vulgaris* even under normal growth conditions, whereas levels in mature leaves, cotyledons, hypocotyls, and roots were low (Yamada *et al.*, 2009). Under the same conditions, CMO accumulates exclusively in old leaves and is difficult to detect in young leaves. Incubation of segments of mature and young leaves with deuterium-labeled choline revealed that the rate of deuterium-labeled-GB synthesis in mature leaves was about 9-fold higher than that in young leaves. The levels of CMO mRNA were higher in mature leaves than in young leaves. Thus, the differential accumulation of CMO was essentially controlled by transcription level. When deuterium-substituted GB was applied exogenously to one of the attached mature leaf, labeled-GB were preferentially detected in young leaves and root, but not in other expanded leaves. These data indicate that GB is primarily synthesized in mature leaves and translocated into young leaves. In response to salt stress, GB levels increased in all tissues, but most significantly increased in young leaves with only a small increase in the levels of CMO. GB functions as a chemical chaperone to protect the inactivation of proteins under high salinity and high temperature. In actively developing cells, the levels of protein synthesis and translocation would be high. GB might facilitate folding and translocation, serving as a chemical chaperone. Another possibility is that GB might be utilized to maintain high cellular pressure. It is known that high cellular pressure is required for cell growth. Higher osmolarity in young leaves than in mature leaves is compatible with this idea. In animal cells, it has been shown that GB plays a role in cell-volume homeostasis.

The high concentration of GB in young leaves with a little *de novo* synthesis of GB suggests the importance of GB transporters in GB accumulation. To date, many amino acid transporter genes have been identified in the plant genomes (Yamada *et al.*, 2009; Lehmann *et al.*, 2010). Proline transporters (ProTs) were first isolated from *Arabidopsis thaliana* as highly selective transporters for proline. However, ProTs from tomato (LeProTs) were subsequently shown to transport GB as well as proline, although tomato is a GB nonaccumulating plant. Hitherto, functional properties of ProTs have been reported on GB nonaccumulating plants; *A. thaliana* (AtProT1-3), tomato (LeProT1-3), and rice (OsProT1), and GB-accumulating plants; mangrove (AmBet/ProT1-2), sugar beet (BvBet/ProT1) and barley (HvProT1-2) (Waditee *et al.*, 2002; Grallath *et al.*, 2005; Yamada *et al.*, 2009; Fujiwara *et al.*, 2010). Among them, the selectivity of rice ProT for GB remains uninvestigated, and the barley

HvProT1 was reported to recognize proline, but not GB. All other transporters mediate transport of both GB and proline. The selectivity of ProTs for proline suggests their specific role in proline homeostasis under stressed and nonstressed conditions (Lehmann *et al.*, 2010). However, the physiological role of GB uptake by Bet/ProTs remains obscure. A homologous gene BvBet/ProT1 was isolated from sugar beet. The fusion protein of green fluorescent protein and BvBet/ProT1 showed that BvBet/ProT1 was localized at the plasma membrane. Levels of mRNA for BvBet/ProT1 were much higher in mature leaves than in young leaves under normal and salt-stress conditions. *In situ* hybridization experiments revealed the localization of BvBet/ProT1 in phloem and xylem parenchyma cells (Yamada *et al.*, 2011). Since GB is transported from the mature leaves to young leaves and the accumulated levels of BvBet/ProT1 gene transcript were higher in mature leaves than in young leaves, BvBet/ProT1 might play a role in the efflux of GB from CMO-expressing cells. The transport properties of Bet/ProTs from GB accumulating (sugar beet, *Amaranthus*, and *Atriplex*) and nonaccumulating (*Arabidopsis*) plants were compared (Yamada *et al.*, 2011). It was shown that all these transporters exhibited a higher affinity for GB than proline. Moreover, these transporters exhibited a higher affinity for choline uptake rather than GB. The uptake of GB and proline was pH dependent. However, uptake of choline by sugar beet BvBet/ProT1 was affected to a much lesser extent by proton gradients. These results suggest that there might be differences in proton binding between choline and GB during transport processes.

## 2.6 Utilization of GB from Sugar Beet

### 2.6.1 Exogenous Application for Crop Production

In many crop plants the natural accumulation of GB is lower than sufficient to ameliorate the adverse effects of dehydration caused by various environmental stresses. Exogenous application of GB to low-accumulating or nonaccumulating plants may help reduce adverse effects of environmental stresses (Makela, 2004; Ashraf and Foolad, 2007). Exogenously applied GB is, at least partially, immediately taken up by plant tissues and is readily translocated to roots, meristems and expanding leaves (Makela *et al.*, 2000). Because GB is metabolically quite inert in plant, it remains in the plant tissue for several weeks. There is now strong evidence that GB plays an important role in tolerance to abiotic stress. Exogenous application of GB to nonaccumulator plants may be a possible alternative approach for tolerance against multiple abiotic stresses.

Foliar application of GB improved salt and drought tolerance in tomatoes. Application of tomato plants in field condition with 3.36 kg ha$^{-1}$ GB during the midflowering period increased fruit yield 36%, as compared to control in salt stress (Makela, 2004). Salt stress decreased net photosynthesis of tomato plant, and application of GB increased the rate of photosynthesis of the salt stressed plants to the same level as adequately watered plants, which did not have GB applied. The increase of net photosynthesis by GB application was also observed in nonstress plants. Stomatal conductance was increased by GB treatment, resulted in efficient gas exchange and reduced photorespiration. GB application protected chloroplast ultrastructure and prevented the decrease of chlorophyll and RuBisCO activity under salt- and drought-stress conditions (Makela *et al.*, 2000). The relative water content of the total shoot of tomato decreased under drought stress and GB had an increasing effect. The positive effect of GB on drought-stress tolerance was more efficient at the vegetative growth stage. Exogenous application of GB increases chilling tolerance in tomato plants. GB levels in tomato plants that took up exogenous GB were high in meristematic tissues such as shoot apices and flower buds. In leaves, GB existed mainly in cytosol with at most 22% in chloroplasts. The transport of GB from cytosol to chloroplasts might be inefficient. However, GB-treated plants exhibited increased levels of photosystem II (PSII) activity compared with control plants. GB-treated plants had significantly greater catalase (CAT) activity and CAT1 gene expression, although their $H_2O_2$ levels remained unchanged. During chilling treatment, the $H_2O_2$ level was lower and catalase activity was higher in GB-treated plants than those in control plants. These results suggest that GB may enhance the induction of antioxidant mechanisms under chilling-stress conditions.

When 28-day-old rice seedlings were exposed to NaCl, relative water content in the leaves drastically decreased. This adverse effect was largely prevented when the seedlings were treated with 15 mM GB before exposure to NaCl stress (Harinasut *et al.*, 1996). GB added in culture medium was taken up by the roots and accumulated in the leaves to reach a concentration of 5.0 μmol g $fwt^{-1}$ before transfer to NaCl stress. This level of GB is comparable with those of barley, a GB accumulator. The level of GB in rice seedlings did not change significantly after transfer to NaCl stress. No apparent difference in relative water content was found between control and GB-treated rice seedling before transfer to salt-stress conditions, changes in relative water content after transfer to salt stress was significantly different. Control rice seedlings showed a large decrease in relative water content (RWC) after transfer to salt stress, the extent of the decrease in RWC was small in GB-treated seedlings. Photosynthetic capacity of control seedlings also drastically decreased, whereas that of GB-treated seedlings was maintained near the value of nonstressed seedlings. Higher water use efficiency in GB-treated rice plants than that in control plants under salt-stress conditions was reported (Cha-um *et al.*, 2006). Foliar-spray of 50 mM GB to a salt-sensitive cultivar of rice enhanced proline accumulation under salt-stress conditions, resulted in maintenance of transpiration efficiency and a high photosynthetic performance. The photosynthetic abilities in plants were positively related to seed-fertility percentage and total seed grain weight after recovery from the stress. This performance was rather inhibited with the high dose of 200 mM GB. Under salt stress, GB-treated rice plants had significantly lower $Na^+$ and higher $K^+$ concentrations in the shoots, compared with untreated plants (Cha-um *et al.*, 2006). Moreover, salt-induced ultrastructural damages in chloroplasts and mitochondria were prevented by GB pretreatment. Although the inhibition of root growth by NaCl was not alleviated, formation of large vacuoles was observed in root cells by GB pretreatment. It was proposed that the mitigation of the NaCl-induced damages in the shoots is due to the enhanced sequestration of $Na^+$ into the vacuoles in root cells by GB pretreatment.

Studies with wheat showed contradictory results. There were reports that exogenous GB delayed the canopy senescence of wheat, but this was not associated with its grain yields after the exposure to drought stress. On the other hand, there are reports that GB had an effect to mitigate the adverse effect of water stress in wheat. There are also reports of the protective effects of GB against NaCl stress and freezing stress. Drought stress reduced plant biomass, grain yield, and phosphate (P) and $K^+$ contents in shoot and root, but foliar-applied GB mitigated the adverse effects of drought stress. It was reported that mitigation of a decrease in photosynthetic rates was not prominent but the transpiration rate was apparently decreased by GB application. The extent of the effect of GB was varied among wheat cultivars. Water stress caused an increase in osmotic pressure in the flag leaves of wheat. The drought-resistant cultivar had higher values of osmotic pressure than the sensitive one. Water stress induced a marked decrease in grain yield, but spraying of 10 mM GB mitigated the effect of drought stress. The application of GB increased the osmotic pressure, the osmolytes concentrations as well as the grain yield. Among the organic osmolytes, proline and keto-acid concentrations were positively correlated with the grain yield. The drought-resistant cultivar had a higher concentration of $K^+$ than the sensitive one and the application of GB caused an accumulation in $K^+$ in both wheat cultivars. Because $K^+$ has a role in stabilization of native proteins, the increase in $K^+$ level may play a role in adaptation of wheat plants to water-stress conditions.

### 2.6.2 Protective Effects and Risks of GB Application

Osmolytes accumulation in plant cells results in a decrease of the cell osmotic potential and thus in maintenance of water absorption and cell turgor pressure, which might contribute to sustain physiological processes, such as stomatal opening, photosynthesis, and growth expansion. Furthermore, the occurrence of osmolytes accumulation at sensitive crop-reproductive stages has been reported to play a constructive role against floral abortion, which results in maintaining grain number under water deficit. Additionally, osmolytes accumulation has also been claimed to facilitate a better translocation of pre-anthesis carbohydrate reserves to the grain during the grain-filling period (Subbarao *et al.*, 2001). Solute accumulation in plant cells creates an intracellular osmotic potential that in the presence of rigid cell walls generates turgor pressure. Maintenance of turgor is necessary for maintenance of growth through cell elongation. GB may not only protect the integrity of the cell membrane from drought-stress damage, but also be involved in osmolytes accumulation in transgenic cotton plants. GB was proved to be a singularly effective in the protection of the photosynthetic oxygen-evolving complex from heat-induced and salt-induced inactivation (Chen and Murata, 2011). Although the exogenous application of GB has effects on crop production, a possible disadvantage should be mentioned that GB is a potent growth substrate for some species of pathogenic fungi and can act as attractant to aphids (Rhodes and Hanson, 1993).

## 2.7 Genetic Engineering of GB Biosynthesis in Plants

Various genes have been used to generate transgenic plants that accumulate GB and exhibit enhanced tolerance to abiotic stress (Chen and Murata, 2011). The various transgenic plants accumulate GB at a variety of levels and exhibit enhanced tolerance to various types of stress. Most of the transgenic plants of nonaccumulators of GB accumulate only low levels of GB compared with the range reported for natural GB accumulators, so the contribution of GB to the total osmotic potential of transformed plant cells is small. However, these transgenic plants often exhibited increases in abiotic stress tolerance. There are two known factors that can limit the accumulation of GB in chloroplasts of transgenic plants: the availability of endogenous choline and the transport of choline across the chloroplast envelope. Exogenous supply of choline showed significantly increase in the level of GB accumulation in transgenic plants. Indeed, the activity of PEAMT was 30 to 100 times lower in tobacco, a GB nonaccumulator, compared with spinach, a GB accumulator. Furthermore, it was revealed that the import of choline, a precursor of GB, into chloroplasts limits GB synthesis.

It has been shown that a halotolerant cyanobacterium *Aphanothece halophytica* synthesis GB by three-step methylation of glycine (Waditee *et al.*, 2003) (Figure 2.4). Two genes are involved in GB synthesis; ApGSMT encodes the methylation of glycine and sarcosine, ApDMT encodes the methylation of dimethylglycine to GB. Genes coding glycine methylation enzymes (ApGSMT + ApDMT) were introduced to *Arabidopsis* plant (Waditee *et al.*, 2005). Arabidopsis plants were also transformed by CMO gene. The *Arabidopsis* plants expressing the glycine methylation genes accumulated GB in all organs, whereas *Arabidopsis* plants expressing the CMO gene accumulated GB exclusively in the roots. GB levels in the roots of transgenic plants expressing the glycine methylation genes were 2-fold higher than that of CMO expressing plants. *Arabidopsis* plants expressing the glycine methylation genes showed improved tolerances for various abiotic stresses more than the CMO-expressing plants. The reason for the accumulation of GB in various tissues of the glycine methylation genes-expressing plants might be the availability of substrates. Glycine and serine are both potential sources of one carbon units (C1 units) in plants. They are readily interconvertible *via* the action of serine hydroxymethyltransferase (SHMT) and glycine decarboxylase enzymes (Figure 2.5). The GB level in the transformed plants expressing the glycine

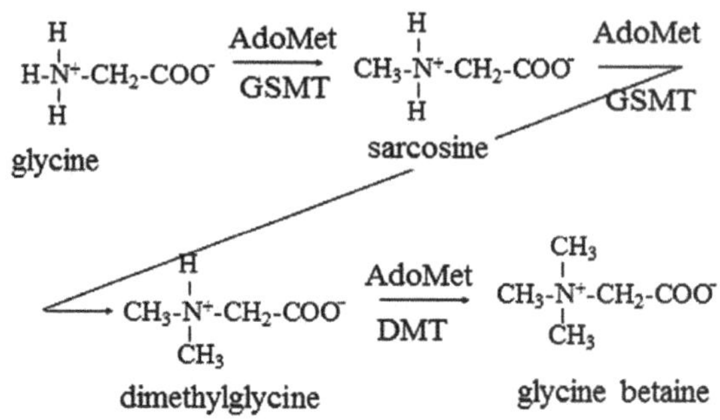


**Figure 2.4** Biosynthetic pathway of GB in a halotolerant cyanobacterium.

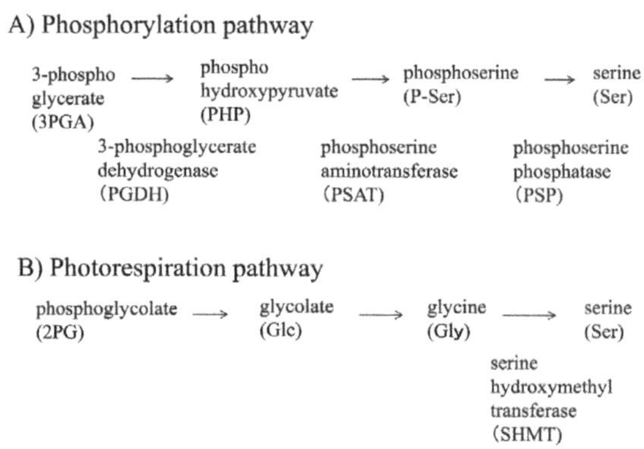


**Figure 2.5** Biosynthetic pathway of serine in plants. (A) Phosphorylation pathway. (B) Photorespiration pathway.

methylation genes increased at high salinity, although their enzyme levels did not change very much. This increase might be due to the increased accumulation of glycine and serine at high salinity, as shown in wild-type plants (Figure 2.6). The high accumulation of serine at high salinity is also reported (Ho and Saito, 2001). Moreover, the induction of AdoMet synthetase expression at high salinity may also increase the AdoMet level. Levels of proline were also higher at high salinity, but these were the same in wild-type and transgenic plants. Supply of exogenous glycine to *Arabidopsis* transformants enhanced the accumulation levels of GB and stress tolerance. Furthermore, a gene-encoding enzyme involved in serine biosynthesis (3-phosphoglycerate dehydrogenase from a halotolerant cyanobacterium *Aphanothece halophytica* (ApPGDH), was introduced into *Arabidopsis* plant expressing the glycine methylation genes (Waditee *et al.*, 2007). Transgenic plants harboring these three genes could accumulate more GB than plants transformed with two methylation enzymes. This demonstrates the importance of the *ApPGDH* gene in precursor supply for biosynthesis of GB. ApPGDH can be expressed in various organs of the *Arabidopsis* plants such as the leaves, stems, and

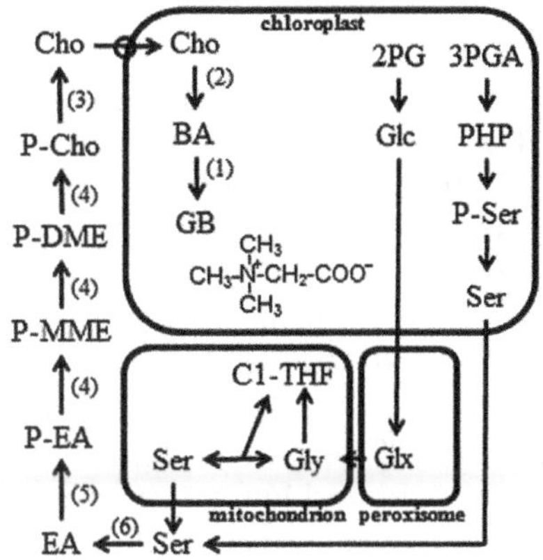


**Figure 2.6** Schematic figure for the GB synthesis in plants. In the figure, the following abbreviations were used. BA, betaine aldehyde; C1-THF, methylenetetrahydrofolate; Cho, choline; EA, ethanolamine; GB, glycinebetaine; Glc, glycolate; Glx, glyoxylate; P-Cho, phospho-choline; P-DME, phospho-dimethylethanolamine; P-EA, phospho-ethanolamine; PHP, phospho-hydroxypyruvate; P-MME, phospho-monomethylethanolamine; 2PG, 2-phosphoglycolate; 3PGA, 3-phosphoglycerate; P-Ser, phosphorserine; Suc, sucrose. (1) Betaine aldehyde dehydrogenase, (2) cholinemono-oxygenase (CMO), (3) phosphocholine phosphatase, (4) *S*-adenosyl-L-methionine: phosphoethanolamine *N*-methyltransferase (PEAMT), (5) ethanolamine kinase, (6) serine decarboxylase.

the root tissues together with the glycine methylation enzymes. *Arabidopsis* plants have both phosphorylation and photorespiratory pathways for serine biosynthesis (Figure 2.5). During stress condition, the photosynthesis would be inhibited due to the closure of the stomata. It is not surprising that the plant has to activate an alternative route for amino acid synthesis under abiotic stress. GB levels in the transgenic plants expressing ApPGDH, and glycine methylation genes were notably higher than those in the transgenic plants expressing the glycine methylation genes. This might be due to a high activity of PGDH, which, in turn, produces more serine in the root tissues. Indeed, levels of serine and glycine in the root tissues were much higher than those in the leaves. These facts support the viewpoint that the simultaneous introduction of the glycine methylation pathway and ApPGDH to non-GB accumulating plant *Arabidopsis* relieves the limitation of the supply of the precursor glycine. The transgenic *Arabidopsis* plants expressing ApPGDH, ApGSMT, and ApDMT exhibited an enhanced tolerance for salt stress compared with those of the wild-type or transgenic plants expressing ApGSMT and ApDMT. This was caused by the elevated levels of GB because of the overexpression of ApPGDH. Although the accumulation levels of GB in transgenic *Arabidopsis* plants were still considerably lower than that in natural GB accumulator, higher levels of GB accumulation resulted in the higher tolerance to abiotic stress.

## Summary Points

- This chapter focuses on betaine in sugar beet.
- Betaine is a potent osmoprotectant in plants.
- Sugar beet is a halophyte plant and accumulates betaine significant amount, however, many important cereal plants such as rice do not accumulate betaine.
- Betaine is synthesized in chloroplast *via* two-step oxidation of choline, choline mono-oxygenase (CMO) and betaine aldehyde dehydrogenase (BADH).
- The accumulation level of betaine is significantly induced by abiotic stress such as salt, drought and low temperature.
- In the tap root of sugar beet, sucrose, betaine and raffinose are the main constituents.
- The importance of long-distance translocation of betaine *via* betaine transporter.
- There is preferential accumulation of betaine in young leaves than in mature leaves.
- Precursor supply such as choline and serine are important for betaine accumulation.
- Genetic engineering of betaine synthesis genes is an important target to improve stress tolerance in plants.

## Key Facts for Betaine in Sugar Beet

1. Salinity is a major factor to decrease the productivity of crop plants.
2. Due to the rapid growth of world population, the improvement of salinity tolerance in the major crop plants is an urgent issue for the international community.
3. Soil salinity depresses plant growth in the form of osmotic stress and ion-specific toxicity.
4. Osmoprotectants or compatible solutes are small molecules that help organisms to survive under extreme osmotic stress such as high salinity and drought.
5. Betaines, amino acids, and the sugar trehalose are example of osmoprotectants that accumulate in cells and balance the osmotic difference between the cell's surroundings and the cytosol.
6. Sugar beet is an important betaine-accumulating plant.
7. Betaine is synthesized from choline by two-step oxidation–reductions *via* betaine aldehyde. The enzymes involved in betaine synthesis are choline mono-oxygenase (CMO) and betaine aldehyde dehydrogenase (BADH).
8. Sugar beet is an important resource for betaine, raffinose and sucrose.
9. Sugar beet is a halophyte and can survive under high salinity, whereas many important crop plants such as rice cannot synthesize betaine and rice is a salt-sensitive plant.

10. Improvement of salinity tolerance of crop plants by introducing betaine synthetic genes is an important target to increase the crop yield in the world.

## Definitions of Words and Terms

***Aphanothece halophytica***: A halotolerant cyanobacterium isolated from the Dead Sea. This cyanobacterium accumulates large amount of betaine that is synthesized by a three-step methylation of glycine.

**Betaine nonaccumulator**: Plants that do not accumulate betaine in nature are betaine nonaccumulators. These plants do not contain betaine-synthesis genes, choline mono-oxygenase (CMO) and betaine aldehyde dehydrogenase (BADH). Sugar beet (*Beta vulgaris*) and spinach (*Spinacia oleracea*) are GB accumulators. Rice, tomato, and *Arabidopsis* are GB nonaccumulators.

**CMO**: An enzyme that catalyzes the rate-limiting step for betaine synthesis from choline in plants. For the enzymatic activity, ferredoxin is required.

**Compatible solutes**: Compatible solutes are small molecules that do not decrease the cellular metabolism even though they accumulate large amounts in the cytoplasm to balance external osmotic pressure.

**Halophyte**: A halophyte is a plant that grows in waters of high salinity. Relatively few plants are halophytes. Sugar beet and mangroves are examples of halophytes. The majority of plants are glycophytes.

**Genetic engineering**: A technique that enables one organism to produce substance(s) not or scarcely contained naturally in that organism by introducing DNA isolated from other organism(s).

**Osmoprotectants**: Osmoprotectants are small molecules that play the role of osmotic balance between inside and outside cells. Examples of osmoprotectants are betaine, potassium ion, proline and trehalose.

**Photosystem II**: Chlorophyll-containing molecular apparatus in plants, which withdraws electrons from water by using light energy to make a strong oxidant in its reaction center.

**Solute potential**: Solute potential determines the direction that solutes move. Solutes totally move from the area with higher potential to the area with lower potential. On the contrary, water molecules move from the area with lower solute potential (less solute = more water) to the area with higher solute potential (more solute = less water).

**Transgenic plants**: Plants whose genome has been altered by the transfer of gene(s) from other organisms or of the DNA fragment(s) whose introduction modifies the expression of native genes in the plants.

**Vacuole**: Intracellular compartment separated by a membrane from cytoplasmic metabolism in plant cells. Usually it occupies nearly 90% of the cell volume.

**Water potential**: Addition of solutes to water lowers the water's potential (to more negative). Water moves from higher water potential to lower water potential area. Water that contains a dissolved salt, has a negative water potential, relative to the pure water.

## List of Abbreviations

| | |
|---|---|
| AdoHcy | S-Adenosyl-homocysteine |
| AdoMet | *S*-Adenosyl-L-methionine |
| Ap | *Aphanothece halophytica* |
| BADH | Betaine aldehyde dehydrogenase |
| BetT | Betaine transporter |
| CAT | Catalase |
| C1 unit | One carbon unit |
| CMO | Choline mono-oxygenase |
| DM | Dry weight |
| DMT | AdoMet-dependent dimethylglycine methyltransferase |
| GB | Glycine betaine |
| GSMT | AdoMet-dependent glycine and sarcosine methyltransferase |
| HAD | Haloacid dehalogenase |
| PGDH | 3-Phosphoglycerate dehydrogenase |
| PEAMT | *S*-Adenosyl-L-methionine: phosphoethanolamine *N*-methyltransferase |
| ProT | Proline transporter |
| PSII | Photosystem II |
| RWC | Relative water content |
| SHMT | Serine hydroxymethyltransferase |

## References

Ashraf, M. and Foolad, M. R., 2007. Roles of glycine betaine and proline in improving plant abiotic stress resistance. *Environmental and Experimental Botany*. 59: 206–216.

Bell, C. I., Milford, G. F. J. and Leigh, R. A., 1996. Photoassimilate distribution in plants and crops: source-sink relationships. In: Zamski, E. and Schaffer, A. A. (ed.) *Sugar Beet*. Marcel Dekker, Inc., New York, pp. 691–707.

Bolen, D. W. and Baskakov, I. V., 2001. The osmophobic effect: natural selection of a thermodynamic force in protein folding. *Journal of Molecular Biology*. 310: 955–963.

Cha-um, S., Supaibulwatana, K. and Kirdmanee, C., 2006. Water relation, photosynthetic ability and growth of Thai jasmine rice (*Oryza sativa L.* ssp. *Indica* cv. KDML 105) to salt stress by application of exogenous glycinebetaine and choline. *Journal of Agronomy and Crop Science*. 192: 25–36.

Chen, T. H. H. and Murata, N., 2011. Glycinebetaine protects plants against abiotic stress: mechanisms and biotechnological applications. *Plant, Cell & Environment*. 34: 1–20.

Draycott, A. P. (ed.), 2006. *Sugar Beet*. Blackwell Publishing, Oxford, UK, 474 pp.

Fujiwara, T., Mitsuya, S., Miyake, H., Hattori, T. and Takabe, T., 2010. Characterization of a novel glycinebetaine/proline transporter gene expressed in the mestome sheath and lateral root cap cells in barley. *Planta*. 232: 133–143.

Grallath, S., Weimar, T., Meyer, A., Gumy, C., Suter-Grotemeyer, M., Neuhaus, J. M. and Rentsch, D., 2005. Compatible solute transporters with similar substrate specificity but differential expression patterns. *Plant Physiology*. 137: 117–126.

Harinasut, P., Tsutsui, K., Takabe, T., Nomura, M., Takabe, T. and Kishitani, S., 1996. Exogenous glycine betaine accumulation and increased salt tolerance in rice seedlings. *Bioscience, Biotechnology and Biochemistry*. 60: 366–368.

Hibino, T., Waditee, R., Araki, E., Ishikawa, H., Aoki, K., Tanaka, Y. and Takabe, T., 2002. Functional characterization of choline monooxygenase, an enzyme for betaine synthesis in plants. *Journal of Biological Chemistry*. 277: 41352–41360.

Ho, C.-L. and Saito, K., 2001. Molecular biology of the plastidic phosphorylated serine biosynthetic pathway in *Arabidopsis thaliana. Amino Acids*. 20: 243–259.

Hoffmann, C. M., 2005. Changes in N composition of sugar beet varieties in response to increasing N supply. *Journal of Agronomy and Crop Science*. 191: 138–145.

Hoffman, C. M., 2013. Adaptive responses of *Beta vulgaris L.* and *Cichorium intybus L.* Root and leaf forms to drought stress. *Journal of Agronomy and Crop Science*. DOI: 10.1111/jac.12051.

Lehmann, S., Funck, D., Szabados, L. and Rentsch, D., 2010. Proline metabolism and transport in plant development. *Amino Acids*. 39: 949–962.

Makela, P., Karkkainen, J. and Somersalo, S., 2000. Effect of glycinebetaine on chloroplast ultrastructure, chlorophyll and protein content, and RuBPCO activities in tomato grown under drought or salinity. *Biologia Plantarum*. 43: 471–475.

Makela, P., 2004. Agro-industrial uses of glycinebetaine. *Sugar Technology*. 6: 207–212.

McNeil S. D., Rhodes D., Russell B. L., Nuccio M. L., Shachar-Hill Y. and Hanson A. D., 2000. Metabolic modeling identifies key constraints on an engineered glycine betaine synthesis pathway in tobacco. *Plant Physiology*. 124: 153–162.

Nuccio, M. L., Ziemak, M. J., Henry, S. A., Weretilnyk, E. A., and Hanson, A. D., 2000. cDNA cloning of phosphoethanolamine *N*-methyltransferase from spinach by complementation in *Schizosaccharomyces pombe* and characterization of the recombinant enzyme. *Journal of Biological Chemistry*. 275: 14095–14101.

Rathinasabapathi, B., Burnet, M., Russell, B. L., Gage, D. A., Liao, P. C., Nye, G. J., Scott, P., Golbeck, J. H. and Hanson, A. D., 1997. Choline monooxygenase, an unusual iron-sulfur enzyme catalyzing the first step of glycine betaine synthesis in plants: prosthetic group characterization and cDNA cloning. *Proceedings of the National Academy of Sciences of the United States of America*. 94: 3454–3458.

Rhodes, D. and Hanson, A. D., 1993. Quaternary ammonium and tertiary sulfonium compounds in higher plants. *Annual Review of Plant Physiology and Plant Molecular Biology*. 44: 357–384.

Rontein, D., Nishida, I., Tashiro, G., Yoshioka, K., Wu, W. I., Voelker D. R., Basset, G. and Hanson A. D., 2001. Plants synthesize ethanolamine by direct decarboxylation of serine using a pyridoxal phosphate enzyme. *Journal of Biological Chemistry*. 276: 35523–35529.

Subbarao, G. V., Wheeler, R. M., Stutte, G. W. and Levine, L. H., 1999. How far can sodium substitute for potassium in red beet? *Journal of Plant Nutrition*. 22: 1745–1761.

Subbarao, G. V., Wheeler, R. M., Levine, L. H. and Stutte, G. W., 2001. Glycine betaine accumulation, ionic and water relations of red-beet at contrasting levels of sodium supply. *Journal of Plant Physiology*. 158: 767–776.

Tabuchi, T., Okada, T., Takashima, Y., Azuma, T., Nanmori, T. and Yasuda, T., 2006. Transcriptional response of glycinebetaine-related genes to salt stress and light in leaf beet. *Plant Biotechnology*. 23: 317–320.

Waditee, R., Hibino, T., Tanaka, Y., Nakamura, T., Incharoensakdi, A., Hayakawa, S., Suzuki, S., Futsuhara, Y., Kawamitsu, Y., Takabe, T. and Takabe, T., 2002. Functional characterization of betaine/proline transporters in betaine-accumulating mangrove. *Journal of Biological Chemistry*. 277: 18373–18382.

Waditee, R., Tanaka, Y., Aoki, K., Hibino, T., Jikuya, H., Takano, J., Takabe, T. and Takabe, T., 2003. Isolation and functional characterization of N-methyltransferases that catalyze betaine synthesis from glycine in a halotolerant photosynthetic organism *Aphanothece halophytica*. *Journal of Biological Chemistry*. 278: 4932–4942.

Waditee, R., Bhuiyan, M. N. H., Rai, V., Aoki, K., Tanaka, Y., Hibino, T.,Suzuki, S., Takano, J., Jagendorf, A. T., Takabe, T. and Takabe, T., 2005. Genes for direct methylation of glycine provide high levels of glycinebetaine and abiotic-stress tolerance in *Synechococcus* and *Arabidopsis*. *Proceedings of the National Academy of Sciences of the United States of America*. 102: 1318–1323.

Waditee, R., Bhuiyan, M. N. H., Hirata, E., Hibino, T., Tanaka, Y., Shikata, M. and Takabe, T., 2007. Metabolic engineering for betaine accumulation in microbes and plants. *Journal of Biological Chemistry*. 282: 34185–34193.

Yamada, N., Promden, W., Yamane, K., Tamagake, H., Hibino, T., Tanaka, Y. and Takabe, T., 2009. Preferential accumulation of betaine uncoupled to choline monooxygenase in young leaves of sugar beet--importance of long-distance translocation of betaine under normal and salt-stressed conditions. *Journal of Plant Physiology*. 166: 2058–2070.

Yamada, N., Sakakibara, S., Tsutsumi, K., Waditee, R., Tanaka, Y. and Takabe, T., 2011. Expression and substrate specificity of betaine/proline transporters suggest a novel choline transport mechanism in sugar beet. *Journal of Plant Physiology*. 168: 1609–1616.

CHAPTER 3

# *Betaine in Food Plants: A Focus on Synthesis and Content*

ROSARIO A. MUÑOZ-CLARES*[a]

[a]Departamento de Bioquímica, Facultad de Química, Universidad Nacional Autónoma de México, Ciudad Universitaria, 04510, México D.F., México
*E-mail: clares@unam.mx

## 3.1 Introduction

Betaine, which I will name hereafter glycine betaine (GB, *N,N,N*-trimethylglycine), is a small, highly soluble, zwitterionic quaternary ammonium compound (Figure 3.1) that *in vivo*, and depending on the organism, may function as: (i) a very efficient osmoprotectant that is accumulated under osmotic stress by both prokaryotic and eukaryotic organisms to balance the osmotic pressure of the cells with that of the environment (Rhodes and Hanson, 1993); (ii) a chemical chaperone that protects proteins against the deleterious effects of salt and temperature (Papageorgiou and Murata, 1995); (iii) a methyl donor for methionine synthesis (du Vigneaud *et al.*, 1946); or (iv) a metabolic intermediate in the catabolism of choline (Cho) in some bacteria that are able to grow in Cho or Cho precursors as the only source of carbon, nitrogen and energy (Velasco-García *et al.*, 2006). This review deals with what is known about the enzymological aspects of the synthesis of GB in plants, discussing the differences between them in their ability to synthesize, and therefore accumulate, this compound, as well as the biotechnological attempts made to engineer this ability in important crops that naturally lack it.

Food and Nutritional Components in Focus No. 7
Betaine: Chemistry, Analysis, Function and Effects
Edited by Victor R Preedy

Published by the Royal Society of Chemistry, www.rsc.org

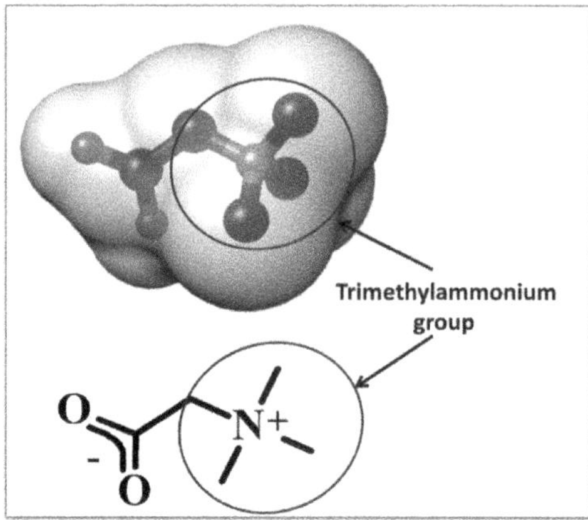


**Figure 3.1** Chemical structure of glycine betaine. The van der Waals surface of the molecule is displayed so that the shape of the molecule can be more fully appreciated. Note the bulky thrimethylammonium group. GB is neutral at physiological pH at which it has a negatively charged carboxyl group. The trimethylammonium group has a positive charge at any pH.

## 3.2 Glycine Betaine is an Important Osmoprotectant in Plants

Plants are sessile organisms that have evolved a great variety of strategies to escape from or contend with the many kinds of abiotic and biotic stresses to which they may be exposed. Among the former, osmotic stress—caused by drought, saline soils or low temperatures—constitutes the major limitation of agricultural production worldwide. To cope with osmotic stress one of the strategies developed, not only by plants but by all kinds of organisms, is the synthesis and/or accumulation of neutral, highly soluble, small organic compounds known as compatible solutes or osmoprotectants. As these compounds can be accumulated by cells up to high concentrations without any toxic effect, they prevent water loss and maintain cell turgor when the cells are in a hypertonic environment (Yancey *et al.*, 1982), and at the same time protect the intracellular proteins from the noxious effects of abnormally high ion concentrations, particularly of sodium, and/or of extreme temperatures (Papageorgiou and Murata, 1995). In plants, GB is the most widely used and effective osmoprotectant (Rhodes and Hanson, 1993). The plants that synthesize it—known as GB accumulators—tolerate osmotic stress much better than the nonaccumulators (Rhodes and Hanson, 1993). Because of the importance of the synthesis of GB for the plant to successfully contend with osmotic stress, and because many important crops are GB nonaccumulators, the engineering of the ability to synthesize this osmoprotectant in these plants has been, and still is, a biotechnological goal (McNeil *et al.*, 1999). Thus, rice, tomato, potato, carrot, alfalfa, tobacco, and cotton plants have been transformed with bacterial or plant genes involved in the GB biosynthetic pathway in an attempt to convert them into GB accumulators. However, although the

**Table 3.1** Glycine betaine-accumulator plants used as food by human or animals.[a]

| Plant genus (vulgar name) | References |
|---|---|
| *Amaranthus* (amaranth)** | Valenzuela -Soto and Muñoz-Clares (1994) |
| *Atriplex* (orach, salt bushes)** | Storey and Wyn Jones (1975), Tipirdamaz *et al.* (2006) |
| *Beta* (sugar beet, beetroot, sea beet)** | Hanson and Wyse (1982), Catusse *et al.* (2008) |
| *Chenopodium* (goosefoot, quinoa)** | Tipirdamaz *et al.* (2006) |
| *Spinacia oleracea* (spinach)** | Weretilnyk *et al.* (1989) |
| *Salicornia* (glasworth, sea beans)** | Tipirdamaz *et al.* (2006) |
| *Suaeda* (seepweed, seblite)** | Storey and Wyn Jones, (1975), Tipirdamaz *et al.* (2006) |
| *Avena* (oats)* | Hitz and Hanson (1980) |
| *Chloris guyana* (Rhodes grass)* | Storey and Wyn Jones (1975) |
| *Hordeum* (barley)** | Hitz and Hanson (1980), Ishitani *et al.* (1993) |
| *Sorghum* (sorghum)* | Grieve and Maas (1984), Ishitani *et al.* (1993) |
| *Seacale* (rye)** | Hitz and Hanson (1980), Ishitani *et al.* (1993) |
| *Triticum* (wheat)** | Hitz and Hanson (1980), Ishitani *et al.* (1993) |
| *Zea* (maize)* | Hitz and Hanson (1980), Ishitani *et al.* (1993) |

[a] The list is not exhaustive. Other GB-accumulator plants are used as food by certain human communities, or as animal forage in particular geographical regions. High GB accumulators are marked with two asterisks; medium GB accumulators are marked with one.

synthesis of GB was increased in the transgenic plants, the levels of GB that they attained when subjected to osmotic stress conditions were much lower that those found in the natural GB-accumulator plants. This outcome stresses the need to get a much fuller understanding of the catalytic properties and regulation of the enzymes involved in GB biosynthesis, and the need to identify and study any other factor that might limit this synthesis. Given the observed health benefits of the intake of GB in humans (Craig, 2004), which is the subject of Chapter 3 of this book, the ability of an edible plant to accumulate GB is important not only from an agricultural but also from a nutritional point of view. Moreover, it has been argued that GB-accumulator plants impart important health benefits to animal that feed on them (Joseph *et al.*, 2013). In fact, GB has been used for many years as a dietary supplement in animal nutrition (references in Craig, 2004). This role of GB in human and animal nutrition undoubtedly adds interest to the mentioned efforts to increase the synthesis of GB in crops or forage plants. The main GB-accumulator plants used as food for humans or animals are given in Table 3.1. The content of GB in these plants under normal growth conditions is highly increased when they are grown under drought, salt or cold conditions.

## 3.3 Biosynthesis of Glycine Betaine in Plants

In plants, GB is formed from choline (Cho) through a short pathway consisting only of two steps: the oxidation of the alcohol group of Cho to betaine aldehyde (BAL) in a reaction catalyzed by Cho mono-oxygenases (E.C.1.14.15.7, CMOs), enzymes that are unique to plants (Brouquisse

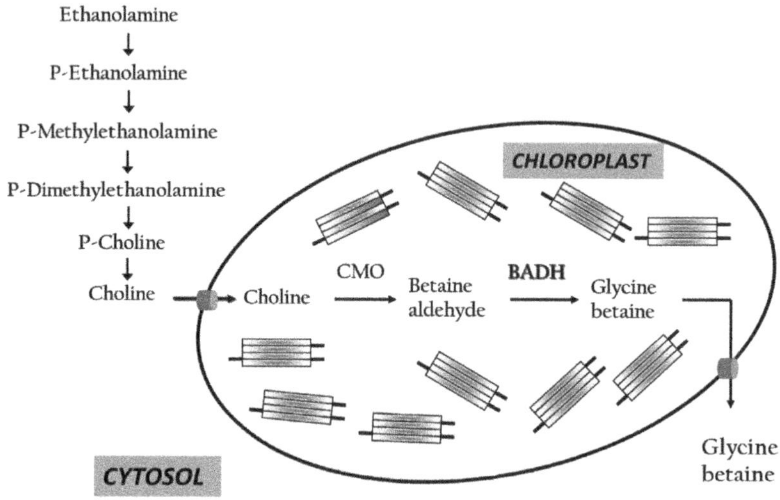


**Figure 3.2** Biosynthetic pathway of glycine betaine from ethanolamine in spinach. Choline is synthesized in the cytosol and must be transported into the chloroplast by a specific transporter located in the membrane. CMO, choline mono-oxygenase; BADH, betaine aldehyde dehydrogenase.

*et al.*, 1989), and the oxidation of the aldehyde group of BAL to the carboxyl group of GB in a reaction catalyzed by betaine aldehyde dehydrogenases (E.C. 1.2.1.8, BADHs) (Weretilnyk and Hanson,1989). No other enzymes with Cho oxidase or BAL oxidase activity have been found in plants so far. In GB accumulators, BADH and CMO enzymes are induced by drought, salt and cold (Weretilnyk and Hanson, 1989; Valenzuela-*Soto* and Muñoz-*Clares*, 1994; Rusell *et al.*, 1998), consistent with the relevance of the accumulation of GB in the plant defense against these adverse environmental conditions. In many of these GB accumulators the synthesis of GB takes place in the chloroplasts since both CMO and BADH are chloroplastic enzymes (Hanson *et al.*, 1985), but it has been reported that other plants have peroxisomal BADH and CMO enzymes (Fujiwara *et al.*, 2008; Mitsuya *et al.*, 2011). Cho is synthesized in the cytosol from ethanolamine, which is phosphorylated to phosphoetanolamine by a kinase present in all plants and then trimethylated by SAM:*N*-methyltransferases producing P-Cho. Cho is then formed by the action of a P-Cho phosphatase (Summers and Weretilnyk, 1993) (Figure 3.2).

For the successful engineering of the biosynthesis of GB, the identification of the factors that may limit this synthesis in nonaccumulator plants is of pivotal importance. The catalytic properties and levels of BADH and CMO enzymes, the synthesis of Cho and the import of Cho into the chloroplast or peroxisomes are the main possible factors, and will be discussed below.

### 3.3.1 Betaine Aldehyde Dehydrogenase (BADH)

BADH enzymes catalyze the irreversible, $NAD^+$-dependent oxidation of BAL to GB (Figure 3.3). Within the aldehyde dehydrogenase (ALDH) superfamily, plant BADHs belong to the family 10 (ALDH10). The plants ALDH10 enzymes exhibit a certain degree of substrate promiscuity since they are ω-aminoaldehyde dehydrogenases that apart from BAL can oxidize small aldehydes possessing ω-primary amine groups, such as 3-aminopropionaldehyde (APAL) and 4-aminobutyraldehyde (ABAL) (Trossat *et al.*, 1997; Vojtěchová *et al.*, 1997), ω-trimethylammonium group, such as 4-trimethylaminobutyraldehyde (TMA-BAL) (Fujiwara *et al.*, 2008), or ω-dimethylsulfonium groups, as 3-dimethylsulfoniopropionaldehyde DMSPAL (Trossat *et al.*, 1997; Vojtěchová *et al.*, 1997) (Figure 3.4). Plant ALDH10 enzymes, therefore, can participate not only in synthesis of GB, if they use BAL as substrate, but also in other important processes as: (1) the synthesis of other osmoprotectants—such as β-alanine betaine from APAL (Rathinasabaphathi *et al.*, 2000), dimethylsulfoniopropionate from DMSPAL (Trossat *et al.*, 1997; Vojtěchová *et al.*, 1997) and 4-aminobutyrate from APAL; (2) polyamines catabolism, where APAL and ABAL are intermediates; and (3) synthesis of carnitine—which in plants not only participates in fatty acid metabolism but is also an osmoprotectant (Charrier *et al.*, 2012)—from TMABAL. Although the so far biochemically characterized plant ALDH10 enzymes oxidize all the above-mentioned ω-aminoaldehydes, only some of them can efficiently use BAL as substrate and, therefore, participate in the synthesis of GB. Studying the enzyme from spinach (*So*BADH)—which is a high GB accumulator—by means of X-ray crystallography, docking of the BAL substrate, and site-directed mutagenesis, it was recently found (Díaz-Sánchez *et al.*, 2012) that only an amino acid residue—Ala441 (*So*BADH numbering) that is located behind one active-site residue—is critical for the enzyme to accept or reject BAL as substrate, because this residue determines the size of the pocket where the bulky trimethylammonium group of BAL binds. If this residue is an alanine or a cysteine, there is enough room in the active site for BAL binding, whereas if it is a bigger residue, an isoleucine, it pushes the active-site residue reducing the size of the cavity where the trimethylammonium group binds, so that the binding of BAL is hindered (Díaz-Sánchez *et al.*, 2012). By correlating the reported level of BADH activity with the presence of either of these residues, the prediction was made that those enzymes that have an isoleucine would not have significant BADH activity, while those that have an alanine or a cysteine would have this activity. Moreover, a perfect correlation was found between the ability of the plant to accumulate GB and the presence of an ALDH10 isoenzyme with probed or predicted—on the basis of the nature of the residue at this critical position—BADH activity. This finding strongly supports the proposal that the absence of an ALDH10 isoenzyme containing alanine or cysteine at the position equivalent to 441 of *So*BADH is the major limitation for the synthesis of GB in nonaccumulator plants (Díaz-Sánchez *et al.*, 2012). Amino acid sequence analysis indicated that: (i) the ancestral plant ALDH10s contained an isoleucine at this critical position and therefore they

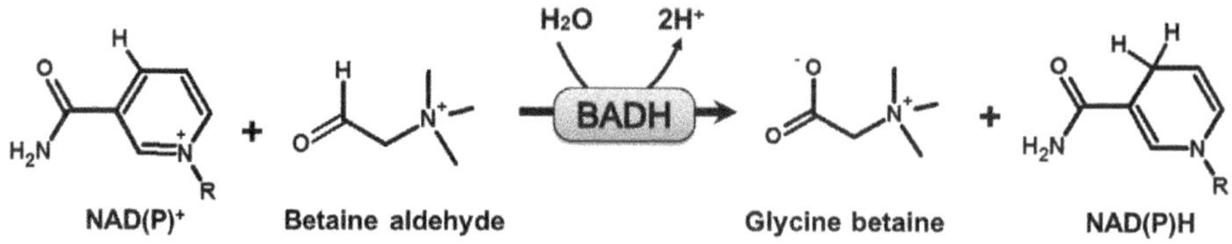


**Figure 3.3** Reaction catalyzed by the plant BADH enzymes. The aldehyde group of BAL is oxidized to the carboxyl group of GB at the same time that the coenzyme NAD+ (oxidized nicotinamide adenine dinucleotide) is reduced to NADH (reduced nicotinamide adenine dinucleotide).

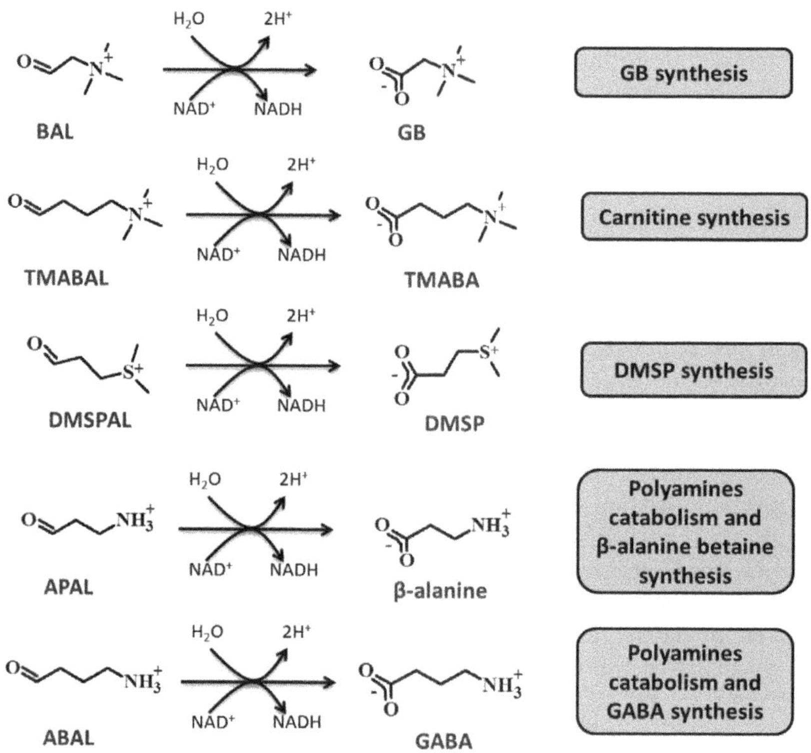


**Figure 3.4** Aldehyde substrates of the plant ALDH10 enzymes. The plant members of the ALDH10 family of the ALDH superfamily are aminoaldehyde dehydrogenase enzymes that *in vitro* can accept several small aminoaldehydes as substrates. *In vivo*, depending of the substrate used, these enzymes may be involved in several important biochemical processes, some of them essential for the plants as the catabolism of polyamines. Only some of the plant ALDH10 enzymes can efficiently oxidize BAL and therefore participate in GB synthesis. BAL, betaine aldehyde; TMABAL, 4-trimethylaminobutyraldehyde; DMSPAL, 3-dimethylsulfoniopropionaldehyde; APAL, 3-aminopropionaldehyde; ABAL, 4-aminobutyraldehyde.

probably were devoid of BADH activity, and (ii) most higher plants have two ALDH10 isoenzymes as a result of a gen duplication event. In some plants, one of the two copies of the gene was mutated so that the isoleucine changed to alanine or cysteine; as a consequence the enzyme acquired BADH activity and the plant the capability to accumulate GB. This specialization event occurred several times during higher plant evolution and in different plant families (Díaz-Sánchez *et al.*, 2012), indicating that it confers a great adaptive advantage to plants, not only to halophytes, which grow in habitats where saline soils are prevalent, but also to mesophytes, which may experience sporadic episodes of water deficit or freezing temperatures (Figure 3.5).

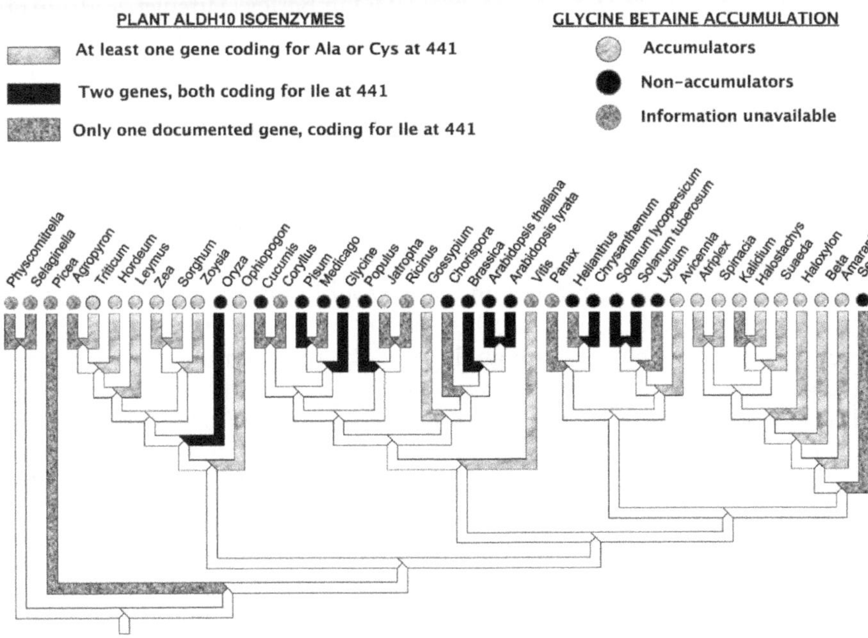


**Figure 3.5** Correlation between the kind of amino acid residue present at position 441 of the ALDH10 isoenzymes and the ability of the plant to synthesize glycine betaine. This figure includes those plants whose ALDH10 enzymes have been sequenced and/or its ability to synthesize GB has been reported. A perfect correlation was found between the plant being a GB accumulator and having one ALDH10 isoenzyme that contains alanine (Ala) or cysteine (Cys) at the position equivalent to 441 of *So*BADH. The nonaccumulators only have ALDH10 isoenzymes that have isoleucine (Ile) at this position. In some of the plants included in the figure, the ALDH10 amino acid sequence or its ability to synthesize GB have not yet been reported, but it can be predicted that those plants that have an Ala-441 or Cys-441-containing ALDH10 are GB accumulators, and *vice versa*, that those plants having only Ile-441-containing ALDH10 isoenzymes are nonaccumulators. Similarly, it can be predicted that if the plant is a GB accumulator it has an Ala-441 or Cys-441-containing ALDH10 isoenzyme, whereas if it is a nonaccumulator both of its ALDH10 isoenzymes are of the Ile-441-containing kind.

### 3.3.2 Choline Mono-Oxygenase (CMO)

Our current knowledge about CMO enzymes mostly comes from studies on the enzyme from spinach (*So*CMO). Although the *cmo* gene has been found in every plant studied so far, only the *So*CMO protein has been purified and its catalytic activity measured. *So*CMO is a chloroplastic nonheme iron mono-oxygenase that contains Rieske-type [2Fe–2S] and mononuclear iron centers (Rathinasabapathi *et al.*, 1997). It catalyzes the irreversible oxidation of Cho, using molecular oxygen and the electrons provided by reduced Fd and forming a water molecule and betaine aldehyde hydrate, which in aqueous solution is in equilibrium with the aldehyde form that is the BADH substrate (Figure 3.6). The report of a peroxisomal CMO in barley (Mitsuya *et al.*, 2011) is at odds with the observation that the activity of *So*CMO requires the electrons provided by reduced ferredoxin (Fd) since it is known that plant Fds are plastidic proteins (Hanke and Mulo, 2013). However, peroxisomal CMO enzymes, which have not been yet biochemically characterized, might use an electron donor different from Fd. Indeed, the current experimental evidence, although scarce, suggests that there are two kinds of CMO enzymes, which could be the chloroplastic and peroxisomal ones. The existence of these two kinds is supported by the finding that the CMO enzymes from mangrove, barley (both GB-accumulator plants) and sunflower (a nonaccumulator) are not recognized by antibodies raised against the *So*CMO (Hibino *et al.*, 2001). Also, a recently reported phylogenetic analysis of the CMO family clearly showed that CMO proteins form two groups (Joseph *et al.*, 2013). However, these findings do not show a correlation between the kind of CMO enzyme that plant possesses and its ability to synthesize GB. The biochemical characterization of these two possible kinds of CMO enzymes will clarify whether they are functionally different or not, and, in the case that they are, it will show which differences exist between them as well as the physiological relevance of these differences. In fact, a functional CMO has been so far found only in plants that also have active BADH enzymes, for instance amaranth (Russell *et al.*, 1998), spinach (Rathinasabapathi *et al.*, 1997), sugar beet (Russell *et al.*, 1998), and barley (Mitsuya *et al.*, 2011), whereas a nonfunctional gene was

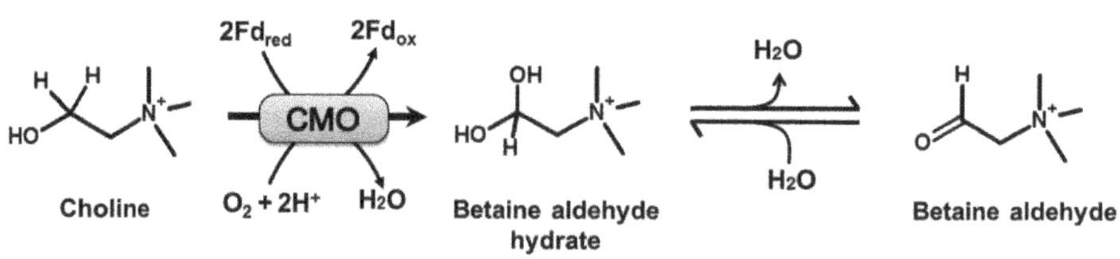


**Figure 3.6** Reaction catalyzed by the *So*CMO enzyme. The alcohol group of Cho is oxidized to the aldehyde group of BAL, while molecular oxygen is reduced to water. The two electrons needed in this reaction are provided by two ferredoxin (Fd) molecules, and the two protons are taken from the aqueous solvent. In chloroplasts, Fd is reduced during photosynthesis. $Fd_{red}$, reduced ferredoxin; $Fd_{ox}$, oxidized ferredoxin.

found in rice (Luo *et al.*, 2007), a nonaccumulator plant, and the recombinant CMO protein from *Arabidopsis*, another nonaccumulator plant, has no activity (Hibino *et al.*, 2002). However, in the latter experiments the lack of CMO activity was inferred from the absence of GB accumulation, but it is possible that the enzyme was active with a substrate different from choline. Given that BAL is highly toxic (Rathinasabapathi *et al.*, 1996), a high CMO activity would be deleterious for the plant if it occurred without an accompanying BADH activity that converts BAL to the corresponding innocuous acid (GB). And since, as mentioned in Section 3.1, GB-nonaccumulator plants are unable to oxidize BAL, or have a very low BADH activity, it can be hypothesized that either (i) their *cmo* genes are not expressed or expressed at a very low level, (ii) their CMO protein are nonfunctional, or (iii) their CMO enzymes have a different catalytic activity—*i.e.* they oxidize a substrate different from Cho. The determination of the substrate specificity of CMO enzymes, as well as of the nature of possible electron donors different from Fd, is lacking and deserves attention. The results obtained with transgenic nonaccumulator plants that have been transformed with highly active BADH enzymes—such as *So*BADH—indicated that the CMO protein of the nonaccumulator plant exists and is active. But its level and/or its activity with Cho could be insufficient to supply enough BAL to the transformant BADH enzyme, thus explaining the small increases in GB content observed in the transgenic plants. Similarly, the transformation of rice, a nonaccumulator plant, with the *cmo* gene from spinach, resulted in low GB accumulation (Shirasawa *et al.*, 2006), a result that could have been expected given that rice has ALDH10 isoenzymes that lack BADH activity. Therefore, the experimental evidence so far does not support the notion that CMO is the primary constraint on GB production in GB-deficient plants (Nuccio *et al.*, 1998). But it is true that the activity of the CMO enzyme cannot exceed that of the BADH even in GB-accumulator plants—given that CMO catalyses an irreversible reaction—otherwise the accumulation of BAL would have deleterious effects for the plant. This is consistent with both enzymes being simultaneously overexpressed by osmotic stress signals in GB-accumulator plants, as already mentioned.

### 3.3.3 Synthesis and Transport of Choline

As discussed above, most of the reported failures of transgenic nonaccumulators to synthesize significant amounts of GB can be explained by the plants being transformed with just one of the two genes, *cmo* or *badh*, needed for GB synthesis, when it is reasonable to expect that nonaccumulator plants should be deficient in both activities. But, it is also possible than in nonaccumulator plants the extent of GB synthesis may be limited by other factors, such as insufficient synthesis of Cho or low import of Cho into the chloroplast. The former has been thought to be the cause behind the failure of tobacco plants transformed with the *cmo* gen from spinach to attain levels of GB similar to those reached by the natural GB-accumulators (Nuccio *et al.*, 1998).

Including Cho or P-Cho into the culture medium significantly increased the synthesis of GB by these plants, but the amount of GB produced was still very low. Cho synthesis may indeed be limiting in nonaccumulator plants, since it has been found that the activity of phosphoethanolamine *N*-methyltransferase, which is the rate-limiting enzyme in the Cho biosynthetic pathway, is much lower in tobacco than in spinach, and it has been proved that tobacco transgenic plants overexpressing this enzyme accumulate higher amounts of GB (McNeil *et al.*, 2001). But he conclusion that Cho synthesis is the limiting factor in the tobacco transgenic plants transformed with *So*CMO was drawn assuming that a significant endogenous BADH activity exists in tobacco, which is not the case. Therefore, BADH activity, not the supply of choline to CMO, was probably the major limiting factor for GB synthesis in these experiments. But in a later study (Nuccio *et al.*, 2000) the tobacco plants were transformed with both *cmo* and *badh* genes targeted to the chloroplast and still the transgenic plants produced low amounts of GB. In another experiment, tobacco plants transformed with the bacterial gen for Cho oxidase—an enzyme that is able to catalyze the two oxidation steps of Cho to GB—synthesized more GB than the plants expressing the chloroplastic *So*CMO and *So*BADH enzymes. Since the Cho oxidase protein was expressed in the cytosol, the authors logically concluded that the transport of Cho into the chloroplasts is not sufficient to support a high GB synthesis in nonaccumulator plants. No Cho transporter in the chloroplast membrane has so far been reported. This is another gap in our current knowledge that should be filled to fully understand and successfully manipulate the GB synthesis process.

## 3.4 Concluding Remarks

The synthesis of GB in plants is a metabolic trait of major adaptive importance given that it greatly increases the plant tolerance to several abiotic stresses that result in water deficit. The acquisition of the ability to synthesize GB during the evolution of terrestrial plants allowed them to grow in very adverse environments, like in coastal highly saline lands, or in dry or cold climates. As has been recently shown, the mutation of a single residue in the ancestral plant ALDH10 enzymes suffices to confer on them BADH activity, which is a *sine qua non* condition for the plant to become a GB accumulator. But the current experimental evidence obtained with wild-type and transgenic plants suggests that the acquisition of the ability to synthesize GB by evolving a BADH activity should have been accompanied by other adaptations, such as the gain of a significant CMO activity, the regulation by osmotic stress signals of the levels of CMO and BADH as well as of the critical enzymes of the choline biosynthetic pathway, and probably also the development of efficient Cho, BAL and GB transport through intracellular membranes.

Finally, the present experimental evidence points to the existence of two kinds of GB-accumulator plants: high accumulators—for instance amaranth, spinach, sugar beet, barley and wheat—and medium accumulators—such as maize and sorghum. For the success of future biotechnological manipulations to improve the accumulation of GB, it is of pivotal importance to fully understand the molecular bases of the differences in the GB biosynthetic pathway not only between accumulators and nonaccumulators plants but also between high and medium accumulators.

## Summary Points

- This chapter focuses on the synthesis and accumulation of glycine betaine (GB) in plants.
- The main role of GB in plants is osmoprotection.
- The accumulation of GB by plants is important not only for agriculture and biodiversity but also for human and animal nutrition.
- Plants can be GB accumulators or GB nonaccumulators.
- The GB-accumulator plants are tolerant to drought, saline soils, and cold.
- Many important crops are GB nonaccumulators.
- GB in plants is synthesized from choline in two reactions catalyzed by choline mono-oxygenase (CMO) and betaine aldehyde dehydrogenase (BADH).
- Only GB-accumulator plants have an enzyme with significant BADH activity.
- CMO activity also appears to be different in GB accumulators than in GB nonaccumulators.
- Transgenic nonaccumulator plants expressing bacterial or plant GB-biosynthetic genes produce GB at much lower concentrations that the natural accumulators.
- In most cases, this negative result was due to the plant being transformed with just one of the two genes, *cmo* or *badh*, needed for GB synthesis.
- Choline supply and/or transport may also be factors that limit GB synthesis in transgenic nonaccumulator plants.
- More basic studies are needed to fully understand the biochemical aspects of GB synthesis in plants.

## Key Facts

### Key Facts of Enzyme Catalysis

- Most biochemical reactions are kinetically impossible; *i.e.*, they would not take place at an appreciable rate without being catalyzed.
- Enzymes are the biocatalysts that accelerate the biochemical reactions.
- Enzymes are exceptionally efficient catalysts, increasing the rate of reactions by many orders of magnitude.
- Another important feature of enzyme catalysis is the enzyme specificity for their substrates.

- This specificity is the result of the enzyme ability to discriminate between molecules, so that it binds only one molecule or a small group of related molecules.
- This is achieved by specific steric and chemical recognition of the substrate by the amino acid residues of the enzyme active site.
- Once the substrate is bound, the enzyme catalyzes only one reaction, even if the substrate may participate in different reactions.

## Definition of Words and Terms

**Amino acid sequence analysis**: Evaluation of the degree of conservation of the amino acid residues of homolog proteins by aligning their amino acid sequences.
**Catalytic activity**: A measure of the ability of an enzyme to increase the rate of a given reaction.
**Coenzyme**: Organic molecule needed for some enzymes to catalyze a given reaction.
**Docking**: Bioinformatic technique that permits the prediction of the interactions made by a molecule with a protein of known three-dimensional structure, by means of modeling the position of the ligand molecule in selected regions of the protein.
**Homologs**: Proteins that have evolved from a common ancestor and share a similar three-dimensional structure and some amino acid sequence identity.
**Osmoprotectant**: Small, neutral, and highly soluble organic molecule that can be accumulated up to very high concentrations inside the cell without any toxic effect, thus counteracting the tendency of water to exit the cell when exposed to an environment with low water availability, high salt concentrations, or ice and at the same time protecting proteins from the noxious effects of salts or dehydration.
**Osmotic stress**: Adverse environmental condition that causes a cell to lose water and volume.
**Phylogenetic family**: Group of homolog proteins that have a high percentage of amino acid sequence identity, usually higher than 40%.
**Site-directed mutagenesis**: Technique that changes selected amino acid(s) residue(s) originating a mutated protein.
**Substrate specificity**: The ability of enzymes to bind only a molecule or related groups of molecules and catalyze only one of various possible reactions in which the molecule(s) may be involved.
**Transgenic plants**: Plants in which one or several genes from a different organism have been introduced, and the proteins coded by these genes are expressed.
**X-ray crystallography**: A powerful technique that allows the determination of the three-dimensional structure of molecules, even of very high molecular weight such as proteins, by diffracting an X-ray beam through a crystal of the studied molecule.

**Transgenic plants**: Plants in which one or several genes from a different organism have been introduced, and the proteins coded by these genes are expressed.

## Abbreviations

| | |
|---|---|
| ALDH10 | Aldehyde dehydrogenase of the family 10 |
| APAL | 3-Aminopropionaldehyde |
| BADH | Betaine aldehyde dehydrogenase |
| BAL | Betaine aldehyde |
| Cho | Choline |
| CMO | Choline mono-oxygenase |
| Fd | Ferredoxin |
| GB | Glycine betaine (betaine) |
| *So*BADH | Spinach betaine aldehyde dehydrogenase |
| *So*CMO | Spinach choline mono-oxygenase. |

## Acknowledgements

Part of the work of the author's research group reported here was financially supported by DGAPA UNAM grant (PAPIIT IN204708). The author acknowledges the help of A.G. Díaz-Sánchez, L.P. Martínez-Castilla and H. Riveros-Rosas in the drawing of the figures.

## References

Brouquisse, R., Weigel, P., Rhodes, D., Yocum, C. F. and Hanson, A. D., 1989. Evidence for a ferredoxin-dependent choline monooxygenase from spinach chloroplast stroma. *Plant Physiology*. 90: 322–329.

Catusse, J., Strub, J.-M., Job, C., Dorsselaer A. V. and Job, D., 2008. Proteome-wide characterization of sugarbeet seed vigor and its tissue specific expression. *Proceedings of the National Academy of Sciences of the United States of America*. 105: 10262–10267.

Charrier, A., Rippa, S., Yu, A., Nguyen, P. J., Renou, J. P. and Perrin, Y., 2012. The effect of carnitine on *Arabidopsis* development and recovery in salt stress conditions. *Planta*. 235: 123–135.

Craig, S. A. S., 2004. Betaine in human nutrition. *American Journal of Clinical Nutrition*. 80: 539–549.

du Vigneaud, V., Simmonds, S., Chandler, J. P. and Cohn, M., 1946. A further investigation of the role of betaine in transmethylation reactions *in vivo*. *Journal of Biological Chemistry*. 165: 639–648.

Díaz-Sánchez, A. G., González-Segura, L., Mújica-Jiménez, C., Rudiño-Piñera, E., Montiel, C., Martínez-Castilla, L. P. and Muñoz-Clares, R. A., 2012. Amino acid residues critical for the specificity for betaine aldehyde of the plant ALDH10 isoenzyme involved in the synthesis of glycine betaine. *Plant Physiology*. 158: 1570–1582.

Fujiwara, T., Horia, K., Ozaki, K., Yokota, Y., Mitsuya, S., Ichiyanagi, T., Hattori, T. and Takabe, T., 2008. Enzymatic characterization of peroxisomal and cytosolic betaine aldehyde dehydrogenases in barley. *Physiologia Plantarum*. 134: 22–30.

Grieve, C. M. and Maas, E. V., 1984. Betaine accumulation in salt-stressed sorghum. *Physiologia Plantarum*. 61: 167–171.

Hanke, G. and Mulo, P., 2013. Plant type ferredoxins and ferredoxin-dependent metabolism. *Plant Cell and Environment*. 36: 1071–1084.

Hanson, A. D. and Wyse, R., 1982. Biosynthesis, translocation, and accumulation of betaine in sugar beet and its progenitors in relation to salinity. *Plant Physiology*. 170: 1191–1198.

Hanson, A. D., May, A. M., Grumet, R., Bode, J., Jamieson, G. C. and Rhodes, D., 1985. Betaine synthesis in chenopods: Localization in chloroplasts. *Proceedings of the National Academy of Sciences of the United States of America*. 82: 3678–3682.

Hibino, T., Meng, Y-.L., Kawamitsu, Y., Uehara, N., Matsuda, N., Tanaka, Y., Ishikawa, H., Baba, S., Takabe, T., Wada, K., Ishii, T. and Takabe, T., 2001. Molecular cloning and functional characterization of two kinds of betaine-aldehyde dehydrogenase in betaine-accumulating mangrove *Avicennia marina* (Forsk.) Vierh. *Plant Molecular Biology*. 45: 353–363.

Hibino, T., Waditee, R., Araki, E., Ishikawa, H., Aoki, K., Tanaka, Y. and Takabe, T., 2002. Functional characterization of choline monooxygenase, an enzyme for betaine synthesis in plants. *Journal of Biological Chemistry*. 277: 41352–41360.

Hitz, W. D. and Hanson, A. D., 1980. Determination of glycine betaine by pyrolysis-gas chromatography in cereals and grasses. *Phytochemistry*. 19: 2371–2374.

Ishitani, M., Arakawa, K., Mizuno, N., Kishitani, S. Takabe, T., 1993. Betaine aldehyde dehydrogenase in the Gramineae: levels in leaves of both betaine-accumulating and nonaccumulating cereal plants. *Plant Cell Physiology*. 34: 493–495.

Joseph, S., Murphy, D. and Bhave, M., 2013. Glycine betaine biosynthesis in saltbushes (*Atriplex spp.*) under salinity stress. *Biologia*. 68: 879–895.

Luo, D., Niu, X., Wang, Y., Zheng, W., Chang, L., Wang, Q., Wei, X., Yu, G., Lu, B.-R. and Liu, Y., 2007. Functional defect at the rice choline monooxygenase locus from an unusual post-transcriptional processing is associated with the sequence elements of short-direct repeats. *New Phytologist*. 175: 439–447.

McNeil, S. D., Nuccio, M. L. and Hanson, A. D., 1999. Betaines and related osmoprotectants. Targets for metabolic engineering of stress resistance. *Plant Physiology*. 120: 945–949.

McNeil, S. D., Nuccio, M. L., Ziemak, M. J. and Hanson, A. D., 2001. Enhanced synthesis of choline and glycine betaine in transgenic tobacco plants that overexpress phosphoethanolamine *N*-methyltransferase. *Proceedings of the National Academy of Sciences of the United States of America*. 98: 10001–10005.

Mitsuya, S., Kuwahara, J., Ozaki K., Saeki, E., Fujiwara, T. and Takabe, T., 2011. Isolation and characterization of a novel peroxisomal choline monooxygenase in barley. *Planta*. 234: 1215–1226.

Nuccio, M. L., Russell, B. L., Nolte, K. D., Rathinasabapathi, B., Gage, D. A. and Hanson, A. D., 1998. The endogenous choline supply limits glycine betaine synthesis in transgenic tobacco expressing choline monooxygenase. *Plant Journal*. 16: 487–496.

Nuccio, M. L., McNeil, S. D., Ziemak, M. J., Hanson, A. D., Jain, R. K. and Selvaraj, G., 2000. Choline import into chloroplasts limits glycine betaine synthesis in tobacco: Analysis of plants engineered with a chloroplastic or a cytosolic pathway. *Metabolic Engineering*. 2: 300–311.

Papageorgiou, G. C. and Murata, N., 1995. The unusually strong stabilizing effects of glycinebetaine on the structure and function in the oxygen evolving photosystem II complex. *Photosynthetica*. 44: 243–252.

Rathinasabapathi, B., McCue, K. F., Gage, D. A. and Hanson, A. D., 1996. Metabolic engineering of glycine betaine synthesis: plant betaine aldehyde dehydrogenases lacking typical transit peptides are targeted to tobacco chloroplasts where they confer betaine aldehyde resistance. *Planta*. 193: 155–162.

Rathinasabapathi, B., Burnet, M., Russell, B. L., Gage, D. A., Liao, P.-C., Nye, G. J., Scott, P., Golbeck, J. H. and Hanson, A. D., 1997. Choline monooxygenase, an unusual iron-sulfur enzyme catalyzing the first step of glycine betaine synthesis in plants: prosthetic group characterization and cDNA cloning. *Proceedings of the National Academy of Sciences of the United States of America*. 94: 3454–3458.

Rathinasabapathi, B., Sigua, C., Ho, J. and Douglas, A. G., 2000. Osmoprotectant β-alanine betaine synthesis in the Plumbaginaceae: *S*-adenosyl-L-methionine dependent *N*-methylation of β-alanine to its betaine is via *N*-methyl and *N,N*-dimethyl β-alanines. *Physiologia Plantarum*. 109: 225–231.

Rhodes, D. and Hanson, H. D., 1993. Quaternary ammonium and tertiary sulfonium compounds in higher plants. *Annual Review of Plant Physiology and Plant Molecular Biology*. 44: 357–384.

Russell, B. L., Rathinasabapathi, B. and Hanson, A. D., 1998. Osmotic stress induces expression of choline monooxygenase in sugar beet and amaranth. *Plant Physiology*. 116: 859–865.

Shirasawa, K., Takabe, T., Takabe T. and Kishitani S., 2006. Accumulation of glycinebetaine in rice plants that overexpress choline monooxygenase from spinach and evaluation of their tolerance to abiotic stress. *Annals of Botany*. 98: 565–571.

Storey R. and Wyn Jones, R. G., 1975. Betaine and choline levels in plants and their relationship to NaCl stress. *Plant Science Letters*, 4: 161–168.

Summers, P. S. and Weretilnyk, E. A., 1993. Choline synthesis in spinach in relation to salt stress. *Plant Physiology*. 103: 1269–1276.

Tipirdamaz, R., Gagneul, D., Duhazé, C., Aïnouche, A., Monnier, C., Özkuma, D. and Larher, F., 2006. Clustering of halophytes from an inland salt marsh in Turkey according to their ability to accumulate sodium and nitrogenous osmolytes. *Environmental and Experimental Botany*. 57: 139–153.

Trossat, C., Rathinasabapathi, B. and Hanson, A. D., 1997. Transgenically expressed betaine aldehyde dehydrogenase efficiently catalyzes oxidation of dimethylsulfoniopropionaldehyde and [omega]-aminoaldehydes. *Plant Physiology*. 113: 1457–1461.

Valenzuela-Soto, E. and Muñoz-Clares, R. A., 1994. Purification and properties of betaine aldehyde dehydrogenase extracted from detached leaves of *Amaranthus hypocondriacus L.* subjected to water deficit. *Journal of Plant Physiology*. 143: 145–152.

Velasco-García, R., Villalobos, M. A., Ramirez-Romero, M. A., Mújica-Jiménez, C., Iturriaga, G. and Muñoz-Clares, R. A., 2006. Betaine aldehyde dehydrogenase from *Pseudomonas aeruginosa*: cloning, over-expression in *Escherichia coli*, and regulation by choline and salt. *Archives of Microbiology*. 185: 14–22.

Vojtěchová, M., Hanson, A. D. and Muñoz-Clares, R. A., 1997. Betaine-aldehyde dehydrogenase from amaranth leaves efficiently catalyzes the NAD-dependent oxidation of dimethylsulfoniopropionaldehyde to dimethylsulfoniopropionate. *Archives of Biochemistry and Biophysics*. 337: 81–88.

Weretilnyk, E. A., Bednarek, S., McCue, K. E., Rhodes, D. and Hanson, A. D., 1989. Comparative biochemical and immunological studies of the glycine betaine synthesis pathway in diverse families of dicotyledons. *Planta*. 178: 342–352.

Weretilnyk, E. A. and Hanson, A. D., 1989. Betaine aldehyde dehydrogenase from spinach leaves: purification, *in vitro* translation of the mRNA, and regulation by salinity. *Archives of Biochemistry and Biophysics*. 271: 56–63.

Yancey, P. H., Clark, M. E., Hand, S. C., Bowles, R. D. and Somero, G. N., 1982. Living with water stress: Evolution of osmolyte system. *Science*. **217**: 1214–1222.

# Section II
# Chemistry and Biochemistry

CHAPTER 4

# The Chemistry of Betaine

PANTELEIMON G. TAKIS[a†], KONSTANTINOS D. PAPAVASILEIOU[b†], ANASTASSIOS N. TROGANIS*[a], AND VASILIOS S. MELISSAS*[b]

[a]Department of Biological Applications and Technology, University of Ioannina, GR–451 10 Ioannina, Greece; [b]Department of Chemistry, School of Physical Sciences, University of Ioannina, GR–451 10 Ioannina, Greece
*E-mail: atrogani@cc.uoi.gr; melissas@chem.auth.gr

## 4.1 Introduction

In the 1860s, a German chemist, Scheibler, isolated a new organic base from sugar beet (*Beta vulgaris*). He coined the name "betaine" and identified its structure. Beet is still the main source of betaine, and commercially available betaine is obtained as a natural byproduct of the sugar beet industry (Lever and Slow, 2010).

Betaine exhibits a multifaceted biological role, since it acts as an organic osmolyte (protecting cells against osmotic stress in various ways) as well as methyl donor for important biological processes such as one-carbon metabolism and DNA methylation. Recent findings stressed betaine's contribution to biomolecules' (proteins, DNA) structure and function stability, modulating the thermostability of the latter *via* their interactions with the water molecules (Bruzdziak *et al.*, 2013).

Betaine in plants and micro-organisms protects cells against osmotic inactivation, where exposure to drought, high salinity, or temperature stress triggers betaine synthesis in mitochondria, which results in its accumulation

[†]Authors equally contributed

Food and Nutritional Components in Focus No. 7
Betaine: Chemistry, Analysis, Function and Effects
Edited by Victor R Preedy

Published by the Royal Society of Chemistry, www.rsc.org

**Table 4.1** Foods with high betaine content (Zeisel *et al.*, 2003; Craig, 2004). This table lists some of the major food sources of betaine (Zeisel *et al.*, 2003; Craig, 2004).

| Nutriment | | Betaine content (mg/100 g of nutriment) |
|---|---|---|
| Cereal | Wheat bran | 1339.35 |
| | Wheat germ | 1240.48 |
| Vegetables | Spinach (cooked) | 645.06 |
| | Beets (canned) | 296.73 |

in the cells. It is a compatible osmolyte that increases the water retention of cells, replaces inorganic salts, and protects intracellular enzymes against osmotically induced or temperature-induced inactivation (Craig, 2004). As a methyl donor, betaine provides a methyl group for the remethylation of homocysteine to methionine, which mainly occurs in the liver (Craig, 2004) through the BHMT enzyme.

The osmotic regulation and methyl donation procedures attributed to betaine are not independent because BHMT is osmoregulated with high tonicity reducing its expression, so that betaine metabolism (and hence the mobilization of methyl groups) decreases when osmolyte concentrations need to be maintained. (Lever and Slow, 2010). Reflecting these two main functions, the highest concentrations of betaine in mammals are found in the kidneys and the liver (Ross *et al.*, 2014).

Humans obtain betaine from foods that contain either betaine or choline-containing compounds (through choline's oxidation in mitochondria) (Ueland *et al.*, 2005). It is present in foods in variable amounts that are generally related to growing and osmotic stress conditions. Some examples of food with high betaine content are shown in Table 4.1 (Craig, 2004). High levels of betaine are also found in animal products, especially shellfish and chicken, with the main betaine source in the normal western diet being flour products like bread and pasta (de Zwart *et al.*, 2003), as well as cereal products (Lever and Slow, 2010). Betaine in foods was found to be heat-stable at 220 °C, and only small losses occurred (<14%) after 30 min at 250 °C. Large losses (60–80%) occur during boiling; however, they can be recovered from the water (de Zwart *et al.*, 2003). Betaine insufficiency is associated with the metabolic syndrome, lipid disorders and diabetes, and may have a role in vascular and other diseases (Lever and Slow, 2010). Betaine is important in development and reproduction, from the embryo preimplantation to infancy (Friesen *et al.*, 2007). Betaine is also considered as an antioxidant and studies show that it could be used as a therapeutic component to control neurological disorders (Amiraslani *et al.*, 2012).

## 4.2 Nomenclature and Generic Properties

Betaine (IUPAC name: *N,N,N*-trimethylammonioacetate, synonym: 1-carboxy-*N,N,N*-trimethylammonium inner salt) is a zwitterionic quaternary ammonium compound that according to the HMDB (Human Metabolome

Database) is also known as (carboxymethyl)trimethylammonium hydroxide inner salt, (trimethylammonio)acetate, 1-carboxy-*N,N,N*-trimethyl-methanaminium, 1-carboxy-*N,N,N*-trimethyl-methanaminium hydroxide, 1-carboxy-*N,N,N*-trimethylmethanaminium inner salt, abromine, alpha-earleine, aminocoat, betafin, betafin BCR, betafin BP, cystadane, ektasolve EE, finnstim, glycine betaine, glycocoll betaine, glycylbetaine, greenstim, loramine AMB-13, lycine, *N,N,N*-trimethylglycine, oxyneurine, rubrine C, trimethylaminoacetate, trimethylaminoacetic acid, trimethylbetaine glycine, trimethylglycine and trimethylglycocoll (Wishart *et al.*, 2009). It is a methyl derivative of the amino acid glycine and it has been characterized as a methylamine because of its three chemically reactive methyl groups (Yancey *et al.*, 1982; Craig, 2004).

In general, betaine and similar zwitterionic compounds were originally derived from other amino acids. By extension, neutral molecules having charge-separated forms with an onium atom, which bears no hydrogen atoms and that is not adjacent to the anionic atom, belong to a general family of compounds called betaines, which cannot be represented without formal charges (McNaught *et al.*, 1997). These are highly polar but neutral zwitterionic compounds and share the feature that their cationic functional groups cannot be deprotonated and retain their charges at high pH. Apart from betaine's role in the human diet, it is also used industrially as a moisturizer in the food and cosmetics industries and can also be found in retail outlets selling dietary supplements (Lever and Slow, 2010).

## 4.3 Physical and Chemical Properties – Chemistry of Important Functions

Betaine's molecular formula is $C_5H_{11}NO_2$ and its molecular weight is equal to 117.15 g mol$^{-1}$. It is an off-white to white powder with a characteristic odor. The structural formula of betaine is illustrated in Figure 4.1. With regard to its melting point, several values are reported in the literature, namely in the wide range of 250–260 °C (EFSA, 2013), 293 °C (Lide, 2005) and 310 °C (O'Neil, 2006). It has a p$K_a$ of 1.83 (at 0 °C) and a bulk density in the range of 0.35–0.50 g cm$^{-3}$ (EFSA, 2013). Betaine is a molecule of low symmetry, since it consists of only two symmetry elements: the identity E and a mirror plane $\sigma_h$. It therefore belongs to the $C_s$ group, as illustrated in Figure 4.1. It is fully soluble in water, soluble in methanol and ethanol as well as sparingly soluble in ether (EFSA, 2013). Consequently, organic synthesis of other amino acids betaines exploits methanol as a substrate (where the corresponding reaction is evolved) and their isolation-purification takes place in water (Goldberg *et al.*, 1990). Collected chemical and physical properties of betaine are listed in Table 4.2.

## 4.4 Chemistry of Important Functions

Betaine's unique osmolytic and protein stabilizing functions stem from its underlying chemistry. Very recent experimental evidence by spectroscopic (IR) and calorimetric (DSC) data confirmed the relationship between the

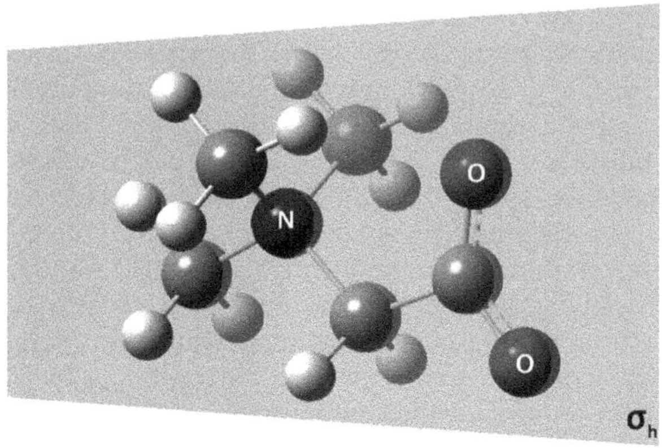


**Figure 4.1** Point group symmetry of zwitterionic betaine. Zwitterionic betaine's molecular structure. Oxygen and nitrogen atoms are selectively labeled, whereas the atom color coding is dark gray for carbon and light gray for hydrogen atoms. Betaine belongs to the $C_s$ symmetry point group. The mirror plane $\sigma_h$ is selectively depicted.

**Table 4.2** Physical and chemical properties of betaine. This table lists collected properties of betaine, including chemical identity information, typical registration codes as well as characteristic chemical and physical properties.

| **Molecular formula** | $C_5H_{11}NO_2$ |
|---|---|
| Molecular weight | 117.15 g mol$^{-1}$ (O'Neil, 2006) |
| Chemical composition | Carbon (51.26%)<br>Hydrogen (9.46%)<br>Nitrogen (11.96%)<br>Oxygen (27.31%) (O'Neil, 2006) |
| CAS number | 107-43-7 (Sigma-Aldrich, 2013) |
| EINECS number | 203-490-6 (Sigma-Aldrich, 2013) |
| Physical state | Solid (Wishart *et al.*, 2009) |
| Color and form | White powder (EFSA, 2013)<br>Deliquescent scales or prisms (O'Neil, 2006) |
| Taste and odor | Sweet taste (O'Neil, 2006)<br>Slight characteristic odor (EFSA, 2013) |
| Melting point | 310 °C (O'Neil, 2006)<br>293 °C (Lide, 2005)<br>250–260 °C (EFSA, 2013) |
| Bulk density | 0.35–0.50 g cm$^{-3}$ (EFSA, 2013) |
| Solubility in water (g/100 g solvent) | 157 g (EFSA, 2013)<br>160 g (O'Neil, 2006) |
| Solubility in organic solvents (g/100 g solvent) | Methanol: 57 (EFSA, 2013)<br>55 (O'Neil, 2006)<br>Ethanol: 8.7 (O'Neil, 2006) (EFSA, 2013)<br>Ether: sparingly soluble<br>Chloroform: slightly soluble (Lide, 2005) |
| Dissociation constant | $pK_a$ = 1.83 (0 °C) (Dawson *et al.*, 1986) |

osmolytes' influence on water structure and their ability to thermally stabilize lysozyme molecules, indicating that structural and energetic characteristics of stabilizing osmolytes' and lysozyme's hydration water molecules appear to be very similar (Bruzdziak *et al.*, 2013). Osmolytes increase lysozyme stabilization in the order bulk water < TMAO < TMG < Gly < DMG < NMG, while the distribution of the H-bond energy of water in the hydration spheres of osmolytes and lysozyme explains preferential hydration of the protein and sheds light on the osmophobic effect (Bruzdziak *et al.*, 2013). NMR spectroscopic data of alpha, beta $^{13}C$ nuclei relaxation times (Figure 4.2A and B) and amino $^{14}N$ nuclei linewidths ($\Delta\nu_{1/2}$) (Figure 4.2C) from six betaines, including

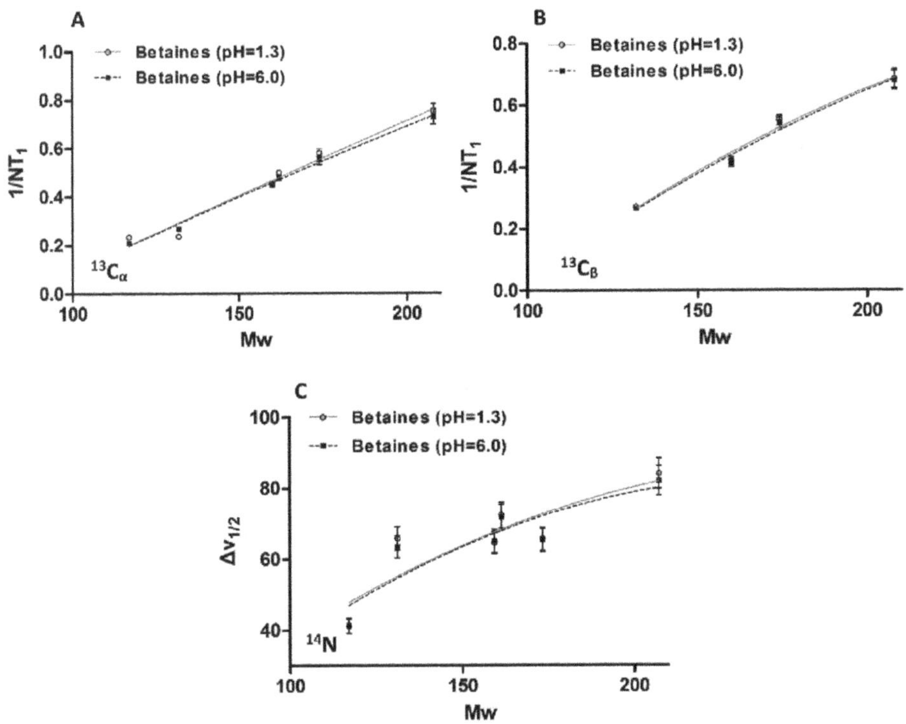


**Figure 4.2** Relaxation rates from: (A) alpha $^{13}C$-NMR of six betaines *vs.* their $M_w$, (B) beta $^{13}C$-NMR of five betaines *vs.* their $M_w$ and (C) $^{14}N$NMR of the six betaines amino groups *vs.* their $M_w$ in their aqueous solutions at two pH values. (A), (B) Spectroscopic $^{13}C_\alpha$ relaxation rates ($1/NT_1$ in $s^{-1}$) results plots of betaines for values in pH = 6.0 (■) and in pH = 1.3 (o), *vs.* their molecular weight ($M_w$). (C) Spectroscopic $^{14}N$ ($\Delta\nu_{1/2}$ in Hz) results plots of betaines in pH = 6.0 (■) and pH = 1.3 (o) *vs.* their molecular weight ($M_w$) respectively (Takis *et al.*, 2012). Statistically testing the nonlinear coincidence of all the above curves indicate that they statistically coincide. Taking into consideration that each curve represents the hydration grade of the understudy betaines in their two ionization states (Takis *et al.*, 2012, 2013), it is concluded that betaines directly interact with the same number of water molecules independent of their ionization state.

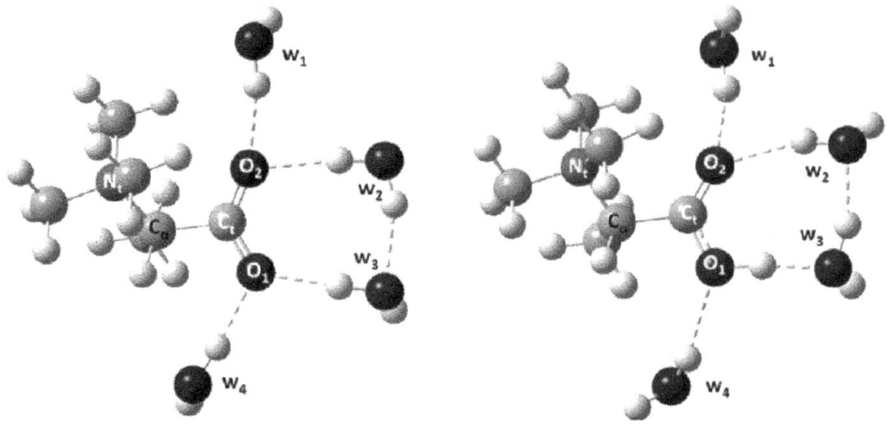


**Figure 4.3** The strongly H-bonded interacting water molecules with zwitterionic and protonated alanine betaine. Fully optimized structures by DFT calculations describing the number of the strongly interacting water molecules with (in the presence of bulk water) the neutral pH and protonated states of alanine betaine through H bonds, at the B3LYP/cc-pVTZ level of theory, applying the IEF-PCM model for bulk water. Dotted lines indicate the maximum number of interacting water molecules through H bonds (Takis *et al.*, 2012).

glycine betaine, in their aqueous solutions at two different pH values (pH = 6.0: zwitterionic betaines and acid pH: protonated betaines) strongly indicated that betaine retains its hydration grade independently of its ionic form, highlighting that its function is the same under acidic and neutral pH conditions in cells, noting its cellular osmotic stress controlling role (Takis *et al.*, 2012). A picture of the strongly H-bonded water molecules with alanine betaine is depicted in Figure 4.3, in agreement with other research studies of betaine (Panuszko *et al.*, 2011).

From the theoretical point of view, the dipole moment magnitude of betaine has been studied with *ab initio* quantum chemical calculations and the calculated value was found to be equal to 11.9 D (Shikata, 2002). The dipole moment vector is illustrated in Figure 4.4. When betaine is hydrated by a single explicitly bound water molecule, this value reduces to 10.6 D, whereas in the case of betaine dihydrate the reduction in magnitude is more significant, within the range of 8.1–8.3 D. These results, in conjunction with dielectric relaxation measurements, showed that betaine is easily and tightly hydrated in aqueous solution, which effectively reduces the total dipole moment magnitude, in comparison to the bare betaine molecule (Shikata, 2002).

Furthermore, molecular dynamics simulations of aqueous betaine solutions showed that water localization increases with increasing solute concentration, and hydrogen-bonded water molecules connect carboxylic oxygen atoms of different solutes. At the highest concentration, approximately twelve water molecules reside within the first hydration shell of each betaine

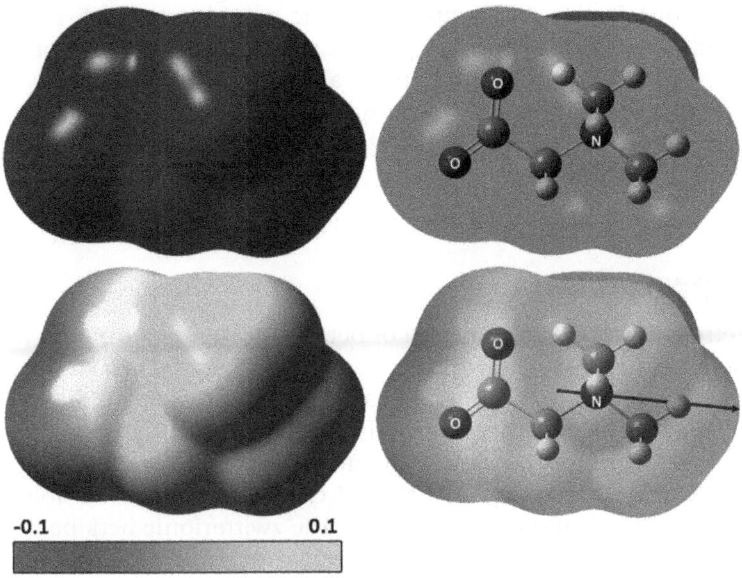


**Figure 4.4** Total electron density surface, electrostatic potential surface and dipole moment of betaine of zwitterionic betaine. Total electron density (isosurface value 0.0004) mapped with the electrostatic potential of betaine, calculated at the MP2/cc-pVTZ (Møller and Plesset, 1934; Dunning, 1989) level of theory. The scale spans the narrow range between 0.1 (dark gray) and 0.1 (light gray) to illustrate betaine's electronegative and electropositive regions. The dipole moment vector is also depicted, with the origin located at the center of mass.

molecule, which is delimited by the water molecules whose oxygen atoms are within 3.5 Å from any atom of the solute (Civera *et al.*, 2003). It was recently found that seven water molecules are sufficient for the preferential interaction of betaine in aqueous phase (Venkatesan and Lee, 2012), as illustrated in Figure 4.5. The interaction of water molecules spans from the carboxylate to the methyl groups, with strong intramolecular forces acting between them, providing additional support for the complex stability when the number of water molecules increases (Venkatesan and Lee, 2012).

Therefore, betaine succeeds in maintaining cell volume when used by cells of numerous water-stressed organisms and tissues, helping their evolutionary adaptation in a dehydrated or concentrated ion solution (Venkatesan and Lee, 2012). In general, organic osmolytes are small solutes used by cells of numerous water-stressed organisms and tissues to maintain cell volume by coordinating water molecules tightly. This results in osmolyte exclusion from hydration layers of peptide backbones, making unfolded protein conformations entropically unfavorable (Yancey, 2001). Hence, cells achieve osmoregulation *via* biosynthesis of betaine with the comparatively low expenditure of metabolic energy (Papageorgiou and Murata, 1995). Being a zwitterion throughout the physiological range of pH values, betaine requires no counterions for electric neutrality in the cytoplasm. As an amphiphilic compound it neutralizes anionic groups situated within hydrophobic protein domains. Moreover, since its tetravalent nitrogen atom lacks the characteristic lone electron pair, it does not interact with electrophilic centers and, therefore, it does not alter their electric properties (Papageorgiou and Murata, 1995).

## 4.5 Spectral Data

The electrospray (ESI) mass spectra of betaine present an ion peak at *m/z* 59, corresponding to $-(CH_3)_3N^+$, which is the base peak in the MS/MS spectrum of glycine betaine. Moreover, an extra peak is generated at *m/z* 118 representing the protonated molecular betaine ion ($[M + H]^+ = m/z\ 118$) (Wood *et al.*, 2002).

The FT-IR spectra of the carboxylic and carboxylate (–CO) band positions from protonated betaine $[(CH_3)_3N^+–CH_2CO_2H,\ Cl^-]$ aqueous solution are 1740 and 1265 $cm^{-1}$, respectively, and for the zwitterionic betaine $[(CH_3)_3N^+–CH_2COO^-]$ 1615 and 1401 $cm^{-1}$. The methyl groups of both protonated and zwitterionic betaine methyl groups $[-(CH_3)_3]$ are located at almost the same region (2993 to 3060 $cm^{-1}$ for the first and 3063 $cm^{-1}$ for the latter) of their IR spectra as well as their $-CH_2$ aliphatic group, namely at 2963 $cm^{-1}$ for the protonated and 2970 $cm^{-1}$ for the zwitterionic betaine (Max and Chapados, 2004).

The $^1H$-NMR spectrum of betaine in aqueous solution (experimental conditions: $T = 298$ K, pH = 7.0, C = 50 mM, 500.13 MHz NMR spectrometer frequency, referenced to DSS) shows that methyl protons resonate at 3.25 ppm and the protons of $-CH_2$ aliphatic group at 3.89 ppm (Wishart *et al.*, 2009). The $^{13}C$-NMR spectrum indicates that $^{13}C$ nuclei of carboxyl group resonate

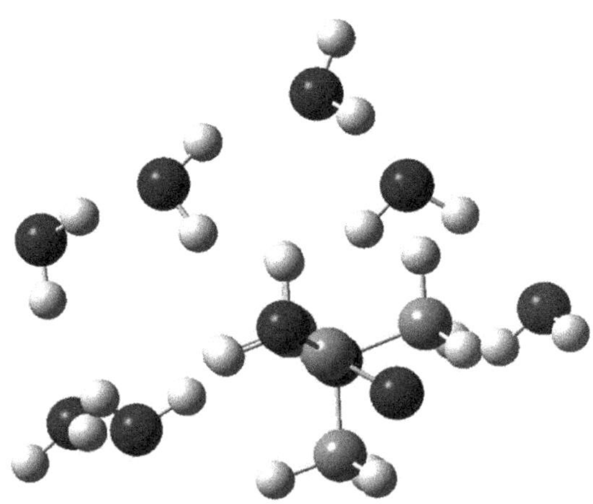

**Figure 4.5** Illustration of betaine's hydration. Illustration of seven water molecules preferentially interacting with betaine in the aqueous phase (Venkatesan and Lee, 2012).

at 171.84 ppm, of [–$(CH_3)_3$] group at 56.08 ppm and of –$CH_2$ group at 68.93 ppm (experimental conditions: $T$ = 298 K, pH = 7.0, C = 50 mM, 100.613 MHz NMR spectrometer frequency, referenced to DSS) (Wishart *et al.*, 2009). Moreover, $^{14}N$ nuclei of betaine trimethylated amino group resonates at 27.48 ppm (experimental conditions: $T$ = 308 K, pH = 7.0, C = 50 mM, 28.91 MHz NMR spectrometer frequency, referenced to nitromethane) (Takis *et al.*, 2012).

## 4.6 Crystal Structure

X-ray diffraction analysis of anhydrous betaine revealed that it crystallizes in the orthorhombic space group *Pnma* with $a$ = 14.544 Å, $b$ = 6.859(3) Å, $c$ = 6.131 Å, $V$ = 6.117(3) $Å^3$ (Figure 4.6). The structure was resolved by direct methods and refined by least-square calculations to $R$ = 0.038 for 531 reflections (Viertorinne *et al.*, 1999). Furthermore, betaine preferentially crystallizes in a highly symmetric form, both in anhydrous and monohydrate crystal structures, retaining its local $C_s$ symmetry. Theoretical calculations attribute this to intramolecular interactions between the carboxylic oxygen and the nearest methyl hydrogens, resulting in internal hydrogen bonding and a slight charge transfer, as indicated by NPA calculations. This is corroborated by the energetically high rotational barrier (9.68 kcal $mol^{-1}$) of the carboxylic C–C bond (Figure 4.6) (Nyrönen *et al.*, 1999).

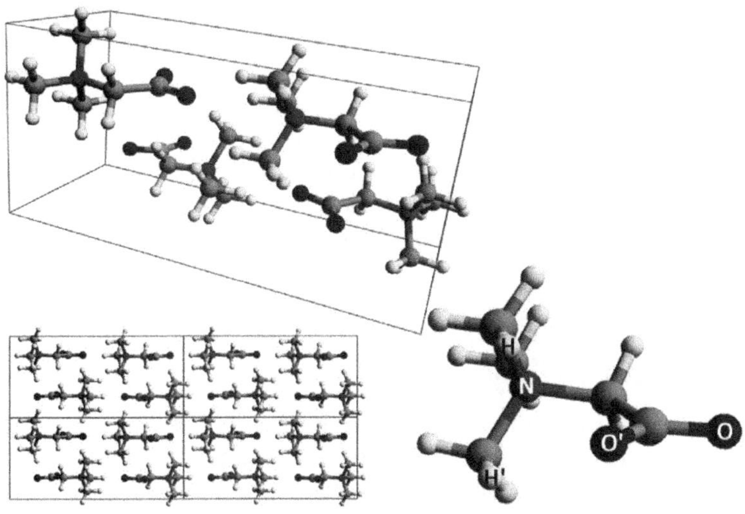


**Figure 4.6** Illustration of the anhydrous betaine unit cell. The unit cell of anhydrous betaine and its replication along $a$- and $b$-axes, based on Xray diffraction data (Viertorinne *et al.*, 1999). Oxygen and nitrogen atoms are selectively labeled on the betaine molecular structure, whereas the atom color coding is dark gray for carbon and light gray for hydrogen atoms. Methyl hydrogens H and H′ are interacting with the carboxylic O′, allowing betaine to conserve its local symmetry (Nyrönen *et al.*, 1999).

## 4.7 Handling and Toxicity

According to a report by EFSA (2013), betaine was found to be genotoxic in a series of *in vitro* and *in vivo* assays. Particle-size data indicate that the additive betaine anhydrous contained about 36% (v/v) particles of diameter <50 μm but no particles of respirable size (<10 μm). No inhalation toxicity studies are available, however, EFSA reports that betaine anhydrous may cause irritation of the respiratory tract and additionally may cause irritation to eyes and skin, thus presenting a health hazard. Considering the dusting potential, the betaine anhydrous may cause concerns for user safety if inhaled. In the absence of data, betaine may be considered an irritant to skin, eyes and mucous membranes and a skin sensitizer. Betaine is widely distributed in nature as an endogenous metabolite in most organisms, including micro-organisms, plants and animals. Consequently, the supplementation of feed with betaine anhydrous and betaine hydrochloride does not pose a risk to the environment (EFSA, 2013).

## Summary Points

- This chapter focuses on betaine chemistry and betaine's physical properties repercussions on its function.
- Betaine is abundant in several plants, from which it is supplied to humans/animals or it is synthesized in mitochondria *via* choline oxidation, acting as a significant methyl donor.
- It is of major importance to understand betaine's osmoregulatory role in cells in order to elucidate osmolytes cellular complex functions.
- It seems that betaine's direct and indirect interactions with water molecules and biomolecules such as proteins and DNA, respectively, play the key role in the understanding of their structure's stabilization/destabilization.
- Betaine and in general osmolytes function and physical origin are still examined largely by physicochemical and biological research studies, both theoretical and experimental, highlighting their so far unexplored nature.

## Key Facts for Betaines and Some of Its Derivatives Function and Utility

- In contrast to betaine, many of its naturally synthesized analogs exhibit controversial activity in different species of the same kingdom. For instance, arsenobetaine and phosphoniobetaine are highly osmoprotective for several bacteria, including *E. Coli* strains, although they are found to be very toxic for *S. meliloti* wild-type bacterial strains and other marine bacteria.
- Because of its "permanent" zwitterionic structure in water, betaine can easily interact with many ions such as $Cl^-$ (after its reaction with hydrochloric acid, HCl) resulting in betaine hydrochloride, whose properties are different compared to betaine.
- Betaine is commercially available as a feed supplement for livestock, as an enrichment agent for human food, and as a biodegradable surfactant.

- Betaine plays a severe role in reducing the elevated serum homocysteine concentrations associated with hyperhomocystinuria—a disease based on a metabolic disorder that is characterized by an excess of the compound homocystine in the urine—and may play a role in epigenetics and athletic performance.

## Definitions of Words and Terms

**Homocysteine**: It is biologically synthesized from amino acid methionine and it is very similar to cysteine. Its high concentration in blood is related to endothelium disorders such as cardiovascular diseases.

**Betaine-homocysteine methyltransferase (BHMT)**: BHMT is a cytosolic zinc metalloenzyme that is highly expressed in the liver and kidneys. BMHT catalyzes methyl transfer from betaine to homocysteine, resulting in methionine and *N,N*-dimethylglycine. It is also found to regulate homocysteine levels in the liver.

**Osmolyte**: The physical phenomenon of osmosis can be regulated by the presence of specific organic compounds, the so-called osmolytes, which are highly soluble in water and exist in cells, tissues or biofluids such as plasma, urine, *etc*.

**Choline**: Choline is a quaternary ammonium salt with the chemical formula $(CH_3)_3N^+(CH_2)_2O^-$ in aqueous solution, being an essential nutrient in diet. Some of its functions include the structure of cell membranes, liver's protection from accumulating fat, and synthesis of the neurotransmitter acetylcholine.

**Nuclear magnetic resonance (NMR) spectroscopy**: NMR spectroscopy is a powerful spectroscopic tool used by physics, chemistry and biology research fields for determining the content, purity, physical properties and molecular structure of a sample in liquid and/or solid state. Its main principle is based on nuclei spin and charge. Consequently, the application of an external magnetic field leads to an energy transfer between the base energy to a higher energy level of spin population distribution. The energy transfer takes place at a wavelength that corresponds to radio frequencies and when the spin population returns to its equilibrium after a sufficient time of period, energy is emitted at the same frequency. The signal that matches this transfer is measured in many ways and processed in order to yield an NMR spectrum for the nucleus concerned.

**Mass spectrometry**: Analytical technique that measures the charge to mass ratio of charged compounds. It is often used coupled to liquid and gas chromatography techniques for identifying the content, as well as the structure of the compounds under study.

**Infrared (Fourier transform, FT) (FT-IR) detection**: Spectroscopic detection is carried out at specific wavelengths. The infrared spectrum exhibits a different pattern for each molecule, depending on its structure, dynamics and general physicochemical properties. The application of

Fourier-transform algorithms for extra mathematical treatment of IR data can provide significant details about molecular studies.

**X-ray crystallography**: It is a method for defining the atomic and molecular structure of a crystal structure, in which the crystalline atoms cause a beam of X-rays to diffract into many specific directions. By measuring the angles and intensities of these diffracted beams, a detailed picture of the crystal's electronic density is acquired, providing the mean positions of the atoms in the crystal, their chemical bonds, as well as the three-dimensional molecular structure.

**Density functional theory (DFT)**: It is a theoretical method aimed at the investigation of many-body systems, with the electron density being the main variable. Instead of using the N-electron wavefuction of the system as the central quantity, DFT methods express the ground state of many-body systems as a functional (function of a function) of the electron density, *i.e.* there is no wavefunction in DFT. DFT methods are computationally rigorous and robust, exhibiting high accuracy in the study of a chemical system's molecular structure, vibrational frequencies, thermochemical, magnetic and other properties in comparison to post-Hartee–Fock methods, especially when the system has a relatively large number of atoms.

**Molecular dynamics (MD)**: It is a computational method which calculates the time-dependent behavior of molecular systems. In general, a system of N particles is considered to behave dynamically under a potential energy function, which is comprised of individual energetic contributions arising from bonded and nonbonded interactions. By employing Newton's equations of motion in conjunction with statistical mechanics, it is possible to deduce the macroscopic properties of a given system (structure, pressure, energy, *etc.*) through microscopic simulations (positions, velocities).

## List of Abbreviations

| | |
|---|---|
| BHMT | Betaine-homocysteine methyl transferase |
| CAS | Chemical Abstract Service Registration Number |
| cc-pVTZ | Correlation consistent polarized valence triple zeta |
| DFT | Density functional theory |
| DMG | Dimethylglycine |
| DSC | Differential scanning calorimetry |
| EFSA | European Food Safety Authority |
| EINECS | European Inventory of Existing Commercial Chemical Substances |
| H-bond | Hydrogen bond |
| IEF-PCM | Integral equation formalism-polarizable continuum model |
| IR | Infrared |
| IUPAC | International Union of Pure and Applied Chemistry |
| MP2 | Second order Møller–Plesset perturbation theory |
| $M_w$ | Molecular weight |

| | |
|---|---|
| NMG | *N*-Methylglycine |
| NMR | Nuclear magnetic resonance |
| NPA | Natural population analysis |
| $pK_a$ | Acid dissociation constant |
| TMAO | Trimethylamine *N*-oxide |
| TMG | Trimethylglycine |

## References

Amiraslani, B., Sabouni, F., Abbasi, S., Nazem, H. and Sabet, M., 2012. Recognition of betaine as an inhibitor of lipopolysaccharide-induced nitric oxide production in activated microglial cells. *Iran Biomedical Journal*. 2: 84–89.

Bruzdziak, P., Panuszko, A. and Stangret, J., 2013. Influence of osmolytes on protein and water structure: a step to understanding the mechanism of protein stabilization. *Journal of Physical Chemistry B*. 46: 11502–11508.

Civera, M., Fornili, A., Sironi, M. and Fornili, S. L., 2003. Molecular dynamics simulation of aqueous solutions of glycine betaine. *Chemical Physics Letters*. 1–2: 238–244.

Craig, S. A., 2004. Betaine in human nutrition. *American Journal of Clinical Nutrition*. 3: 539–549.

Dawson, R. M. C., Elliott, D. C., Elliott, W. H. and Jones, K. M., 1986. *Data for biochemical research*. 3rd edn. Clarendon Press, Oxford, pp. 8–9.

de Zwart, F. J., Slow, S., Payne, R. J., Lever, M., George, P. M., Gerrard, J. A. and Chambers, S. T., 2003. Glycine betaine and glycine betaine analogues in common foods. *Food Chemistry*. 2: 197–204.

Dunning, T. H., 1989. Gaussian basis sets for use in correlated molecular calculations. I. The atoms boron through neon and hydrogen. *Journal of Chemical Physics*. 2: 1007–1023.

EFSA, 2013. Scientific Opinion on the safety and efficacy of betaine anhydrous as a feed additive for all animal species based on a dossier submitted by Danisco Animal Nutrition. *EFSA Journal*. 5: 20.

Friesen, R. W., Novak, E. M., Hasman, D. and Innis, S. M., 2007. Relationship of dimethylglycine, choline, and betaine with oxoproline in plasma of pregnant women and their newborn infants. *Journal of Nutrition*. 12: 2641–2646.

Goldberg, Y., Ābele, E., Bremanis, G., Trapenciers, P., Gaukhman, A., Popelis, J., Gomtsyan, A., Kalviņš, I., Shymanska, M. and Lukevics, E., 1990. Betaines derived from amino and hydrazino acids as phase transfer catalysts. *Tetrahedron*. 6: 1911–1922.

Lever, M. and Slow, S., 2010. The clinical significance of betaine, an osmolyte with a key role in methyl group metabolism. *Clinical Biochemistry*. 9: 732–744.

*CRC Handbook of Chemistry and Physics 2005–2006 a Ready-Reference Book of Chemical and Physical Data*, ed. Lide, D. R., 2005 ed. D. R. Lide. 86th edn. CRC Press Taylor & Francis Group, Boca Raton New York Washington.

Max, J.-J. and Chapados, C., 2004. Infrared spectroscopy of aqueous carboxylic acids: comparison between different acids and their salts. *Journal of Physical Chemistry A*. 16: 3324–3337.

McNaught, A. D., Wilkinson, A. and International Union of Pure and Applied Chemistry, 1997. *Compendium of Chemical Terminology: IUPAC Recommendations*. 2nd edn. Blackwell Science, Oxford Oxfordshire; Malden, MA.

Møller, C. and Plesset, M. S., 1934. Note on an Approximation Treatment for Many-Electron Systems. *Physical Review*. 7: 618–622.

Nyrönen, T. H., Suontamo, R. and Pitkänen, I., 1999. Why betaine crystallizes in high local Cs symmetry. An *ab initio* MO and DFT study of anhydrous betaine and betaine monohydrate. *Theoretical Chemistry Accounts*. 1–3: 209–214.

O'Neil, M. J., 2006. *The Merck index : an encyclopedia of chemicals, drugs, and biologicals*. Merck, Whitehouse Station, N.J., 14th edn.

Panuszko, A., Smiechowski, M. and Stangret, J., 2011. Fourier transform infrared spectroscopic and theoretical study of water interactions with glycine and its N-methylated derivatives. *Journal of Chemical Physics*. 11: 115104.

Papageorgiou, G. C. and Murata, N., 1995. The unusually strong stabilizing effects of glycine betaine on the structure and function of the oxygen-evolving Photosystem-II complex. *Photosynthesis Research*. 3: 243–252.

Ross, A. B., Zangger, A. and Guiraud, S. P., 2014. Cereal foods are the major source of betaine in the Western diet – Analysis of betaine and free choline in cereal foods and updated assessments of betaine intake. *Food Chemistry*. 859–865.

Shikata, T., 2002. Dielectric relaxation behavior of glycine betaine in aqueous solution. *Journal of Physical Chemistry A*. 34: 7664–7670.

Sigma-Aldrich, 2013. *Betaine safety data sheet*. Sigma-Aldrich Chemie GmbH, Steinheim.

Takis, P. G., Melissas, V. S. and Troganis, A. N., 2012. A "hidden" role of amino and imino groups is unveiled during the micro-solvation study of three biomolecule groups in water. *New Journal of Chemistry*. 9: 1866–1878.

Takis, P. G., Papavasileiou, K. D., Peristeras, L. D., Melissas, V. S., Troganis, A. N., 2013. Probing micro-solvation in "numbers": the case of neutral dipeptides in water. *Physical Chemistry Chemical Physics*. 15: 7354–7362.

Ueland, P. M., Holm, P. I. and Hustad, S., 2005. Betaine: a key modulator of one-carbon metabolism and homocysteine status. *Clinical Chemistry and Laboratory Medicine*. 10: 1069–1075.

Venkatesan, S. and Lee, S. L., 2012. Computational investigation on microsolvation of the osmolyte glycine betaine [GB $(H_2O)_{1-7}$]. *Journal of Molecular Modeling*. 12: 5017–5028.

Viertorinne, M., Valkonen, J., Pitkanen, I., Mathlouthi, M. and Nurmi, J., 1999. Crystal and molecular structure of anhydrous betaine, $(CH_3)_3NCH_2CO_2$. *Journal of Molecular Structure*. 1–3: 23–29.

Wishart, D. S., Knox, C., Guo, A. C., Eisner, R., Young, N., Gautam, B., Hau, D. D., Psychogios, N., Dong, E., Bouatra, S., Mandal, R., Sinelnikov, I., Xia, J., Jia, L., Cruz, J. A., Lim, E., Sobsey, C. A., Shrivastava, S., Huang, P., Liu, P., Fang, L., Peng, J., Fradette, R., Cheng, D., Tzur, D., Clements, M., Lewis, A., De Souza, A., Zuniga, A., Dawe, M., Xiong, Y., Clive, D., Greiner, R., Nazyrova, A., Shaykhutdinov, R., Li, L., Vogel, H. J. and Forsythe, I., 2009. HMDB: a knowledgebase for the human metabolome. *Nucleic Acids Research*. Database issue: D603–D610.

Wood, K. V., Bonham, C. C., Miles, D., Rothwell, A. P., Peel, G., Wood, B. C. and Rhodes, D., 2002. Characterization of betaines using electrospray MS/MS. *Phytochemistry*. 7: 759–765.

Yancey, P. H., 2001. Water stress, osmolytes and proteins. *American Zoologist*. 4: 699–709.

Yancey, P. H., Clark, M. E., Hand, S. C., Bowlus, R. D. and Somero, G. N., 1982. Living with water stress: evolution of osmolyte systems. *Science*. 4566: 1214–1222.

Zeisel, S. H., Mar, M.-H., Howe, J. C. and Holden, J. M., 2003. Concentrations of choline-containing compounds and betaine in common foods. *Journal of Nutrition*. 5: 1302–1307.

CHAPTER 5

# Chemical Interactions of Betaine with Folate and Other Biomolecules

MICHAEL R. DUFF JR[a] AND ELIZABETH E. HOWELL*[a]

[a]Biochemistry, Cellular and Molecular Biology Department, University of Tennessee, 1414 Cumberland Avenue, Knoxville, TN 37996-0840, USA
*E-mail: lzh@utk.edu

## 5.1 Introduction to Osmolytes and Models of Osmolyte Action

Osmolytes are small molecules produced by cells in response to altered water conditions in the environment. Their production allows organisms to withstand harsh conditions such as heat, dehydration, and high salt content (Yancey *et al.*, 1982). Three classes of compounds serve as osmolytes (Bolen, 2001). The first class includes proline, taurine, glutamic acid and other amino acids. The second class contains polyols such as glycerol, sucrose, trehalose and other related molecules. The third class comprises methylamines such as trimethylamine oxide (TMAO) and glycine betaine (GB).

Osmolytes can also be differentiated based on their effect on protein function. The above osmolytes are protecting osmolytes, while an example of a nonprotecting osmolyte is urea (Bolen, 2001). Urea can accumulate in mammalian kidney cells and also in sharks, rays, and coelacanths (Yancey *et al.*, 1982). Since urea can lead to protein unfolding, methylamines are often produced to counteract the negative effects of urea.

Food and Nutritional Components in Focus No. 7
Betaine: Chemistry, Analysis, Function and Effects
Edited by Victor R Preedy

Published by the Royal Society of Chemistry, www.rsc.org

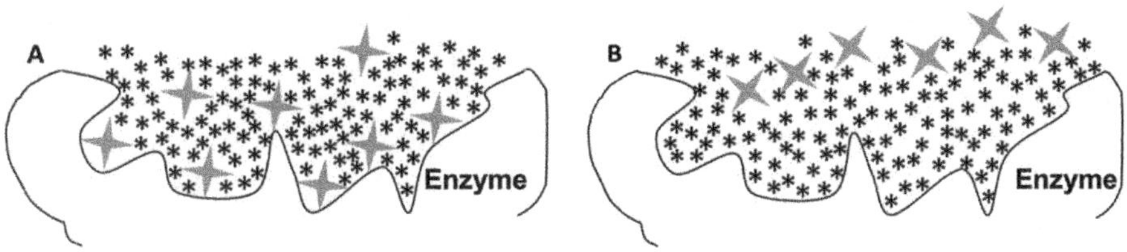


**Figure 5.1** Models describing preferential interaction and preferential exclusion of solutes from a protein surface. Panel A: model showing preferential interaction of osmolyte (×) with a slice of the protein surface. Waters are shown by ∗. Various osmolytes may interact differently and yield variable effects on ligand binding. Panel B: due to a preference of the protein surface to interact with water, osmolytes are excluded. For this situation, all osmolytes show identical behavior.

To describe how osmolytes work, several models have been proposed. Figure 5.1 shows the preferential exclusion model, where osmolytes are excluded from protein surfaces (Parsegian *et al.*, 1995, 2000). For glycine betaine (GB), the source of the exclusion is likely its inability to be a hydrogen-bond donor. Rephrased, the protein prefers to interact with water rather than osmolyte. In this scenario, different osmolytes have identical effects on ligand binding as they are all similarly excluded from the protein surface. Two examples of this result include addition of NADPH to R67 dihydrofolate reductase (DHFR) (Chopra *et al.*, 2008) and binding of DNA by lac repressor (Fried *et al.*, 2002).

A second model describing osmotic-pressure effects is preferential interaction (Figure 5.1). Here, the osmolytes act as a cosolvent and interact with the protein or ligand surface, albeit weakly. The different interaction abilities of the various osmolytes result in different osmotic-pressure effects. This is a more typical result, see for example differential interaction of osmolytes with bovine serum albumin (Courtenay *et al.*, 2000) and the binding of nonspecific (star sequence) DNA by BamH1 (Sidorova *et al.*, 2006).

## 5.2 Osmolytes Function as Chemical Chaperones

How does osmolyte addition affect protein folding? Initial studies by the Bolen lab used a ribonuclease $T_1$ (RNase $T_1$) that was reduced and modified by iodoacetamide, yielding an unfolded protein as assessed by circular dichroism (Baskakov *et al.*, 1999). The enzyme activity of this modified species, RCAM-$T_1$, was decreased by ~250-fold. RCAM-$T_1$ was induced to fold by addition of trimethylamine oxide (TMAO). The full circular dichroism signal was restored at 2.7 M TMAO and at this TMAO concentration, RCAM-$T_1$ enzyme activity increased to ~20% of the wild-type value. These results indicate addition of TMAO induces folding. Lin and Timasheff (1994) studied wildtype RNase-$T_1$ folding in TMAO and the denaturant, urea. They find urea stabilizes both the native and unfolded states by interactions with the protein backbone such that the energy difference between the two states is smaller; this situation leads to protein denaturation. Addition of TMAO destabilizes both the native and unfolded states by its exclusion from the peptide backbone, however, the unfolded state is more affected, resulting in a larger energy difference between the native and unfolded states. This leads to protein folding (see Figure 5 in Lin and Timasheff, 1994). In the words of Lin and Timasheff: "What determines the effect of a cosolvent on the stability of a protein relative to water? It is the difference between its interactions with the protein in the denatured and native states." As betaine and TMAO are excluded from the protein surface, their addition somewhat destabilizes the native state with a larger destabilization of the unfolded state. As the water concentration is concurrently decreased; this also favors the state with the lowest surface area, which is the native state.

More recently, chemical-chaperone studies have used GB and similar results have been observed. Examples include positive effects of GB addition on folding of α-chymotrypsin (Venkatesu *et al.*, 2009) and catalase (Belluzo *et al.*, 2011).

*In vivo* effects of betaine acting as a chemical chaperone have also been observed. For example, chemical mutagenesis was used to generate a library of mutant *E. coli* that were screened for growth on high salt minimal media plus GB (Bourot *et al.*, 2000). A *LysA* mutant strain of *E. coli* was obtained with a S384F mutation in the diaminopimelate decarboxylate gene. *In vivo*, the strain could not grow on minimal media plus 0.5 M salt, however, addition of 1 mM GB allowed growth. *In vitro*, the enzyme activity of this mutant could be partially restored by addition of GB. Further, fluorescence and FTIR studies suggest the mutant protein is not well folded and addition of GB leads to a folded, albeit altered, state.

The above discussion raises the question of whether osmolality varies *in vivo?* Yes, for example, an isosmotic solution for humans is 300 mOsm/kg $H_2O$, however, kidney cells can accommodate variations from 160 (hypotonic) to ~1700 mOsm (hypertonic) (Roy *et al.*, 1992). Less dramatic variations occur in the intestine. Table 1 in Burg *et al.* (2007) lists how the concentrations of many molecules are affected by small to large changes in osmolality and Table 2 in Neuhofer (2010) indicates pathologies associated with local inflammation and hypertonicity.

## 5.3 Preferential Interactions of Betaine with Macromolecules

In 2000, the Record lab developed the use of vapor pressure osmometry (VPO) to monitor the preferential interaction of osmolytes with bovine serum albumin (BSA) (Courtenay *et al.*, 2000). For example, they monitored the solution osmolality as the betaine concentration was increased. Upon addition of a fixed concentration of BSA, if osmolyte is excluded from the protein surface, the bulk concentration of betaine will be higher than expected. Alternatively, if betaine is excluded from the protein surface, water preferentially solvates the protein, which in turn decreases the bulk water concentration. The latter is measured with an osmometer. The difference in slope from a GB only *versus* GB + BSA osmolality plot indicates the degree of osmolyte exclusion from the protein surface. For BSA, Courtenay *et al.* find betaine is highly excluded from the protein surface, followed by proline, TMAO, trehalose, potassium glutamate, and glycerol. They conclude betaine is the most excluded osmolyte.

In an effort to deconvolute how betaine interacts with BSA and develop a predictive approach for GB interaction with macromolecules, the Record lab determined partition coefficients ($K_p$), which describe the interaction of GB with the surface of an atom as compared to bulk solution (Capp *et al.*, 2009). A $K_p$ of 0 describes an atom that totally excludes osmolyte. A partition coefficient rather than a binding constant is used as no evidence of saturation of the surface by osmolytes has been found. The 23 small model compounds studied include glycerol, urea, amino acids, cyclic peptides, sodium benzoate, various salts and other molecules. These compounds include various atom types such as aliphatic and aromatic carbons, hydroxyl oxygens, carboxylate and phosphate oxygens, amide nitrogens and oxygens, and cationic nitrogens. Capp *et al.* find GB interacts with amide nitrogens but is excluded from amide oxygens. Also, GB is weakly excluded from polyols and strongly excluded from anionic oxygens. Furthermore, GB interacts with cationic nitrogens as well as the aromatic benzene ring. Using changes in accessible surface area calculations, they propose the interactions between GB and the various atom types are additive. Thus, the various experimental effects can be deconvoluted using changes in accessible surface area associated with the different atom surfaces. They find GB is totally excluded from phosphate oxygen, with a $K_p$ of 0 (see Table 2 in Capp *et al.*, 2009). This exclusion most likely occurs as water can be a hydrogen-bond donor, while GB cannot. The strongest interaction occurs between GB and aromatic carbons, with a $K_p$ of 1.7. A similar osmometry study found urea interacts weakly with aliphatic carbons and moderately with amide oxygens (Guinn *et al.*, 2011). These differences in solute partitioning result in urea interaction with backbone atoms and the hydrophobic core of proteins, which leads to denaturation.

## 5.4 Betaine Association with Folate and Its Derivatives

When GB acts as a chemical chaperone, it affects protein folding. However, osmolytes can additionally affect ligand binding. As an example, we present several studies performed by our lab, where the effects of betaine on ligand binding to three different, nonhomologous dihydrofolate reductase (DHFR) enzymes – type-I chromosomal DHFR, type-II R67 DHFR and FolM, a pteridine reductase homolog (see Figure 5.2 for protein structures) - were assessed using steady-state kinetics and ligand binding assays (Chopra *et al.*, 2008; Grubbs *et al.*, 2011; Bhojane *et al.*, 2014). The NADPH cofactor bound

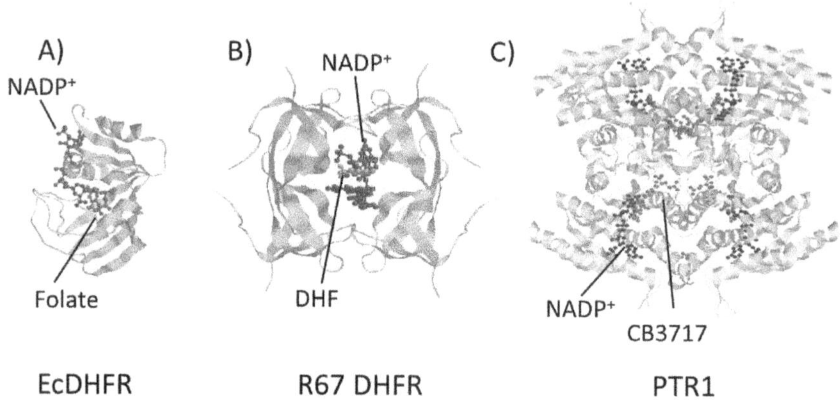


**Figure 5.2** Structures of various dihydrofolate reductases. (A) Structure of *E. coli* chromosomal DHFR (Protein Data Bank code 1RA2) (Sawaya and Kraut, 1997). $NADP^+$ is shown in dark gray, while folate is depicted in light gray. (B) Structure of tetrameric R67 DHFR (PDB code 1VIF) (Krahn *et al.*, 2007). The central pore is the active site. $NADP^+$ and DHF are colored dark and light gray, respectively. (C) Structure of *Leishmania major* pteridine reductase (PDB code 2BFA) (Schuttelkopf *et al.*, 2005). Bound NADPH and 10-propargyl-5,8-dideazafolate (CB3717) are shown in dark gray and light gray. Reprinted with permission from Bhojane, P. P., Duff, M. R., Jr., Patel, H. C., Vogt, M. E., and Howell, E. E., 2014, Investigation of Osmolyte Effects on FolM: Comparison with Other Dihydrofolate Reductases. *Biochemistry 53,* 1330–1341. Copyright 2014 American Chemical Society.

more tightly to all three DHFRs in the presence of betaine. These results are consistent with release of water upon cofactor binding to DHFR. From the slope in Figure 5.3A, NADPH binding to R67 DHFR is accompanied by the net release of $38 \pm 6$ water molecules. This result is consistent with conventional wisdom, *i.e.* desolvation conditions tighten binding (Parsegian *et al.*, 1995).

In contrast, osmolyte effects on dihydrofolate (DHF) binding "break the rules." (Table 5.1 lists a few key facts about folate.) DHF binding by all three DHFRs is weakened upon osmolyte addition and is sensitive to osmolyte identity. The results for R67 DHFR are shown in Figure 5.3B. (Control plots of solution viscosity or solution dielectric do not support their involvement (Chopra *et al.*, 2008)). How can opposite results for the NADPH and DHF ligands be understood? As R67 DHFR possesses 222 symmetry, each binding site is symmetry related and can accommodate either NADPH or DHF (Krahn *et al.*, 2007). As Figure 5.3A shows that all osmolytes display identical effects on NADPH binding to R67, this leads to the "preferential exclusion" mechanism (Parsegian *et al.*, 1995) where water prefers to interact with the protein and the osmolytes are excluded (see Figure 5.1). This result allows use of NADPH binding as an internal control, *i.e.* if osmolytes are excluded from the cofactor binding site, then from the 222 symmetry of the R67 active site

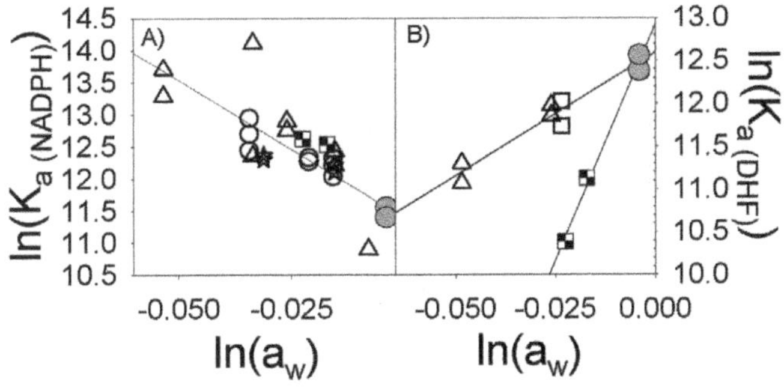


**Figure 5.3** The effects of osmolytes on ligand binding to R67 dihydrofolate reductase. Panel A shows an increase in ln $K_a$ for NADPH binding to R67 DHFR with a decrease in water activity (or an increase in osmolyte concentration) as measured by isothermal titration calorimetry. Panel B shows the decrease in ln $K_a$ for DHF binding to the R67 DHFR-NADP$^+$ complex with decreasing water activity. Symbols in the plots describe buffer (●), betaine (Δ), DMSO (o), sucrose (□), ethylene glycol (☆) and PEG400 (◨). This research was originally published by Chopra, S., Dooling, R. M., Horner, C. G., and Howell, E. E., 2006. "A Balancing Act between Net Uptake of Water during Dihydrofolate Binding and Net Release of Water upon NADPH Binding in R67 Dihydrofolate Reductase," *Journal of Biological Chemistry*. 283: 4690–4698. © the American Society for Biochemistry and Molecular Biology.

**Table 5.1** Folate key facts. Importance of folate to human health. Folate metabolism allows transfer of single carbons; an example is formation of thymidylate that is incorporated into DNA. Thus, antifolates can be used to treat various diseases, including cancer.

1. Folate is vitamin B9.
2. Bacteria synthesize folate *de novo*, while humans obtain folate from their diet.
3. Foods rich in folate include dark leafy greens, citrus and lentils/peas/beans.
4. A folate deficiency can lead to fetal neural tube defects and macrocytic anemia as well as other conditions.
5. Fortification of foods with folic acid can prevent folate deficiency.
6. The active forms of folate are based on different redox states of tetrahydrofolate that can transfer single carbons in metabolic reactions.
7. Antifolates such as methotrexate are used to treat cancer as they inhibit production of tetrahydrofolate in rapidly dividing cells.
8. Methotrexate inhibits the key enzyme, dihydrofolate reductase.

pore, osmolytes must also be excluded from the substrate binding site. Elimination of preferential binding of osmolytes to R67 DHFR during DHF binding leaves the remaining possibility of osmolyte effects on free DHF. This model is presented in Figure 5.4. If the DHF–osmolyte interaction(s) is more difficult to break than the DHF–$H_2O$ interaction(s), then weaker binding of

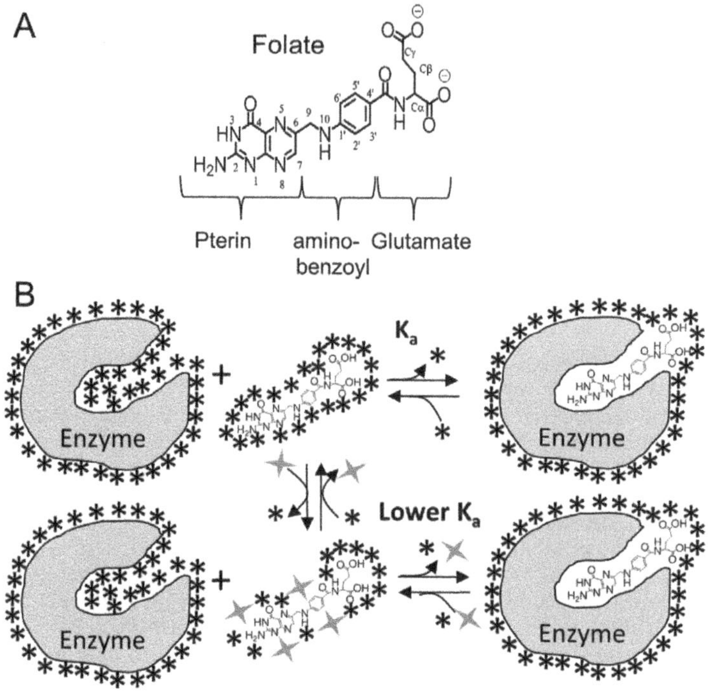


**Figure 5.4** Model describing the effects of the dihydrofolate-(osmolyte)$_n$ interaction on binding to a dihydrofolate reductase enzyme. Panel A, structure of folate. Panel B, binding of DHF to DHFR is accompanied by water release from the ligand and binding site surfaces (top row). When osmolytes are added, they interact with free DHF (middle column). The osmolyte–DHF interaction shifts the DHF–DHFR binding equilibrium towards the unbound state (bottom row). Reprinted with permission from Timson, M. J., Duff, M. R., Jr., Dickey, G., Saxton, A. M., Reyes-De-Corcuera, J. I. and Howell, E. E., 2013. Further Studies on the Role of Water in R67 Dihydrofolate Reductase. *Biochemistry* 52, 2118–2127. Copyright 2013 American Chemical Society.

substrate to DHFR results. This is a solvent-substitution scenario with the osmolytes acting as a solute or cosolvent (Schellman, 1994).

Additional evidence for interaction of betaine with DHF was noted in stopped-flow studies monitoring ligand binding to *E. coli* chromosomal DHFR (EcDHFR) (Grubbs *et al.*, 2011). These rapid kinetics experiments find that betaine mostly affects DHF binding *via* changes in $k_{on}$, while betaine mostly affects NADPH binding *via* changes in $k_{off}$. The decrease in dissociation rate arises as the ligand needs to be rehydrated upon dissociation from the enzyme, however, due to osmolyte addition, the bulk water concentration has been reduced. In a parallel fashion, ligands must also release water upon binding. When the association rate was measured, the $k_{on}$ rate for NADPH binding to EcDHFR was unaffected by betaine, while the association rate for DHF decreased about 5-fold upon addition of betaine. This result supports our model in Figure 5.4 as associated betaine molecules would need to be removed prior to DHF binding to EcDHFR.

These interactions are weak and may also be called "soft" associations or nonspecific interactions or 2nd virial coefficients. Another way to envision these pairings suggests that osmolytes act as an alternate solvent as water prefers to hydrate osmolytes and/or proteins rather than the folate molecules. This model suggests that the typical desolvation events that occur upon ligand binding also apply to other solutes.

The betaine–folate interaction(s) was additionally studied by proton nuclear magnetic resonance (NMR) spectroscopy (Duff *et al.*, 2012). Folate dimerization decreased two-fold in 20% betaine compared to buffer, indicating betaine shifts the equilibrium towards the 2 monomers due to their larger surface area. Also, the most notable nuclear Overhauser effects (NOEs) were between the protons on the C7 and C9 atoms, between the C7 and C3′/C5′ protons and between the C9 and the C3′/C5′ protons (see Figure 5.4 for atom numbers). Furthermore, the sign of the NOE between the C9 and C3′/C5′ protons in osmolyte solutions is opposite to that in buffer alone (see Figure 3 in Duff *et al.*) A change in sign indicates a change in the rotational rate for the *p*-amino-benzoic acid ring, consistent with an increased size due to association with osmolyte. As a control, NOE spectra were recorded for the DHFR cofactor, NADPH, in buffer and in 20% betaine. No significant differences in NOEs for NADPH were noted.

To understand which parts of folate may preferentially interact with betaine, VPO experiments using folate fragments such as pterin-6-carboxylate and *p*-aminobenzoyl-glutamate (Figure 5.4A) were performed. Our results (Duff *et al.*, 2012) indicated these compounds possess an equal preference for water *versus* betaine. As both these folate fragments possess an aromatic ring system as well as several amines, these are the likely interaction sites for betaine.

### 5.4.1 Do These *In Vitro* Observations Translate into *In Vivo* Effects?

Yes, *in vivo* osmolality effects on R67 DHFR function in bacteria are observed. In these experiments, exogenous sorbitol was added to raise the solution osmolality (Chopra *et al.*, 2008). The bacteria respond by making their own osmoprotectants, which include betaine, proline and glutamate (Record *et al.*, 1998). As Figure 5.3B shows that an increasing betaine concentration decreases the catalytic efficiency of R67 DHFR, cell growth can be titrated and ultimately blocked by adding sorbitol. This observation applies to both the ability of R67 DHFR to provide resistance to trimethoprim to the DH5α strain of *E. coli* as well as R67 DHFR's ability to rescue a DHFR minus strain of *E. coli (ΔFolA::kan)* from Fol end product auxotrophy. Control sorbitol titrations show DH5α can grow in high-osmolality media without trimethoprim; also the *ΔFolA::kan* strain can grow in high-osmolality media containing added Fol end products.

As well as folate-osmolyte effects, crowding agents such as high molecular weight PEGs and macromolecules (BSA, lysozyme, casein) interfere with DHF binding to R67 DHFR and EcDHFR upon small changes in osmolality. These results indicate *in vivo* effects may be mediated by both osmolytes and macromolecules. Even small changes can have a substantial effect on DHF capture.

## 5.5 Cation–π Interactions Provide a Handle to Bind Betaine

If GB prefers to be excluded from protein surfaces, how can proteins bind it for synthesis or transport purposes? To address this question, (Schiefner *et al.*, 2004) determined the crystal structure of ProX. Under conditions of osmotic stress and when GB is available in the environment, GB passively diffuses through the outer membrane of *E. coli* using the OmpC and OmpB porins. Once in the periplasmic space, ProX binds GB with a dissociation constant of ~1 μM. ProX then delivers GB to another protein, ProW, in the cytoplasmic membrane and GB is transported into the cell, helping to restore cell turgor. Figure 5.5 shows the contacts associated with the GB-ProX complex. Tryptophans 65, 140 and 188 form an "aromatic cage," allowing both van der Waals and cation–π interactions with the quaternary amine of GB. The carboxylate group of GB also forms hydrogen bonds with nearby residues. To evaluate whether cation–π interactions are involved in GB binding, site directed

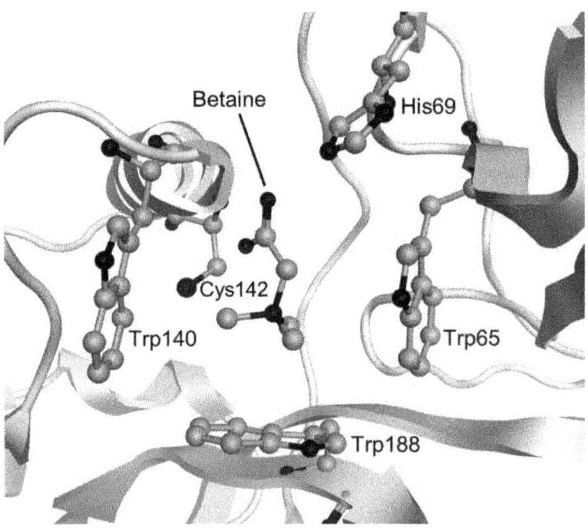


**Figure 5.5** Structure of ProX, a GB transporter (PDB file 1R9L). The ProX structure is described in Schiefner *et al.* (2004). Glycine betaine is shown near the middle of the figure. The "aromatic box" is composed of Tryptophans 65, 140, and 188. Hydrogen bonds are indicated by dashed lines. The figure was constructed using MOE v.2010 (Chemical Computing Group, Montreal, Canada).

mutants were constructed at all 3 tryptophans. W180 provides contacts to all three methyl groups of GB and forms the "floor" of the aromatic box; this residue appears most important as only substitutions to aromatic residues (phenylalanine or tyrosine) still bound GB. Mutations at W65 and W140 showed more tolerance to amino acid substitution, however, a double W65A + W140A mutant lost the ability to bind GB, indicating all three aromatic groups play an important role in binding GB. The ProX structure was the first to show a solution to the puzzle of binding GB. Other proteins have since been found to use an aromatic cage to bind GB and related molecules, including choline binding protein (Cheng *et al.*, 2013) and an aldehyde dehydrogenase from spinach that reduces betaine aldehyde to GB using $NAD^+$ (Diaz-Sanchez *et al.*, 2012).

In summary, while GB is normally excluded from protein surfaces as it cannot form hydrogen bonds with backbone amines, proteins have found a way to bind it. The GB handle that allows binding appears to be the quaternary amine group, which interacts with the π-electron cloud associated with aromatic groups.

## 5.6 Conclusions

The chemical chaperone function of GB derives from its exclusion from protein surfaces as GB is not a hydrogen-bond donor. However, GB can bind to synthesis and transport proteins *via* interactions with aromatic rings and amines. An unexpected consequence of these types of interaction is a weak ability to associate with folate. This weak interaction impedes binding of folate to DHFR, as depicted in Figure 5.4. Does this model only describe the DHFR–DHF/folate interaction or can it be expanded to provide a general description of weak interactions in related molecules? We propose this model will apply to the binding of antifolates as our experiments with FolM found weaker binding of methotrexate in the presence of osmolytes, including betaine (Bhojane *et al.*, 2014). The structure of methotrexate is very similar to that of folate, with an amine off C4 substituting for a carbonyl in folate and a methyl on N10 compared to a hydrogen. The model also likely applies to enzymes of folate-mediated one-carbon metabolism as methylene-tetrahydrofolate (for example) also possesses the pterin and *p*-aminobenzoate rings. Thus, the ability of betaine to interact with related molecules should be similar. This in turn should affect proteins that use pterins as cofactors (for example, molybdopterins in xanthine oxidase and tetrahydrobiopterin in several nitric oxide synthases as well as phenylalanine hydroxylase), proteins that use folates as allosteric effectors and as traps for reactive formaldehyde. Thus, this phenomenon may play a more general role than previously expected.

## Summary Points

- Osmolytes are small molecular weight compounds such as glycine betaine, sucrose and proline.
- Crowding agents are larger-sized compounds such as polyethylene glycols or proteins.

- Osmolytes can either be excluded from protein and ligand surfaces and/or they can interact weakly with these surfaces, acting as an alternate solvent to water.
- Glycine betaine cannot act as a hydrogen-bond acceptor. It prefers to interact with aromatic rings and cationic and amide nitrogens.
- The exclusion of glycine betaine from protein surfaces lets it act as a chemical chaperone, enabling protein folding.
- Glycine betaine weakly interacts with folate, which impairs binding of folate to dihydrofolate reductase.
- In living cells, functional ligand–protein interactions must compete with a large number of nonspecific interactions. Typically, the high specificity and affinity of cognate interactions leads to their being preferred, however, high concentrations of osmolytes and/or crowding agents can weaken the pairwise formation of the binding interaction both *in vitro* and *in vivo*.
- Addition of sorbitol to *E. coli* growth media results in production of osmoprotectants inside the cell. This leads to weaker binding of dihydrofolate to dihydrofolate reductase and blockage of cell growth when dihydrofolate reductase activity is limiting.

## Definitions and Explanation of Key Terms

**Antifolates**: Drugs that target enzymes utilizing various redox states of folate.

**Cation–π interaction**: Benzene, phenol and indol are simplified models of the phenylalanine, tyrosine and tryptophan side chains found in proteins. While these small molecules are neutral, their aromatic rings possess delocalized π electrons above and below the plane of the ring, which allows cations to interact with the ring face.

**Chemical chaperone**: Small molecules such as glycine betaine and trimethylamine oxide help proteins to fold by being excluded from the protein surface. The exclusion occurs as the protein prefers to interact with water. In conditions where the water content is low, the protein species with the lowest surface area is preferred. This is the native state rather than the unfolded state.

**Dihydrofolate reductase**: An essential enzyme in folate metabolism that catalyzes the reduction of dihydrofolate to tetrahydrofolate using NADPH as a cofactor. Chromosomal dihydrofolate reductase is inhibited by the antibacterial drug, trimethoprim, the antimalarial drug, pyrimethamine, and the anticancer drugs, methotrexate and pemetrexed.

**Osmolytes**: Small molecules that are soluble in water.

**Osmoprotectants**: Compatible osmolytes produced by the cell in response to osmotic stress.

**Osmotic stress**: The contents of a cell are separated from the external environment by a membrane. When the external solution osmolality is altered, equilibrium is re-established by rapid movement of water across the cell membrane. This process is osmosis.

**Preferential exclusion**: A model of osmolyte action where the small molecule is excluded from the protein surface.

**Preferential inclusion**: A model of osmolyte action where the small molecule interacts with compatible motifs on the protein surface.

**R67 Dihydrofolate reductase**: Resistance to the antibacterial drug trimethoprim has been observed; one resistance mechanism involves production of a type II DHFR that is not homologous in sequence or structure to chromosomal DHFR. R67 DHFR is an example of this novel enzyme.

**Vapor pressure osmometry**: If a solute is dissolved in pure water, the solution osmolality increases. An osmometer measures the dew point temperature depression of a sample solution and provides a measure of the solution osmolality.

## Abbreviations

| | |
|---|---|
| BSA | Bovine serum albumin |
| DHF | Dihydrofolate |
| DHFR | Dihydrofolate reductase |
| EcDHFR | Chromosomal DHFR from *E. coli* |
| GB | Glycine betaine |
| NADPH/($NADP^+$) | Reduced and oxidized states of nicotinamide adenine Dinucleotide phosphate |
| NOE | Nuclear Overhauser effect |
| PEG | Polyethylene glycol |
| RNase $T_1$ | Ribonuclease $T_1$ |
| RCAM-$T_1$ | Reduced and carboxyamidated ribonuclease $T_1$ |
| TMAO | Trimethylamine *N*-oxide |
| VPO | Vapor pressure osmometry. |

## References

Baskakov, I. V., Kumar, R., Srinivasan, G., Ji, Y. S., Bolen, D. W. and Thompson, E. B., 1999. Trimethylamine N-oxide-induced cooperative folding of an intrinsically unfolded transcription-activating fragment of human glucocorticoid receptor. *Journal of Biological Chemistry*. 274: 10693–10696.

Belluzo, S., Boeris, V., Farruggia, B. and Pico, G., 2011. Influence of stabilizers cosolutes on catalase conformation. *International Journal of Biological Macromolecules*. 49: 936–941.

Bhojane, P., Duff Jr., M. R., Patel, H. C., Vogt, M. E. and Howell, E. E., 2014. Investigation of osmolyte effects on FolM: comparison with other dihydrofolate reductases. *Biochemistry*. 53: 1330–1341.

Bolen, D. W., 2001. Protein stabilization by naturally occurring osmolytes. *Methods in Molecular Biology*. 168: 17–36.

Bourot, S., Sire, O., Trautwetter, A., Touze, T., Wu, L. F., Blanco, C. and Bernard, T., 2000. Glycine betaine-assisted protein folding in a lysA mutant of *Escherichia coli*. *Journal of Biological Chemistry*. 275: 1050–1056.

Burg, M. B., Ferraris, J. D. and Dmitrieva, N. I., 2007. Cellular response to hyperosmotic stresses. *Physiological Reviews*. 87: 1441–1474.

Capp, M. W., Pegram, L. M., Saecker, R. M., Kratz, M., Riccardi, D., Wendorff, T., Cannon, J. G. and Record, M. T, Jr., 2009. Interactions of the osmolyte glycine betaine with molecular surfaces in water: thermodynamics, structural interpretation, and prediction of m-values. *Biochemistry*. 48: 10372–10379.

Cheng, J., Goldstein, R., Gershenson, A., Stec, B. and Roberts, M. F., 2013. The cation-pi box is a specific phosphatidylcholine membrane targeting motif. *Journal of Biological Chemistry*. 288: 14863–14873.

Chopra, S., Dooling, R., Horner, C. G. and Howell, E. E., 2008. A balancing act: net uptake of water during dihydrofolate binding and net release of water upon NADPH binding in R67 dihydrofolate reductase. *Journal of Biological Chemistry*. 283: 4690–4698.

Courtenay, E. S., Capp, M. W., Anderson, C. F. and Record Jr., M. T., 2000. Vapor pressure osmometry studies of osmolyte-protein interactions: implications for the action of osmoprotectants in vivo and for the interpretation of "osmotic stress" experiments *in vitro*. *Biochemistry*. 39: 4455–4471.

Diaz-Sanchez, A. G., Gonzalez-Segura, L., Mujica-Jimenez, C., Rudino-Pinera, E., Montiel, C., Martinez-Castilla, L. P. and Munoz-Clares, R. A., 2012. Amino acid residues critical for the specificity for betaine aldehyde of the plant ALDH10 isoenzyme involved in the synthesis of glycine betaine. *Plant Physiology*. 158: 1570–1582.

Duff Jr., M. R., Grubbs, J., Serpersu, E. H. and Howell, E. E., 2012. Weak interactions between folate and osmolytes in solution. *Biochemistry*. 51: 2309–2318.

Fried, M. G., Stickle, D. F., Smirnakis, K. V., Adams, C., MacDonald, D. and Lu, P., 2002. Role of hydration in the binding of lac repressor to DNA. *Journal of Biological Chemistry*. 277: 50676–50682.

Grubbs, J., Rahmanian, S., Deluca, A., Padmashali, C., Jackson, M., Duff, M. R. and Howell, E. E., 2011. Thermodynamics and solvent effects on substrate and cofactor binding in *Escherichia coli* chromosomal dihydrofolate reductase. *Biochemistry*. 50: 3673–3685.

Guinn, E. J., Pegram, L. M., Capp, M. W., Pollock, M. N. and Record Jr., M. T., 2011. Quantifying why urea is a protein denaturant, whereas glycine betaine is a protein stabilizer. *Proceedings of the National Academy of Sciences of the United States of America.* 108: 16932–16937.

Krahn, J., Jackson, M., DeRose, E. F., Howell, E. E. and London, R. E., 2007. Structure of a type II dihydrofolate reductase ternary complex: use of identical binding sites for unrelated ligands. *Biochemistry*. 46: 14878–14888.

Lin, T. Y. and Timasheff, S. N., 1994. Why do some organisms use a urea-methylamine mixture as osmolyte? Thermodynamic compensation of urea and trimethylamine N-oxide interactions with protein. *Biochemistry*. 33: 12695–12701.

Neuhofer, W., 2010. Role of NFAT5 in inflammatory disorders associated with osmotic stress. *Current Genomics*. 11: 584–590.

Parsegian, V. A., Rand, R. P. and Rau, D. C., 1995. Macromolecules and water: probing with osmotic stress. *Methods in Enzymology*. 259: 43–94.

Parsegian, V. A., Rand, R. P. and Rau, D. C., 2000. Osmotic stress, crowding, preferential hydration, and binding: a comparison of perspectives. *Proceedings of the National Academy of Sciences of the United States of America*. 97: 3987–3992.

Record Jr., M. T., Courtenay, E. S., Cayley, D. S. and Guttman, H. J., 1998. Responses of *E. coli* to osmotic stress: large changes in amounts of cytoplasmic solutes and water. *Trends in Biochemical Sciences*. 23: 143–148.

Roy, D. R., Layton, H. E. and Jamison, R. L., 1992. Countercurrent mechanism and its regulation. In: Seldin, D. W. and Giebisch, G. (ed.) The Kidney: Physiology and Pathophysiology. Raven, New York. pp. 1649–1692.

Sawaya, M. R. and Kraut, J., 1997. Loop and subdomain movements in the mechanism of *Escherichia coli* dihydrofolate reductase: crystallographic evidence. *Biochemistry*. 36: 586–603.

Schellman, J. A., 1994. The thermodynamics of solvent exchange. *Biopolymers*. 34: 1015–1026.

Schiefner, A., Breed, J., Bosser, L., Kneip, S., Gade, J., Holtmann, G., Diederichs, K., Welte, W. and Bremer, E., 2004. Cation-pi interactions as determinants for binding of the compatible solutes glycine betaine and proline betaine by the periplasmic ligand-binding protein ProX from *Escherichia coli*. *Journal of Biological Chemistry*. 279: 5588–5596.

Schuttelkopf, A. W., Hardy, L. W., Beverley, S. M. and Hunter, W. N., 2005. Structures of *Leishmania major* pteridine reductase complexes reveal the active site features important for ligand binding and to guide inhibitor design. *Journal of Molecular Biology*. 352: 105–116.

Sidorova, N. Y., Muradymov, S. and Rau, D. C., 2006. Differences in hydration coupled to specific and nonspecific competitive binding and to specific DNA Binding of the restriction endonuclease BamHI. *Journal of Biological Chemistry*. 281: 35656–35666.

Venkatesu, P., Lee, M. J. and Lin, H. M., 2009. Osmolyte counteracts urea-induced denaturation of alpha-chymotrypsin. *Journal of Physical Chemistry B*. 113: 5327–5338.

Yancey, P. H., Clark, M. E., Hand, S. C., Bowlus, R. D. and Somero, G. N., 1982. Living with water stress: evolution of osmolyte systems. *Science*. 217: 1214–1222.

CHAPTER 6

# Betaine Uptake with a Focus on the Betaine/GABA Transporter

Y. ZHOU*[a], S. A. KEMPSON[b], AND N. C. DANBOLT[a]

[a]The Neurotransporter Group, Department of Anatomy, Institute of Basic Medical Sciences, University of Oslo, Oslo, Norway; [b]Department of Cellular & Integrative Physiology, Indiana University School of Medicine, Indianapolis, Indiana, USA
*E-mail: yun.zhou@medisin.uio.no; www.neurotransporter.org

## 6.1 Introduction

Betaine (glycine betaine or *N,N,N*-trimethylglycine) was first discovered in the 19th century, and it is derived from two sources. (A) Endogenous betaine is synthesized *de novo* from choline (essential nutrient in mammals) in two steps catalyzed by choline dehydrogenase (CHDH; EC 1.1.99.1) and then by betaine aldehyde dehydrogenase (EC 1.2.1.8) (Craig, 2004). (B) Exogenous (or dietary) betaine comes directly from intestinal uptake of betaine (Clow *et al.*, 2008; Slow *et al.*, 2009; see Figure 6.2). Certain foods, *e.g.* whole grains, spinach, shrimp and beet, are rich in betaine (de Zwart *et al.*, 2003) although the exact betaine content varies greatly between sources (*e.g.*, de Zwart *et al.*, 2003). Dietary betaine is absorbed from the proximal small intestine (Kettunen *et al.*, 2001) and carried to the systemic circulation *via* the portal vein (Figure 6.1). Human blood plasma typically contains 20–60 μmol $L^{-1}$ betaine depending on the betaine intake (Schwab *et al.*, 2006). Several organs, including the liver (Kettunen *et al.*, 2001; Kim *et al.*, 2003), the kidneys (Kim *et al.*, 2003) and the testis (Johnson *et al.*, 2010) accumulate betaine from

Food and Nutritional Components in Focus No. 7
Betaine: Chemistry, Analysis, Function and Effects
Edited by Victor R Preedy

Published by the Royal Society of Chemistry, www.rsc.org

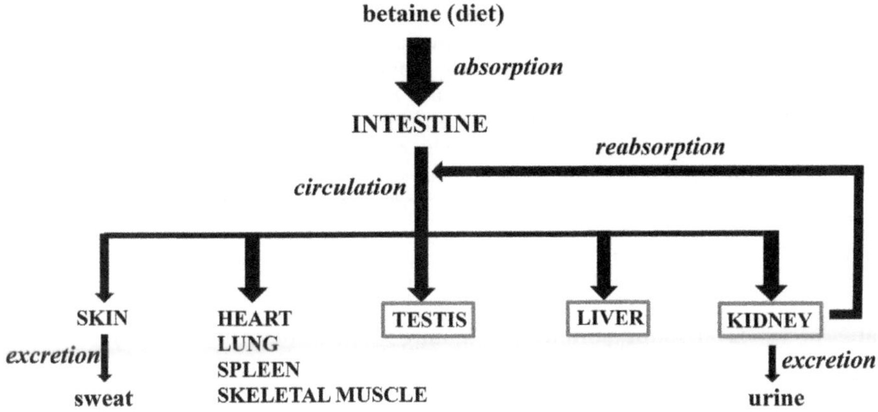


**Figure 6.1** The routes of dietary betaine in the body. Dietary betaine is absorbed in small intestine and reaches the systemic circulation *via* the portal vein. Several organs, including liver, kidney, heart, lung, spleen and testis, accumulate betaine from the blood. Liver, kidney and testis contains substantial amounts of betaine; some of which is synthesized from choline (indicated by square). A small portion of betaine is excreted in the form of urine and sweat.

blood (Figure 6.1). Some of the betaine is consumed by betaine-homocysteine *S*-methyltransferase (BHMT1), while the rest is excreted in urine and sweat (Craig, 2004).

## 6.2 Betaine Acts as a Compatible Osmolyte and as a Methyl Donor

The principal roles of betaine are to act as a compatible osmolyte (Burg and Ferraris, 2008) and as a methyl donor (Craig, 2004). The osmolyte role is particularly important in the kidney medulla where the environment is rather hostile due to hyperosmolarity (up to about 1000 mosmol $kg^{-1}$) and significant concentrations of the denaturing agent urea. To adapt and maintain normal cell functions, cells compensate by changing their content of osmolytes, including betaine (Burg and Ferraris, 2008). In the liver, which is usually slightly hypertonic (about 330–335 mosmol $kg^{-1}$; Go *et al.*, 2004), betaine is primarily linked to one-carbon metabolism (see details in Chapter 7 of this book).

## 6.3 Identification of the Betaine-GABA Transporter (BGT1; slc6a12)

With the discovery of betaine uptake activities in kidney-derived Madin–Darby canine kidney (MDCK) cells, a betaine transporter cDNA was cloned (Yamauchi *et al.*, 1992). The nucleotide sequence turned out to exhibit high

**Table 6.1** The reported organ distribution of BGT1 in different species. The BGT1 expression may vary among species. For example, in rat and mouse BGT1 is most abundant in the liver, whereas in dog it is reported to be more abundant in the kidney. In man, BGT1 mRNA was in heart and skeletal muscle, but not in the lung. In the rat, BGT1 was detected in lung, but not in heart and skeletal muscle. All species express BGT1 in liver, kidney and brain.

| Species | Localization | References |
|---|---|---|
| Man | Kidney, brain, liver, placenta, heart and skeletal muscle (not lung and pancreas) | Rasola *et al.* (1995) |
| Rat | Liver, kidney, brain, spleen, lung (not heart, skeletal muscle, testis) | Burnham *et al.* (1996) |
| Mouse | Liver, kidney, brain (not small intestine, testis) | López-Corcuera *et al.* (1992), Zhou *et al.* (2012a) |
| Dog | Kidney, brain, liver (not ileum, epididymis) | Yamauchi *et al.* (1992), Takenaka *et al.* (1995) |

similarity to the brain transporters for GABA and noradrenaline (Yamauchi *et al.*, 1992). Because the novel transporter was able to transport not only GABA, but also betaine, it was named the betaine-GABA transporter (BGT1; slc6a12). Two other research groups, in parallel, cloned BGT1 homologs from mouse (López-Corcuera *et al.*, 1992) and human (Borden *et al.*, 1995) brains based on homology to GABA transporter 1 (GAT1; slc6a1; Guastella *et al.*, 1990). Soon afterwards, BGT1 was cloned from human kidney (Rasola *et al.*, 1995) and rat liver (Burnham *et al.*, 1996). Table 6.1 summarizes the reported organ distribution of BGT1 in different species. Because the mouse isoform was initially named mouse GAT2 (López-Corcuera *et al.*, 1992), it is important to avoid confusing BGT1 (slc6a12) with rat/human GAT2 (slc6a13), which has no capacity in transporting betaine.

## 6.4 BGT1 is Expressed in the Brain, Kidney and Liver

### 6.4.1 The Roles of BGT1 in the Kidney

The kidney contains a high amount of betaine (Slow *et al.*, 2009), and a significant fraction of this is *de novo* synthesized from reabsorbed choline as elimination of choline dehydrogenase activity leads to a marked reduction of renal betaine levels (Johnson *et al.*, 2010). This, however, does not explain the distribution pattern of betaine along the corticopapillary axis (increase from the inner medulla to the papillary tip), because the highest choline dehydrogenase activity is in the renal cortex (Miller *et al.*, 1996). Therefore, an additional mechanism has to be postulated. The *in situ* rat model suggests that the newly synthesized betaine in the cortex accumulates in the medulla by means of $Na^+$- and $Cl^-$-dependent betaine transporters (Moeckel *et al.*, 1997). This fits with the expression pattern of BGT1 (Moeckel *et al.*, 1997; Zhou *et al.*, 2012a).

BGT1 is believed to act as a protector against osmolarity changes in the kidney (Burg and Ferraris, 2008), and the functions of BGT1 have been extensively studied in the MDCK cells. Under the isotonic condition BGT1 is mainly in the cytoplasm and the betaine transport capacity of the cells is low (Kempson and Montrose, 2004). Hypertonicity elevates the transcriptional activity of tonicity-responsive enhancer binding protein (TonEBP/NFAT5; Burg and Ferraris, 2008). Activation of transcription of the downstream genes, for example the BGT1 gene, can be achieved by the specific binding of TonEBPs to the tonicity-responsive enhancer sequences (TonE1 and TonE2) that are present in the promotor region of the BGT1 gene (Burg and Ferraris, 2008). This leads to increased abundance of BGT1 mRNA more than 10-fold, accompanying the sequential 5–10-fold increase of the transport capacity during hypertonic stress (Uchida *et al.*, 1993). Recent studies (Schweikhard *et al.*, 2014) suggest that phosphorylation of the Thr 40 residue of BGT1 protein is essential for accurate trafficking to the plasma membrane. Up- and downregulation of BGT1 mRNA transcription in response to changes in medullary tonicity have been confirmed also *in vivo* (Kaneko *et al.*, 1997). Hypertonic upregulation, however, is a relatively slow process that requires up to 20 h for completion (Kempson, 1998). To allow more acute control of plasma membrane abundance of BGT1 protein, a post-translational downregulation of plasma membrane BGT1 by intracellular calcium, extracellular ATP/adenosine or activators of protein kinase C was identified (Kempson and Montrose, 2004).

Even though BGT1 has been suggested for more than two decades to play an important role in osmolyte accumulation and survival of the renal medullary cells, the localization of BGT1 protein *in vivo* was not reported until recently (Zhou *et al.*, 2012a). BGT1 protein was found at low levels in the renal outer medulla, and at very high concentration in the renal papilla (Zhou *et al.*, 2012a). Mice lacking BGT1 were viable (Lehre *et al.*, 2011) and, unexpectedly, they concentrated urine and showed no obvious defects after 72 h of salt drinking (Zhou *et al.*, 2012a). No detailed examination on the kidney has been done but, nevertheless, this was in contrast to the severe phenotype of TonEBP knockout mice (Lopez-Rodriguez *et al.*, 2004). TonEBP does not regulate the BGT1 gene exclusively, it also regulates several other osmolyte transporter genes including the sodium/myo-inositol cotransporter 1 (SMIT1; slc5a3) and the taurine transporter (TAUT; slc6a6), as well as aldose reductase (AR), which converts glucose to sorbitol (Lopez-Rodriguez *et al.*, 2004). Deletion of one of the osmolyte transporters or the enzyme AR alone caused only mild defects in renal function (Aida *et al.*, 2000; Huang *et al.*, 2006). These observations can perhaps be explained by the *in vitro* observation that the other osmolytes (*e.g.* myo-inositol and sorbitol) compensate for the loss (Moriyama *et al.*, 1991), but more investigations are required to draw a firm conclusion.

In fact, renal handling of betaine involves more than one active transport mechanism because BGT1 is not expressed in the cortex and is present at the basolateral membrane (Zhou *et al.*, 2012a). Thus, BGT1 cannot account for the significant reabsorption of betaine in proximal tubules (Wunz and

Wright, 1993) and cannot explain why so little betaine is excreted in the urine in adults (Patterson *et al.*, 2011). There is another $Na^+$- and $H^+$-dependent betaine transporter in the brush-border in the renal cortex, and this transporter is $Cl^-$ independent and can be inhibited by proline and glycine (Wunz and Wright, 1993). The identity has not been unequivocally determined, but is believed to be SIT1 (slc6a20, also called XT3s1) which may also be the primary transporter responsible for the absorption of betaine from the intestine (Romeo *et al.*, 2006; Patterson *et al.*, 2011). In some patients with diabetes mellitus urinary betaine excretion is elevated because of defective reabsorption by SIT1 (Patterson *et al.*, 2011). Part of the absorbed betaine probably contributes to the renal homocysteine clearance (Delgado-Reyes *et al.*, 2001).

### 6.4.2 The Role of BGT1 in the Liver

The betaine levels in rodent liver are usually not very high (approximately1–2 μmol $g^{-1}$), but when betaine is supplemented in the diet, betaine levels can increase up to 5-fold to about 10 μmol $g^{-1}$ (Clow *et al.* 2008). The accumulation of betaine is believed to be primarily by BGT1 and amino acid transport system A (Craig, 2004; see Figure 6.2). In fact, most of the BGT1 in the body is in liver, at least in rodents (Zhou *et al.,* 2012a). The liver lobes are composed of hepatocytes (often referred to as parenchymal cells) and nonparenchymal cells comprising the endothelial cells lining the sinusoids, the Kupffer cells (liver macrophages) and hepatic stellate cells. The literature on BGT1 localization is confusing in part because earlier immunocytochemical studies were hampered due to lack of a good negative control (see detailed discussion in Holmseth *et al.*, 2012) and because betaine uptake activities were attributed to BGT1 before it was realized that they may be more than one betaine transporter. The first immunocytochemistry study where the labeling specificity was validated using BGT1-deficient mice, concluded that BGT1 is expressed in the plasma membrane of hepatocytes (Zhou *et al.*, 2012a). This conclusion has recently been confirmed by using isolated mouse hepatocytes (Kempson *et al.*, 2013; Y. Zhou and N.C. Danbolt, unpublished data). Considering (Delgado-Reyes *et al.*, 2001) that hepatocytes also express the betaine metabolizing enzyme BHMT1 (EC 2.1.1.5) it seems legitimate to suggest that the role of betaine is to act as methyl donor for removing homocysteine. A disturbance of hepatic one-carbon metabolism can lead to fatty liver, steatosis and the development of hepatocellular carcinoma (Corbin and Zeisel, 2012). Betaine supplements protect the liver from injuries induced by a variety of toxins, for example, carbon tetrachloride, alcohol (see Chapter 17) and lipopolysaccharide, and improve liver function in nonalcoholic fatty liver disease (Craig, 2004), but the mechanisms behind these observations remain obscure.

Deficiency in choline (the betaine precursor) leads to multiple consequences, for example, liver dysfunction, muscle damage and defects in brain function (Corbin and Zeisel, 2012). Choline is needed for several purposes one of which is synthesis of betaine. Thus, lack of betaine can be compensated by increased consumption of choline, but not the other way around.

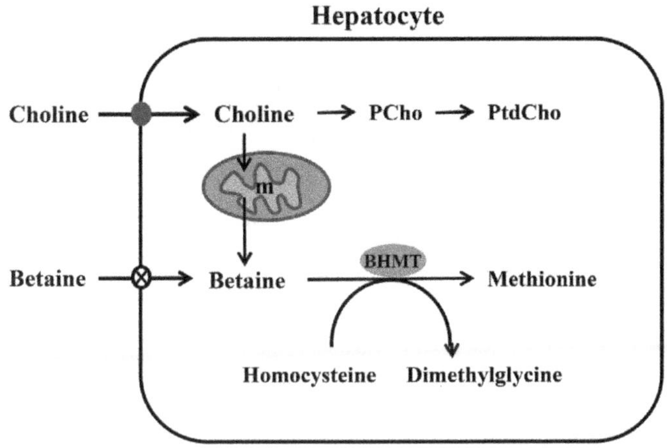


**Figure 6.2** Schematic diagram of the source of hepatic betaine and the choline-betaine pathway in hepatocytes. In addition to phosphatidylcholine synthesis, a major fate of choline is to be converted to betaine. Hepatic betaine can also come from an exogenous source, and this is mediated by transporter proteins (*e.g.*, BGT1). Betaine-homocysteine methyltransferase (BHMT; E.C.2.1.1.5) catalyzes the transfer of one methyl group from betaine to convert homocysteine to methionine. Abbreviations: m, mitochondria; PCho, phosphocholine; PtdCho, phosphatidylcholine.

Exogenous betaine can restore hepatic betaine concentrations and correct *S*-adenosylmethionine/*S*-adenosylhomocysteine ratio, but cannot prevent development of fatty liver induced by choline-deficient diet (Shin *et al.*, 1997). The observations that betaine supplement reduces total homocysteine in homocystinurics and attenuates the total homocysteine increase after methionine loading in healthy subjects are clear (Holm *et al.*, 2004). Plasma homocysteine can be removed by methylating it to form methionine through the action of BHMT1 or the methionine synthase (MS) (Figure 6.3). Betaine serves as methyl donor of BHMT1. In mice the BHMT1 pathway is a major route for the removal of homocysteine, whereas the MS pathway shows little excess capacity to methylate homocysteine (Corbin and Zeisel, 2012). When homocysteine flux is high (*e.g.*, induced by methionine loading), the choline-betaine-dependent pathway becomes important (Corbin and Zeisel, 2012). The addition of betaine from diet allows a marked reduction (up to 50% in chicken) in the total dietary choline requirement (Dilger *et al.*, 2007).

Further characterization of betaine transporters in the liver is needed in order to fully understand the function of hepatic betaine. As the other GABA transporter subtype (GAT2; slc6a13) is also in hepatocytes with even higher affinity to GABA ($K_m$ = 18 μM) (Liu *et al.*, 1993; Zhou *et al.*, 2012b), BGT1 transport activity in hepatocytes has to be characterized using betaine as the substrate. Kempson and coworkers showed that about 70% of [$^{14}$C]betaine transport activity by isolated hepatocytes was sodium dependent (Kempson *et al.*, 2013). Furthermore, the transport was partly inhibited by quinidine

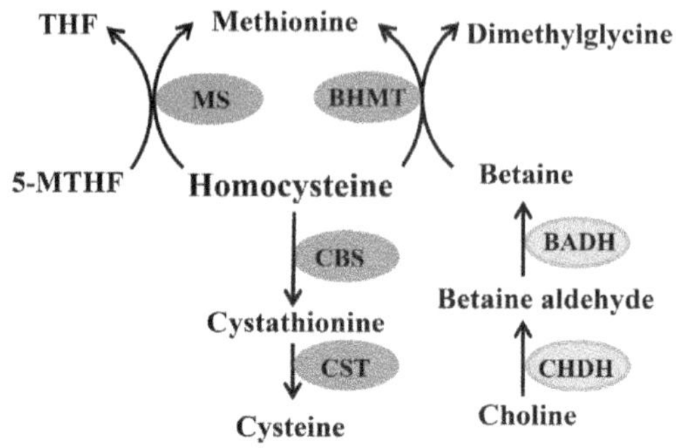


**Figure 6.3** Schematic diagram of the pathways for homocysteine removal. There are three different pathways for homocysteine removal. It can be degraded to cysteine by the sequential actions of cystathionine β-synthase (CBS; E.C.4.2.1.22) and γ-cystathionase (CST; E.C.4.4.1.1). Alternatively, it can be converted by methylation to methionine through the action of BHMT1 or by the action of methionine synthase (MS). In the BHMT pathway betaine is the methyl donor, whereas 5-methyltetrahydrofolate (5-MTHF) is the methyl donor in MS pathway. Betaine can be synthesized from choline oxidation by choline dehydrogenase (CHDH; E.C.1.1.99.1) and betaine aldehyde dehydrogenase (BADH; E.C.1.2.1.8).

and nipecotic acid, similar to the activity of the renal BGT1 transporter, and was partly inhibited by α-methylaminoisobutryic acid (an inhibitor for system A amino acid transporter) and carnitine (Kempson *et al.*, 2013). Thereby, non-BGT1 transporters appear to be involved in betaine accumulation. Among the three system A subtypes (SNAT1, slc38a1; SNAT2, slc38a2; SNAT4, slc38a4), only SNAT2 shows significant [$^{14}$C]betaine uptake (Nishimura *et al.*, 2014), but SNAT2 is hardly expressed in the liver (Mackenzie and Erickson, 2004). This suggests that other transporters are involved. Slc-family 38 comprises 11 members not all of which are characterized with respect to a potential role as liver betaine transporters. Alternatively, there are potential candidates in the organic cation/carnitine transporter family (*e.g.* slc22a27).

### 6.4.3 The Role of BGT1 in the Brain

Because BGT1 also transports GABA (see Table 6.2), it was assumed that BGT1 might be involved in the inactivation of GABA in the brain. GABA is the major inhibitory neurotransmitter in the brain, and it was hypothesized that BGT1 might be a useful target for development of anticonvulsive drugs (for review see: Zhou and Danbolt, 2013). To address this question, BGT1 knockout mice were generated and subjected to a number of seizure threshold tests, including pentylentetrazol (PTZ) seizure threshold test, minimal clonic seizure threshold test, the 6 Hz seizure threshold tests and minimal tonic extension threshold test (Lehre *et al.*, 2011). However, no differences in seizure thresholds were noted between BGT1 knockout and wild-type mice (Lehre *et al.*, 2011). The reason for this is probably the low expression levels of BGT1 in the

**Table 6.2** Key facts of GABA.

1. The major inhibitory neurotransmitter in the brain
2. An amino acid
3. Synthesis from glutamate by glutamic acid decaroxylases
4. Exerts its signaling action from the extracellular fluid by binding to receptors in the plasma membrane.
5. GABA receptor activation is limited by removal of GABA from the extracellular fluid by means of cellular uptake catalyzed by GABA transporter proteins

brain relative to the GAT1 (slc6a1) and GAT3 (slc6a11) GABA transporter subtypes. GAT1 and GAT3 also have higher affinity for GABA, and together this implies that BGT1 is unlikely to influence the GABA levels around synapses (see detailed discussion in Lehre *et al.*, 2011; Zhou *et al.*, 2012a; Zhou and Danbolt, 2013). Secondly, the cycling of these transporters is relatively slow and the diffusion of transmitters is fast. As a result, a high number of neurotransmitter transporter molecules are required to provide enough binding sites. This explains why the expression levels of EAAT2 and GAT1 are so high in the brain (see the detailed discussion in Zhou and Danbolt, 2013).

BGT1 is also unlikely to play any significant role in maintaining cellular volume in the brain because brain betaine levels are also very low (0.02–0.1 μmol $g^{-1}$), and because betaine is believed to come from the blood (Schwahn *et al.*, 2004) rather than being produced from choline as the brain lacks the enzyme choline dehydrogenase (Barlow and Marchbanks, 1985; Johnson *et al.*, 2010). In contrast, taurine and myo-inositol, which are other osmolytes, are abundant in brain tissue (5–10 μmol $g^{-1}$ and 5–7 μmol $g^{-1}$, respectively; *e.g.* Heilig *et al.*, 1989), and thereby are present in sufficient quantities to affect osmolarity. Further, these osmolytes are substrates for the SMIT and TAUT transporters that are abundantly expressed in the brain (Zhou *et al.*, 2012a; Zhou *et al.* 2012b). The notion that BGT1 and betaine do not play any role in osmolyte regulation in the brain is further strengthened by the observations that neither salt loading nor water deprivation caused changes in brain betaine content (Heilig *et al.*, 1989) and that acute and chronic salt loading induced no significant increase of BGT1 mRNA (Bitoun and Tappaz, 2000). To maintain the low betaine levels in the brain, there may be an efflux mechanism because the brain has little capacity for betaine catabolism (Delgado-Reyes *et al.*, 2001). BGT1 may play a role in betaine disposal because it is highly expressed in the leptomeninges (Zhou *et al.*, 2012a) and because the leptomeninges have recently been identified as a key contributor to the clearance of interstitial solute (Iliff *et al.*, 2012).

### 6.4.4 The Role of Betaine in the Testis

In addition to the kidney and liver, high levels of betaine have been observed in the testis. Remethylation of homocysteine to methionine is probably essential for the spermatogenesis, and this is mediated *via* two pathways. One is the choline-betaine pathway because deletion of choline dehydrogenase in mice caused depletion of betaine in several organs including testis ((Johnson *et al.*, 2010). This did not cause liver dysfunction, but impaired fertility. The other pathway is the folate pathway using the methyl donor 5-methyltetrahydrofolate. Deletion of the methylenetetrahydrofolate reductase (MTHFR) gene resulted in abnormal spermatogenesis and infertility (Kelly *et al.*, 2005). Betaine supplements alleviate the symptoms (Kelly *et al.*, 2005; Johnson *et al.*, 2010), but the responsible transporter remains to be identified. It is probably not BGT1 as BGT1 is not present in the testis, at least in mice (Zhou *et al.*, 2012a).

## Summary Points

- This chapter focused on betaine uptake.
- The accumulation of betaine in the body is mediated by several different betaine transporters.
- The most studied betaine transporter is the $Na^+$- and $Cl^-$-dependent betaine/GABA transporter (BGT1, slc6a12).
- In mice, BGT1 is predominantly expressed in the liver and at lower levels in the kidney and brain.
- The primary role of BGT1 in hepatocytes is probably to supply the methyl donor betaine.
- BGT1 cannot contribute to GABA inactivation around synapses and does not affect seizure thresholds due to both lower expression levels and lower affinity than the other GABA transporters.

## Definitions of Words and Terms

**Alkaloid:** Alkaloid is a large group of nitrogenous basic substances found in plants. Many are pharmacologically active. Examples are atropine, caffeine and nicotine.

**Hepatocellular carcinoma:** Hepatocellular carcinoma is the most common type of liver cancer.

**Homocysteine:** Homocysteine is a sulfur-containing amino acid ($C_4H_9NO_2S$) that is produced by the demethylation of methionine.

**Homocystinurics:** Excretion of excess homocystine in the urine. The condition is usually genetic (inherited).

**Leptomeninges:** Leptomeninges is the two innermost layers (pia mater and arachnoid) of the meninges covering the brain.

**Nonalcohol fatty liver disease:** Fatty liver not caused by excessive alcohol use.

**Osmoprotectant:** Osmoprotectant is a substance that is beneficial for cell survival under osmotic stress.

**Pentylentetrazol:** Pentylentetrazol is a compound that can chemically induce seizures.

**Seizure:** A sudden surge of abnormal electrical activities in the brain.

**Steatohepatitis:** Fatty liver often accompanying with inflammation.

**Steatosis:** Abnormal retention of fat in cells. This is also referred to as fatty change or fatty degeneration as large accumulations may disrupt the cells.

## List of Abbreviations

| | |
|---|---|
| BGT1 | Sodium/chloride-dependent betaine-GABA transporter (slc6a12) |
| BHMT1 | Betaine-homocysteine-*S*-methyltransferase (E.C. 2.1.1.5) |
| BADH | Betaine aldehyde dehydrogenase (E.C.1.2.1.8) |
| CBS | Cystathionine β-synthase (E.C.4.2.1.22) |
| CST | γ-Cystathionase (E.C.4.4.1.1) |
| CHDH | Choline dehydrogenase (E.C.1.1.99.1) |
| EF1502 | (*N*-[4,4-bis(3-methyl-2-thienyl)-3-butenyl]-4-(methylamino-4,5,6,7-tetrahydrobenzo[d]isoxazol-3-ol |
| EAAT | Excitatory amino acid transporter |
| GABA | Gamma-aminobutyric acid |
| GAT | GABA Transporter |
| Hcy | Homocysteine |
| THF | Tetrahydrofolate |
| MDCK | Madin–Darby canine kidney |
| 5-MTHF | 5-Methyltetrahydrofolate |
| MS | Methionine synthase |
| PCho | Phosphocholine |
| PtdCho | Phosphatidylcholine |
| SMIT1 | Sodium-dependent myo-inositol cotransporter-1 (slc5a3) |
| TAUT | Sodium/chloride-dependent taurine transporter (slc6a6) |
| TonEBP | Tonicity-responsive enhancer binding protein, also known as NFAT5, nuclear factor of activated T Cells-5 |

## Acknowledgements

This work was supported by private funds and by the University of Oslo (stimulation funds to NCD).

## References

Aida, K., Ikegishi, Y., Chen, J., Tawata, M., Ito, S., Maeda, S. and Onaya, T., 2000. Disruption of aldose reductase gene (Akr1b1) causes defect in urinary concentrating ability and divalent cation homeostasis. *Biochemical and Biophysical Research Communications*. 277: 281–286.

Barlow, P. and Marchbanks, R. M., 1985. The effects of inhibiting choline dehydrogenase on choline metabolism in mice. *Biochemical Pharmacology*. 34: 3117–3122.

Bitoun, M. and Tappaz, M., 2000. Gene expression of taurine transporter and taurine biosynthetic enzymes in brain of rats with acute or chronic hyperosmotic plasma. A comparative study with gene expression of myo-inositol transporter, betaine transporter and sorbitol biosynthetic enzyme. *Molecular Brain Research*. 77: 10–18.

Borden, L. A., Smith, K. E., Gustafson, E. L., Branchek, T. A. and Weinshank, R. L., 1995. Cloning and expression of a betaine/GABA transporter from human brain. *Journal of Neurochemistry*. 64: 977–984.

Burg, M. B. and Ferraris, J. D., 2008. Intracellular organic osmolytes: function and regulation. *Journal of Biological Chemistry*. 283: 7309–7313.

Burnham, C. E., Buerk, B., Schmidt, C. and Bucuvalas, J. C., 1996. A liver-specific isoform of the betaine/GABA transporter in the rat: cDNA sequence and organ distribution. *Biochimica et Biophysica Acta (BBA) - Biomembranes*. 1284: 4–8.

Clow, K. A., Treberg, J. R., Brosnan, M. E. and Brosnan, J. T., 2008. Elevated tissue betaine contents in developing rats are due to dietary betaine, not to synthesis. *Journal of Nutrition*. 138: 1641–1646.

Corbin, K. D. and Zeisel, S. H., 2012. Choline metabolism provides novel insights into nonalcoholic fatty liver disease and its progression. *Current Opinion in Gastroenterology*. 28: 159–165.

Craig, S. A., 2004. Betaine in human nutrition. *American Journal of Clinical Nutrition*. 80: 539–549.

Delgado-Reyes, C. V., Wallig, M. A. and Garrow, T. A., 2001. Immunohistochemical detection of betaine-homocysteine S-methyltransferase in human, pig, and rat liver and kidney. *Archives of Biochemistry and Biophysics*. 393: 184–186.

Dilger, R. N., Garrow, T. A. and Baker, D. H., 2007. Betaine can partially spare choline in chicks but only when added to diets containing a minimal level of choline. *Journal of Nutrition*. 137: 2224–2228.

Go, W. Y., Liu, X., Roti, M. A., Liu, F. and Ho, S. N., 2004. NFAT5/TonEBP mutant mice define osmotic stress as a critical feature of the lymphoid microenvironment. *Proceedings of the National Academy of Sciences of the United States of America*. 101: 10673–10678.

Guastella, J., Nelson, N., Nelson, H., Czyzyk, L., Keynan, S., Miedel, M. C., Davidson, N., Lester, H. A. and Kanner, B. I., 1990. Cloning and expression of a rat brain GABA transporter. *Science*. 249: 1303–1306.

Heilig, C. W., Stromski, M. E., Blumenfeld, J. D., Lee, J. P. and Gullans, S. R., 1989. Characterization of the major brain osmolytes that accumulate in salt-loaded rats. *American Journal of Physiology*. 257: F1108–F1116.

Holm, P. I., Bleie, O., Ueland, P. M., Lien, E. A., Refsum, H., Nordrehaug, J. E. and Nygard, O., 2004. Betaine as a determinant of postmethionine load total plasma homocysteine before and after B-vitamin supplementation. *Arteriosclerosis, Thrombosis, and Vascular Biology*. 24: 301–307.

Holmseth, S., Zhou, Y., Follin-Arbelet, V. V., Lehre, K. P., Bergles, D. E. and Danbolt, N. C., 2012. Specificity controls for immunocytochemistry: the antigen pre-adsorption test can lead to inaccurate assessment of antibody specificity. *Journal of Histochemistry and Cytochemistry*. 60: 174–187.

Huang, D. Y., Boini, K. M., Lang, P. A., Grahammer, F., Duszenko, M., Heller-Stilb, B., Warskulat, U., Haussinger, D., Lang, F. and Vallon, V., 2006. Impaired ability to increase water excretion in mice lacking the taurine transporter gene TAUT. *Pflugers Archiv - European Journal of Physiology*. 451: 668–677.

Iliff, J. J., Wang, M., Liao, Y., Plogg, B. A., Peng, W., Gundersen, G. A., Benveniste, H., Vates, G. E., Deane, R., Goldman, S. A., Nagelhus, E. A. and Nedergaard, M., 2012. A paravascular pathway facilitates CSF flow through the brain parenchyma and the clearance of interstitial solutes, including amyloid beta. *Science Translational Medicine*. 4: 147ra111.

Johnson, A. R., Craciunescu, C. N., Guo, Z., Teng, Y.-W., Thresher, R. J., Blusztajn, J. K. and Zeisel, S. H., 2010. Deletion of murine choline dehydrogenase results in diminished sperm motility. *FASEB Journal*. 24: 2752–2761.

Kaneko, T., Takenaka, M., Okabe, M., Yoshimura, Y., Yamauchi, A., Horio, M., Kwon, H. M., Handler, J. S. and Imai, E., 1997. Osmolarity in renal medulla of transgenic mice regulates transcription via 5'-flanking region of canine BGT1 gene. *American Journal of Physiology*. 272: F610–F616.

Kelly, T. L., Neaga, O. R., Schwahn, B. C., Rozen, R. and Trasler, J. M., 2005. Infertility in 5,10-methylenetetrahydrofolate reductase (MTHFR)-deficient male mice is partially alleviated by lifetime dietary betaine supplementation. *Biology of Reproduction*. 72: 667–677.

Kempson, S. A., 1998. Differential activation of system A and betaine/GABA transport in MDCK cell membranes by hypertonic stress. *Biochimica et Biophysica Acta*. 1372: 117–123.

Kempson, S. A. and Montrose, M. H., 2004. Osmotic regulation of renal betaine transport: transcription and beyond. *Pflugers Archiv - European Journal of Physiology*. 449: 227–234.

Kempson, S. A., Vovor-Dassu, K. and Day, C., 2013. Betaine transport in kidney and liver: use of betaine in liver injury. *Cellular Physiology and Biochemistry*. 32: 32–40.

Kettunen, H., Tiihonen, K., Peuranen, S., Saarinen, M. T. and Remus, J. C., 2001. Dietary betaine accumulates in the liver and intestinal tissue and stabilizes the intestinal epithelial structure in healthy and coccidia-infected broiler chicks. *Comparative Biochemistry and Physiology Part A: Molecular & Integrative Physiology*. 130: 759–769.

Kim, S. K., Choi, K. H. and Kim, Y. C., 2003. Effect of acute betaine administration on hepatic metabolism of S-amino acids in rats and mice. *Biochemical Pharmacology*. 65: 1565–1574.

Lehre, A. C., Rowley, N. M., Zhou, Y., Holmseth, S., Guo, C., Holen, T., Hua, R., Laake, P., Olofsson, A. M., Poblete-Naredo, I., Rusakov, D. A., Madsen, K. K., Clausen, R. P., Schousboe, A., White, H. S. and Danbolt, N. C., 2011. Deletion of the betaine-GABA transporter (BGT1; slc6a12) gene does not affect seizure thresholds of adult mice. *Epilepsy Research*. 95: 70–81.

Liu, Q. R., López-Corcuera, B., Mandiyan, S., Nelson, H. and Nelson, N., 1993. Molecular characterization of 4 pharmacologically distinct gamma-aminobutyric acid transporters in mouse brain (vol. 268, Pg 2106, 1993). *Journal of Biological Chemistry*. 268: 9156.

Lopez-Rodriguez, C., Antos, C. L., Shelton, J. M., Richardson, J. A., Lin, F., Novobrantseva, T. I., Bronson, R. T., Igarashi, P., Rao, A. and Olson, E. N., 2004. Loss of NFAT5 results in renal atrophy and lack of tonicity-responsive gene expression. *Proceedings of the National Academy of Sciences of the United States of America*. 101: 2392–2397.

López-Corcuera, B., Liu, Q. R., Mandiyan, S., Nelson, H. and Nelson, N., 1992. Expression of a mouse brain cDNA encoding novel gamma-aminobutyric acid transporter. *Journal of Biological Chemistry*. 267: 17491–17493.

Mackenzie, B. and Erickson, J. D., 2004. Sodium-coupled neutral amino acid (system N/A) transporters of the SLC38 gene family. *Pflugers Archiv - European Journal of Physiology*. 447: 784–795.

Miller, B., Schmid, H., Chen, T. J., Schmolke, M. and Guder, W. G., 1996. Determination of choline dehydrogenase activity along the rat nephron. *Biological Chemistry Hoppe-Seyler*. 377: 129–137.

Moeckel, G. W., Lai, L. W., Guder, W. G., Kwon, H. M. and Lien, Y. H., 1997. Kinetics and osmoregulation of Na(+)-and Cl(−)-dependent betaine transporter in rat renal medulla. *American Journal of Physiology*. 272: F100–F106.

Moriyama, T., Garcia-Perez, A., Olson, A. D. and Burg, M. B., 1991. Intracellular betaine substitutes for sorbitol in protecting renal medullary cells from hypertonicity. *American Journal of Physiology*. 260: F494–F497.

Nishimura, T., Yagi, R., Usuda, M., Oda, K., Yamazaki, M., Suda, S., Takahashi, Y., Okazaki, F., Sai, Y., Higuchi, K., Maruyama, T., Tomi, M. and Nakashima, E., 2014. System A amino acid transporter SNAT2 shows subtype-specific affinity for betaine and hyperosmotic inducibility in placental trophoblasts. *Biochimica et Biophysica Acta*. 1838(5): 1306–1312.

Patterson, A. D., Bonzo, J. A., Li, F., Krausz, K. W., Eichler, G. S., Aslam, S., Tigno, X., Weinstein, J. N., Hansen, B. C., Idle, J. R. and Gonzalez, F. J., 2011. Metabolomics reveals attenuation of the SLC6A20 kidney transporter in nonhuman primate and mouse models of type 2 diabetes mellitus. *Journal of Biological Chemistry*. 286: 19511–19522.

Rasola, A., Galietta, L. J., Barone, V., Romeo, G. and Bagnasco, S., 1995. Molecular cloning and functional characterization of a GABA betaine transporter from human kidney. *FEBS Letters*. 373: 229–233.

Romeo, E., Dave, M. H., Bacic, D., Ristic, Z., Camargo, S. M., Loffing, J., Wagner, C. A. and Verrey, F., 2006. Luminal kidney and intestine SLC6 amino acid transporters of B0AT-cluster and their tissue distribution in Mus musculus. *American Journal of Physiology: Renal Physiology*. 290: F376–F383.

Schwab, U., Torronen, A., Meririnne, E., Saarinen, M., Alfthan, G., Aro, A. and Uusitupa, M., 2006. Orally administered betaine has an acute and dose-dependent effect on serum betaine and plasma homocysteine concentrations in healthy humans. *Journal of Nutrition*. 136: 34–38.

Schwahn, B. C., Laryea, M. D., Chen, Z., Melnyk, S., Pogribny, I., Garrow, T., James, S. J. and Rozen, R., 2004. Betaine rescue of an animal model with methylenetetrahydrofolate reductase deficiency. *Biochemical Journal*. 382: 831–840.

Schweikhard, E. S., Kempson, S. A., Ziegler, C. and Burckhardt, B. C., (2014). Mutation of a single threonine in the cytoplasmic NH2 terminus disrupts trafficking of renal betaine-GABA transporter 1 during hypertonic stress. *American Journal of Physiology: Renal Physiology*. 307(1): F107–F115. DOI: 10.1152/ajprenal.00085.2014.

Shin, O. H., Mar, M. H., Albright, C. D., Citarella, M. T., da Costa, K. A. and Zeisel, S. H., 1997. Methyl-group donors cannot prevent apoptotic death of rat hepatocytes induced by choline-deficiency. *Journal of Cellular Biochemistry*. 64: 196–208.

Slow, S., Lever, M., Chambers, S. T. and George, P. M., 2009. Plasma dependent and independent accumulation of betaine in male and female rat tissues. *Physiological Research*. 58: 403–410.

Uchida, S., Yamauchi, A., Preston, A. S., Kwon, H. M. and Handler, J. S., 1993. Medium tonicity regulates expression of the Na(+)- and Cl(−)-dependent betaine transporter in Madin-Darby canine kidney cells by increasing transcription of the transporter gene. *Journal of Clinical Investigation*. 91: 1604–1607.

Wunz, T. M. and Wright, S. H., 1993. Betaine transport in rabbit renal brush-border membrane vesicles. *American Journal of Physiology*. 264: F948–F955.

Yamauchi, A., Uchida, S., Kwon, H. M., Preston, A. S., Robey, R. B., Garcia-Perez, A., Burg, M. B. and Handler, J. S., 1992. Cloning of a $Na^+$-dependent and $Cl^-$-dependent betaine transporter that is regulated by hypertonicity. *Journal of Biological Chemistry*. 267: 649–652.

Zhou, Y. and Danbolt, N. C., 2013. GABA and Glutamate Transporters in Brain. *Frontiers in Endocrinology*. 4: 165.

Zhou, Y., Holmseth, S., Guo, C., Hassel, B., Hofner, G., Huitfeldt, H. S., Wanner, K. T. and Danbolt, N. C., 2012b. Deletion of the gamma-aminobutyric acid transporter 2 (GAT2 and SLC6A13) gene in mice leads to changes in liver and brain taurine contents. *Journal of Biological Chemistry*. 287: 35733–35746.

Zhou, Y., Holmseth, S., Hua, R., Lehre, A. C., Olofsson, A. M., Poblete-Naredo, I., Kempson, S. A. and Danbolt, N. C., 2012a. The betaine-GABA transporter (BGT1, slc6a12) is predominantly expressed in the liver and at lower levels in the kidneys and at the brain surface. *American Journal of Physiology - Renal Physiology*. 302: F316–F328.

de Zwart, F. J., Slow, S., Payne, R. J., Lever, M., George, P. M., Gerrard, J. A. and Chambers, S. T., 2003. Glycine betaine and glycine betaine analogs in common foods. *Food chemistry*. 83: 197–204.

CHAPTER 7

# Betaine-Homocysteine S-Methyltransferase

TIMOTHY A. GARROW*[a]

[a]1201 West Gregory Drive, Room 258, Urbana, Illinois 61801, USA
*E-mail: tagarrow@illinois.edu

## 7.1 Introduction

BHMT is an enzyme of choline and sulfur amino acid metabolism that supports the biosynthesis of Met (Figure 7.1) and *S*-adenosylmethionine, the latter of which is required for over 200 different methylation reactions in humans (Petrossian and Clarke, 2011). There have been numerous studies that have investigated the tissue and subcellular distribution of BHMT in mammalian species. Although relatively low levels of activity, protein or mRNA can be detected in many tissues, the majority of BHMT in most mammals resides in the liver, with secondary sites of significant expression limited to kidney and pancreas. In liver, BHMT represents 0.5–2% of the soluble protein. In human and pig liver, the expression of BHMT is evenly distributed throughout the lobule, but in rat liver there is a gradient of expression with periportal hepatocytes expressing more that centrilobular hepatocytes. In pigs, guinea pigs and primates, the enzyme is abundantly expressed in kidney where it is has been specifically localized to the proximal tubules of the cortex. A third location where high expression has been observed is in the pancreas of sheep and guinea pig. In terms of subcellular localization, BHMT expression has so far only been localized to the cytoplasm, and a portion of which has been found to be specifically associated with microtubules.

Food and Nutritional Components in Focus No. 7
Betaine: Chemistry, Analysis, Function and Effects
Edited by Victor R Preedy

Published by the Royal Society of Chemistry, www.rsc.org

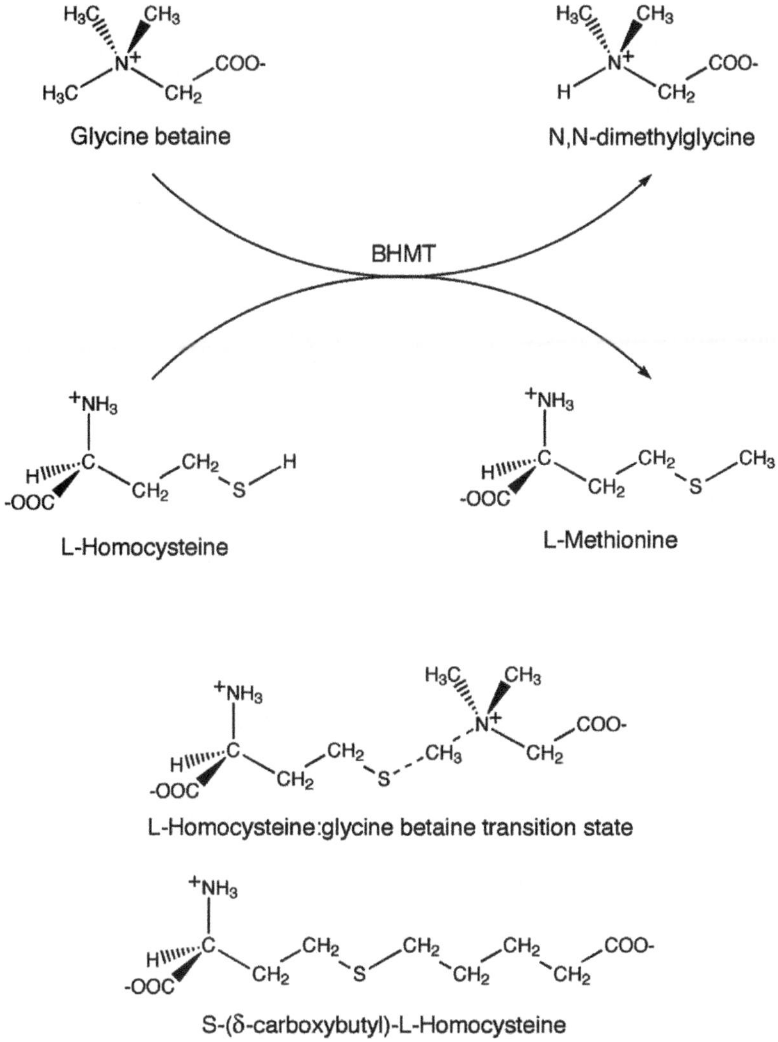


**Figure 7.1** Reactants, products, the transition state and a bisubstrate analog of the BHMT-catalyzed reaction. The BHMT catalyzed reaction; the L-homocysteine:glycine betaine transition state; and the bisubstrate inhibitor *S*-(δ-carboxybutyl)-L-homocysteine (CBHcy). Reprinted (adapted) with permission from Castro *et al.* (2004) Dissecting the catalytic mechanism of betaine-homocysteine *S*-methyltransferase by use of intrinsic tryptophan fluorescence and site-directed mutagenesis. *Biochemistry*. 43: 5341–5351. Copyright 2004 American Chemical Society.

Over the last 20 years there has been considerable interest in the role of BHMT in the regulation of plasma Hcy, an amino acid thought to be a graded and independent risk factor for vascular diseases and thrombosis (Williams and Schalinske, 2010). Recently, *in vivo* studies using *S*-(δ-carboxybutyl)-L-homocysteine (CBHcy), a specific and potent inhibitor of BHMT that mimics the transition state (Figure 7.1), and the development of *bhmt* $^{-/-}$ mice have confirmed that BHMT has a major role in tissue Hcy homeostasis (Collinsova *et al.*, 2006; Strakova *et al.*, 2010, 2011; Teng *et al.*, 2011). A lack of BHMT function causes Hcy and Bet to increase nearly ten-fold in liver and plasma. Besides being a methyl donor for BHMT, Bet is an important intracellular osmolyte. The expression BHMT has been shown to be sensitive to dietary sodium, choline and sulfur amino acid intake. For example, Met restriction in combination with surfeit choline (or betaine) elevates BHMT expression in liver, whereas excess dietary sodium decreases BHMT expression in both liver and kidney.

To date, there are three enzymes in mammals that are known to methylate Hcy to form Met; BHMT, Met synthase (cobalamin dependent: EC 2.1.1.13), and BHMT-2 (EC 2.1.1.10). Met synthase uses 5 *N*-methyltetrahydrofolate as the methyl donor. BHMT-2 was more recently discovered and it was named only knowing its cDNA sequence, which is highly homologous to BHMT. However, when recombinant BHMT-2 was overexpressed and purified it was discovered that the enzyme does not use Bet as a methyl donor but rather *S*-methylmethionine (Szegedi *et al.*, 2008), a compound only produced in plants and fungi. Unless specifically indicated, BHMT-2 is not discussed in this chapter.

## 7.2 Early Studies of Bet-Dependent Methyltransferase Activity

The Bet-dependent methylation of Hcy was first reported by Borsook and Dubnoff (1947) using liver slices. Ericson (1960) studied the methyl donor specificity of homogeneous pig liver BHMT and showed that the enzyme could use three structurally related substrates as methyl donors; Bet, dimethylsulfonioacetate (dimethylacetothetin) and dimethylsulfoniopropionate (dimethylpropiothetin). Bet is derived from choline *in vivo* and both are relatively abundant in the human diet. In addition, choline can be synthesized by animals from serine. Dimethylsulfoniopropionate is a naturally occurring compound that is produced at high levels in marine algae and at much lower levels in terrestrial plants, but it is not synthesized in animals. Dimethylsulfonioacetate is not known to be a natural product.

Early kinetic studies by Fromm and Nordlie (1959) using purified rat liver BHMT indicated that the reaction proceeds through a ternary enzyme–substrate complex. Consistent with that result, initial rate kinetics using partially purified enzyme from rat liver in the absence and presence of products indicated that BHMT catalyzes an ordered bi bi reaction with Hcy being the first substrate to bind, and Met as the last product off (Finkelstein *et al.*, 1972). The latter study also confirmed previous observations that DMG is a potent inhibitor of BHMT activity. Using purified BHMT isolated from human liver, Awad *et al.* (1983) developed a bisubstrate inhibitor of the enzyme called CBHcy. The fact that a bisubstrate mimic could potently inhibit BHMT added support that the reaction mechanism of BHMT was indeed sequential rather than ping-pong. These early studies also highlight a notable feature of all BHMT enzymes, which is that they have very slow turnover numbers (<100 $min^{-1}$).

## 7.3 Studies Using Recombinant BHMT Enzymes

The ability to perform detailed biophysical studies with BHMT was greatly facilitated when cDNAs encoding the enzyme were isolated. The first cDNAs encoding BHMT were described in one report where the cloning and sequencing of a partial pig liver cDNA, and a full-length human liver cDNA were described (Garrow, 1996). Since then, other full-length cDNAs encoding BHMT have been isolated and sequenced. The amino acid sequences of mammalian BHMT enzymes have striking amino acid homology. For example, the mouse, rat, pig and human enzymes share greater than 90% amino acid identity. Their genes have been sequenced and each has a similar arrangement; eight exons interrupted by seven introns.

The human BHMT cDNA was the first to be overexpressed (*E. coli*) and purified to homogeneity (Millian and Garrow, 1998). It was shown to have kinetic properties that were indistinguishable from enzyme that was partially purified from human liver. Before the first crystal structures of BHMT were solved, biochemical studies with the human recombinant enzyme and derived mutants provided significant information about the enzyme's catalytic properties.

When human recombinant BHMT was first purified it was observed that zinc (Zn) coeluted with homogenous enzyme during the final chromatographic step (Millian and Garrow, 1998). Using the reversible and thiol-specific modification reagent, methyl methanethiosulfonate, it was determined that Zn binds to the enzyme through one or more thiol groups (Cys residues). The metal is tightly bound to the enzyme and circular dichroism indicated that Zn is not needed for the protein to retain its overall structure. However, Zn is absolutely required for catalysis, and specifically for Hcy binding and activation. By extrapolating from a series of elegant studies with the related enzyme, Met synthase (Goulding and Matthews, 1997; Peariso *et al.*, 1998), it was proposed that as the thiol of Hcy binds to the Zn atom it is simultaneously deprotonated to generate the more potent nucleophile, the thiolate anion. The thiolate then attacks the nitronium pole of Bet to initiate methyl transfer.

At the time the deduced amino acid sequence of human BHMT was obtained, the protein sequence databases were relatively small, and the only proteins found to have any homology to BHMT were Met synthase and a few proteins in bacteria and yeast. Using the limited homology between these proteins, the two most conserved regions were targeted for further study: G[ILV]NCX$^{218}$ and [ALV]X(2)[ILV]GGCCX(3)PX(2)I$^{307}$. Using site-specific, semirandom and random mutagenesis (Breksa and Garrow, 1999; Breksa and Garrow, 2002), it was discovered that Gly214, Cys217, Cys299 and Cys300 were essential for BHMT function, and that the positions held by Gly297, Gly298 and Gly301 could only accommodate amino acids with small side chains, such as Ala or Ser. BHMT was unable or severely hindered in its ability to bind Zn and catalyze methyl transfer if any of the Cys residues in those regions were changed, and the Gly residue at position 214 was absolutely essential; no other amino acid could substitute at that position. The fact that Zn binding to the enzyme was sensitive to methyl methanethiosulfonate, and that substitution of Cys217, Cys299 or Cys300 with Ala or Ser abolished or nearly abolished Zn binding and enzyme activity led to the conclusion that these Cys residues were Zn-binding ligands. The role of the Gly residues at positions 214, 297, 298 and 301 were surmised to have a role in creating the Zn-binding pocket. It was also noted that positions 216 and 218, which are occupied by Asn and His in the wild-type enzyme, could accommodate amino acid substitutions but that enzyme activity was enhanced when proton donors occupied those positions. These data suggested that Asn218 and His218 could be second sphere ligands to Zn and affect the net charge at the Zn catalytic center and influence activity.

As protein databases became larger and more proteins were discovered that had significant amino acid homologies and identities to BHMT, these highly related proteins were group together into a family designated Pfam02574.

## 7.4 BHMT Structure and Characterization of Ligand Binding

There have been two reports describing the crystal structure of BHMT. The first structure was the human enzyme with the bisubstrate analog CBHcy bound to the active site (Evans *et al.*, 2002). The second structure was of the rat liver enzyme without any ligands bound (Gonzalez *et al.*, 2004). Other reports (Breksa and Garrow, 1999; Breksa and Garrow, 2002; Gonzalez *et al.*, 2003; Castro *et al.*, 2004) have probed the function of specific residues in substrate, product and CBHcy binding, and catalysis, and have identified key conformational changes the enzyme undergoes during its catalytic cycle.

BHMT is a tetramer composed of identical subunits with a quaternary structure that is best described as a dimer of dimers. Each subunit is composed of 406 (human) or 407 (rat) amino acids that have two functional domains. The *N*-terminal domain, encoded by residues 1-318, is a $(\beta/\alpha)_8$ barrel that contains the active site Zn and nearly all the residues that are essential for substrate and product binding and catalysis. The C-terminal domain contains residues 319–406/7 and is composed of three substructures that are important for oligomerization. The first substructure is the "dimerization arm" that is encoded by residues 319–370. In addition to being intimately involved in the dimerization of two monomers, this region appears to help stabilize the tetramer as well. The dimerization arm is followed by a second substructure that is simply a flexible linker (residues 371–380) that connects the dimerization arm to the third and final substructure of the C-terminal domain, an alpha helix encoded by residues 381–406/7. This C-terminal alpha helix is the dominant structure that binds the dimers together to form a stable tetramer. A more detailed description of the secondary, tertiary and quaternary structure of BHMT can be found in the original works (Evans *et al.*, 2002; Gonzalez *et al.*, 2004) while this chapter will focus primarily on the key residues required for ligand binding and catalysis.

Prior to the elucidation of any crystal structure of BHMT, the most highly conserved regions across Pfam02574 members; G[ILV]NCX$^{218}$ and [ALV]X(2)[ILV]GGCCX(3)PX(2)I$^{307}$, had already undergone considerable study (Breksa and Garrow, 1999; Breksa and Garrow, 2002). As proposed in a study that used reversible chemical modification to deplete and reconstitute the enzyme of Zn (Millian and Garrow, 1998), and a second study that used site directed mutagenesis to probe Zn binding (Breksa and Garrow, 1999), the crystal structures confirmed that the sulfur atoms of Cys217, Cys299 and Cys300 are ligands to the catalytic Zn (Figure 7.2). The structure of the enzyme with the bisubstrate analog CBHcy bound, which contains an intact Hcy moiety, confirmed that the sulfur of Hcy binds directly to the Zn atom and that catalysis must therefore proceed by a tetrathiolate intermediate (Evans *et al.*, 2002). As suggested by random mutagenesis (Breksa and Garrow, 2002), the absence of side chains at the positions held by Gly214, Gly297, Gly298 and Gly301 is necessary for the tight packing that is required to form the Zn-binding pocket. Finally, as surmised by mutagenesis experiments (Breksa and Garrow, 2002), Asn216 and His218 are second sphere ligands to Zn by virtue of being hydrogen-bond donors to Cys299 and Cys217, respectively.

In addition to the crystal structures offering clarification to previous mutagenesis experiments, the model structures implicated other residues that were likely involved in ligand binding and catalysis. Besides showing the interaction of Hcy with Zn (Figure 7.2A), the human BHMT-CBHcy binary structure suggested that Hcy specifically interacts with Phe29, Val30 and Glu159 (Evans *et al.*, 2002). In that structure the carboxylate of Hcy is positioned to form bidentate hydrogen bonds to the amide nitrogens of Phe29 and Val30, and because these are backbone amides, it is not surprising that these residues are not strictly conserved in Pfam02574. However, Glu159 is an invariant residue within this family of enzymes. The gamma carboxyl group of this Glu residue is in position to form a hydrogen bond with the amino group of L-Hcy. Using steady-state intrinsic tryptophan fluorescence to monitor ligand binding, Castro *et al.* (2004) studied the impact of mutagenesis at position 159 and found that changing it to Gln increased the $K_{dHcy}$ and $K_{mHcy}$ 400- and 28-fold, respectively, and reduced the $V_{max}/K_m$ by 94%. Similarly, mutating Glu159 in the rat enzyme to either Gly or Lys resulted in a greater than 98% decrease in $k_{cat}$ (Gonzalez *et al.*, 2003).

The human BHMT-CBHcy model structure (Figure 7.2B) implicated residues Trp44, Try77 and Tyr160 in Bet binding (Evans *et al.*, 2002). Trp44 and Tyr77 are within hydrogen-bonding distance to the carboxybutyl moiety (Bet mimic) of CBHcy, while Tyr160 is tightly packed against Tyr77. Changing Trp44 to Ala and Tyr160 to Phe did not disrupt Hcy binding, but significantly

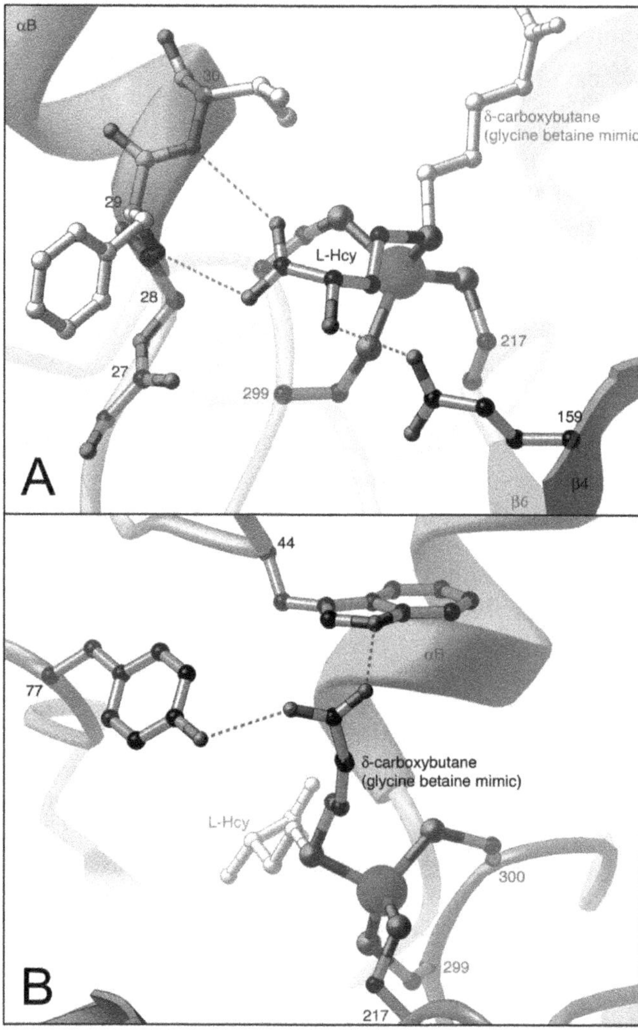


**Figure 7.2** The active site of human BHMT. CBHcy bound to the active site of human BHMT. Panel A shows the sulfur atom of the Hcy moiety of CBHcy interacting with the catalytic Zn (gray); the carboxylate oxygen atoms of Hcy interacting with the amide nitrogen atoms of Phe29 and Val30; and the amino nitrogen of Hcy interacting with Glu159. Panel B shows the carboxylate of the Bet mimic interacting with Typ44 and Tyr77. The carboxylate of the Bet mimic also interacts with Tyr160 (not shown). Both panels show the Zn ion bound to the enzyme by thiolates derived from Cys217, Cys299, and Cys300. Reprinted (adapted) with permission from Evans *et al.* (2002) Betaine-homocysteine methyltransferase: zinc in a distorted barrel. *Structure*. 10: 1159–1171. Copyright 2002 American Chemical Society.

impaired DMG and CBHcy binding, increased the $K_{mBet}$, and had no effect on $k_{cat}$ (Castro *et al.*, 2004). These results confirmed a role for these residues in Bet and DMG binding. Mutating Tyr77 provided unexpected results. Changing Tyr77 to Ala in the rat enzyme (Gonzalez *et al.*, 2003) or Phe in the human enzyme (Castro *et al.*, 2004) abolished enzyme activity. Although in the human BHMT-CBHcy structure this residue appeared to only be involved in Bet/DMG binding, directly measuring ligand binding using intrinsic fluorescence showed that the enzyme could not bind Hcy (Castro *et al.*, 2004). These results were clarified when Gonzalez *et al.* (2004) published the structural details of ligand-free rat BHMT. This structure shows that Tyr160 was the forth ligand to the catalytic Zn, and comparing the human BHMT-CBHcy structure to the ligand-free rat BHMT structure, they proposed that Phe76 and Tyr77 first have a key role orienting Hcy as it binds to the enzyme, which results in displacing Tyr160 off the Zn, which in turn causes a conformational change that moves Tyr77 and Tyr160 into positions as Bet binding determinants. These crystal structures, and the fact that measuring ligand affinity by changes in steady-state intrinsic tryptophan fluorescence depends upon conformational changes in enzyme structure, strongly support the hypothesis that Tyr77 functions as a switch that triggers conformational change upon Hcy binding and thereby creates the Bet binding site. Using steady-state intrinsic tryptophan fluorescence Castro *et al.* (2004) were also able to prove that BHMT catalyzes an ordered bi bi reaction, and that DMG is a very potent inhibitor of BHMT activity because it binds to the BHMT-Hcy binary complex to form a tight-binding ternary complex, as depicted in Figure 7.3.

It is interesting to note that BHMT is the only member of Pfam02574 that oligomerizes. Not including the most distantly related member, Met synthase,

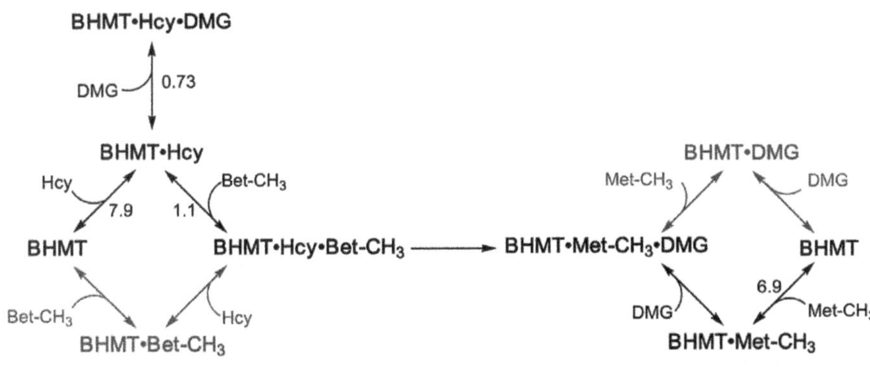


**Figure 7.3** The kinetic mechanism of BHMT. The BHMT reaction sequence. The allowed sequence of substrate binding and product release is shown in black, including $k_d$ values (micromolar). The forbidden sequence of substrate binding and product release is shown in grey. Reprinted (adapted) with permission from Castro *et al.* (2004) Dissecting the catalytic mechanism of betaine-homocysteine *S*-methyltransferase by use of intrinsic tryptophan fluorescence and site-directed mutagenesis. *Biochemistry*. 43: 5341–5351. Copyright 2004 American Chemical Society.

the other family members are missing most or all of what is referred to here as the C-terminal domain (residues 319–406/7). In light of the studies described above on the structure and function of BHMT, it was peculiar when it was observed that removing all of the C-terminal domain by truncating BHMT at residue 325 resulted in an inactive protein (Szegedi and Garrow, 2004). This was surprising since the entire sequence that encodes the *N*-terminal $(\beta/\alpha)_8$ barrel, which appeared to include the entire active site, was intact. Using a truncation mutant that left the dimerization arm but removed the C-terminal residues encoding the linker and alpha helix (truncate at residue 371), it was found that BHMT had residual activity. Although the oligomeric state of these truncated proteins were not evaluated, it can be assumed that the Ser325 truncation mutant was monomeric, and that the Asp371 truncation mutant was either dimeric or in a monomer–dimer equilibrium. These mutants suggested that some oligomerization, at least dimerization, might be required for BHMT activity (Szegedi and Garrow, 2004). Hence, further analysis of the human BHMT-CBHcy structure revealed that the His at position 338 within the dimerization arm is projecting very near the Bet binding site of its dimeric partner. Mutating that residue to Ala reduced the catalytic efficiency of the enzyme by 95%. This work showed that the active site of the enzyme is composed primarily of residues within the *N*-terminal domain of a given monomer, but that the complete active site is formed by the participation of at least one residue (His338) from the dimerization arm of its partner (Szegedi and Garrow, 2004).

Having the crystal structure of the human enzyme and the ability to create and characterize mutant BHMT proteins provides a robust platform to study the effect of any mutation that has already or might be discovered within the coding region of BHMT. Although several mutations and one common polymorphism within the human BHMT have been described, their links to disease have been absent or tenuous, and none of these mutant enzymes have been purified and subjected to advanced biochemical and biophysical analyses.

## Summary Points

- This chapter focuses on the structure and function of BHMT.
- BHMT is expressed at high levels in the liver, and in some species the kidney cortex and pancreas are secondary sites.
- In addition to Bet, BHMT can use dimethylsulfonioacetate and dimethylsulfoniopropionate as methyl donors to methylate homocysteine.
- BHMT is a tetramer composed of identical subunits composed of approximately 400 residues.
- BHMT contains a catalytic Zn.
- Structure and function studies have identified key residues required for substrate, product and inhibitor binding and catalysis.
- In rodents, lack of BHMT function causes hyperhomocysteinemia, a condition associated with several human diseases.

## Key Facts

1. Hcy is the first substrate to bind to BHMT. As it enters the active site, Hcy triggers a conformational change, mediated in part by Phe76 and Tyr77. The conformational change causes Tyr160 to become displaced off the active site Zn, while at the same time exposing the thiol of Hcy to Zn, which leads to its activation (conversion to a thiolate anion) as it binds to the metal. Other residues critical for Hcy binding include Phe29, Val30 and Glu159.
2. Bet is the second substrate to bind to BHMT. The conformational change triggered by Hcy binding creates the Bet binding site. Residues important for Bet binding include Trp44, Tyr77, Tyr160 and His338.
3. When both substrates are bound, methyl transfer occurs and product release is ordered; DMG comes off first, followed by Met.
4. DMG is a potent inhibitor of the BHMT reaction because it can participate in the formation of a tight-binding ternary complex with BHMT and Hcy (BHMT-Hcy-DMG).

## Definitions of Words and Terms

**Amino acid homology**: The degree of similarity between amino acid sequences of proteins.
**Amino acid identity**: The degree of identical amino acid sequences between proteins.
**Binary complex**: Two molecules interacting by noncovalent forces.
**Exons**: A nucleotide sequence within a gene that encodes amino acid sequence.
**Introns**: A nucleotide sequence within a gene that does not code for amino acid sequence and is removed from heteronuclear RNA to form mRNA.
**Random mutagenesis**: A procedure that allows substitution of any amino acid at a specific position within a protein.
**Semirandom mutagenesis**: A procedure that allows a limited number of amino acids to be substituted at a specific position within a protein.
**Site-directed mutagenesis**: A procedure that only allows one amino acid to be substituted for another at a specific position within a protein.
**Ternary complex**: Three molecules interacting by noncovalent forces.
**Thiol**: A compound that contains a carbon-bonded sulfhydryl group.
**Thiolate**: A deprotonated thiol.
**Thrombosis**: The formation of a blood clot in an artery or vein that obstructs flow.

## List of Abbreviations

| | |
|---|---|
| Bet | Betaine |
| BHMT | Betaine-homocysteine *S*-methyltransferase |
| cDNA | Complementary DNA |

| | |
|---|---|
| DMG | Dimethylglycine |
| Hcy | Homocysteine |
| Met | Methionine |
| mRNA | Messenger ribonucleic acid |
| Pfam | Protein family |
| Zn | Zinc |

## References

Awad, W. M., Whitney, P. L. and Skiba, W. E., 1983. Evidence for direct methyl transfer in betaine:homocysteine methyltransferase. *Journal of Biological Chemistry*. 258: 12790–12792.

Borsook, H. and Dubnoff, J. W., 1947. Methionine formation by transmethylation *in vitro*. *Journal of Biological Chemistry*. 169: 247–258.

Breksa, A. P. and Garrow, T. A., 1999. Recombinant human betaine-homocysteine methyltransferase: identification of three cysteine residues critical for zinc binding. *Biochemistry*. 38: 13991–13998.

Breksa, A. P. and Garrow, T. A., 2002. Random mutagenesis of the zinc-binding motif of betaine-homocysteine methyltransferase reveals that Gly 214 is essential. *Archives of Biochemistry and Biophysics*. 399: 73–80.

Castro, C., Gratson, A. A., Evans, J. C., Jiracek, J., Collinsova, M., Ludwig, M. L. and Garrow, T. A., 2004. Dissecting the catalytic mechanism of betaine-homocysteine *S*-methyltransferase by use of intrinsic tryptophan fluorescence and site-directed mutagenesis. *Biochemistry*. 43: 5341–5351.

Collinsova, M., Strakova, J., Jiracek, J. and Garrow, T. A., 2006. Inhibition of betaine-homocysteine *S*-methyltransferase causes hyperhomocysteinemia in mice. *Journal of Nutrition*. 136: 1493–1497.

Ericson, L. E., 1960. Betaine-homocysteine methyltransferases: the methyl donor specificity of the transferase isolated from pig liver. *Acta Chemica Scandinavica*. 14: 2127–2134.

Evans, J. C., Huddler, D. P., Jiracek, J., Castro, C., Millian, N. S., Garrow, T. A. and Ludwig, M. L., 2002. Betaine-homocysteine methyltransferase. Zinc in a distorted barrel. *Structure*. 10: 1159–1071.

Finkelstein, J. D., Harris, B. J. and Kyle, W. E., 1972. Methionine metabolism in mammals: kinetic study of betaine-homocysteine methyltransferase. *Archives of Biochemistry and Biophysics*. 153: 320–324.

Fromm, H. J. and Nordlie, R. C., 1959. On the purification and kinetics of rat liver thetin-homocysteine transmethylase. *Archives of Biochemistry and Biophysics*. 81: 363–376.

Garrow, T. A., 1996. Purification, kinetic properties, and cDNA cloning of mammalian betaine-homocysteine methyltransferase. *Journal of Biological Chemistry*. 271: 22831–22838.

Gonzalez, B., Campillo, N., Garrido, F., Gasset, M., Sanz-Aparicio, J. and Pajares, M. A., 2003. Active-site mutagenesis study of rat liver betaine homocysteine *S*-methyltransferase. *Biochemical Journal*. 370: 945–952.

Gonzalez, B., Pajares, M. A., Martinez-Ripoll, M., Blundell, T. L. and Sanz-Aparicio J., 2004. Crystal structure of rat liver betaine homocysteine *S*-methyltransferase reveals new oligomerization features and conformational changes upon substrate binding. *Journal of Molecular Biology*. 338: 771–782.

Goulding, C. W. and Matthews, R. G., 1997. Cobalamin-dependent methionine synthase from *Escherichia coli*: involvement of zinc in homocysteine activation. *Biochemistry*. 36: 15749–15757.

Millian, N. S. and Garrow, T. A., 1998. Human betaine-homocysteine methyltransferase is a zinc metalloenzyme. *Archives of Biochemistry and Biophysics*. 356: 93–98.

Peariso, K., Goulding, C. W., Huang, S., Matthews, R. G. and Penner-Hahn, J. E., 1998. Characterization of the zinc binding site in methionine synthase enzymes of *Escherichia coli*: The role of zinc in the methylation of homocysteine. *Journal of the American Chemical Society*. 120: 8410–8416.

Petrossian, T. C. and Clarke, S. G., 2011. Uncovering the human methyltransferasome. *Molecular & Cellular Proteomics*. 10: M110.000976.

Strakova, J., Williams, K. T., Gupta, S., Schalinske, K. L., Kruger, W. D., Rozen, R., Jiracek, J., Li, L. and Garrow, T. A., 2010. Dietary intake of S-(alpha-carboxybutyl)-DL-homocysteine induces hyperhomocysteinemia in rats. *Nutrition Research*. 30: 492–500.

Strakova, J., Gupta, S., Kruger, W. D., Dilger, R. N., Tryon, K. and Garrow, T. A., 2011. Inhibition of betaine-homocysteine *S*-methyltransferase in rats causes hyperhomocysteinemia and reduces liver cystathionine β-synthase activity and methylation capacity. *Nutrition Research*. 31: 563–571.

Szegedi, S. S. and Garrow, T. A., 2004. Oligomerization is required for betaine-homocysteine *S*-methyltransferase function. *Archives of Biochemistry and Biophysics*. 426: 32–42.

Szegedi, S. S., Castro, C. C., Koutmos, M. and Garrow, T. A., 2008. Betaine-homocysteine S-methyltransferase-2 is an *S*-methylmethionine-homocysteine methyltransferase. *Journal of Biological Chemistry*. 283: 8939–8945.

Teng, Y. W., Mehedint, M. G., Garrow, T. A. and Zeisel S. H., 2011. Deletion of betaine-homocysteine *S*-methyltransferase in mice perturbs choline and 1-carbon metabolism, resulting in fatty liver and hepatocellular carcinomas. *Journal of Biological Chemistry*. 286: 36258–36267.

Williams, K. T. and Schalinske, K. L., 2010. Homocysteine metabolism and its relation to health and disease. *BioFactors*. 36: 19–24.

# Section III
# Analysis

CHAPTER 8

# Assays of Betaines using $^1H$ NMR Spectroscopy

PANTELEIMON G. TAKIS[a], KONSTANTINOS D. PAPAVASILEIOU[b], ANASTASSIOS N. TROGANIS*[a], AND VASILIOS S. MELISSAS*[c]

[a]Department of Biological Applications and Technology, University of Ioannina, GR–451 10, Ioannina, Greece; [b]Department of Chemistry, University of Ioannina, GR–451 10, Ioannina, Greece; [c]Department of Chemistry, School of Physical Sciences, University of Ioannina, GR–451 10, Ioannina, Greece
*E-mail: atrogani@cc.uoi.gr; melissas@chem.auth.gr

## 8.1 Introduction

Betaine and betaines in general possess a catalytic role in cells and functions in living organisms that are of major importance for a large portion of the scientific research community (Yancey *et al.*, 1982). First, it is a prerequisite to define the terms: betaine and betaines. The term betaine is commonly used to describe glycine betaine, while in its plural number (*i.e.* betaines) it encompasses the majority of methylamine compounds (*N*-methylated) and metabolites (Lee *et al.*, 2006). This terminology is followed throughout this short review.

As described in previous chapters, betaine is characterized by a plethora of unique physicochemical properties such as its ability to interact with many anions, its existence as a zwitterion in aqueous solutions leading to cell

Food and Nutritional Components in Focus No. 7
Betaine: Chemistry, Analysis, Function and Effects
Edited by Victor R Preedy

Published by the Royal Society of Chemistry, www.rsc.org

osmoprotection (Yancey *et al.*, 1982), its role as a thermodynamic stabilizer of proteins structures in water (Bruździak *et al.*, 2013), *etc.* Moreover, betaine participates in the one-carbon metabolism and DNA methylation and its clinical importance is also highlighted by its role in reducing the elevated serum homocysteine concentrations associated with hyperhomocystinuria (Ueland *et al.*, 2005).

Apart from betaine, betaines and generally *N*-methylated compounds as well as betaine's metabolites comprise a large ensemble of compounds exhibiting significant biological contributions. An added proof of this is the continuous attempt for their detection and quantification in almost all biological fluids such as blood (Psychogios *et al.*, 2011) and urine (Bouatra *et al.*, 2013). Table 8.1 describes betaine and the *N*-methylated compounds that will be discussed in detail, as well as their human blood and urine concentrations which were determined mainly by NMR spectroscopy. Their structural formulas are illustrated in Figure 8.1.

Considering the cellular, biological and clinical importance of all the above-mentioned *N*-methylated compounds, the focus of the scientific community was directed to developing new identification and analytical tools with emphasis on the detection and quantification of those compounds in complex biological fluids, without being limiting to simple mixtures (*i.e.* blood, urine, food). The ultimate goal is to establish protocols for identifying specific diseases (Lodi *et al.*, 2013), while developing highly selective drugs for their cure (Kim *et al.*, 2012) and to explore food-intake repercussions to human metabolism (Heinzmann *et al.*, 2010). Amongst the thus applied and commonly used analytical techniques of high-pressure liquid chromatography (HPLC) and mass spectrometry (MS) (Holm *et al.*, 2003) *etc.*), lies yet another powerful one, NMR spectroscopy (Wishart, 2008; Zhang *et al.*, 2010).

Since the 1970s, NMR has proved outstanding as an analytical tool, since its employment allowed quantification and detection of metabolites concentrations in biofluids and/or complex mixtures effectively and in a precise way

**Table 8.1** Collected *N*-methylated compounds-metabolites' concentrations in urine and blood serum biofluids. This table lists collected *N*-methylated compounds and their concentrations in blood and urine biofluids, of healthy subjects, both male and female, above 18 years old, determined mainly by NMR.

| *N*-methylated compounds | Blood (Psychogios *et al.*, 2011) (μM) | Urine (Bouatra *et al.*, 2013) (μM mmol$^{-1}$ creatinine) |
|---|---|---|
| Betaine | 51.6–94.4 | 2.7–24.7 |
| Carnitine | 34.1–57.3 | 0.7–16.4 |
| Choline | 9.2–19.8 | 1.4–6.1 |
| Phosphorylcholine | 1.2–3.2[a] | 0.7–3.0 |
| Proline betaine | 0.0–50.0[a] | 0.8–4.0 |
| TMA | 0.3–0.5[a] | 0.3–19.4 |
| TMAO | 17.4–58.2[a] | 4.8–509.0 |
| Trigonelline | Expected/not quantified | 5.5–109.3 |

[a]These concentrations are determined by other analytical techniques (Wishart *et al.*, 2009).

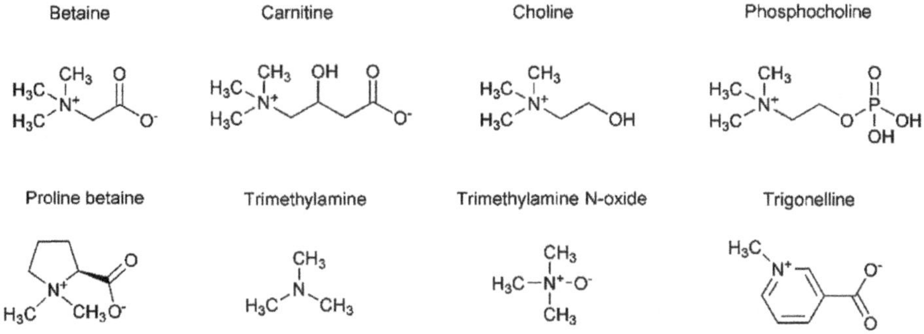


**Figure 8.1** The structures of here reported *N*-methylated compounds. Betaine's and other similar *N*-methylated compounds' structures in aqueous solutions are depicted. It is noted that all these metabolites are present in both blood and urine biofluids.

(Wilson *et al.*, 1974). In particular, it has been shown that $^1$H-NMR one-dimensional (1D) spectra contain valuable information not only with regard to the compound's molecular structure but also their concentration in high complexity mixtures. Consequently, it is reasonable that prefix q- (*i.e.* quantitative) was added before the NMR initials, resulting in qNMR, whose applications are multiplied exponentially, in diverse research areas, ranging from food chemistry and biomedicine to natural and pharmaceutical product analysis at relatively low operational and financial cost (Van Duynhoven *et al.*, 2013).

The employment of all the above-mentioned capabilities of NMR spectroscopy led to new research areas flourishing, such as NMR-based metabolomics, which in combination with advanced chemometrics tools, evolved as a perfect technique to discover new biomarkers, measure, separate or distinguish with high accuracy metabolites in human biofluids, food extracts as well as to detect the effect of the latter to human metabolism (Zhang *et al.*, 2010).

Given, on the one hand, the recent advances and optimization of $^1$H-NMR pulses along with NMR spectrometer's increasing sensitivity (cryoprobes development, higher magnetic fields) and betaine's high biological, food-nutritional and pharmaceutical importance and specific structure characteristics (*i.e.* the trimethylation of their amino group) on the other, was only for betaines to have been brought under the prism of numerous research $^1$H-NMR spectroscopy studies.

## 8.2 $^1$H-NMR Detection and Quantification of Betaines in Biofluids

$^1$H-NMR spectroscopy effectiveness and rapidity concerning the identification and quantification of specific compounds in complex mixtures, led to its extensive employment as an analytical tool for betaines mostly in human biofluids (Wishart *et al.*, 2009; Zhang *et al.*, 2010; Bouatra *et al.*, 2013). It should be mentioned that in several cases laboratory animal biofluids are exploited (for example mice and pigs, *etc.*) in order to detect biomarkers concerning medicinal or nutritional studies (Yde *et al.*, 2012; Jiang *et al.*, 2013). Apart from the conventional 1D $^1$H-NMR experiments, the most commonly applied NMR experiments for complex biofluids such as serum, plasma, urine, *etc.* are described in Table 8.2, whilst some of them have been utilized in several studies for the characterization and the detection of the concentration variance levels of betaine, betaine's metabolites and other *N*-methylated compounds, usually in conjunction with chemometrics.

With regard to compound quantification, 1D proton NMR spectra usually include an internal quantitative standard such as TSP (trimethylsilyl propanoic acid) or DSS (4,4-dimethyl-4-silapentane-1-sulfonic acid), which are variants of TMS (tetramethylsilane) but they exhibit higher solubility in water. However, if the corresponding samples contain binding factors such as proteins in blood, these internal standards interact with the latter, rendering them inappropriate for internal calibration usually due to their $^1$H-NMR peaks linewidth broadening (Ross *et al.*, 2007). Thus, artificial signal (ERETIC, Electronic Reference To access *In vivo* Concentrations) is exploited as a concentration reference (Martínez-Bisbal *et al.*, 2009), where an external

**Table 8.2** The most common one-dimension $^1$H-NMR experiments used for biomarkers detection in biofluids. This table lists the most common $^1$H-NMR experiments that are employed in $^1$H-NMR based metabolomics studies. Due to large water amounts in all biofluids, all the above-mentioned NMR experiments pulses are combined with suppression of $^1$H-NMR water signal.

| $^1$H-NMR | Blood | Urine | CSF | Description |
|---|---|---|---|---|
| 1D NOESY | + | + | + | Includes information about both small size metabolites and large macromolecules |
| 1D CPMG | + | − | + | Suppression of large biomolecules proton NMR signals, emphasizing small to medium-sized metabolites |
| 1D DIFFUSION | + | − | + | Includes only proton NMR signals of macromolecules (*i.e.* proteins, lipids, *etc.*) |

signal is generated by the rf electronics and provides a reference to analytically calculate the compound concentrations under study.

### 8.2.1 Selected Glycine Betaine Assays by $^1$H-NMR

Recently, it was showed that $^1$H-NMR spectroscopy at several high magnetic fields led to the fast detection and quantification of more than 40 metabolites, amongst which glycine betaine, yielded high accuracy in human serum samples of healthy subjects (Psychogios *et al.*, 2011). In Figure 8.2, the equivalent nine *N*-methyl protons of the resonating betaine molecule in the typical 1D $^1$H–NMR spectrum of human serum sample are selectively pointed out. Despite human serum mixture complexity, it is methyl protons' resonances of glycine betaine are clearly detectable (Figure 8.2).

The same results are drawn upon urine complex mixtures (Bouatra *et al.*, 2013), where $^1$H-NMR spectra reveal betaine's presence, offering information about betaine's concentration levels (Figure 8.3). Consequently, Diaz *et al.*, while trying to detect possible metabolic biomarkers of prenatal disorders by $^1$H-NMR spectroscopy (Diaz *et al.*, 2011), observed that fetal malformations altered betaine concentrations of second trimester pregnant women plasma, which are closely correlated to distortions of homocysteine–methionine conversion.

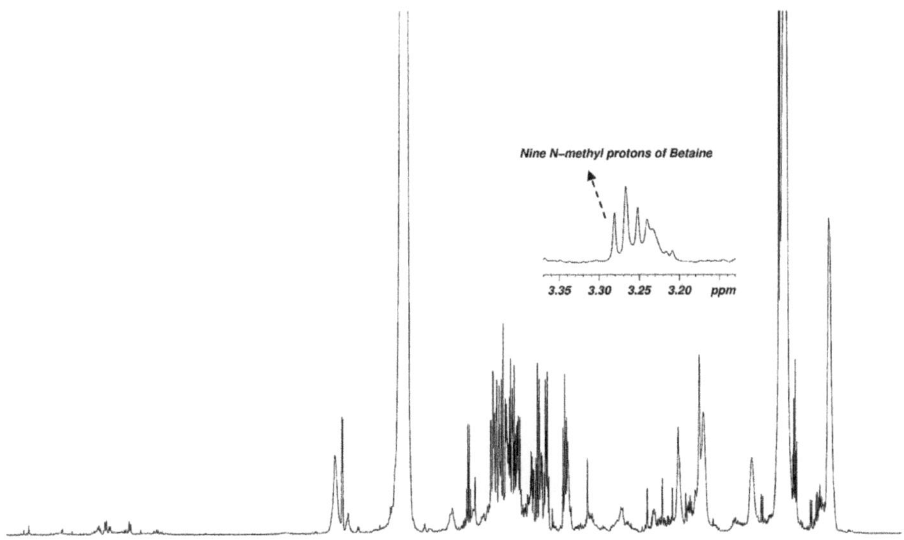


**Figure 8.2** The $^1$H-NMR spectrum of human serum biofluids. A common $^1$H-NMR CPMG spectrum of human serum biofluid is depicted (the proton frequency is 500.13 MHz, $T = 313$ K, 64 scans). In addition, the region where the *N*-methyl protons of betaine resonate is highlighted. It is obvious that those protons are quite detectable by $^1$H-NMR spectroscopy.

Recent studies (Barter *et al.*, 2010), among which a $^{1}$H-NMR spectroscopic approach (Kang *et al.*, 2011), showed quite significant correlation of increased betaine levels in plasma with atherosclerosis, heart failure, and other cardiovascular diseases. On the basis of these results, a recent $^{1}$H-NMR study (Geamanu *et al.*, 2013) indicated that the intake of the biologically active fraction of ProAlgaZyme (BaP) (*i.e.* a fermentation product of different freshwater organisms such as algae) reduces betaine levels of plasma profiles on the subjects, thus lowering the risk of cardiovascular diseases.

Moreover, a $^{1}$H-NMR metabonomic study in human urine metabolic profiles of patients experiencing immunoglobulin A nephropathy (IAN), showed that betaine is involved in tumor necrosis factor alpha (TNF-α) regulation and it is differentially modulated in those patients (*i.e.* increased concentration levels) compared to the healthy ones (Del Coco *et al.*, 2013). Similarly, another NMR-based metabonomic study, exploiting rat urine, showed that chronic unpredictable mild stress has a direct impact to betaine's concentration levels in urine (*i.e.* a significant decrease) (Tian *et al.*, 2013).

In addition, quite recently, $^{1}$H-NMR studies of catheterized sows' plasma revealed a relationship between betaine absorption and elevated creatine concentrations attributed to the quantity of fiber-containing food, namely the betaine uptake (Yde *et al.*, 2012), while previous studies highlighted the

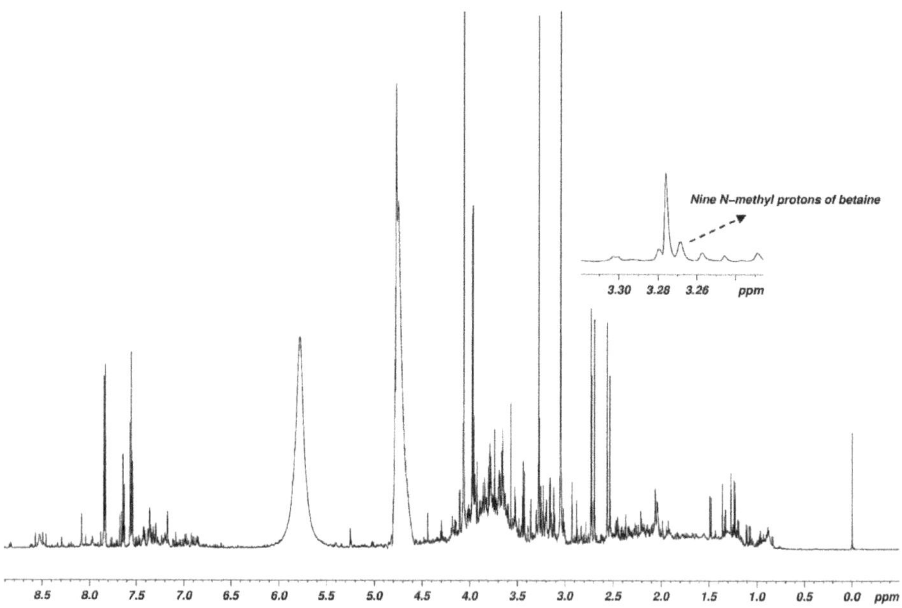


**Figure 8.3** The $^{1}$H-NMR spectrum of human urine biofluid. In this figure a $^{1}$H-NMR NOESY spectrum of human urine biofluid is depicted (the proton frequency is 500.13 MHz, $T$ = 303 K, 64 scans). The NMR peak of the betaine's *N*-methyl protons is extracted and highlighted above the whole expanded spectrum.

biochemical effects of rye *vs.* wheat-based diet in hypercholesterolemic pigs, based upon the observed elevated betaine's concentration in their plasma specimens (Bertram *et al.*, 2009).

Piglet urine $^1$H-NMR studies allowed for betaine's involvement in hypoxia (*i.e.* increase of betaine concentration in hypoxic subjects), establishing it as a useful biomarker for the identification of hypoxic animals (Skappak *et al.*, 2013), since the increase of urine betaine levels was reported to decrease the inflammatory response of adipocytes stressed under hypoxic conditions(Olli *et al.*, 2013).

### 8.2.2 Selected Other Betaines' and *N*-Methylated Compounds Assays by $^1$H-NMR

As mentioned above, except for glycine betaine there are a lot of *N*-methylated metabolites and other compounds that exhibit a significant biological role and for which a large number of $^1$H-NMR studies have been performed. The following paragraphs highlight the $^1$H-NMR spectroscopy contribution to the detection and quantification of some *N*-methylated metabolites in biological specimens.

Betaine-homocysteine methyltransferase (BHMT) is a zinc metalloenzyme expressed in liver and kidney tissue and catalyzes the methylation of homocysteine from glycine betaine forming *N*-dimethyl-glycine (DMG) and methionine (Millian & Garrow, 1998). Lee *et al.* introduced a 1D $^1$H-NMR assay as an excellent tool for exploring interactions between BHMT and glycine betaine analogs and for measuring BHMT catalytic activity in an accurate and non-time-consuming fashion (Lee *et al.*, 2004).

Carnitine consists of another important *N*-methylated metabolite, generally located in tissues but also appearing in blood, urine, cerebrospinal (CSF) biofluids in various concentrations (Wishart *et al.*, 2009). In particular, carnitine's detection and quantification in serum samples is a prerequisite, since carnitine's levels in blood are important biomarkers (Lodi *et al.*, 2013) for numerous severe metabolic disorders such as in lipid metabolism, amino acid degradation, while low serum carnitine levels may result in liver (Hoppel, 2003), heart, or brain malfunction and damage (Möder *et al.*, 2005). Recently, a very promising $^1$H-NMR methodology was developed for accurate and rapid carnitine determination in human serum samples by employing $^1$H-NMR single quantum coherence filtering (SQCF) spectra, succeeding in the distinction of the *N*-methyl proton signal of carnitine that is greatly overlapped mostly by *beta* ($C_2$H) proton of glucose in the conventional $^1$H-NMR spectrum of serum (Figure 8.4) (Tsiafoulis *et al.*, 2011).

$^1$H NMR spectra indicated that choline's and -hosphorylcholine's levels in serum were elevated after the ischemic stroke – modeled by middle cerebral artery occlusion rats – probably because it is a basic component of membranes used in the organism as a self-repair process after the ischemic stroke (Wang *et al.*, 2014).

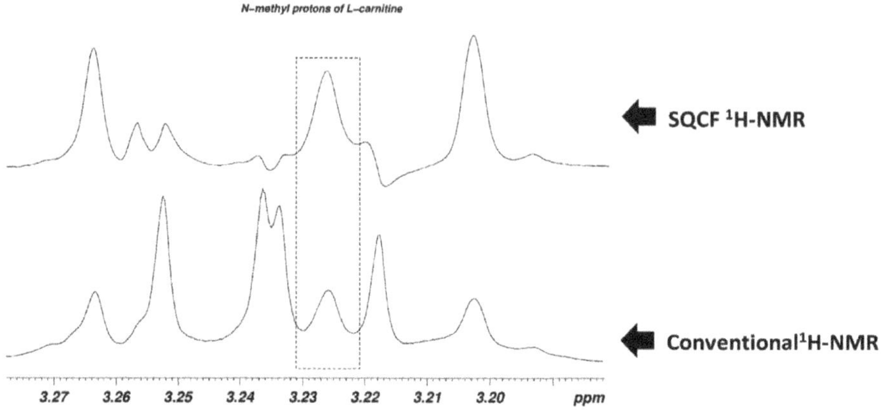


**Figure 8.4** A comparison between a typical $^1$H-NMR spectrum of human serum sample and its corresponding $^1$H-NMR SQCF. This figure displays a selected region, where carnitine's *N*-methyl protons resonate and they are pointed out in the dashed box, of a typical $^1$H-NMR spectrum of a human serum sample (below spectrum) using solvent suppression. In this spectrum, it is apparent that the L-carnitine signal cannot be accurately quantified. In the above spectrum of this figure, the same sample was measured with the SQC filtering $^1$H-NMR technique; the L-carnitine signal is exposed and can be properly quantified (Tsiafoulis *et al.*, 2011).

Another analog of glycine betaine is proline betaine, commonly found in urine where it provides an osmoprotective role for the kidney (Atkinson *et al.*, 2007). Moreover, $^1$H-NMR studies of human urine specimens characterized proline betaine as an important nutritional biomarker, since it is contained in citrus foods and their consumption leads to increased concentrations in urine biofluid (Atkinson *et al.*, 2007; Heinzmann *et al.*, 2010).

The malfunction of the conversion of trimethylamine (TMA) to trimethylamine-*N*-oxide (TMAO) – two important *N*-methyl metabolites – leads to trimethylaminuria (or fish-odor syndrome), which is an inherited metabolic disorder that causes a strong, repulsive odor of breath, sweat and urine (Ayesh *et al.*, 1993). Accordingly, previous studies showed that a conventional $^1$H-NMR spectrum is adequate for the rapid and simple diagnosis of trimethylaminuria by comparing the $^1$H-NMR urine fingerprint of healthy subjects with those experiencing the fish-odor syndrome, detecting the TMA/TMAO concentration differences (Maschke *et al.*, 1997).

Betaine nicotinate or trigonelline is another betaine metabolite. Trigonelline is present in coffee and human urine. Except for a potential nutritional biomarker, very recent $^1$H-NMR based metabolomics showed that it can be used as a medicinal biomarker for human gastric adenocarcinoma as well as TMA and TMAO (Kim *et al.*, 2013). Trigonelline and TMA levels in urine increase significantly, whereas TMAO appears to be decreased. In addition, urine $^1$H-NMR spectra showed that those methylamines could be used as drug-efficiency markers, since their proton NMR peaks reverted to normal levels after the medication with specific drugs against gastric adenocarcinoma (Kim *et al.*, 2013). It should be noted that trigonelline was identified as a biomarker for lung cancer through $^{1}$H-NMR spectroscopy, where in this case it appeared to be reduced in lung cancer patients' urine metabolic profile (Stretch *et al.*, 2012).

## Summary Points

- This chapter focuses on glycine betaine's as well as other *N*-methylated biomolecules' biological significance, and their unravelled exclusivity by $^{1}$H-NMR spectroscopy, in particular.
- $^{1}$H-NMR spectroscopy revealed and quantified betaine in human and animal complex biofluids such as blood and urine, where its concentration variance is found to be directly correlated to atherosclerosis, heart failure and other cardiovascular diseases as well as animal hypoxia, *etc.*
- A single $^{1}$H-NMR spectrum quantifies carnitine's levels in human serum samples, while conventional proton NMR spectra reveal that choline's concentration increases after an ischemic stroke.
- Proline betaine becomes an established nutritional biomarker following citrus fruit consumption, whereas TMAO, TMA and trigonelline led to the identification of human gastric adenocarcinoma.
- $^{1}$H-NMR spectroscopy has eluded its traditional applications on molecular structure studies by transforming to a new analytical tool leading to new directions in medicine regarding diagnostics, improvement of selectivity of drug design and in food industry though the food products development for our diet.

## Key Facts for NMR Contribution to Metabolites Function Revelation

- Except for one-dimensional (1D) proton NMR spectra, homonuclear as well as heteronuclear two-dimensional (2D) NMR is employed in several cases in order to accurately predict the structure or the quantity of metabolites in complex mixtures, since they provide extra details for each mixture component's structure, especially when 1D proton NMR signals overlap.
- Even though NMR spectroscopy can quite easily predict the metabolites concentration variance between different samples, further treatment of NMR data is required for more accurate results, especially in the case of numerous samples. Consequently, statistics-chemometrics have become a complementary, prerequisite tool for the detection of biomarkers in complex biofluids.
- More and more research studies emerge proposing new medical and nutritional biomarkers by $^{1}$H-NMR spectroscopy. However, biochemical and other biological experiments have to take place in order to verify which metabolites have the leading role in each case studied.

- Despite the large number of metabonomic nutritional studies, many betaine metabolites' functions still remain unknown; betonicine, which is commonly found in herbs and spices is a prominent example.
- $^1$H-NMR spectroscopy can provide a wide variety of information for understanding metabolism, but its main drawback remains its lack of sensitivity in comparison to other analytical tools such as mass spectrometry, gas chromatography. As a result, the combination of all those techniques maybe essential for the elucidation of all of an organism's metabolism complicated function and response to diseases or nutrition.
- Constantly, the pursue for discovering important biomarkers results to multiple assignments, whose appearance/disappearance/concentration variance are possibly correlated to each case studied. Those assumptions justify that metabolism is characterized by a multivariate function and behavior, so studies should not be limited to specific metabolic pathways but rather treat them as a whole.

## Definitions of Words and Terms

**Serum biofluid**: Blood serum is a primary carrier of small molecules in the body. Not only does this biofluid play a critical role in transporting dissolved gases, nutrients, hormones and metabolic wastes, but it also plays a key role in the regulation of the pH of interstitial fluids, the restriction of fluid losses at injury sites, the defence against toxins and pathogens and the stabilization of body temperature.

**Plasma biofluid**: Plasma is a light yellow liquid in which blood cells are suspended and is often confused with serum. Plasma is extracted from blood *via* anticoagulants and centrifugation. Namely, blood plasma is the supernatant liquid of the centrifuged blood sample plus the anticoagulants, whereas serum is the supernatant fluid of clotted blood if no anticoagulant addition takes place. Both serum and plasma are used for biomarker detection.

**Urine biofluid**: The kidneys filter the wastes from the bloodstream, water, and other compounds, as a result urine fluid is produced, which is a transparent, sterile fluid containing mainly urea (from amino acid metabolism), inorganic salts (chloride, sodium, and potassium), creatinine, ammonia, organic acids, toxins and urobilin.

**Cerebrospinal biofluid (CSF)**: The central nervous system produces CSF (colorless fluid) for filling the ventricles and the subarachnoid space of the brain and spinal column. CSF contains nutrients and several compounds from the blood and transfers them to the subarachnoid compartment and the lymphatic system, while it provides brain protection.

**Tumor necrosis factor alpha (TNF-a)**: Tumor necrosis factor is an adipokine protein, member of cytokines protein group. It is produced by macrophages during inflammation.

**Immunoglobulin A (IgA) nephropathy**: This disease is characterized by the inflammation of the glomeruli part of the kidney, which at first causes arthritis and constant abdominal pain and if it is not treated leads to chronic renal failure.

**Proton nuclear magnetic resonance ($^1$H-NMR) spectroscopy**: This field of NMR spectroscopy corresponds to the application of NMR spectroscopy with respect to hydrogen-1 nuclei ($^1$H). This nucleus is the most common and abundant within almost all organic molecules. Consequently, $^1$H-NMR spectroscopy becomes the most powerful tool for compound-structure determination as well as their quantification in complex mixtures.

**Mass spectrometry**: Analytical technique that measures the charge to mass ratio of charged compounds and is one of the most valuable techniques for identifying the content as well as the structure of the compounds under study, especially in the case of very low concentration metabolites, in complex mixtures. This is the reason for its wide use in metabolomics and proteomics.

**High-pressure liquid chromatography (HPLC)**: This analytical tool is widely applied in analytical chemistry for separating, identifying, quantifying the components of a mixture. HPLC function principles rely on pumps that circulate a liquid solvent containing the sample mixture through a specific separation column. Depending on its physicochemical characteristics, each component interacts with the column in a specific way, causing different flow rates for the different components, leading to their separation.

**Gastric adenocarcinoma**: It is the second most common cancer worldwide; stomach cancer, and ninety per cent of all tumors are malignant.

## List of Abbreviations

| | |
|---|---|
| 1D | One Dimensional |
| 2D | Two Dimensional |
| BHMT | Betaine-Homocysteine Methyl Transferase |
| BaP | Biologically active ProAlgaZyme |
| CPMG | Carr–Purcell–Meiboom–Gill |
| CSF | Cerebrospinal Fluid |
| DMG | Dimethylglycine |
| DSS | 4,4-Dimethyl-4-Silapentane-1-Sulfonic acid |
| ERETIC | Electronic Reference To access *In vivo* Concentrations |
| HPLC | High-Pressure Liquid Chromatography |
| IAN, | Immunoglobulin A Nephropathy |
| MS | Mass Spectrometry |
| NMR | Nuclear Magnetic Resonance |
| NOESY | Nuclear Overhauser Effect spectroscopy |
| qNMR | Quantitative Nuclear Magnetic Resonance |
| SQCF | Single Quantum Coherence Filtering |
| TMAO | Trimethylamine N-oxide |
| TMA | Trimethylamine |
| TMS | TetraMethylSilane |
| TNF-α | Tumor Necrosis Factor alpha |
| TSP | Trimethylsilyl Propanoic acid |

## References

Atkinson, W., Downer, P., Lever, M., Chambers, S. and George, P., 2007. Effects of orange juice and proline betaine on glycine betaine and homocysteine in healthy male subjects. *European Journal of Nutrition*. 46: 446–452.

Ayesh, R., Mitchell, S. C., Zhang, A. and Smith, R. L., 1993. The fish odour syndrome: biochemical, familial, and clinical aspects. *British Medical Journal*. 307: 655–657.

Barter, P. J., Brandrup-Wognsen, G., Palmer, M. K. and Nicholls, S. J., 2010. Effect of statins on HDL-C: a complex process unrelated to changes in LDL-C: analysis of the VOYAGER database. *Journal of Lipid Research*. 51: 1546–1553.

Bertram, H. C., Malmendal, A., Nielsen, N. C., Straadt, I. K., Larsen, T., Knudsen, K. E. B. and Laerke, H. N., 2009. NMR-based metabonomics reveals that plasma betaine increases upon intake of high-fiber rye buns in hypercholesterolemic pigs. *Molecular Nutrition and Food Research*. 53: 1055–1062.

Bouatra, S., Aziat, F., Mandal, R., Guo, A. C., Wilson, M. R., Knox, C., Bjorndahl, T. C., Krishnamurthy, R., Saleem, F., Liu, P., Dame, Z. T., Poelzer, J., Huynh, J., Yallou, F. S., Psychogios, N., Dong, E., Bogumil, R., Roehring, C. and Wishart, D. S., 2013. The human urine metabolome. *PLoS One*. 8: e73076–e73103.

Bruździak, P., Panuszko, A. and Stangret, J., 2013. Influence of osmolytes on protein and water structure: a step to understanding the mechanism of protein stabilization. *Journal of Physical Chemistry B*. 117: 11502–11508.

Del Coco, L., Assfalg, M., D'Onofrio, M., Sallustio, F., Pesce, F., Fanizzi, F. and Schena, F., 2013. A proton nuclear magnetic resonance-based metabolomic approach in IgA nephropathy urinary profiles. *Metabolomics*. 9: 740–751.

Diaz, S. O., Pinto, J., Graça, G., Duarte, I. F., Barros, A. S., Galhano, E., Pita, C., Almeida, M. D. C., Goodfellow, B. J., Carreira, I. M. and Gil, A. M., 2011. Metabolic biomarkers of prenatal disorders: an exploratory NMR metabonomics study of second trimester maternal urine and blood plasma. *Journal of Proteome Research*. 10: 3732–3742.

Geamanu, A., Goja, A., Saadat, N., Khosla, P. and Gupta, S. V., 2013. ProAlgaZyme subfraction improves the lipoprotein profile of hypercholesterolemic hamsters, while inhibiting production of betaine, carnitine, and choline metabolites. *Nutrition & Metabolism*. 10: 55–65.

Heinzmann, S. S., Brown, I. J., Chan, Q., Bictash, M., Dumas, M.-E., Kochhar, S., Stamler, J., Holmes, E., Elliott, P. and Nicholson, J. K., 2010. Metabolic profiling strategy for discovery of nutritional biomarkers: proline betaine as a marker of citrus consumption. *American Journal of Clinical Nutrition*. 92: 436–443.

Holm, P. I., Ueland, P. M., Kvalheim, G. and Lien, E. A., 2003. Determination of choline, betaine, and dimethylglycine in plasma by a high-throughput method based on normal-phase chromatography-tandem mass spectrometry. *Clinical Chemistry*. 49: 286–294.

Hoppel, C., 2003. The role of carnitine in normal and altered fatty acid metabolism. *American Journal of Kidney Diseases*. 41: S4–S12.

Jiang, C.-Y., Yang, K.-M., Yang, L., Miao, Z.-X., Wang, Y.-H. and Zhu H.-B., 2013. A 1H NMR-based metabonomic investigation of time-related metabolic trajectories of the plasma, urine and liver extracts of hyperlipidemic hamsters. *PLoS One*. 8: e66786–e66805.

Kang, S.-M., Park, J.-C., Shin, M.-J., Lee, H., Oh, J., Ryu D. H., Hwang, G.-S. and Chung, J. H., 2011. 1H nuclear magnetic resonance based metabolic urinary profiling of patients with ischemic heart failure. *Clinical Biochemistry*. 44: 293–299.

Kim, K.-B., Yang, J.-Y., Kwack, S. J., Kim, H. S., Ryu, D. H., Kim, Y.-J., Bae J. Y., Lim, D. S., Choi, S. M., Kwon, M. J., Bang, D. Y., Lim, S. K., Kim, Y. W., Hwang, G.-S. and Lee, B.-M., 2013. Potential metabolomic biomarkers for evaluation of adriamycin efficacy using a urinary $^1$H-NMR spectroscopy. *Journal of Applied Toxicology*. 33: 1251–1259.

Lee, M. B., Blunt, J. W., Lever, M. and George, P. M., 2004. A nuclear-magnetic-resonance-based assay for betaine–homocysteine methyltransferase activity. *Analytical Biochemistry*. 330: 199–205.

Lee, M. B., Storer, M. K., Blunt, J. W. and Lever, M., 2006. Validation of 1H NMR spectroscopy as an analytical tool for methylamine metabolites in urine. *Clinica Chimica Acta*. 365: 264–269.

Lodi, A., Tiziani, S., Khanim, F. L., Günther, U. L., Viant, M. R., Morgan, G. J., Bunce, C. M. and Drayson, M. T., 2013. Proton NMR-based metabolite analyses of archived serial paired serum and urine samples from myeloma patients at different stages of disease activity identifies acetylcarnitine as a novel marker of active disease. *PLoS One*. 8: e56422–e56433.

Martínez-Bisbal, M. C., Monleon, D., Assemat, O., Piotto, M., Piquer, J., Llácer, J. L. and Celda, B., 2009. Determination of metabolite concentrations in human brain tumour biopsy samples using HR-MAS and ERETIC measurements. *NMR in Biomedicine*. 22: 199–206.

Maschke, S., Wahl, A., Azaroual, N., Boulet, O., Crunelle, V., Imbenotte, M., Foulard, M., Vermeersch, G. and Lhermitte, M., 1997. $^1$H-NMR analysis of trimethylamine in urine for the diagnosis of fish-odour syndrome. *Clinica Chimica Acta*. 263: 139–146.

Millian, N. S. and Garrow, T. A., 1998. Human betaine–homocysteine methyltransferase is a zinc metalloenzyme. *Archives of Biochemistry and Biophysics*. 356: 93–98.

Möder, M., Kießling, A. and Löster, H., 2005. Current methods for determination of l-Carnitine and Acylcarnitines. *Monatshefte für Chemie/Chemical Monthly*. 136: 1279–1291.

Olli, K., Lahtinen, S., Rautonen, N. and Tiihonen, K., 2013. Betaine reduces the expression of inflammatory adipokines caused by hypoxia in human adipocytes. *British Journal of Nutrition*. 109: 43–49.

Psychogios, N., Hau, D. D., Peng, J., Guo, A. C., Mandal, R., Bouatra, S., Sinelnikov, I., Krishnamurthy, R., Eisner, R., Gautam, B., Young, N., Xia, J., Knox, C., Dong, E., Huang, P., Hollander, Z., Pedersen, T. L., Smith, S. R., Bamforth, F., Greiner, R., McManus, B., Newman, J. W., Goodfriend, T. and Wishart, D. S., 2011. The human serum metabolome. *PLoS One*. 6: e16957–e16579.

Ross, A., Schlotterbeck, G., Dieterle, F. and Senn, H., 2007. NMR spectroscopy techniques for application to metabonomics. In: Lindon, J. C., Nicholson J.K. (ed.) *Holmes EBT-TH of the Handbook of Metabonomics and Metabolomics*. Elsevier Science B.V., Amsterdam, pp. 55–112.

Skappak, C., Regush, S., Cheung, P.-Y. and Adamko, D. J., 2013. Identifying hypoxia in a newborn piglet model using urinary NMR metabolomic profiling. *PLoS One*. 8: e65035–e65041.

Stretch, C., Eastman, T., Mandal, R., Eisner, R., Wishart, D. S., Mourtzakis, M., Prado, C. M. M., Damaraju, S., Ball, R. O., Greiner, R. and Baracos, V. E., 2012. Prediction of skeletal muscle and fat mass in patients with advanced cancer using a metabolomic approach. *The Journal of Nutrition*. 142: 14–21.

Tian, J.-S., Shi, B.-Y., Xiang, H., Gao, S., Qin, X.-M. and Du, G.-H., 2013. $^1$H-NMR-based metabonomic studies on the anti-depressant effect of genipin in the chronic unpredictable mild stress rat model. *PLoS One*. 8: e75721–e75731.

Tsiafoulis, C. G., Exarchou, V., Tziova, P. P., Bairaktari, E., Gerothanassis, I. P. and Troganis, A. N., 2011. A new method for the determination of free l-carnitine in serum samples based on high field single quantum coherence filtering $^1$H-NMR spectroscopy. *Analytical and Bioanalytical Chemistry*. 399: 2285–2294.

Ueland, P. M., Holm, P. I. and Hustad, S., 2005. Betaine: a key modulator of one-carbon metabolism and homocysteine status. *Clinical Chemistry and Laboratory Medicine*. 43: 1069–1075.

Van Duynhoven, J., Van Velzen, E. and Jacobs, D. M., 2013. Quantification of complex mixtures by NMR. *Annual Reports on NMR Spectroscopy*. 80: 181–236.

Wang, P.-R., Wang, J.-S., Yang, M.-H. and Kong, L.-Y., 2014. Neuroprotective effects of Huang-Lian-Jie-Du-Decoction on ischemic stroke rats revealed by 1H NMR metabolomics approach. *Journal of Pharmaceutical and Biomedical Analysis*. 88: 106–116.

Wilson, D. M., Burlingame, A. L., Cronholm, T. and Sjövall, J., 1974. Deuterium and carbon-13 tracer studies of ethanol metabolism in the rat by $^2$H, $^1$H-decoupled $^{13}$C nuclear magnetic resonance. *Biochemical and Biophysical Research Communications*. 56: 828–835.

Wishart, D. S., 2008. Quantitative metabolomics using NMR. *Trends in Analytical Chemistry*. 27: 228–237.

Wishart, D. S., Knox, C., Guo, A. C., Eisner, R., Young, N., Gautam, B., Hau, D. D., Psychogios, N., Dong, E., Bouatra, S., Mandal, R., Sinelnikov, I., Xia, J., Jia, L., Cruz, J. A., Lim, E., Sobsey, C. A., Shrivastava, S., Huang, P., Liu, P., Fang, L., Peng, J., Fradette, R., Cheng, D., Tzur, D., Clements, M., Lewis, A., De Souza, A., Zuniga, A., Dawe, M., Xiong, Y., Clive, D., Greiner, R., Nazyrova, A., Shaykhutdinov, R., Li, L., Vogel, H. J. and Forsythe, I., 2009. HMDB: a knowledgebase for the human metabolome. *Nucleic Acids Research*. (*Database issue*): D603–D610.

Yancey, P. H., Clark, M. E., Hand, S. C., Bowlus, R. D. and Somero, G. N., 1982. Living with water stress: evolution of osmolyte systems. *Science*. 4566: 1214–1222.

Yde, C. C., Westerhuis, J. A., Bertram, H. C. and Bach Knudsen, K. E., 2012. Application of NMR-based metabonomics suggests a relationship between betaine absorption and elevated creatine plasma concentrations in catheterised sows. *British Journal of Nutrition*. 107: 1603–1615.

Zhang, S., Nagana Gowda, G. A., Ye, T. and Raftery, D., 2010. Advances in NMR-based biofluid analysis and metabolite profiling. *Analyst*. 135: 1490–1498.

CHAPTER 9

# Determination of Betaines in Vegetable Oils by Capillary Electrophoresis-Ion Trap Mass Spectrometry

LAURA SÁNCHEZ-HERNÁNDEZ[a,b], MARIA LUISA MARINA ALEGRE[a], AND ANTONIO L. CREGO NAVAZO*[a]

[a]Department of Analytical Chemistry, Physical Chemistry and Chemical Engineering, Faculty of Biology, Environmental Sciences and Chemistry, University of Alcalá, Ctra. Madrid-Barcelona Km. 33.600, 28871, Alcalá de Henares (Madrid), Spain; [b]Department of Analytical Chemistry, Faculty of Chemistry, Aalen University, Beethovenstraße 1, 73430, Aalen, Germany
*E-mail: antonio.crego@uah.es

## 9.1 Introduction

Betaines are molecules defined as quaternary amino derivatives of amino acids that exist in zwitterionic form at neutral pH with both a permanent positive charge of the amino group and a negatively charged derived from a carboxylic group. This is why most of these compounds are classified within the group of N-substituted nonprotein amino acids, *e.g.*, β-alanine betaine, glycine betaine, carnitine or acyl-carnitines (Hunt, 1985). However, other betaines are classified as alkaloids, *i.e.* compounds synthesized generally from amino acids and having heterocyclic nitrogen. Examples of this type of

Food and Nutritional Components in Focus No. 7
Betaine: Chemistry, Analysis, Function and Effects
Edited by Victor R Preedy

Published by the Royal Society of Chemistry, www.rsc.org

betaine include proline betaine, which is a pyrrolidine alkaloid, and trigonelline, which is a pyridine alkaloid.

Different studies have confirmed that betaines are available to humans through endogenous synthesis and food intake. They are known to be in a large number of foods (Zeisel *et al.*, 2003; de Zwart *et al.*, 2003) and also to be natural osmolytes in plants in which they confer enzymatic protection and higher stability to the membranes under environmental stress conditions (Zhang *et al.*, 2002). Due to their beneficial effects described on health (acting as osmorregulating compounds in most tissues and maintaining the tertiary structure of the macromolecules) (Lang, 2007; Lever and Slow, 2010), betaines are considered important nutrients in foods and therefore interesting compounds to evaluate food quality. In this regard, the quality and authenticity of foods are currently of great concern for researchers, consumers, and industries being important to provide relevant information about food processing, contamination, alterations or adulterations, and to assess that legislation is fulfilled. There are some regulations from agencies as the European Food Safety Authority and the Food and Drug Administration, such as Regulation 178/2002 (European Communities, 2002) stating that foods should be safe, both quantitatively and qualitatively, and indicating that, consequently, only those foods that have been proven to be safe may be marketed in order to protect consumer's health. Moreover, the detailed information about the composition of food products has to be labeled properly in order to promote and guarantee their quality and to help consumers to be aware of what they acquire and its nutritional value (European Communities, 2011).

Accordingly, the regulatory agencies require highly selective and sensitive analytical methods providing the data required to assess the control of foodstuffs. From this point of view, mass spectrometry (MS) has become a powerful detection tool to characterize foods because it combines sensitivity, selectivity, and versatility, being capable of gently ionizing a wide range of molecules, especially polar ones, when soft ionization methods such as electrospray ionization (ESI) are used. Moreover, the combination of MS with separation techniques such as gas chromatography (GC), liquid chromatography (LC) or capillary electrophoresis (CE) allows for fingerprinting of complex food matrices, and provides structural information on the separated compounds. LC continues to be the preferred separation technique in routine laboratories due to its high precision and robustness. However, CE has been shown to be a separation technique with a high potential offering simplicity, speed, highly efficient separations and minimum solvent and reagent consumption. Although this technique is still not well established in food analysis, it is of great interest to investigate its possibilities in the field of food characterization and quality control (Piñero *et al.*, 2011).

A large number of methodologies by CE have been reported in the literature for the determination of betaines. However, the majority of them only describe the determination of carnitine and its acyl derivatives in biological samples using UV absorption detection, laser-induced fluorescence or contactless conductivity detectors (Mansour *et al.*, 2013) and very few articles were devoted to the analysis of foods, being focused on the determination of carnitine (Dabrowska and Starek, 2014). Note that there are only four studies dealing with the determination of different kind of betaines. Two of them were based on the use of UV detection after derivatization of the betaines: one was aimed at the separation of different standards of betaines (2-naphthacyl ester derivatives of glycine betaine, carnitine, acetylcarnitine, octanoylcarnitine, and proline betaine) (Storer *et al.*, 2006) and in the other, different betaines were determined in plants (*p*-bromophenacyl esters of glycine betaine, trigonelline, proline betaine, γ-butyro betaine, and carnitine) (Zhang *et al.*, 2002). However, in spite of carrying out a previous derivatization, the sensitivity achieved in both cases was very low, only allowing detection of the presence of glycine betaine in leaf extract of plants (Zhang *et al.*, 2002). As a consequence, the use of MS detectors is very attractive in order to achieve the sensitive determination of betaines. Recently, one method enabling the analysis of three kinds of betaines (glycine betaine, carnitine and acetyl-carnitine) has been described for the metabolomic fingerprinting of rat serum samples (Naz *et al.*, 2013). Nevertheless, the method was focused on the characterization of metabolic profiles but not on the separation of betaines. Considering all these aspects, this chapter describes a recent work (Sánchez-Hernández *et al.*, 2011a) based on the use of capillary electrophoresis coupled to mass spectrometry (CE/MS) as a powerful tool for the separation of several betaines (glycine betaine, proline betaine, trigonelline and carnitines) (see Section 9.2) and their determination in food such as vegetable oils, and presents the innovative application of evaluating the potential of betaines as novel quality markers for olive oils (see Section 9.3).

## 9.2 Capillary Electrophoresis Coupled to Mass Spectrometry for the Determination of Betaines

The choice of the analytical technique is a relevant point when investigating the quality of foods. Selective, sensitive, accurate and precise analytical techniques are necessary for the separation, identification and quantitation of target compounds. With this in mind, there is no doubt that the use of MS as a detector coupled to a high-performance separative technique as CE can solve the identification problems associated with the determination of unknown compounds and the sensitivity problems associated to the low concentrations detected with on-column UV absorption detection, limited by the path length of the capillaries used in CE to carry out the separation (usually 50–75 μm). In addition, the high sensitivity that can be achieved by CE/MS using the specificity and sensitivity of MS/MS or $MS^2$ experiments certainly opens up promising possibilities for the determination of target compounds in complex food matrices.

The development of a CE/MS analytical method requires the optimization of the conditions of both separation by CE and detection by MS. This section describes the experimental considerations to be taken into account in the development and validation of a fast and sensitive method for the determination of a group of betaines (glycine betaine, trigonelline, proline betaine, carnitine, acetyl-carnitine and palmitoyl-carnitine) by CE/MS using a previous derivatization of them with butanol.

### 9.2.1 Derivatization of Betaines

In order to make possible the determination of analytes present at low concentration in complex matrices, an increase in the sensitivity may be required. In this sense, the derivatization of the target compounds can be an advantage even when MS detection is employed and in spite of the increase in the analysis time derived from the pretreatment of the sample. A possible derivatization reaction for betaines previous to their determination by MS is the butylation of their acid group. Butanol as a derivatizing reagent allows for considerable improvement in the ionization efficiencies and, hence the analytical sensitivity (Chace, 2001), and supposes an increase in the size of the betaines (>150 mass-to-charge ratio, *m/z*). For these higher mass-to-charge ratio values, the MS background noise is usually lower and obtaining a higher number of product ions in the fragmentation of precursor ions is favored when $MS^2$ experiments are employed.

The optimization of the derivatization conditions is always necessary previously to CE/MS analysis in order to obtain the highest sensitivity and robustness in the analysis. The easiest way to perform this optimization is measuring by CE with UV detection the percentage of derivatization achieved for a betaine presenting a chromophore group such as trigonelline. Thus, a butilation method described in the literature (Chace *et al.*, 2009) was employed, in which the butanol derivatizing reagent (3 N HCl in butanol) is added to the evaporated extract of betaines and the reaction is carried out in an oven at 60 °C for 20 min. Under these conditions, two peaks corresponding to derivatized and underivatized trigonelline were obtained (see Figure 9.1a), resulting in a percentage of derivatization of only 62%. In order to achieve the complete derivatization of the betaine, both the temperature and the time of derivatization must be increased to 80 °C (butanol boiling point, 118 °C) and 30 min, respectively (see Figure 9.1b). It is noteworthy that shorter analysis time is obtained for the butyl ester derivative (from 14 to 7 min for trigonelline) showing the potential of this derivatization reaction for betaines.

Under these conditions, acetyl-carnitine and palmytoil-carnitine, as well as other acyl derivatives of carnitine, are degraded to carnitine during the derivatization reaction with butanol. Consequently, the total content of carnitines is determined by this method because the differentiation between carnitine and other acyl-carnitines is not possible.

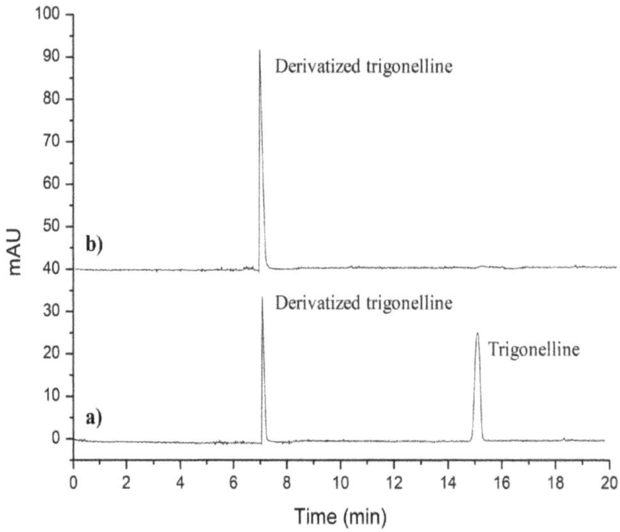


**Figure 9.1** Effect of the temperature on the derivatization of the betaine trigonelline. The electropherograms show the effect of an increase in the temperature and time of derivatization for trigonelline from (a) 60 °C and 20 min to (b) 80 °C and 30 min. CE conditions: BGE, 0.1 M formic buffer (pH 2.0); uncoated fused-silica capillary, 50 μm ID × 68.5 cm; injection by pressure at 50 mbar × 15 s; applied voltage, 25 kV; temperature, 25 °C; detection at 195 nm. Standard of 50 μg $mL^{-1}$ in water. Data from Sánchez-Hernández *et al.* (2011a), with permission from the Publishers.

### 9.2.2 Separation of Betaines by CE/MS

Capillary zone electrophoresis (CZE) is the most common separation mode of CE employed in food analysis due to its ease of operation and because the separation of the compounds is based only on their mass-to-charge ratio (Piñero *et al.*, 2011). The separation and migration of compounds by CZE depends upon the selection of a suitable background electrolyte (BGE), the temperature of the capillary during the analysis, the separation voltage, and the capillary length. Thus, betaines are separated by CZE in acidic media using a 0.1 M formic acid buffer at pH 2.0, which makes negligible the electro-osmotic flow and the interaction of these cationic analytes with the negatively charged capillary wall that takes place when the pH is higher than 3.0. Moreover, an ion trap (IT) is chosen as an analyzer for the detection of betaines by MS using ESI for the CE/MS coupling in order to take advantage of the selectivity and sensitivity of $MS^2$ experiments.

The base peak electropherogram (BPE) shows the baseline separation of the group of betaines investigated in about 10 min using a temperature of 25 °C, a voltage of 25 kV and a capillary with a total length of 60 cm (see Figure 9.2a). The identification of each peak was possible from its mass spectra through the *m/z* value for the protonated molecule, $[M + H]^+$, with the *m/z*

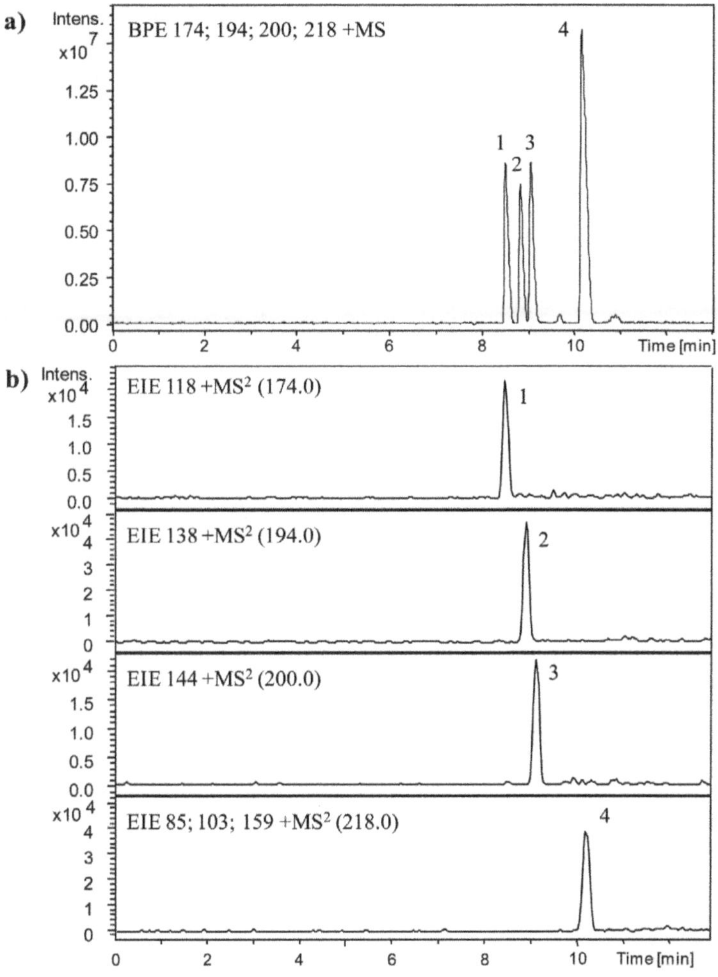


**Figure 9.2** Separation of four betaines by CE/MS and CE/MS$^2$. The electropherograms show the separation of a standard betaines mixture (a) CE/MS BPE and (b) CE/MS$^2$ EIEs. CE Conditions: BGE, 0.1 M formic buffer (pH 2.0); uncoated fused-silica capillary, 50 μm ID × 60 cm; injection by pressure at 50 mbar × 15 s; applied voltage, 25 kV; temperature, 25 °C. ESI conditions: positive ion mode; spray voltage, 4.5 kV; sheath liquid, isopropanol/water (50/50 v/v) with 0.1% formic acid at 3.3 μL min$^{-1}$; drying gas flow, 3 L min$^{-1}$; drying temperature, 300 °C; nebulizer pressure, 2 psi. Ion trap conditions: maximum accumulation time, 300 ms; averages, 1; scan, 50–280 *m*/*z*. MS$^2$ transitions in pseudo-MRM mode with width, 4 *m*/*z*; fragmentation amplitude, 1.00 and fragmentation time, 10 ms. Peak 1. Glycine betaine, 2. Trigonelline, 3. Proline betaine, 4. Carnitines. Standards dissolved in acetonitrile/water (40:60, v/v) at 5 μg mL$^{-1}$ each one. Data are from Sánchez-Hernández *et al.* (2011a), with permission from the Publishers.

ratios of 174 for glycine betaine, 194 for trigonelline, 200 for proline betaine, and 218 for carnitine. Note that the peak #4 is the largest of the four peaks because it includes carnitine and the other two acyl-carnitines analyzed (acetyl- and palmitoyl-carnitine). Then, to obtain the highest sensitivity and selectivity with an IT analyzer is necessary to work in the pseudomultiple reaction monitoring (MRM) mode, *i.e.*, to develop a CE/$MS^2$ method, where the precursor ion of each analyte is selected to obtain its $MS^2$ spectra from the called product scan. Moreover, the fragmentation conditions should be set. Thus, using 300 ms for the maximum accumulation time and 10 ms for the fragmentation time, the best signal-to-noise ratio (S/N) with enough precision (about 10% and more than 12 points per peak) are obtained for all betaines. Figure 9.2b illustrates the extracted ion electropherograms (EIEs) for glycine betaine, trigonelline, proline betaine, and total content of carnitines under $MS^2$ optimal conditions.

The $MS^2$ spectra of the protonated molecular ions and the proposed fragmentation structures of the derivatized betaines are shown in Figure 9.3. For all betaines, the neutral loss of the derivatizing agent with $m/z$ 56 is observed. For carnitine, as well, the neutral loss of $m/z$ 59 of the quaternary ammonium group and of $m/z$ 18 of a water molecule is produced. This CE/$MS^2$ assay enables the unequivocal identification of carnitine and almost unequivocal identification of glycine betaine, trigonelline and proline betaine, and could contribute to their specific determination in different complex matrices.

### 9.2.3 Validation of the CE/$MS^2$ Method

One crucial aspect to be taken into account when an analytical approach is developed is the validation of the methodology in order to demonstrate that it is suitable for its intended purpose. The International Conference on Harmonization (ICH) guidelines on validation of analytical procedures delineate the guidance and methodology to validate their characteristics in terms of selectivity, linearity, accuracy, precision, limit of detection (LOD), and limit of quantification (LOQ) (ICH, 2005). The method should be validated using standard and samples. The validation of CE/$MS^2$ method was carried out in order to evaluate its suitability for the determination of betaines in vegetables oils. The results obtained are summarized in Table 9.1.

Excellent selectivity (almost specificity) is achieved, since at least the baseline resolution is obtained for all betaines and also, at least one transition of the precursor ion of betaines to a product ion can be monitored due to the potential of $MS^2$ experiments. Even for carnitine, monitoring two transitions of a precursor ion to two product ions is possible, allowing its unequivocal identification.

Precision is the most critical component in a validation because it allows the claims of accuracy and linearity to be made. To assess precision, a sample of oil was selected (sunflower oil) and the instrumental and method repeatability, and intermediate precision were evaluated. Glycine betaine, trigonelline

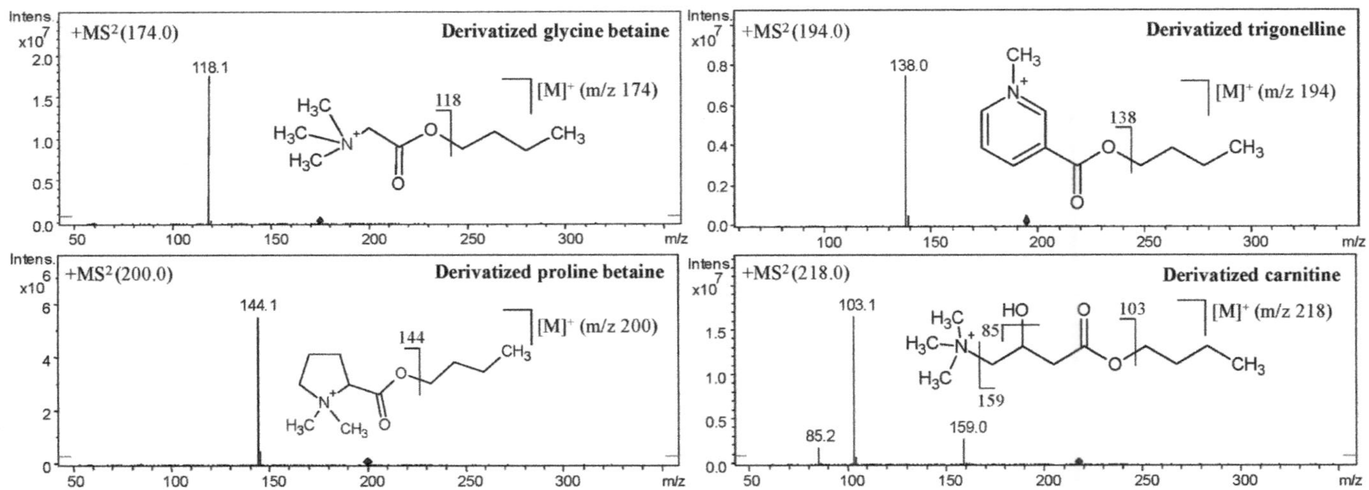


**Figure 9.3** Spectra and structures of four betaines by MS² experiments. The MS² spectra and structure of precursor ions for each betaine are shown. In the structures is shown the fragmentation pattern when the MS² experiments are carried out. CE conditions: uncoated fused-silica capillary, 50 μm ID × 85 cm. Ion trap conditions: averages, 3; scan, 50–350 *m/z*; fragmentation time, 40 ms. Other CE conditions, ESI conditions, and IT conditions as in Figure 9.2. Data from Sánchez-Hernández *et al.* (2011a), with permission from the Publishers.

**Table 9.1** Analytical characteristics of the CE/MS$^2$ method for the determination of betaines Data from Sánchez-Hernández *et al.* (2011a), with permission from the Publishers. ND: not detected (concentration < LOD).

| Analytical characteristics | Compounds | | | |
|---|---|---|---|---|
| | ***Glycine betaine*** | ***Trigonelline*** | ***Proline betaine*** | ***Carnitine*** |
| **Precision** (RSD, %) | | | | |
| ***Instrumental repeatability***[a] | | | | |
| *Retention Time* | 5.5 | 5.9 | ND | 7.7 |
| *Peak Area corrected* | 4.8 | 5.8 | ND | 5.4 |
| ***Method repeatability***[b] | | | | |
| *Retention Time* | 6.8 | 8.9 | ND | 8.9 |
| *Peak Area corrected* | 7.7 | 7.4 | ND | 7.9 |
| ***Intermediate precision***[c] | | | | |
| *Retention Time* | 11.2 | 10.2 | ND | 12.5 |
| *Peak Area corrected* | 10.7 | 9.5 | ND | 9.1 |
| **Linearity**[d] | | | | |
| ***R*** | 0.996 | 0.994 | 0.997 | 0.994 |
| ***Intercept*** | 7.5 (±7.6) × 10$^3$ | 12.8 (±15.8) × 10$^3$ | 1.9 (±9.4) × 10$^3$ | 8.2 (±12.0) × 10$^3$ |
| ***Slope*** | 109.9 (±8.3) | 148.3 (±30.3) | 75.6 (±10.6) | 241.9 (±22.8) |
| **Recovery**[e] | | | | |
| *Low level* | 99 ± 2 | 92 ± 5 | 80 ± 5 | 96 ± 1 |
| *High level* | 94 ± 1 | 88 ± 2 | 88 ± 2 | 99 ± 1 |
| **LOD** (pg g$^{-1}$) | 38 | 25 | 38 | 25 |
| **LOQ** (pg g$^{-1}$) | 125 | 83 | 125 | 83 |

[a]Obtained from six consecutive injections of a refined sunflower oil sample in the same day ($n$ = 6).
[b]Obtained from three individual samples of refined sunflower oil injected by triplicate in the same day ($n$ = 3).
[c]Assessed from three individual samples of refined sunflower oil injected in triplicate in three consecutive days ($n$ = 9).
[d]Six standard solutions at different concentration levels (LOQ-100LOQ) injected in triplicate during three days. The calibration plot was represented using the average of the triplicate injections for each day. Values in parentheses are confidence intervals at 95%: $\pm t \times s_{\text{intercept}}$; $\pm t \times s_{\text{slope}}$.
[e]Recovery for a sample of hojiblanca extra virgin olive oil spiked at low level (5 ng of each compound) and at high level (50 ng of each compound). Average ±standard deviation ($n$ = 3).

and carnitine were detected and their percentage values of relative standard deviations (RSDs) for corrected peak areas (Ac, peak area to migration time ratio) and retention times (RT) were lower than 8% for repeatability and lower than 13% for intermediate precision. These values are higher than those obtained by LC/MS methodologies; however, they are usual values for CE/MS systems that can offer good possibilities in repeatability, with values around 5% of RSD, but with higher values in terms of intraday reproducibility.

On the other hand, linearity was evaluated by analyzing six concentration levels of standard solutions. Each point of the calibration graph corresponded to the mean value from three individual injections of the standard mixture. The obtained data were used to build the calibration curves by taking the Ac from the EIEs of each compound *versus* the concentrations using a least-squares linear regression model. Good linearity for the calibration curves was achieved for all the studied compounds at the concentrations checked with adequate correlation coefficients ($r > 0.99$) and with all the confidence intervals at 95% for the intercept including the zero value as can be seen in Table 9.1.

The evaluation of the accuracy was carried out as the recovery obtained for each compound when a representative extra virgin olive oil sample was spiked with standard solutions at concentration levels of 5 and 50 ng of each betaine. Three injections of each sample were accomplished. Values of recovery ranged from 80% to 99% with RSDs ≤ 5%.

LODs and LOQs for betaines were calculated as the minimum analyte concentration yielding an S/N ratio equal to 3 and 10 times, respectively. LOD values of 38 pg $g^{-1}$ for glycine betaine and proline betaine and of 25 pg $g^{-1}$ for trigonelline and total carnitines were achieved. Therefore, the method offers an excellent sensitivity (at ppt level) and an adequate precision and accuracy for all the betaines, and allows their determination in complex samples as vegetable oils to be carried out.

## 9.3 Application to the Analysis of Vegetable Oils

A wide number of powerful techniques have been employed for the analysis of vegetable oils and can be classified into nonseparative and separative techniques. Spectrometric methods without separation, such as MS, nuclear magnetic resonance, Raman spectroscopy, infrared spectroscopy, fluorescence, and chemiluminiscence, among others, have broadly been reported (Arvanitoyannis and Vlachos, 2007; Frankel, 2010). Among the spectrometric techniques, MS has become a powerful technique to characterize vegetable oils. In addition, the injection of the sample can easily be automated, particularly when using robotized sample injection systems such as direct infusion or flow injection analysis (FIA), permitting very high throughput. The most commonly used mass analyzers for this purpose are tandem mass analyzers. Thus, the use of direct infusion in MS has been reported employing IT or quadrupole time-of-flight (QTOF) mass spectrometers, whereas FIA has been used with triple quadrupole (QqQ) or QTOF. Separative methods for the analysis of several vegetable oils have widely been described being based on LC or GC (Aparicio and Aparicio-*Ruíz*, 2000; Cert *et al.*, 2000; Cserháti *et al.*, 2005; Frankel, 2010), and more recently on CE (Sánchez-Hernández *et al.*, 2011a, 2011b; Monasterio *et al.*, 2013) with MS detection or other detectors.

The compounds studied in both separative and spectrometric techniques were free fatty acids, triacylglycerols, diacylglycerols, and monoacylglycerols as major components and aliphatic alcohols, triterpene alcohols, tocopherols, or other phenolic compounds, and nonprotein amino acids as minor components.

Regarding the analysis of betaines, several approaches have been reported by LC with UV or MS detection and CE with UV detection to quantify glycine betaine, proline betaine, trigonelline and carnitine in a wide range of foods (Zeisel *et al.*, 2003; de Zwart *et al.*, 2003; Sánchez-Hernández *et al.*, 2010; Dabrowska and Starek, 2014). However, only the presence of glycine betaine was detected in olive oil at a high concentration of 1.1 mg $kg^{-1}$ (Zeisel *et al.*, 2003), whereas proline betaine and trigonelline were not detected (de Zwart *et al.*, 2003; Sánchez-Hernández *et al.*, 2010).

This section describes the application of the CE/$MS^2$ methodology described above (Section 9.2) to the analysis of several vegetable oils.

### 9.3.1 Sample Preparation

Vegetable oils are complex samples to be analyzed directly so a previous treatment of the sample used to be necessary. Moreover, an important aspect to take into account in CE is the very low volumes injected in the system (in the nL range) which implies that the sensitivity can be affected compared with other separation techniques such as LC (with volumes injected in the µL range). Therefore, a sample treatment including a preconcentration of the betaines in the sample is required.

Many treatment methods have been described in the analysis of oils depending on the compounds under study, such as liquid–liquid extraction (LLE), solid-phase extraction (SPE), or liquid–liquid microextraction (LLME), among others. Betaines are minor compounds in the unsaponificable fraction of oils. To extract them, a LLE procedure was carried out in order to simplify the treatment. This technique allows the isolation, concentration and clean-up of the sample. The preconcentration is obtained by evaporating the sample extracted to dryness and dissolving the residue in a smaller volume of solvent. The selectivity and efficiency of the extraction process in LLE depends mainly on the selection of the immiscible solvents.

The treatment procedure consists in weighing directly 40 g of the oil in polypropylene copolymer tubes of 500 mL. Betaines are extracted with a mixture of methanol and chloroform (160 mL methanol/chloroform at 2:1 v/v) and left at −20 °C overnight. The mixture is centrifuged for 15 min at 4 °C and the upper phase is collected. Then, the bottom phase is washed with a mixture of methanol/chloroform/water (100 mL at 2:1:0.8, v/v/v) and the new upper phase is combined with the previous one. The mixed fractions are washed with chloroform (40 mL) and water (100 mL) and centrifuged (4000 × g, 15 min, 4 °C). The volume of these solvents was optimized because a higher proportion of water led to a cleaner aqueous phase. Afterwards, the aqueous phase is separated and evaporated (at 80 °C). Finally, the residue is derivatized with butanol and reconstituted in acetonitrile/water (300 µL of 40:60, v/v) before injection in the CE system.

### 9.3.2 Determination of Betaines in Vegetable Oils

The determination of betaines was performed in three different samples for each kind of refined seed oil, such as soy oils, sunflower oil and corn oil, and nine different samples of extra virgin olive oil of three different varieties:

**Table 9.2** Concentration of betaines in vegetable oils from different botanical origin Data from Sánchez-Hernández *et al.* (2011a), with permission from the Publishers. ND: not detected (concentration < LOD).

| *Origin* | *Samples* | *Concentration (ng g⁻¹)* | | | |
|---|---|---|---|---|---|
| | | *Glycine betaine* | *Trigonelline* | *Proline betaine* | *Carnitines* |
| ***Sunflower oil*** | *#1* | 9 ± 1 | 5.5 ± 0.1 | ND | 2.1 ± 0.5 |
| | *#2* | 12 ± 1 | 6.7 ± 0.1 | ND | 4.4 ± 0.3 |
| | *#3* | 11 ± 1 | 7.7 ± 0.6 | ND | 2.2 ± 0.3 |
| ***Corn oil*** | *#1* | 4 ± 1 | 0.3 ± 0.1 | ND | 0.3 ± 0.1 |
| | *#2* | 8 ± 1 | 1.1 ± 0.7 | ND | 0.8 ± 0.2 |
| | *#3* | 5 ± 2 | 0.5 ± 0.1 | ND | 1.0 ± 0.3 |
| ***Soybean oil*** | *#1* | 4 ± 1 | 0.7 ± 0.1 | ND | 1.1 ± 0.2 |
| | *#2* | 9.7 ± 0.8 | 1.2 ± 0.3 | ND | 1.0 ± 0.3 |
| | *#3* | 5.4 ± 0.4 | 0.6 ± 0.2 | ND | 0.8 ± 0.1 |
| ***Hojiblanca olive oil*** | *#1* | 0.16 ± 0.01 | < LOQ | ND | < LOQ |
| | *#2* | 0.12 ± 0.01 | < LOQ | ND | ND |
| | *#3* | < LOQ | < LOQ | ND | ND |
| ***Arbequina olive oil*** | *#1* | 0.13 ± 0.02 | 0.085 ± 0.007 | ND | ND |
| | *#2* | 0.24 ± 0.01 | < LOQ | ND | < LOQ |
| | *#3* | 0.19 ± 0.01 | 0.089 ± 0.001 | ND | ND |
| ***Picual Olive oil*** | *#1* | < LOQ | < LOQ | ND | < LOQ |
| | *#2* | 0.12 ± 0.02 | < LOQ | ND | ND |
| | *#3* | 0.14 ± 0.01 | < LOQ | ND | ND |

*hojiblanca*, *arbequina* and *picual*. Table 9.2 summarizes the results for the content of each betaine in these samples.

The strategy employed to carry out the quantification of betaines was the single-point standard addition calibration. Only two solutions for each oil sample have to be injected, a sample solution and the sample solution spiked with 0.1 μg mL$^{-1}$ of each betaíne. The results shown in Table 9.2 confirm the absence of proline betaine and the presence of glycine betaine, trigonelline and carnitines in soybean, corn, and sunflower oils, being the glycine betaine content the highest in all these seed oils, while the content of trigonelline is similar to the carnitines in soybean and corn oils but about 2 times higher in sunflower oils.

Furthermore, it is noteworthy that the carnitines are not detected in extra virgin olive oils, while the contents in trigonelline and glycine betaine are around 30 and 40 times lower than the amounts obtained in seed oils, respectively.

These results can be observed in Figure 9.4, which shows the EIEs obtained by CE/MS$^2$ for glycine betaine, trigonelline and total content of carnitines in a seed oil sample and an olive oil sample. This figure also shows the MS$^2$ spectra employed to carry out the identification of each betaine in the seed oil.

In view of the obtained data, the total content of carnitines can be considered as a marker to detect adulterations of olive oils with seed oils where percentages ≥ 2% (w/w) of soybean oil in olive oil may be detected (Sánchez-Hernández *et al.*, 2011a). These results show the high potential of the developed method to assess the quality and authenticity of olive oils. The importance of olive oil and in particular the extra virgin olive oil has been correlated with its high

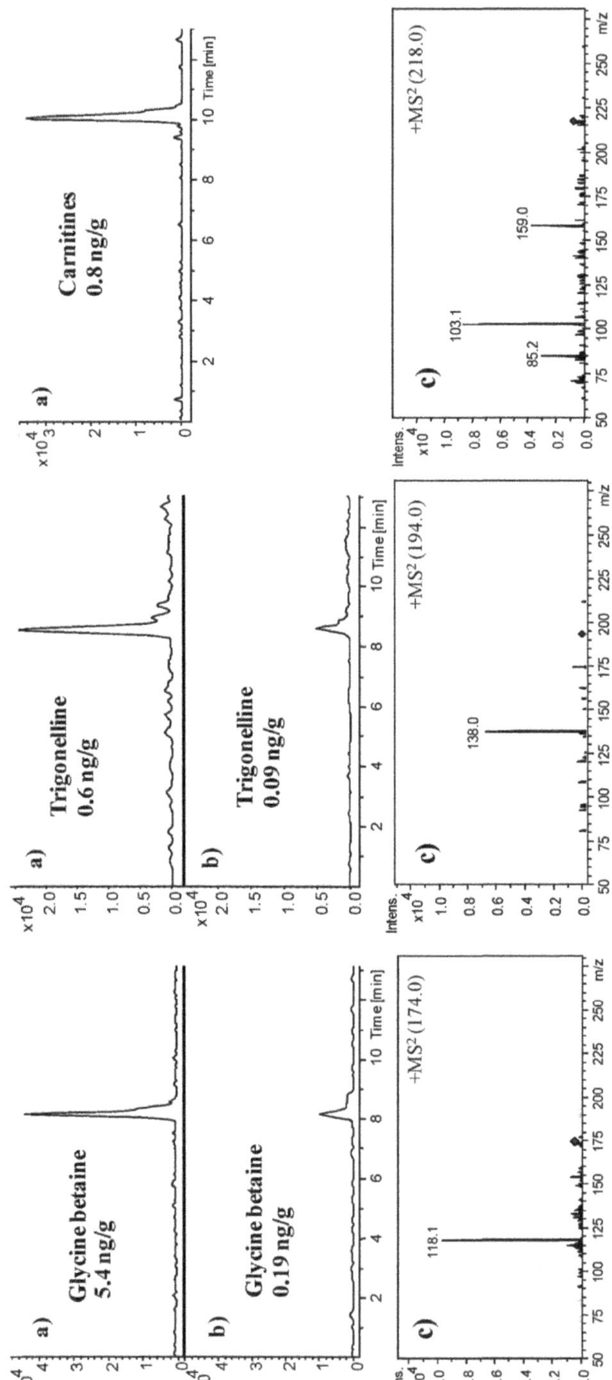


**Figure 9.4** Detection by CE/MS$^2$ of three betaines in seed and olive oils. CE/MS$^2$ EIEs for glycine betaine, trigonelline and total content of carnitines in (a) seed oil sample (soybean oil#3), (b) olive oil sample (Arbequina olive oil#3), and (c) MS$^2$ spectra for the peaks obtained in (a). All other experimental conditions were as in Figure 9.2. Data from Sánchez-Hernández *et al.* (2011a), with permission from the Publishers.

nutritional value and with its benefits in human health (Martin-Moreno *et al.*, 1994; Bruckner, 2000; Owen *et al.*, 2000; Fasciotti and Pereira Netto, 2010). In addition, because of its high price, extra virgin olive oils have become the object of adulterations for economical purposes. The most common method for adulteration is doping olive oils with cheaper oils such as seed oils (sunflower, corn, or soy oils), which are the most widely employed for this purpose. However, according to European regulations, the definition of virgin olive oils excludes the mixture with other kind of oils (European Communities, 2004). For these reason, the genuineness of olive oils is a very important aspect from the point of view of quality and marketing.

Finally, note that very few compounds have been validated so far as markers for specific foods, therefore, the proposal of new markers of foods remain a challenge.

## Summary Points

- This chapter focuses on a very sensitive determination of betaines using a capillary electrophoresis coupled to mass spectrometry method.
- The previous derivatization of betaines with butanol has proven be very useful to carry out their determination by mass spectrometry with adequate sensitivity and selectivity.
- Satisfactory separation among four betaines with short analysis times (about 10 min) is achieved and their identification is possible through their $MS^2$ spectra.
- The methodology shows good sensitivity and adequate precision and accuracy for all the betaines allowing their determination at low concentrations in food samples with adequate selectivity.
- Sample preparation permits the extraction and preconcentration of betaines to yield higher sensitivity in the analysis of vegetable oils.
- The analysis of betaines in different commercial seed oils (soybean oils, sunflower oils and corn oils) reveals the presence of glycine betaine, trigonelline and carnitines in these oils.
- The results obtained in the analysis of betaines in different commercial extra virgin olive oils show the absence of carnitines enabling them to be proposed as markers for detecting adulterations of extra virgin olive oils with seed oils.
- This methodology offers a sensitive and rapid fingerprint of olive oils for quality control purposes.

## Key Facts

### Key Facts for Betaines

- Betaines are quaternary amino derivatives of amino acids.
- Betaines are found mainly in plants and foods.
- Betaines have beneficial effects on human health.

## Key Facts for the Extraction Procedure

- The LLE technique allows the isolation, concentration of analytes and clean-up of the sample.
- The preconcentration is obtained by evaporating the sample extracted to dryness and by dissolving the residue in a smaller volume of solvent.
- The selectivity and efficiency of the extraction process in LLE depends mainly on the selection of the immiscible solvents.

## Key Facts for the Separation Technique

- CE is an analytical technique for the separation of mixtures of compounds.
- MS is an analytical detector that uses the *m/z* ratio of ionized compounds.
- MS allows quantitative and qualitative (chemical and structural) information based on the distinctive fragmentation patterns of the molecules.

# Definitions of Words and Terms

**Accuracy:** Validation parameter defined as how close the determined value is to the true value.

**Capillary electrophoresis (CE):** Separation technique based on the different migration rate of analytes charged into a capillary tube under the action of an electric field.

**Capillary electrophoresis/mass spectrometry (CE/MS):** Hybridized analytical technique based on the coupling between capillary electrophoresis and mass spectrometry that is used for qualitative/quantitative analyses.

**Derivatization:** Chemical reaction between an analyte and a suitable reagent to convert the analyte into a new compound to increase the sensitivity and selectivity of its detection.

**Electrospray ionization (ESI):** Ionization method most used in the coupling between a liquid-phase separative technique and mass spectrometry due to the wide spectrum of molecules capable of ionizing, such as polar molecules, ionic or ionizable, which are within the scope of CE separations.

**Liquid chromatography (LC):** Separation technique using a liquid phase as the mobile phase based on the different retention of the analytes on a chromatographic packing when dragged by the mobile phase.

**Linearity:** this is the ability to obtain results within a given range that are directly proportional to the concentration of analyte in the sample.

**Limit of detection (LOD):** The lowest concentration of the analyte that can be distinguished from the noise. In general, the LOD is defined as the concentration corresponding to a signal-to-noise ratio of 3.

**Limit of quantification (LOQ):** The lowest amount of analyte that can be quantified with acceptable accuracy and precision. In general, the LOQ is defined as the concentration corresponding to a signal-to-noise ratio of 10.

**Liquid–liquid extraction (LLE)**: Sample treatment technique for clean-up and signal enhancement in the routine analysis based on the distribution of the analyte between two immiscible liquid phases.
**Mass-to-charge ratio (m/z)**: The *m/z* value corresponding to the mass and charge of an ion.
**Multiple reaction monitoring (MRM)**: is a mode of acquisition in mass spectrometry based on monitoring multiple product ions obtained from one to several precursor ions after a MS/MS or $MS^2$ experiment. It is a tool of high sensitivity and selectivity.
**Precision**: determines the degree of closeness of a set of values obtained in given conditions. There are three levels: repeatability, intermediate precision and reproducibility.
**Selectivity**: expresses the ability of the method to identify and/or measure analytes of interest in the presence of other compounds of the sample.
**MS/MS or $MS^2$ experiment**: is a working mode in mass spectrometry to isolate one or more characteristic ions of the analyte (named precursor ions) for further fragmentation in order to obtain a series of fragment ions (named product ions) that aid in identification of the analyte.
**Validation**: Systematic study of the quality parameters of an analytical method to confirm if an analytical methodology is suitable for the purpose for which it was designed ("fitness for purpose").

# List of Abbreviations

| | |
|---|---|
| Ac | Corrected peak area |
| BGE | Background electrolyte |
| BPE | Base peak electropherogram |
| CE | Capillary electrophoresis |
| CZE | Capillary zone electrophoresis |
| EIE | Extracted ion electropherogram |
| ESI | Electrospray ionization |
| FIA | Flow injection analysis |
| GC | Gas chromatography |
| IT | Ion trap |
| LLE | Liquid–liquid extraction |
| LLME | Liquid–liquid microextraction |
| LC | Liquid chromatography |
| LOD | Limit of detection |
| LOQ | Limit of quantification |
| $[M + H]^+$ | Protonated molecule |
| MRM | Multiple reaction monitoring |
| MS | Mass spectrometry |
| $MS^2$ | Tandem mass spectrometry |
| *m/z* | Mass-to-charge ratio |
| QqQ | Triple quadrupole |

| | |
|---|---|
| QTOF | Quadrupole time-of-flight |
| r | Correlation coefficient |
| RSD | Relative standard deviation |
| RT | Retention time |
| S/N | Signal-to-noise ratio |
| SPE | Solid-phase extraction |
| UV | Ultraviolet |

## Acknowledgements

The authors thank the Spanish Ministry of Science and Innovation (project CTQ2009-09022) and the Comunidad Autónoma of Madrid (Spain) and European funding from FEDER programme (project S2013/ABI-3028, AVANSECAL). They also thank the University of Alcalá (Spain) and the Comunidad Autónoma of Madrid (Spain) for project CCG10-UAH/AGR-5950.

## References

Aparicio, R. and Aparicio-Ruíz, R., 2000. Authentication of vegetable oils by chromatographic techniques. *Journal of Chromatography A*. 881: 93–104.

Arvanitoyannis, I. S. and Vlachos, A., 2007. Implementation of physicochemical and sensory analysis in conjunction with multivariate analysis towards assessing olive oil authentication/adulteration. *Critical Reviews in Food Science and Nutrition*. 47: 441–498.

Bruckner, G., 2000. *Fatty Acids in Foods and Their Health Implications*, Chow, C. K, (ed.) Marcel Deckker, New York, p. 843.

Cert, A., Moreda, W. and Pérez-Camino, M. C., 2000. Chromatographic analysis of minor constituents in vegetable oils. *Journal of Chromatography A*. 881: 131–148.

Chace, D. H., 2001. Mass spectrometry in the clinical laboratory. *Chemical Reviews*. 101: 445–477.

Chace, D. H., Lim, T., Hansen, C. R., De Jesus, V. R. and Hannon, W. H., 2009. Improved MS/MS analysis of succinylacetone extracted from dried blood spots when combined with amino acids and acylcarnitine butyl esters. *Clinica Chimica Acta*. 407: 6–9.

Cserháti, T., Forgács, E., Deyl, Z. and Miksik, I., 2005. Chromatography in authenticity and traceability tests of vegetable oils and dairy products: a review. *Biomedical Chromatography*. 19: 183–190.

Dabrowska, M. and Starek, M., 2014. Analytical approaches to determination of carnitine in biological materials, foods and dietary supplements. *Food Chemistry*. 142: 220–232.

European Communities. 2002. Regulation 178/2002. *Official Journal of the European Communities: Legislation*. 31: 1–24.

European Communities. 2004. Regulation 865/2004. *Official Journal of the European Communities: Legislation*. 206: 37–50.

European Communities. 2011. Regulation 1169/2011. *Official Journal of the European Communities: Legislation*. 304: 18–63.

Fasciotti, M. and Pereira Netto, A. D., 2010. Optimization and application of methods of triacylglycerol evaluation for characterization of olive oil adulteration by soybean oil with HPLC-APCI-MS-MS. *Talanta*. 81: 1116–1125.

Frankel, E. N., 2010. Chemistry of extra virgin olive oil: adulteration, oxidative stability, and antioxidants. *Journal of Agricultural and Food Chemistry*. 58: 5991–6006.

Hunt, S., 1985. *The Non-protein Amino Acids in "Chemistry and Biotechnology of Amino Acids"*. Barret, G. C. (ed.), Chapman & Hall LTD, Chapter 4.

ICH, 2005. International Conference on Harmonization. *Validation of analytical procedures: text and methodology Q2 (R1)*.

Lang, F., 2007. Mechanisms and significance of cell volume regulation. *Journal of the American College of Nutrition*. 26: 613S–623S.

Lever, M. and Slow, S., 2010. The clinical significance of betaine, an osmolyte with a key role in methyl group metabolism. *Clinical Biochemistry*. 43: 732–744.

Mansour, F. R., Wei W. and Danielson N. D., 2013. Separation of carnitine and acylcarnitines in biological samples: a review. *Biomedical Chromatography*. 27: 1339–1353.

Martin-Moreno, J. M., Willett, W. C., Gorgojo, L., Banegas, J. R., Rodriguez-Artalejo, F., Fernandez-Rodriguez, J. C., Maisonneuve, P. and Boyle, P., 1994. Dietary-fat, olive oil intake and breast-cancer risk. *International Journal of Cancer*. 58: 774–780.

Monasterio, R. P., de los Ángeles Fernández, M. and Fernanda Silva, M., 2013. Olive oil by capillary electrophoresis: characterization and genuineness. *Journal of Agricultural and Food Chemistry*. 61: 4477–4496.

Naz, S., Garcia, A., Rusak, M. and Barbas, C., 2013. Method development and validation for rat serum fingerprinting with CE–MS: application to ventilator-induced-lung-injury study. *Analytical and Bioanalytical Chemistry*. 405: 4849–4858.

Owen, R. W., Giacosa, A., Hull, W. E., Haubner, R., Wurtele, G., Spiegelhalder, B. and Bartsch, H., 2000. Olive-oil consumption and health: the possible role of antioxidants. *The Lancet Oncology*. 1: 107–112.

Piñero, M.-Y., Bauza, R. and Arce, L., 2011. Thirty years of capillary electrophoresis in food analysis laboratories: potential applications. *Electrophoresis*. 32: 1379–1393.

Sánchez-Hernández, L., Puchalska, P., García-Ruiz, C., Crego, A. L. and Marina, M. L., 2010. Determination of trigonelline in seeds and vegetable oils by capillary electrophoresis as a novel marker for the detection of adulterations in olive oils. *Journal of Agricultural and Food Chemistry*. 58: 7489–7496.

Sánchez-Hernández, L., Castro-Puyana, M., Marina, M. L. and Crego, A. L., 2011a. Determination of betaines in vegetable oils by capillary electrophoresis tandem mass spectrometry- application to the detection of olive oil adulteration with seed oils. *Electrophoresis*. 32: 1394–1401.

Sánchez-Hernández, L., Marina, M. L. and Crego, A. L., 2011b. A capillary electrophoresis–tandem mass spectrometry methodology for the determination of non-protein amino acids in vegetable oils as novel markers for the detection of adulterations in olive oils. *Journal of Chromatography A*. 1218: 4944–4951.

Storer, M. K., McEntyre, C. J. and Lever, M., 2006. Separation of cationic aracyl derivatives of betaines and related compounds. *Journal of Chromatography A*. 1104: 263–271.

Zeisel, S. H., Mar, M.-H., Howe, J. C. and Holden, J. M., 2003. Concentrations of choline-containing compounds and betaine in common foods. *Journal of Nutrition*. 133: 1302–1307.

Zhang, J., Nishimura, N., Okubo, A. and Yamazaki, S., 2002. Development of an analytical method for the determination of betaines in higher plants by capillary electrophoresis at low pH. *Phytochemical Analysis*. 13: 189–194.

de Zwart, F. J., Slow, S. R., Payne, J., Lever, M., George, P. M., Gerrard, J. A. and Chambers, S. T., 2003. Glycine betaine and glycine betaine analogs in common foods. *Food Chemistry*. 83: 197–204.

CHAPTER 10

# *Hydrophilic Interaction Liquid Chromatography for Determination of Betaine*

JONATHAN M. CURTIS*[a] AND SI MI[b]

[a]Department of Agricultural, Food and Nutritional Science, Faculty of Agricultural, Life and Environmental Sciences, University of Alberta, 3-60D South Academic Building, Edmonton, AB, Canada T6G 2P5; [b]Department of Agricultural, Food and Nutritional Science, Faculty of Agricultural, Life and Environmental Sciences, University of Alberta, 3-24 South Academic Building, Edmonton, AB, Canada T6G 2P5
*E-mail: jcurtis1@ualberta.ca

## 10.1 Introduction

### 10.1.1 Basics of HILIC

Hydrophilic interaction liquid chromatography (HILIC) was first introduced by Andrew Alpert in 1990. The HILIC method was originally proposed as an alternative to the more conventional normal-phase (NP) and reversed-phase (RP) LC methods to separate extremely polar and hydrophilic compounds (Guo and Gaiki, 2011). Since its development, HILIC has been widely used in many research fields including food, environmental, pharmaceutical and clinical analyses (Holm *et al.*, 2003; Bruce *et al.*, 2010; Kirsch *et al.*, 2010; Zhao *et al.*, 2011; Xiong *et al.*, 2012). Through these applications, HILIC has been shown to be a valuable separation method for the analysis of both polar and ionic compounds (McCalley, 2007).

Food and Nutritional Components in Focus No. 7
Betaine: Chemistry, Analysis, Function and Effects
Edited by Victor R Preedy

Published by the Royal Society of Chemistry, www.rsc.org

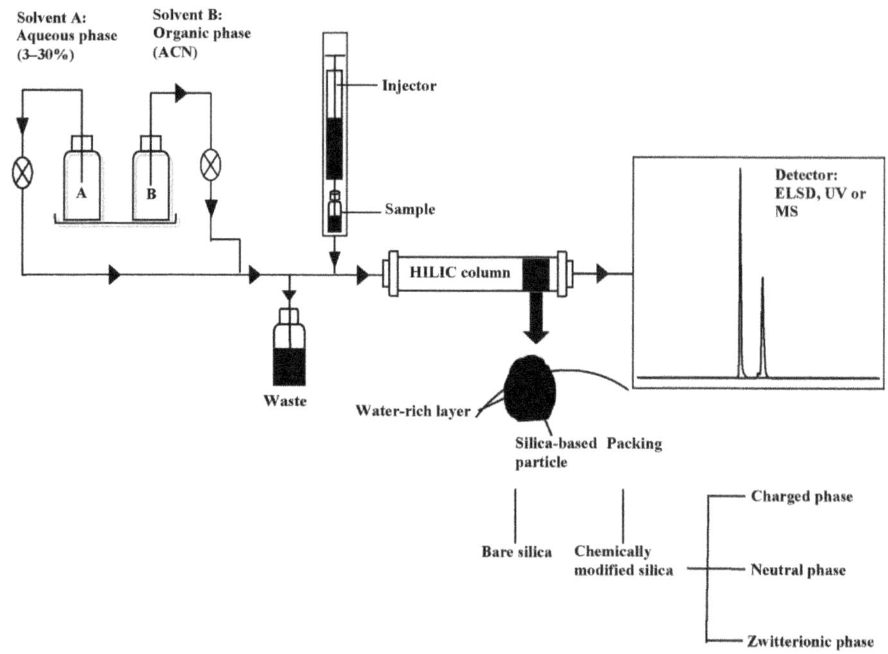


**Figure 10.1** The HILIC chromatography apparatus. This figure shows the LC components used in HILIC separations. A binary system of water (solvent A) and ACN (solvent B) is mixed as indicated. The mobile phase and injected sample is passed into a HILIC column that is packed with the stationary-phase particles. Under HILIC conditions, a water-enriched layer forms around each particle as shown. This layer plays a crucial role in the HILIC separation.

A typical HILIC system is usually comprised of a hydrophilic stationary phase and a binary mobile phase system (Figure 10.1). In comparison with the widely used RPLC techniques, HILIC has the advantage of providing high retention of highly hydrophilic and polar molecules but with less requirement for ion-pair reagents. The typical elution order is similar to that found in NPLC with the less polar compounds eluting ahead of the more polar ones.

#### *10.1.1.1 Separation Mechanisms of HILIC*

Although HILIC chromatography has been developed and widely applied over several decades, the relevant mechanisms of separation are still not well clarified. Since the pattern of separation in HILIC mode may be a result of multiple types of interactions between the analyte and stationary phase, it can be influenced by many chromatographic conditions, such as the stationary phase type, the mobile phase composition and the structural properties of the target solutes (Gama *et al.*, 2012). There are two accepted retention mechanisms that are believed to operate under HILIC conditions.

First, as with RPLC, partition is considered to be one of the primary separation mechanisms in HILIC. This mechanism was proposed by Andrew Alpert (1990) who suggested that the retention on HILIC columns is based upon the partitioning of analytes between a water-enriched layer adsorbed onto the surface of the hydrophilic stationary phase and the relatively less polar organic eluent. In this model, the partially immobilized water-rich layer is formed *via* attractive interactions between the polar functional groups on the stationary phase and water molecules in the mobile phase. Thus, hydrophilic analytes carried in the mobile phase preferentially partition into the immobilized water-rich layer leading to strong retention on the column. This explains the elution order in HILIC since the less polar compounds elute from the column earlier than the more polar ones. However, it should be noted that there is a saturation limit for the amount of water that can be adsorbed on the surface of the stationary phase (Orth and Engelhardt, 1982; Gama *et al.*, 2012). Beyond this limit, excess water molecules would stay in the organic-enriched eluent, which makes the water-rich layer and the organic bulk phase more similar to each other. Consistent with this idea, Hemström and Irgum (2006) discussed the results obtained by Orth and Engelhardt (1982), which noted that the retention of solutes decreased along with the decrease of the difference in polarity between the two phases. Therefore, there is a defined range of water content required for the HILIC mobile phase. A lower limit of 3% water is generally sufficient to guarantee the formation of a water layer around the stationary phase particles, whereas an upper limit of 30% water ensures that there is sufficient retention of polar solutes. The partition mechanism has been discussed extensively and is supported experimentally in many publications (Dejaegher *et al.*, 2008; Dejaegher and Vander Heyden, 2010; Gama *et al.*, 2012; Cavazzini and Felinger, 2013).

However, some separation results in HILIC cannot be explained by the partition mechanism alone. For example, different retention properties of four HILIC stationary phases were observed for analysis of the same group of model compounds under identical chromatographic conditions (Guo and Gaiki, 2005). This suggests that other retention mechanisms besides simple partitioning must also occur in HILIC. It was reported that the elution pattern of peptides achieved in HILIC was similar to that in NPLC (Hemström and Irgum, 2006). The retention mechanism in NPLC is known to be controlled by surface adsorption (Yoshida, 2004) and hence the adsorptive retention model may also play a role in HILIC separations. The adsorption mechanism requires direct contact or interaction between the analytes and the stationary phase (Hao *et al.*, 2008). These interactions have been investigated in many publications and mostly involve hydrogen bonding, electrostatic forces and dipole–dipole interactions (Yoshida, 2004; Hemström and Irgum, 2006; McCalley, 2007; Cavazzini and Felinger, 2013). Notably, the adsorption mechanism usually involves multiple weak interactions between the analytes and the stationary phase, hence in some publications it is also referred to as a multimodal mechanism (Cavazzini and Felinger, 2013).

In some cases, the predominant separation mechanism may even vary during the separation process as chromatographic conditions change (Hao *et al.*, 2008).

### 10.1.1.2 Components of HILIC–Mobile and Stationary Phases

**10.1.1.2.1 Mobile Phases.** In general, the mobile phase in HILIC chromatography is composed of a binary system of water with a higher percentage of a water-miscible polar organic solvent. Thus, compared with NPLC, the aqueous–organic mobile phase in HILIC provides greater solubility for the polar and hydrophilic analytes (Dejaegher *et al.*, 2008). Importantly, the polar mobile phase makes HILIC highly compatible with the electrospray-mass spectrometry (ESI-MS) detection allowing for high ionization efficiency and sensitivity.

Acetonitrile (ACN) is commonly employed, being an aprotic organic solvent with low viscosity, so that separations can be optimized at low back-pressures (Gama *et al.*, 2012). Other water-miscible solvents including methanol, ethanol, isopropanol (IPA) and tetrahydrofunan (THF) have also been utilized as components of the HILIC mobile phase. However, it is worth mentioning that alcohols are both protic and stronger eluting solvents than ACN (Bernal *et al.*, 2011; Gama *et al.*, 2012). These characteristics may result in lower retention and decreased separation efficiency. Moreover, since THF is a better hydrogen-bond acceptor than ACN, its inclusion in place of ACN will result in competition with water for the active sites on stationary phase. This will affect the formation of the immobilized water layer on the surface of stationary phase (Li and Huang, 2004). Hence, in general it is found that ACN is the preferred choice of organic modifier in HILIC mobile phase.

The aqueous phase, which may contain volatile salts or weak acids (*e.g.* ammonium acetate, ammonium formate, trifluoroacetic acid), is the stronger eluting solvent (Jian *et al.*, 2010; Kahsay *et al.*, 2013) in the HILIC mobile phase. The salts or weak acids are added in low amounts to manipulate the separation (Lehnert *et al.*, 2013) or the ionization efficiency for MS detection. Differences in both selectivity and retention due to choice of additive have been investigated in detail (McCalley, 2007) and been shown to be important for optimizing HILIC separations.

**10.1.1.2.2 Stationary Phases.** In recent years, a very large number of stationary phases have been employed in HILIC experiments. Generally, these can be divided into silica-based stationary phases and polymer-based stationary phases. Of these two kinds of stationary phases, silica-based columns are far more commonly used in HILIC, either bare silica columns or chemically modified silica columns with functionalities including amide, cyano, sulfobetaine and diol. Chemically modified silica columns can be divided into categories depending on the surface charge properties of the silica. Thus neutral phases carry no net charge over the pH range of the mobile phase used for HILIC as may be the case for amide, cyano, diol and cyclodextrin modified silica (Guo and Gaiki, 2011). Phases carrying a net positive or negative charge over the mobile phase pH range for HILIC include the widely used amino phase (Guo and Gaiki, 2011). Examples of zwitterionic phases used in HILIC separations include sulfobetaine phases and phosphocholine phases (Guo and Gaiki, 2011). In addition, many other novel HILIC phases have been developed, such as those modified with polysaccharides (Lehnert *et al.*, 2013).

### 10.1.2 Summary and Comparison to Other Analytical Techniques Used for Analysis of Betaine

Betaine is present in many samples of both plant and animal origins, as described elsewhere in this book. Consequently, a wide variety of analytical techniques have been developed for the measurement of betaine in various matrices.

RPLC is an efficient and widely used separation technique, but betaine is poorly retained due to its high polarity. The use of RPLC for betaine analysis in *Lycium chinense* was reported by Lee *et al.* (2011). In order to obtain sufficient retention, an ion-pairing reagent, in this case perfluoropentanoic acid (PFPA), was used. A disadvantage of this approach is that many ion-pairing reagents are not compatible with MS detection due to potential ion-suppression effects and/or possible ion-source contamination. NPLC separations have been discussed for betaine analysis (Zhao *et al.*, 2011) in the context of a method to analyze choline-containing compounds in food samples. The use of typical normal phase organic solvents is less desirable in a laboratory setting, less compatible with ESI-MS compared to RP-LC or HILIC mobile phases and not ideal for polar analytes such as betaine.

Ion chromatography (IC) with different detectors has been tested for betaine measurement. Multiple ion exchange columns were tested to separate betaine in feed ingredients (Chendrimada *et al.*, 2002). However, good separation of betaine from other impurities present in the feed and subsequent quantification of betaine could only be achieved after the development of suitable extraction procedures and further clean-up using ion-exchange resins. The application of gas chromatography (GC) in betaine analysis requires the conversion of nonvolatile betaine into more volatile derivatives. For example, Hitz and Hanson (1980) described the pyrolytic conversion of betaine from plants to trimethylamine (TMA), whereas Allen *et al.* (1993) measured betaine in serum by enzymatically converting it to dimethylglycine (DMG) including a stable isotope dilution procedure for quantification by GC/MS. The requirement for derivatization in GC analysis may increase the uncertainty in quantitative analysis. Capillary electrophoresis (CE) is another possible separation technique for betaine that has been used for quantification in plants (Zhang *et al.*, 2002) and in various seed oils

(Sánchez-Heránndez *et al.*, 2011). Prior esterification of betaine was used to reduce interferences from other compounds and to provide a chromophore for UV detection.

The use of the HILIC method in betaine analysis may overcome many of the problems associated with the analytical methods described above. For example, the polar stationary phases employed can strongly retain betaine molecules without the need for ion-pairing reagents or derivatization. At the same time, the HILIC mobile phase is typically a mixture of water with a high content of ACN, which can solubilize betaine and is compatible with ESI-MS detection.

## 10.2 The Use of HILIC in the Determination of Betaine Concentrations.

### 10.2.1 Betaine in Standard Mixtures

The analysis of a mixture of quaternary amine standards of biological and environmental importance (acetylcholine, choline, betaine, chlormequat and mepiquat) by HILIC/MS was described by Guo (2005). Two HILIC columns were compared for this separation, a YMC-Pack $NH_2$ (YMC Co.), containing aminopropyl groups attached to the silica surface and a TSKgel Amide-80 (Tosoh Bioscience), an amide phase containing amide groups bound to silica *via* a hydrocarbon chain. The effects of ACN content, column temperature, buffer salt type (ammonium acetate or formate) and its concentration on the retention behavior were investigated for both columns. Under most conditions betaine was more strongly retained on the amide column compared to the amino column. All five of the model compounds gave rise to peak tailing on the amide phase, which was explained as a result of interactions between the positively charged amines and negatively charged surface silanols. Increasing the ACN content of the mobile phase gave rise to greater betaine retention. However, plots of ln $k'$ (the capacity factor) against ACN content of betaine deviated from linearity indicating that secondary interactions contribute to retention in addition to HILIC mode. With the amino column, use of low concentrations of ammonium formate (~10 mM) resulted in better peak shapes than with similar concentrations of ammonium acetate in the mobile phase. Although betaine quantification was not described, under optimized HILIC conditions using the YMC-Pack $NH_2$ column and coupled with ESI-MS in SIM mode, the LOD for choline was 0.4 fmol ($S/N = 4$). This is considerably lower than the LOD of 30 fmol previously reported for a RPLC-MS method (Dunphy and Burinsky, 2003). This difference in sensitivity is attributed to avoiding the use of ion-pairing reagents and the use of higher levels of organic solvent in the mobile phase that is closer to optimal for ESI. The HILIC separation had a run time of less than 5 min and exhibited at least 2 orders of magnitude of linearity. Although this discussion of sensitivity refers to choline, it can be assumed that similar arguments would apply to betaine analysis.

**Table 10.1** The determination of betaine from HILIC-based methods. A summary of published HILIC-based methods used for betaine analysis in a variety of samples.

| Matrix | Target compound (s) | Stationary phase | Mobile phase | Detection | References |
|---|---|---|---|---|---|
| Plasma, cereals, cereal products | Betaine, free choline | UHPLC HILIC (waters) (2.1 mm × 150 mm, 1.7 μm) | 10 mM aqueous ammonium acetate-0.005% acetic acid + ACN; gradient | ESI-MS/MS | Bruce *et al.* (2010) |
| Mixture of standards | Acetylcholine, choline, betaine, chlormequat, mepiquat | YMC-Pack $NH_2$ (waters) (150 mm × 4.6 mm, 3 μm) TSKgel Amide-80 (Tosoh Bioscience) (250 mm × 4.6 mm, 5 μm) | ACN + 10 mM aqueous ammonium acetate (81 : 19, v/v); isocratic | ESI-MS/MS | Guo (2005) |
| Plasma | Choline, betaine, dimethylglycine | Hypersil silica column (100 mm × 2.1 mm, 5 μm) | 15 mM ammonium formate (pH = 3.5) + ACN; gradient | ESI-MS/MS | Holm *et al.* (2003) |
| EDTA plasma, urine, amniotic fluid, cerebrospinal fluid | Acetylcholine, choline, betaine, dimethylglycine | Acquity UPLC BEH HILIC (100 mm × 2.1 mm, 1.7 μm) | 15 mM ammonium formate (pH 3.5) + ACN; gradient | ESI-MS/MS | Kirsch *et al.* (2010) |
| Lycium chinense fruits | Betaine | YMC-Pack NH2 (250 mm × 4.6 mm, 5 μm) | water + ACN (25 : 75, v/v); isocratic | ESI-MS/MS | Shin *et al.* (1999) |
| Lycii fructus | Betaine | Phenomenex Kinetex HILIC (2.1 mm × 100 mm, 2.6 μm) Waters Atlantis HILIC (3.0 mm × 100 mm, 5 μm) HALO HILIC (2.1 mm × 100 mm, 2.7 μm) | ACN + 10 mM ammonium formate (pH 3.0)-ACN (90/10, v/v); isocratic | ELSD | Shin *et al.* (2012) |
| Refined sugar, wine | Betaine | Amino-bonded silica (150 mm × 4.7 mm, 10 μm) | water + ACN (25 : 75, v/v); isocratic | Refractive index detection | Vialle *et al.* (1981) |

(*continued*)

**Table 10.1** (*continued*)

| Matrix | Target compound (s) | Stationary phase | Mobile phase | Detection | References |
|---|---|---|---|---|---|
| Egg yolk extracts, human diets, rat livers, stomach contents | Acetylcholine, betaine, choline, glycerophosphocholine, lysophosphatidylcholine, lysophosphatidylethanolamine, phosphatidylcholine, phosphatidylethanolamin, phosphatidylinositol, phosphocholine, sphingomyelin, cytidine diphosphocholine, phosphatidylserine, phosphatidylglycerol | Ascentis Express HILIC (150 mm × 2.1 mm, 2.7 µm) | ACN + 10 mM aqueous ammonium formate (pH = 3); gradient | ESI-MS/MS | Xiong *et al.* (2012) |
| Egg yolk extracts | Acetylcholine, betaine, choline, glycerophosphocholine, lysophosphatidylcholine, lysophosphatidylethanolamine, phosphatidylcholine, phosphatidylethanolamin, phosphatidylinositol, phosphocholine, sphingomyelin | Ascentis Express HILIC (150 mm × 2.1 mm, 2.7 µm) | ACN + 10 mM aqueous ammonium formate (pH = 3); gradient | ESI-MS/MS | Zhao *et al.* (2011) |
| Lycii fructus | Betaine | Atlantis HILIC silica (4.6 mm × 150 mm, 5 µm) Phenomenex Luna HILIC (4.6 mm × 150 mm, 5 µm) | 30 mM ammonium acetate + ACN (20 : 80, v/v); isocratic | ELSD | Zhao *et al.* (2013) |

### 10.2.2 Betaine in Samples of Plant Origin

Fructus Lycii is a commonly used traditional Chinese herbal medicine that is said to have many functional properties as a tonic including as an antioxidant and antitumor agent. Betaine is reported to be a major bioactive component and therefore quantitative determination of betaine is of importance in quality control of products derived from Fructus Lycii. Furthermore, both the Korean and Chinese Pharmacopoeia specify the required minimum betaine concentrations. Three HILIC-based methods for the quantification of betaine in Fructus Lycii have been reported (Shin *et al.*, 1999; Shin *et al.*, 2012; Zhao *et al.*, 2013).

In 1999, Shin *et al.* extracted the dried *Lycium chinense* fruit samples at room temperature with methanol using sonication, followed by centrifugation. The supernatant was dried under nitrogen and reconstituted with water. The final extract was then directly injected for HILIC analysis on a YMC $NH_2$-phase column (250 mm × 4.6 mm I.D., 5 μm in particle size) with an isocratic mobile phase composition of water and ACN (25 : 75, v/v) followed by both UV (210 nm) and ESI-MS detection in both selected ion monitoring (SIM) and full scan modes. Protonated molecular ions of betaine at *m/z* 118 were observed but no molecular species could be seen in negative-ion mode. Although the HILIC LC-UV trace was complex and the betaine peak was not completely resolved, there were no apparent interferences seen for the betaine SIM trace at *m/z* 118. A linear betaine calibration curve was constructed over the 0.02 to 1 mg $mL^{-1}$ range and used to demonstrate an LOD ($S/N = 3$) of 0.2 ng $mL^{-1}$ for ESI-MS (SIM) compared to 0.1 μg $mL^{-1}$ by LC-UV. The HILIC LC-MS method was validated and showed acceptable levels of precision and accuracy. Levels of betaine in *Lycium chinense* fruits were found to lie at between 0.15 and 0.21%. In summary, the HILIC LC-MS (SIM) method described in this work is a simple and effective way of measuring the high betaine levels present in Fructus Lycii but may not be sufficiently specific for use in other cases where much lower concentrations of betaine are present.

A later publication by Shin *et al.* in 2012 also described a quantitative assay for betaine in Fructus Lycii but made use of evaporative light-scattering detection (ELSD) instead of MS. Extraction was simply carried out by sonication in 70% methanol (25 °C, 40 min) followed by centrifugation, filtration and the addition of a solution of the internal standard stachydrine in methanol. Three HILIC columns were tested including a Phenomenex Kinetex HILIC (2.1 mm × 100 mm, 2.6 μm), a Waters Atlantis HILIC (3.0 mm × 100 mm, 5 μm) and a HALO HILIC (2.1 mm × 100 mm, 2.7 μm). The optimal separation was achieved using the Phenomenex Kinetex HILIC column with a gradient elution of (A) ACN and (B) 10 mM ammonium formate (pH 3.0)/ACN (90/10, v/v). The gradient used was 15% B at 0–1 min, 25% B at 16 min and 40% B at 17 min. It is apparent from the chromatograms in the paper that a final hold to 35 min was used, although this was not specified in the experimental section of the paper. The ELSD was operated at an optimized temperature and nebulizing gas pressure of 50 °C and 3.0 bar. The method LOQ ($S/N = 10$) and LOD

($S/N = 3$) were 10 and 3 μg mL$^{-1}$, respectively. Hence, the HILIC LC-ELSD method is less sensitive than the HILIC LC-MS (SIM) method described by Shin *et al.* (1999), but this is unimportant due to the high content of betaine in Fructus Lycii samples. Replacement of the more sophisticated and specific MS method by ELSD detection has the advantages of lower cost and simplicity. The HILIC LC-ELSD method achieved recoveries of greater than 96.5%, intra- and inter-day precision of less than 5.3% ($n = 5$) and accuracy in the range of 94–107%. The validated HILIC LC-ELSD method was successfully applied to 14 Fructus Lycii samples (one originated from *L. barbarum* and 13 from *L. chinense*) that contained betaine at weight percentages of between ~0.5–1.0%.

The most recent HILIC-based method for betaine quantification in Fructus Lycii samples was published by Zhao *et al.* in 2013. In this case betaine was extracted directly into water containing the β-alanine internal standard. Water was said to give higher extraction amounts compared to 70% methanol or 50 or 70% ethanol, with 60 min sonication at room temperature in each case. Extraction into water was followed by filtration prior to analysis by HILIC LC-ELSD. Two HILIC columns were tested: Atlantis silica HILIC column 100 Å (4.6 mm × 150 mm, 5 μm), and Phenomenex Luna HILIC column 200 Å (4.6 mm × 150 mm, 5 μm). The elution was carried out under isocratic conditions with a mobile phase composition of 30 mM ammonium acetate buffer and ACN (20 : 80, v/v), at a flow rate of 1.0 mL min$^{-1}$ for 30 min. For this particular separation, the Atlantis HILIC column was selected since it was shown to have slightly better performance, including higher plate count and capacity factor. Using this simple isocratic HILIC LC-ELSD method an LOD ($S/N = 3$) and LOQ ($S/N = 10$) of 2 and 7 μg mL$^{-1}$ were obtained, similar to the earlier study by Shin *et al.* (2012). However, this study had the advantage of an isocratic separation, thereby avoiding the need for column re-equilibration. Using this method, 92 Fructus Lycii samples (29 from *L. barbarum* L. and 63 from *L. chinense* Miller) with betaine weight percentages of ~0.6–0.8% were quantitatively analyzed. This simple and stable method was suggested to be used as an official analytical method for Fructus Lycii in the Korean Pharmacopoeia (K.P.) and the results used in specifying the content regulation of betaine for Lycii Fructus.

### 10.2.3 Betaine in Samples of Animal Origin

Betaine is present in a wide variety of animal tissues and fluids including plasma, urine and liver. Betaine is produced by the enzymatic oxidation of choline and acts as a methyl donor in the homocysteine–methionine reaction in the presence of betaine-homocysteine methyltransferase (BHMT). The quantification of betaine in tissues and other biological fluids has great significance in clinical studies. This requires a sensitive and specific method due to the low content of betaine in these highly complex matrices.

A method based on LC-MS/MS for the combined determination of free choline, betaine and dimethylglycine (DMG) in human EDTA plasma and serum was proposed by Holm *et al.* in 2003. The LC separation in this work was conducted on a normal-phase Hypersil silica column (100 mm × 2.1 mm, 5 μm).

The target analytes were eluted from the column at a flow rate of 0.6 mL $min^{-1}$ with the mobile phase consisting of 15 mM ammonium formate (pH 3.5) and ACN, in gradient. Thus, the starting conditions (25% aqueous) could be considered as being HILIC conditions although the rapid gradient to high aqueous content indicates normal phase separation. This allows for a short run time but with little or no separation between betaine and DMG. It is probable that a superior separation could be achieved purely in HILIC mode but at the expense of run time. MS detection was performed on a triple-quadrupole tandem mass spectrometer with an ESI interface operated in positive-ion mode. Human plasma and serum samples were treated with three volumes of ACN containing the stable isotopically labeled internal standards ($d_9$-choline and $d_9$-betaine), to precipitate protein. The samples were centrifuged and the supernatant was then directly used for analysis by LC-MS/MS. Betaine was quantified by multiple reaction monitoring (MRM) using the transition *m/z* 118 ($[M + H]^+$)→*m/z* 59 ($[M + H\text{–}CH_2COOH]^+$) and the corresponding transition for the $d_9$ labeled internal standard (*i.e. m/z* 127 → *m/z* 68). Linearity was demonstrated by standard additions to dialyzed plasma over the range of concentrations 0.01–400 μmol $L^{-1}$ with an LOD for betaine of 0.3 μmol $L^{-1}$ ($S/N = 5$). Similar results were obtained for the other two analytes. The method achieved recoveries and precision for betaine of 93–103% and <6%, respectively. This was claimed to be the first published method enabling the simultaneous analysis of three target compounds from a single preparation procedure. The method is suitable for large-scale clinical and epidemiologic studies due to the short analysis time of 6 min (including equilibration) and the simple extraction procedure that is amenable to robotic automation. The method was used to determine the choline, betaine and DMG levels in plasma and serum of 60 fasting and nonfasting donors. The betaine levels in the combined groups were found to be in the range of 27–47 μmol $L^{-1}$ in plasma and 28–48 μmol $L^{-1}$ in serum; thus plasma and serum betaine levels did not differ significantly.

To better understand choline metabolism, Kirsch *et al.* (2010) described a stable-isotope dilution ultraperformance liquid chromatography-mass spectrometry (UPLC-MS) method for the determination of choline and related metabolites including betaine, DMG and acetylcholine (ACh). This allowed the presence of these compounds in EDTA plasma, urine, amniotic fluid (AF), and cerebrospinal fluid (CSF) to be measured. Sample pretreatment included protein precipitation by ACN for all samples proceeded by a 5-fold dilution with water in the case of urine samples due to the presence of a high concentration of biological salts. The chromatographic separation used an Acquity UPLC BEH HILIC column (100 mm × 2.1 mm I.D., 1.7 μm) at 30 °C. A run time of only 6.5 min was used with a gradient of between 12.5% and 20% 15 mmol $L^{-1}$ ammonium formate (pH 3.5) and the remainder ACN. Similar to the earlier method (Holm *et al.*, 2003), positive-ion ESI-MS/MS in MRM mode was used for the quantification of the four target compounds and their corresponding stable-isotope labeled internal standards ($d_9$-ACh, $d_9$-choline, $d_6$-DMG and $d_9$-betaine). The LODs in this study ranged from 0.11 to 0.34 μmol $L^{-1}$, which was similar to that obtained by Holm *et al.* (2003), but the range of tested materials and analytes was extended. Analysis of biological fluids in volumes of <100 μL was demonstrated, suggesting that the method is feasible both when sample size is a limitation (*e.g.* CSF) and for high-throughput analysis in clinical studies.

In 2010, a closely related UPLC-MS/MS method was reported by Bruce *et al.* for the determination of betaine and free choline in plasma as well as in cereal and cereal-related products that are discussed in Section 10.2.4. In this work, the sample-preparation procedure was optimized based on previously reported methods (Koc *et al.*, 2002; Holm *et al.*, 2003; De Zwart *et al.*, 2003). For plasma samples, this involved protein precipitation and extraction into methanol (in place of ACN in other methods) followed by centrifugation. LC separation used a UHPLC HILIC column (2.1 mm × 150 mm, 1.7 μm) with a mobile phase gradient of 10 mM ammonium acetate in water containing 0.005% acetic acid and ACN. However, since the gradient used was between 20% and 100% aqueous, this method is not strictly in HILIC mode. In fact, the LC eluent only flowed to the mass spectrometer at between 1 and 3.65 min at which points in time the aqueous content of the mobile phase was ~50% and 100%, respectively. Nevertheless, the published chromatograms do show separation between the choline and betaine analytes that each eluted at around 2 min. The triple-quadrupole MS/MS system was operated in positive ESI in the MRM mode, monitoring the two transitions for each analyte (*m/z* 118 → 58, 59 for betaine and *m/z* 104 → 60, 58 for choline). Quantitative results were achieved by relying on the signal ratios between the analytes in the sample and the corresponding isotope-labeled standards ($d_{11}$-betaine and $d_9$-choline). This method was applied to the plasma of 47 subjects who were found to have average plasma levels of betaine of 40 μmol $L^{-1}$.

### 10.2.4 Betaine in Food Samples

Interest in the analysis of betaine in foods has been increased considerably in the past few years because of the potential health benefits associated with betaine intake. Many types of food, such as eggs, milk, cheese, whole grains and yogurt, are known to be important sources of betaine (Patterson *et al.*, 2008) and in addition, betaine is produced *via* enzymatic oxidation of choline. A dietary deficiency of betaine might result in an increased risk of diseases such as fatty liver disease and bone abnormalities (Zeisel *et al.*, 2003). Hence, the measurement of betaine and related compounds in food is important in order to track dietary intake.

In 1981, Vialle and Kolosky published a paper in which betaine in commercial refined sugar and wine was measured by refractive index detection using amino-bonded silica as the stationary phase and a mobile phase of ACN and water (75 : 25, v/v), as is customary for sugar analysis. In this way, a mixture of betaine and several sugars was separated with betaine eluting first in the chromatogram ahead of the monosaccharides fructose and glucose. Before chromatographic separation, sample preparation was performed using two ion-exchange columns. Acidified solutions of commercial refined sugar were passed through a cation-exchange resin to concentrate betaine, which was subsequently eluted by an ammonia solution. Wine samples were additionally passed through a second column (a mixed bed of cation- and anion-exchange resins) for further clean-up. Using quantification against an external standard calibration, it was found that the betaine content of refined beet sugar was in the range of 30–37 mg $kg^{-1}$. Wine samples were found to contain 0.8–5.7 mg betaine per liter, which may reflect the level of addition of beet sugar to the wine. Although this paper predates the first description of HILIC, its use of a hydrophilic stationary phase and a mobile phase of 3–30% water plus ACN, indicates that the chromatography was operating in HILIC mode. Thus, amongst all of the papers cited in this chapter, this was the earliest one concerning the application of HILIC to betaine analysis.

Betaine and free choline measurement in a broad range of cereal flours and cereal food products was reported by Bruce *et al.* (2010) using the UPLC-MS/MS method employing a HILIC column, as described above in Section 10.2.3. The range of betaine concentration in the 8 white wheat flours analyzed was 166–326 μg $g^{-1}$, while for 6 whole-grain wheat flours it was 747–1503 μg $g^{-1}$. In contrast, corn and rice flours contained barely measurable amounts, while whole grain oats and barley contained amounts in the range of 114–760 μg $g^{-1}$. Some correlation was found between betaine and free choline levels in cereals.

There have been many reports of the use of HILIC chromatography in phospholipid class separations. Zhao *et al.* in 2011 reported on an LC-MS/MS experiment using multiple scan modes to simultaneously identify and quantify 11 choline-related compounds – acetylcholine (AcCho), betaine (Bet), choline (Cho), glycerophosphocholine (GPC), lysophosphatidylcholine (LPC), lysophosphatidylethanolamine (LPE), phosphatidylcholine (PC), phosphatidylethanolamine (PE), phosphatidylinositol (PI), phosphocholine (PCho) and sphingomyelin (SM). This work used a modification to the Bligh and Dyer method of lipid extraction to ensure collection of the more polar components. In this paper, HILIC separation with an Ascentis Express HILIC column (150 mm × 2.1 mm I.D., 2.7 μm) was compared with normal-phase HPLC separation with an Ascentis silica column (150 mm × 2.1 mm I.D., 3 μm). An important difference between the 2 chromatographic modes is the simplicity of the HILIC method that used an ACN/water (10 mM ammonium formate, pH 3) gradient compared to the normal phase method that used a gradient of two complicated solvent mixtures, each containing ACN/water/ethanol/1 M ammonium acetate/glacial acetic acid in different proportions. It was found that the HILIC mode generally resulted in better peak shapes and higher ionization efficiency in LC-MS, along with a shorter run time and better sensitivity. This applied to both phospholipids and to betaine analysis, where the $m/z$ 118 → $m/z$ 58 MRM transition was monitored and choline $d_9$ was used as the internal standard. Although the focus of this work was to quantify total choline in foods from all sources (*i.e.* free choline and bound), it does demonstrate how in a single HILIC LC run, betaine can be measured alongside many other diverse species.

Later, this work was extended to include 14 compounds or compounds classes comprised of 6 aqueous choline-related compounds (Cho, Bet,

AcCho, PCho, GPC, CDP-Cho) and 8 phospholipid classes (PC, LPC, PE, LPE, SM, PI, PS and PG) (Xiong *et al.*, 2012). For betaine, a working linear dynamic range of 0.05–5 μg $mL^{-1}$, an LOD of 0.02 μg $mL^{-1}$ ($S/N > 3$) and an LOQ of 0.05 μg $mL^{-1}$ ($S/N > 10$) were established. However, it should be noted that these values could be improved on in a method that focused only on the analysis of compounds similar to betaine, rather than one developed to quantify a wide range of phospholipids and choline-related compounds in a single run. The method recoveries were excellent (99–102%) for betaine spiked into food at 20–40 mg/100 g levels. Intraday triplicate measurement precision for betaine quantified in standards and egg yolk was <10%, which is adequate for food and tissue analysis. Thus, betaine analysis as part of a wider nutritional analysis that measures phospholipids and other choline-related metabolites can be achieved by HILIC LC-MS/MS methods.

## 10.3 Conclusions and Future Perspectives

The HILIC method is becoming increasingly popular for the analysis of polar, hydrophilic and ionic compounds. In this chapter, a number of HILIC-based methods that include betaine determination are summarized. In general, these methods were developed to analyze a range of related or unrelated compounds in addition to betaine. The methods discussed employ the HILIC technique in conjunction with several types of detectors including UV, ELSD, refractive index, MS or MS/MS for betaine analysis. Some key developments in the use of the HILIC method in betaine characterization include the simplification of sample preparation procedures, the optimization of mobile-phase composition and selection of the HILIC column from the continually increased number of commercially available columns. Since HILIC chromatography can involve multimodal interactions, the selection of the stationary phase to achieve an optimal separation may vary depending on the nature of the analytes and sample matrix.

Of the ten papers reviewed in this chapter that describe the use of HILIC methods for betaine analyses, six of them were published in the past five years. Hence, this is still an emerging area of study and it is expected that improvements in sensitivity and selectivity will be forthcoming based on newer HILIC phases and more targeted analyses. This would be desirable, especially for biomedical type analyses where the use of limited sample amounts and betaine concentrations are important.

## Summary Points

- This chapter summarizes published work on the use of hydrophilic interaction liquid chromatography (HILIC) for betaine analysis.
- The separation mechanism in HILIC is described as being multimodal, involving the partition of analytes into a water-rich layer around stationary phase particles plus other surface interactions.

- ESI-MS and ELSD are the two most commonly used detection techniques for betaine analysis in HILIC mode.
- Quantification of betaine by HILIC methods is described for food products and other samples of plant and animal origin.
- HILIC-based betaine analysis has been achieved following simple extractions into aqueous acetonitrile (ACN) or methanol.
- 10 papers related to the application of HILIC for betaine determination are discussed, 6 of which were published in the last 5 years.

## Key Facts of Betaine Analysis by HILIC

- Betaine is ubiquitous – it is present in biological fluids and tissues, food, plants and animals.
- Betaine is a small water soluble molecule that carries both positive and negative charges at neutral pH.
- In the past, betaine has been separated from other compounds and quantified by many chromatographic techniques such as normal- or reversed-phase liquid chromatography, gas chromatography, ion chromatography and capillary electrophoresis.
- HILIC chromatography is used for the analysis of highly polar and hydrophilic compounds.
- When ESI-MS/MS is combined with HILIC chromatography betaine can be measured with high sensitivity in the presence of other compounds.
- Betaine levels in samples including biological fluids and tissues, cereals, food and natural health products have been measured using HILIC LC-MS or HILIC LC-ELSD.

## Definitions of Words and Terms

**Hydrophilic interaction liquid chromatography (HILIC)**: A chromatographic technique using a hydrophilic stationary phase and a binary mobile phase system of water, sometimes containing salts or weak acids, with higher-content of organic solvent, usually 70–97% acetonitrile. Used for the analysis of highly polar, hydrophilic and ionic compounds.

**Normal-phase liquid chromatography (NPLC)**: An LC method usually characterized by a polar stationary phase and a nonpolar mobile phase. The separation of the components in a mixture in NPLC mode is based on the specific interaction between the analyte and the stationary phase.

**Reversed-phase liquid chromatography (RPLC)**: Refers to a separation technique that employs a nonpolar stationary phase and an aqueous polar mobile phase. The analytes under RPLC conditions are largely separated according to their hydrophobic properties.

**Mobile phase**: Mobile phase is the term used in the chromatography to describe either a liquid or a gas that functions as a carrier to move the analytes through the stationary phase.

**Stationary phase**: In chromatography, the analytes partition between the flowing mobile phase and a stationary phase, which is a solid or a liquid that is fixed, for example within an LC column, on the walls of a GC column or on the surface of a TLC plate.

**Electrospray ionization (ESI)**: An atmospheric pressure ionization technique used in mass spectrometry (MS) that results in the formation of molecular ions with little fragmentation. In ESI, the analyte solution is nebulized into a high-potential electric field where the droplets are desolvated resulting in ion formation.

**Multiple reaction monitoring (MRM)**: A scan mode in tandem MS with high sensitivity and specificity used for quantification. In this mode, the fragmentation of a single precursor ion to a single fragment ion (*e.g.* $M_p^+ \rightarrow M_f^+$) is monitored.

**Liquid chromatography-tandem mass spectrometry (LC-MS/MS)**: Liquid chromatography directly coupled to tandem mass spectrometry. This combines the LC separation of components from a mixture with the use of MS/MS techniques that provide structural information and quantification.

**Limit of detection (LOD)**: Indicates the lowest quantity of an analyte that can be detected. The LOD value is often defined by the signal-to-noise ($S/N$) ratio at which the height of the analyte peak is ~3 times higher than the baseline noise ($S/N \geq 3$).

**Limit of quantification (LOQ)**: Indicates the lowest amount of an analyte that can be quantified with acceptable precision using the test analytical method. The LOQ value is often defined by the signal-to-noise ($S/N$) ratio at which the height of the analyte peak is > 10 times higher than the baseline noise level ($S/N \geq 10$).

## List of Abbreviations

| | |
|---|---|
| AcCho | Acetylcholine |
| ACN | Acetonitrile |
| AF | Amniotic fluid |
| Bet | Betaine |
| BHMT | Betaine-homocysteine methyltransferase |
| CDP-Cho | Cytidine diphosphocholine |
| CE | Capillary electrophoresis |
| Cho | Choline |
| CSF | Cerebrospinal fluid |
| DMG | Dimethylglycine |
| EDTA | Ethylenediaminetetraacetic acid |
| ELSD | Evaporative light-scattering detection |
| ESI | Electrospray ionization |
| ESI-MS | Electrospray ionization-mass spectrometry |
| GC | Gas chromatography |
| GPC | Glycerophosphocholine |
| HILIC | Hydrophilic interaction liquid chromatography |

| | |
|---|---|
| I.D. | Internal diameter |
| IC | Ion chromatography |
| IPA | Isopropanol |
| LC | Liquid chromatography |
| LC-ELSD | Liquid chromatography-evaporative light scattering detection |
| LC-MS | Liquid chromatography-mass spectrometry |
| LC-MS/MS | High-performance liquid chromatography-tandem mass spectrometry |
| LC-UV | Liquid chromatography-ultraviolet detection |
| LOD | Limit of detection |
| LOQ | Limit of quantification |
| LPC | Lysophosphatidylcholine |
| LPE | Lysophosphatidylethanolamine |
| MRM | Multiple reaction monitoring |
| MS | Mass spectrometry |
| MS/MS | Tandem mass spectrometry |
| NPLC | Normal-phase liquid chromatography |
| PC | Phosphatidylcholine |
| PCho | Phosphocholine |
| PE | Phosphatidylethanolamine |
| PFPA | Perfluoropentanoic acid |
| PG | Phophatidylserine |
| PI | Phosphatidylinositol |
| PS | Phosphatidylglycerol |
| RPLC | Reversed-phased liquid chromatography |
| RSD | Relative standard deviation |
| *S/N* | Signal-to-noise ratio |
| SIM | Selected ion monitoring |
| SM | Sphingomyelin |
| THF | Tetrahydrofunan |
| TMA | Trimethylamine |
| UPLC-MS | Ultraperformance liquid chromatography-mass spectrometry |
| UV | Ultraviolet |

# References

Allen, R. H., Stabler, S. P. and Lindenbaum, J., 1993. Serum betaine, *N,N*-dimethylglycine and *N*-methylglycine levels in patients with cobalamin and folate deficiency and related inborn errors of metabolism. *Metabolism: Clinical and Experimental*. 11: 1448–1460.

Alpert, A. J., 1990. Hydrophilic-interaction chromatography for the separation of peptides, nucleic acids and other polar compounds. *Journal of Chromatography A*. 0: 177–196.

Bernal, J., Ares, A. M., Pól, J. and Wiedmer, S. K., 2011. Hydrophilic interaction liquid chromatography in food analysis. *Journal of Chromatography A*. 42: 7438–7452.

Bruce, S. J., Guy, P. A., Rezzi, S. and Ross, A. B., 2010. Quantitative measurement of betaine and free choline in plasma, cereals and cereal products by isotope dilution LC-MS/MS. *Journal of Agricultural and Food Chemistry*. 4: 2055–2061.

Cavazzini, A. and Felinger, A., 2013. *Liquid Chromatography: Fundamentals and Instrumentation*. Elsevier Science Publishers, Amsterdam, Netherlands, 520 pp.

Chendrimada, T. P., Neto, M. G., Pesti, G. M., Davis, A. J. and Bakalli, R. I., 2002. Determination of the betaine content of feed ingredients using high-performance liquid chromatography. *Journal of the Science of Food and Agriculture*. 13: 1556–1563.

Dejaegher, B., Mangelings, D. and Vander Heyden, Y., 2008. Method development for HILIC assays. *Journal of Separation Science*. 9: 1438–1448.

Dejaegher, B. and Vander Heyden, Y., 2010. HILIC methods in pharmaceutical analysis. *Journal of Separation Science*. 6–7: 698–715.

De Zwart, F. J., Slow, S., Payne, R. J., Lever, M., George, P. M., Gerrard, J. A. and Chambers S. T., 2003. Glycine betaine and glycine betaine analogues in common foods. *Food Chemistry*. 2: 197–204.

Dunphy, R. and Burinsky, D. J., 2003. Detection of choline and acetylcholine in a pharmaceutical preparation using high-performance liquid chromatography/electrospray ionization mass spectrometry. *Journal of Pharmaceutical and Biomedical Analysis*. 5: 905–915.

Gama, M. R., da Costa Silva, R. G., Collins, C. H. and Bottoli, C. B. G., 2012. Hydrophilic interaction chromatography. *Trends in Analytical Chemistry*. 37: 48–60.

Guo, Y., 2005. Analysis of quaternary amine compounds by hydrophilic interaction chromatography/mass spectrometry HILIC/MS. *Journal of Liquid Chromatography and Related Technologies*. 4: 497–512.

Guo, Y. and Gaiki, S., 2005. Retention behavior of small polar compounds on polar stationary phases in hydrophilic interaction chromatography. *Journal of Chromatography A*. 1–2: 71–80.

Guo, Y. and Gaiki, S., 2011. Retention and selectivity of stationary phases for hydrophilic interaction chromatography. *Journal of Chromatography A*. 35: 5920–5938.

Hao, Z., Xiao, B. and Weng, N., 2008. Impact of column temperature and mobile phase components on selectivity of hydrophilic interaction chromatography HILIC. *Journal of Separation Science*. 9: 1449–1464.

Hemström, P. and Irgum, K., 2006. Hydrophilic interaction chromatography. *Journal of Separation Science*. 12: 1784–1821.

Hitz, W. D. and Hanson, A. D., 1980. Determination of glycine betaine by pyrolysis-gas chromatography in cereals and grasses. *Phytochemistry*. 11: 2371–2374.

Holm, P. I., Ueland, P. M., Kvalheim, G. and Lien, E. A., 2003. Determination of choline, betaine, and dimethylglycine in plasma by a high-throughput method based on normal-phase chromatography-tandem mass spectrometry. *Clinical Chemistry*. 2: 286–294.

Jian, W., Edom, R. W., Xu, Y. and Weng, N., 2010. Recent advances in application of hydrophilic interaction chromatography for quantitative bioanalysis. *Journal of Separation Science*. 6–7: 681–697.

Kahsay, G., Song, H., Van Schepdael, A., Cabooter, D. and Adams, E., 2013. Hydrophilic interaction chromatography HILIC in the analysis of antibiotics. *Journal of Pharmaceutical and Biomedical Analysis*. (in press).

Kirsch, S. H., Herrmann, W., Rabagny, Y. and Obeid, R., 2010. Quantification of acetylcholine, choline, betaine, and dimethylglycine in human plasma and urine using stable-isotope dilution ultra performance liquid chromatography-tandem mass spectrometry. *Journal of Chromatography B: Analytical Technologies in the Biomedical and Life Sciences*. 32: 3338–3344.

Koc, H., Mar, M. H., Ranasinghe, A., Swenberg, J. A. and Zeisel, S. H., 2002. Quantitation of choline and its metabolites in tissues and foods by liquid chromatography/electrospray ionization-isotope dilution mass spectrometry. *Analytical Chemistry*. 18: 4734–4740.

Lee, S. M., Park, C. K., Cho, B. G., Cho, K. S., Min, B. S. and Bae, K. H., 2011. A convenient HPLC/ELSD method for the quantitative analysis of betaine in lycium chinense. *Natural Product Sciences*. 2: 104–107.

Lehnert, P., Douša, M. and Lemr, K., 2013. Underivatized amylose and cellulose as new stationary phases for hydrophilic interaction chromatography. *Journal of Separation Science*. 20: 3345–3350.

Li, R. and Huang, J., 2004. Chromatographic behavior of epirubicin and its analogues on high-purity silica in hydrophilic interaction chromatography. *Journal of Chromatography A*. 1–2: 163–169.

McCalley, D. V., 2007. Is hydrophilic interaction chromatography with silica columns a viable alternative to reversed-phase liquid chromatography for the analysis of ionisable compounds? *Journal of Chromatography A*. 1–2: 46–55.

Orth, P. and Engelhardt, H., 1982. Separation of sugars on chemically modified silica gel. *Chromatographia*. 2: 91–96.

Patterson, K. Y., Bhagwat, S. A., Williams, J. R., Howe, J. C. and Holden, J. M., 2008. *USDA Database for the Choline Content of Common Foods*. Available at: http://www.ars.usda.gov/SP2UserFiles/Place/12354500/Data/Choline/Choln02.pdf. Accessed 23 December 2013.

Sánchez-Hernández, L., Castro-Puyana, M., Luisa Marina, M. and Crego, A. L., 2011. Determination of betaines in vegetable oils by capillary electrophoresis tandem mass spectrometry - application to the detection of olive oil adulteration with seed oils. *Electrophoresis*. 11: 1394–1401.

Shin, H. D., Suh, J. H., Kim, J., Lee, H., Eom, H. Y., Kim, U., Yang, D. H., Han, S. B. and Youm, J. R., 2012. Determination of betaine in fructus lycii using hydrophilic interaction liquid chromatography with evaporative light scattering detection. *Bulletin of the Korean Chemical Society*. 2: 553–558.

Shin, Y. G., Cho, K. H., Kim, J. M., Park, M. K. and Park, J. H., 1999. Determination of betaine in lycium chinense fruits by liquid chromatography-electrospray ionization mass spectrometry. *Journal of Chromatography A*. 1–2: 331–335.

Vialle, J., Kolosky, M. and Rocca, J. L., 1981. Determination of betaine in sugar and wine by liquid chromatography. *Journal of Chromatography A*. 429–435.

Xiong, Y. P., Zhao, Y. Y., Goruk, S., Oilund, K., Field, C. J., Jacobs, R. L. and Curtis J. M., 2012. Validation of an LC-MS/MS method for the quantification of choline-related compounds and phospholipids in foods and tissues. *Journal of Chromatography B*. 911: 170–179.

Yoshida, T., 2004. Peptide separation by hydrophilic-interaction chromatography: a review. *Journal of Biochemical and Biophysical Methods*. 3: 265–280.

Zeisel, S. H., Mar, M. H., Howe, J. C. and Holden, J. M., 2003. Concentrations of choline-containing compounds and betaine in common foods. *Journal of Nutrition*. 5: 1302–1307.

Zhang, J., Nishimura, N., Okubo, A. and Yamazaki, S., 2002. Development of an analytical method for the determination of betaines in higher plants by capillary electrophoresis at low pH. *Phytochemical Analysis*. 13: 189–194.

Zhao, B. T., Jeong, S. Y., Hwangbo, K., Moon, D. C., Seo, E. K., Lee, D., Lee, J. H., Min, B. S., Ma, E. S., Son, J. K. and Woo, M. H., 2013. Quantitative analysis of betaine in lycii fructus by HILIC-ELSD. *Archives of Pharmacal Research*. 10: 1231–1237.

Zhao, Y. Y., Xiong, Y. P. and Curtis, J. M., 2011. Measurement of phospholipids by hydrophilic interaction liquid chromatography coupled to tandem mass spectrometry: The determination of choline containing compounds in foods. *Journal of Chromatography A*. 32: 5470–5479.

CHAPTER 11

# *Characterization of Betaine from Seeds*

PRITI PATIL[a] AND PRATIMA TATKE*[a]

[a]Department of Pharmaceutical Chemistry, C. U. Shah College of Pharmacy, S.N.D. T. Women's University, Juhu Campus, Juhu Road, Santacruz(W), Mumbai-400 049, India
*E-mail: pratima.tatke@cushahpharmacy.sndt.ac.in

## 11.1 Introduction

Betaine is an alkaloid that is found in micro-organisms, plants, and animals. It is present in significant amounts in foods such as spinach, sugar beets wheat, and shellfish. Betaine is a zwitterionic quaternary ammonium compound that is also known as trimethylglycine, glycine betaine, lycine, and oxyneurine. It is a methyl derivative of the amino acid glycine with a formula of $(CH3)_3N^+CH_2COO^-$ and a molecular weight of 117.2, and it has been characterized as a methylamine because of its 3 chemically reactive methyl groups.

Betaine has been reported to be present in seeds of many medicinal plants (Table 11.1). Betaine was isolated as hydrochloride salt from Guayule shrub by Murray and Walter (1945). Betaine was isolated from the epigeal part of *Chenopodium botrys* L. (Jerusalem oak) using extraction with 10% $NH_4OH$ and methanol followed by column chromatography using alumina as stationary phase and methanol as eluant. Isolated betaine was recrystallized using *n*-butanol (Rustembekova *et al.*, 1973). It was also isolated along with

Food and Nutritional Components in Focus No. 7
Betaine: Chemistry, Analysis, Function and Effects
Edited by Victor R Preedy

Published by the Royal Society of Chemistry, www.rsc.org

choline in anthers and wheat germ (Strange *et al.*, 1974). Elyakov *et al.* isolated betaine from brown algae using extraction with ethanol. Ethanol extract was treated with water and the aqueous portion was chromatographed on silica column using ethanol as eluent. Fractions containing alkaloid was rechromatographed on silica using 50% alcohol as eluent to get betaine (Chumak *et al.*, 1978). It was also isolated from aerial parts of *Lycium barbarum* Linn. The aerial parts were extracted in methanol. The methanol extract was partitioned between water and ethyl acetate. The aqueous layer was passed through cation exchanger. The column was then eluted with 2 N $NH_4OH$ and eluent was evaporated, taken in methanol and filtered. The residue was further purified using preparative TLC using methanol as solvent. The colorless crystalline material thus obtained was converted to hydrochloride and recrystallized using methanol (Ahmad and Sultan, 1980).

**Table 11.1** List of medicinal plants containing betaine. Betaine derivative/s have been reported to be present in the seeds of many medicinal plants.

| Sr. No | Plant | Family | Part of plant | Medicinal uses | Active constituents |
|---|---|---|---|---|---|
| 1 | *Achyranthes aspera* Linn | Amaranthaceae | Seeds | Astringent, ashes of the plant used in asthma, diuretic, hepatoprotective, emmenagogue | Achyranthine, betaine (50 mg per 100 g) |
| 2 | *Amaranthus tricolor* Linn. | Amaranthaceae | Seeds | Astringent, in menorrhagia, leucorrhoea, dysentery, diarrhoea, haemorrhagic colitis | Amarantin, isoamarantin, betaine (80–100 mg per 100 g) |
| 3 | *Beta vulgaris* Linn. | Chenopodiacae | Seeds | In burns and bruises, for diseases of spleen and liver. expectorant. diuretic. anti-inflammatory. analgesic. | Betaine (250 mg per 100 g) |
| 4 | *Trigonella corniculata* Linn(Mehrafarin,*et al.*, 2010) | Subfamily: Papilionaceae; Family: Fabaceae. | Seeds | Applied to swellings and bruises. | Pyridine alkaloids, mainly trigonelline, (a methyl betaine derivative, 0.2–0.36%), choline, gentianine, carpaine, the flavonoids apigenin, luteolin, orientin, vitexin quercetin, *etc.* |

Glycine-betaine was isolated from human urine by precipitation using ammonium reineckate at acidic pH followed by gel filtration and high-pressure liquid chromatography (Chambers and Kunnin, 1987). Betaine was estimated in sugar and wine using an amino-bonded silica gel column (150 mm × 4.7 mm, 10 μm) as stationary phase and acetonitrile: water (75:25, v/v) as mobile phase. The flow rate of mobile phase was 1 mL $min^{-1}$ (Vialle and Kolosky, 1981). Gorham separated betaine and their sulfur analogs by HPLC using strong cation exchange column (Partisil l0-SCX, 250 mm × 5 mm) as stationary phase and 50 mM of $KH_2PO_4$ buffer containing 5% methanol (pH 4.6) as mobile phase pumped at 1.5 mL $min^{-1}$. Detection was carried out at 190–200 nm. Betaine eluted at 5.47 min with total run time of 10 min (Gorham, 1984). Wurtman developed a normal-phase HPLC method for separation of choline metabolites using silica column, Rainin Microsorb (250 mm × 4.6 mm, 5 μm) and gradient elution with buffer. The flow rate and temperature were kept at 2.7 mL $min^{-1}$ and 45 °C, respectively. The retention time of betaine was 10.1 min with total run time of 60 min (Liscovitch *et al.*, 1985).

HPLC-radioenzymatic assay was used to determine betaine aldehyde and betaine in rat liver mitochondria using a silica column. Chromatographic elution was carried out using a buffer containing 800 mL acetonitrile, 68 mL ethanol, 5 mL of 1 M ammonium acetate: glacial acetic acid buffer (3 : 2, v/v), 127 mL water and 10 mL of 0.1 M $KH_2PO_4$. The flow rate of mobile phase was 1.5 mL $min^{-1}$ (Zhang *et al.*, 1992). The betaine content was estimated in oriental medicine, Jikoppi, which is powdered root-skin of *Lycium chinense* using Kaseisorb LC $NH_2$ super column (250 mm × 4.6 mm, 5 μm) and mobile phase containing acetonitrile: water (75 : 25, v/v) containing 0.1% of Brij 35. The flow rate of mobile phase and the detection wavelength were kept at 1 mL $min^{-1}$ and 210 nm, respectively (Kikuchi *et al.*, 1993). Betaine was determined in biological samples by derivatizing using 4′-bromo-phenacyl triflate and quantitated by UV (Mar *et al.*, 1995). Betaine and *N,N*-dimethylglycine in human urine were determined in urine by HPLC using column Supelcosil LC-SCX (250 mm × 4.6 cm, 5 μm) and eluent $KH_2PO_4$ (1 $mML^{-1}$) and HCl (0.12 $mML^{-1}$). The flow rate was 1 mL $min^{-1}$ and detection was carried out at 195 nm. The limit of detection (LOD) was found to be 0.9 μg $mL^{-1}$ The percentage recovery of the betaine was greater than 85% (Laryea *et al.*, 1994). Betaine and *N,N*-dimethylglycine were analyzed in blood and urine by precolumn derivatization using *p*-bromophenacyl bromide and crown ether as catalyst. Column used was Supelcosil LC-SCX (250 mm × 4.6 cm, 5 μm) and mobile phase used was acetonitrile: water (9 : 1, v/v) containing 22 $mML^{-1}$ choline. The flow rate was 1.5 mL $min^{-1}$ and the detection was carried out at 254 nm. Retention times of the phenacyl bromide esters of dimethyl glycine and betaine were 12.7 and 14.8 min, respectively. The LOD was found was to be 5 mM $L^{-1}$ (Laryea *et al.*, 1998). Chendrimada *et al.* determined betaine content in several feed ingredients like alfalfa, wheat, wheat middling, poultry meal using cation exchange column (Partisil SCX-10) and buffer containing 50 mM $KH_2PO_4$. Detected was carried out at 200 nm. The flow rate of mobile phase was kept at 1.5 mL $min^{-1}$. The method was found to be linear within a range of 15 to 650 μg $mL^{-1}$. The LOD was found to be 5 μg $mL^{-1}$ (Chendrimada *et al.*, 2002). Aryl derivatives of betaine were separated using different strong cation-exchange columns, normal-phase columns and reversed-phase columns. Detection was carried out at 249 nm and the flow rate of mobile phase was maintained at 1 mL min$^{-1}$. Column oven temperature was maintained at 40 °C (Storer *et al.*, 2006). Betaine was estimated in whole plant of *Achyranthes aspera* Linn using reversed phase HPLC using a Zorax column (250 mm × 4.6 mm, 5 μm) and mobile phase consisting of methanol: 50 mM $KH_2PO_4$ buffer pH 4.6 (50:50, v/v) at flow rate of 0.8 mL min$^{-1}$. Detection was carried out at 200 nm. Betaine eluted at 3.492 min. The method was linear in the range of 1–40 μg mL$^{-1}$. LOD and LOQ were 0.29 and 0.88 μg mL$^{-1}$, respectively (Khan *et al.*, 2011). It was determined in 70% methanolic extract of *Fructus lycii* using hydrophilic interaction liquid chromatography with evaporative light-scattering detection using Kinetex HILIC column (100 mm × 2.1 mm, 2.6 μm) and mobile phase consisting of acetonitrile and 10 mM ammonium formate (pH 3.0): acetonitrile (90:10, v/v) gradient. The flow rate of mobile phase was maintained at 0.7 mL min$^{-1}$ and the column was equilibrated set at 27.5 °C. Linearity was observed in the range of 10–250 μg mL$^{-1}$. LOD and LOQ was found to be 3 μg mL$^{-1}$ and 10 μg mL$^{-1}$, respectively. Betaine eluted at 15 min (Shin *et al.*, 2012). Betaine was analyzed in Chinese Lycium by HPLC coupled with an evaporative light-scattering detector. The column used was HILIC (250 mm × 4.6 mm, 5 μm) and mobile phase used was methanol: 10 mM ammonium acetate buffer pH 6 (75:25, v/v). The flow rate of mobile phase was kept at 1 mL min$^{-1}$ (Bullock, 2010).

However, a literature survey reveals that there is no adequate data available on its isolation and characterization from these various medicinal plants especially from seeds. The authors have isolated and characterized betaine from the seeds of *Achyranthus aspera* (Desai *et al.*, 2013). The chapter also includes isolation and characterization of betaine derivatives from two plant sources, Kancolla seeds and Cannabis seeds reported by other authors.

## 11.2 Isolation of Betaine from *Achyranthes aspera* Seeds

*Achyranthes aspera* Linn is an annual to perennial herb belonging to the family Amaranthaceae. Whole plant, leaves, seeds, roots, flowers and fruits are used for their various medicinal values. The plant is used as a diuretic, carminative, digestive, expectorant, bitter tonic, anti-inflammatory and a killer of intestinal worms. The plant contains alkaloids betaine and achyranthine. Betaine is present in considerable amount and can be used as marker for standardization of extracts and formulations containing *Achyranthes aspera.*

### 11.2.1 Procurement of Plant Material

The seeds of *Achyranthes aspera* were obtained from three geographical sources in the month of November 2009 from different parts of India as mentioned below:

S1: Seeds procured from Valsad district of Gujarat, India.
S2: Seeds procured from Virudunagar district of Tamilnadu, India.
S3: Seeds procured from a local market of Mumbai, Maharashtra, India.

The different herbaria of S1, S2 and S3 were deposited at the Botanical Survey of India, Pune under names AAGSOD4, AAGSOD5 and AAGSOD6, respectively.

### 11.2.2 Extraction of Alkaloids from *Achyranthes aspera*

20 g of powdered seeds were extracted with 200 mL of acidified ethanol (pH 4 was made by HCl) by heating at 70 °C for 5 h. It was then filtered. To this ammonia solution was added (25% Extrapure, specific gravity 0.91) to a pH of 10 to get the precipitate of the alkaloids. The precipitate of the crude alkaloid was filtered and dried (Figure 11.1).

### 11.2.3 Isolation of Alkaloids from Seeds of *Achyranthes aspera*

The crude alkaloid thus obtained was sonicated with 5 × 50 mL of distilled water for 20 min. The alkaloid being water soluble gets extracted in water. The aqueous layer showed some colored impurities that were removed using a charcoal column. The charcoal column was washed with 5 × 25 mL of

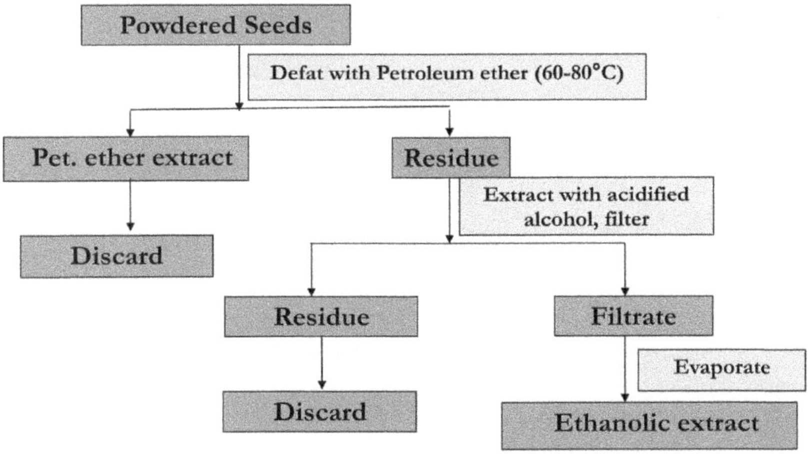


**Figure 11.1** Schematic representation of extraction of alkaloids from *Achyranthes aspera*. Pet. ether (petroleum ether), Powdered seeds of *Achyrathes aspera* were treated with Pet. Ether to remove fatty matter. The defatted seeds were extracted with acidified ethanol and filtered. The filtrate was evaporated to obtain ethanol extract. The figure appears in Desai *et al.* (2013).

distilled water. The clear water extracts were pooled and evaporated to dryness under reduced pressure to yield crude alkaloid.

### 11.2.4 Thin-Layer Chromatographic Studies

To check the purity of isolated alkaloid, TLC studies were carried out on precoated plates of silica gel $GF_{254}$ using a twin-trough chamber (10 × 10 cm), CAMAG Switzerland for development of TLC plates. A CAMAG TLC visualizer was used to visualize TLC plates. Isolated compound was dissolved in methanol and then spotted on TLC plates. Several mobile phases were tried and methanol: chloroform in the ratio of (7 : 3) was selected, which gave good resolution. Isolated compound was seen as a well-resolved orange band at an $R_f$ of 0.82 after spraying with Dragendorrf's reagent, indicating the presence of alkaloid.

### 11.2.5 Purification of Alkaloid

The crude alkaloid was purified by preparative TLC using methanol: chloroform (7 : 3) as mobile phase followed by repeated recrystallization with methanol (Figure 11.2).

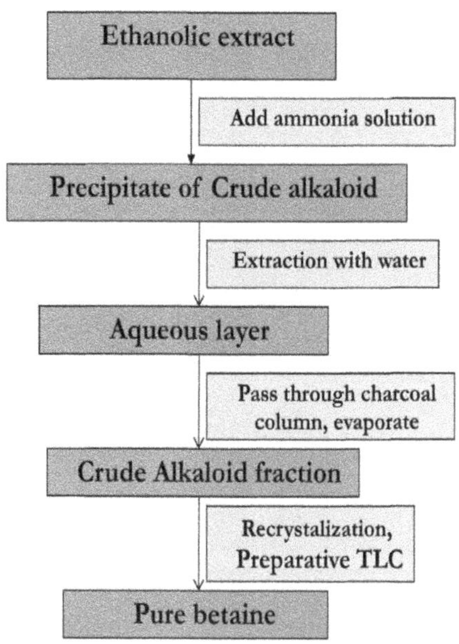


**Figure 11.2** Schematic representation of isolation and purification of betaine from *Achyranthes aspera*. The ethanol extract of seeds of *Achyrathes aspera* was treated with ammonia solution and a precipitate of crude alkaloids was obtained. An aqueous solution of the precipitate was purified over a charcoal column. Betaine was separated by preparative TLC and purified by repeated recrystallization with methanol. The figure appears in Desai *et al.* (2013).

### 11.2.6 Characterization of Alkaloid as Betaine

The isolated pure alkaloid was characterized by chemical (melting point), chromatographic and spectral studies (IR, $^1$HNMR and mass spectroscopy). The NMR spectrum was recorded using a nuclear magnetic resonance spectrophotometer, a Bruker Advance 300 MHz having programme Zg30. The probe used was 5 mm BBO probe. Data were analyzed by Xwin NMR version 1.3. The mass spectrum was recorded using a mass spectrophotometer, a Bruker Daltanics, Model Microtof Q. All the data were analyzed by Hystar software version 3.4. The ionization technique used was the electron spray ionization technique.

### 11.2.7 Analysis of Betaine

The isolated compound was compared with a reference standard of betaine by TLC and HPLC studies. The $R_f$ values and mean retention times ($R_T$) for isolated compound was found to be 0.82 and 2.79 min, respectively, which were the same as that of standard betaine (Figure 11.3). The melting point of isolated compound was found to be 290 °C. IR, NMR and mass spectra of the isolated compound were obtained. The IR spectrum confirmed the presence of $COO^-$ group and C–N stretching (Figure 11.4; Table 11.2). $^1$HNMR showed the presence of nine equivalent methyl protons and two methylene protons (Figure 11.5, Table 11.3). The mass spectrum confirmed the presence of an M + H peak at *m*/*z* value of 118, confirming the molecular weight of the compound (Figure 11.6; Table 11.4). Based on chemical, chromatographic and spectroscopic studies the structure of the isolated alkaloid was confirmed as betaine (Figure 11.7). Thus, betaine can be used as a marker compound for standardization of extracts or formulations containing *Achyranthes aspera* (Desai *et al.*, 2013).

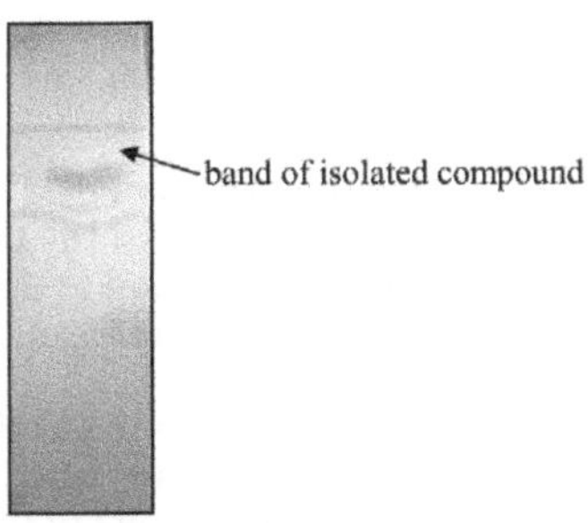


**Figure 11.3** TLC profile of isolated compound. TLC profile shows a band of isolated compound (glycine betaine) after spraying Drangerdorff's reagent, indicating the presence of alkaloid. The figure appears in Desai *et al.* (2013).

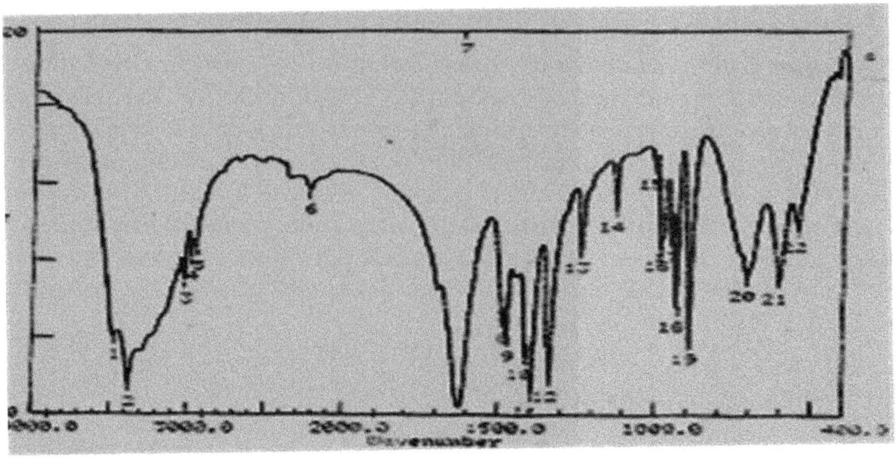

**Figure 11.4** IR spectrum of isolated compound. Peak 4 (2980.29 $cm^{-1}$) and peak 5 (2947.50 $cm^{-1}$) are due to C–H stretching. Peak 7 (1624.21 $cm^{-1}$) shows carboxylate anion stretching. Peak 13 (1238.41 $cm^{-1}$) is due to C–N vibration. The figure appears in Desai *et al.* (2013).

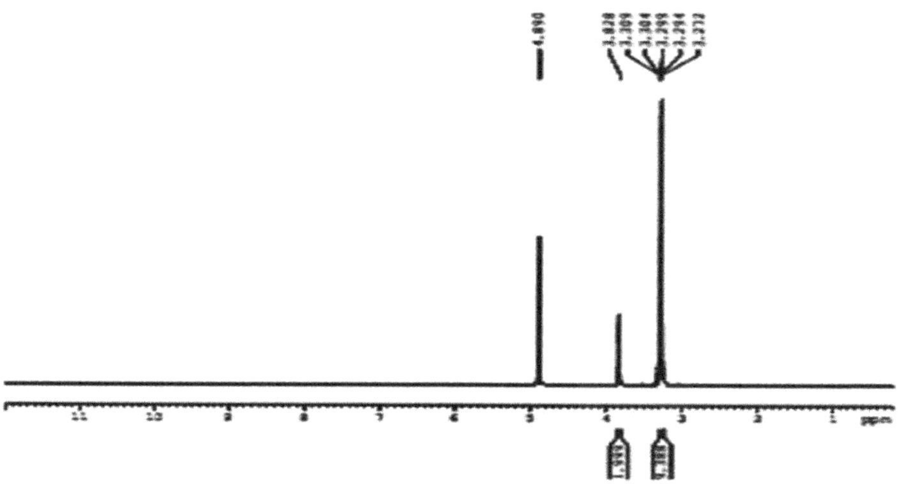

**Figure 11.5** NMR spectrum of isolated compound. $^{1}$HNMR shows presence of nine equivalent methyl protons (δ 3.27, from the three methyl groups) and two methylene protons (δ 3.80) attached to the quaternary nitrogen of glycine betaine. The figure appears in Desai *et al.* (2013).

**Table 11.2** Interpretation of IR spectrum of glycine betaine. IR spectrum of isolated compound (glycine betaine) confirmed the presence of $COO^-$ group (1624.21 $cm^{-1}$) and C–N stretching (1238.41 $cm^{-1}$).

| Peak No. | Wave number ($cm^{-1}$) | Inference |
|---|---|---|
| 4, 5 | 2980.29, 2947.5 | C–H stretching |
| 7 | 1624.21 | Carboxylate anion stretching |
| 13 | 1238.41 | C–N vibration |

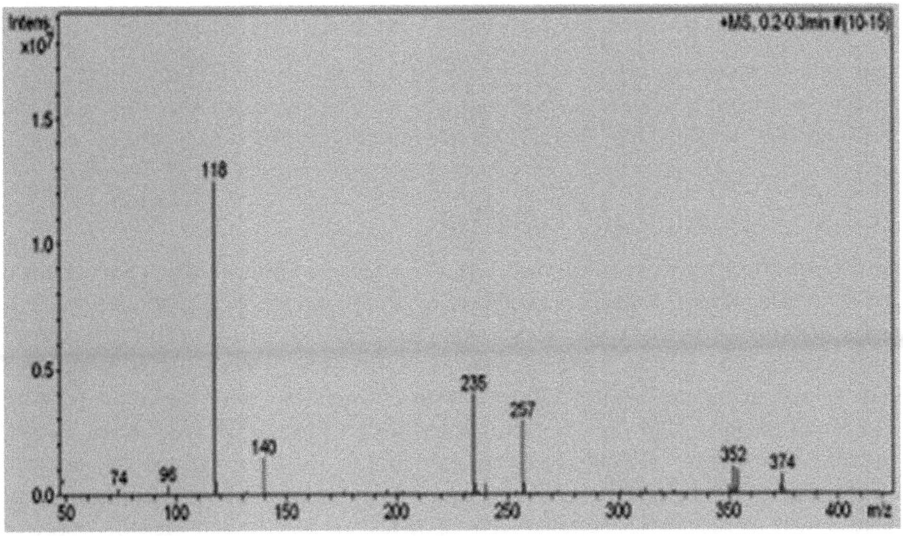


**Figure 11.6** Mass spectrum of isolated compound. *m*/*z*: mass over charge value, Mass spectrum confirms presence of M + H peak (*m*/*z* value 118) and M + Na peak (*m*/*z* 140) confirms the molecular weight of the compound, 117. The figure appears in Desai *et al.* (2013).

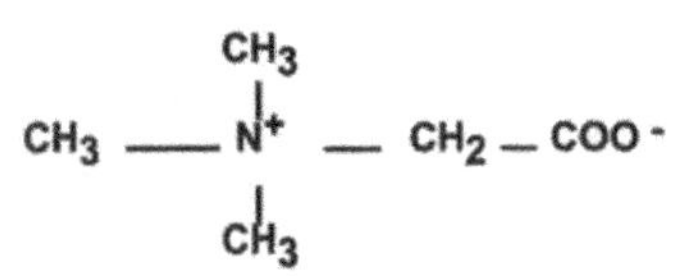


**Figure 11.7** Structure of glycine betaine. The figure appears in Desai *et al.* (2013).

**Table 11.3** Interpretation of H NMR spectrum of glycine betaine. $^1$HNMR spectrum of isolated compound (glycine betaine) showed the presence of nine equivalent methyl protons (3.27) and two methylene protons (3.80).

| Sr. No. | δ Values | Inference |
|---|---|---|
| 1. | 3.27 | (–3$CH_3$, s) |
| 2. | 3.80 | (–$CH_2$, s) |

**Table 11.4** Interpretation of mass spectrum. Mass spectrum of isolated compound (glycine betaine) showed presence of M + H peak (*m*/*z* value 118) which confirms the molecular weight of the compound.

| Sr. No. | *m*/*z* values | Inference |
|---|---|---|
| 1. | 140 | $[M + Na]^+$ peak |
| 2. | 118 | $[M + H]^+$ peak |
| 3. | 74 | $[M + H\text{–}COO]^+$ peak |

## 11.3 Isolation of Betaine from Kancolla Seeds

Kancolla is a sweet variety of *Chenopodium quinoa* Willd. (quinoa) (Chenopodiaceae family), used as a food plant. It is a highly nutritious food and the main edible parts are the seeds (Jancurova *et al.*, 2009).

### 11.3.1 Extraction and Isolation

Dini *et al.*(2006) carried out extraction, isolation and characterization of betaine derivatives from Kankolla seeds using the methods described below.

The whole flour from the seeds was extracted with methanol. The methanol extract was partitioned between *n*-butanol and water. The *n*-butanol extract was evaporated and defatted with chloroform. The residual fraction was chromatographed on a Sephadex LH-20 column, with methanol as eluent. Fractions were collected and checked by TLC [Silica gel plates, *n*-butanol: acetic acid: water (60 : 15 : 25)]. Fractions 1–36 were pooled together and evaporated and were further separated on a silica gel column using chloroform: methanol: water (15 : 5:0.6) as eluent. Fractions 41–50 and fractions 86–189 were chromatographed by reverse-phase HPLC with methanol: water (40 : 60) to yield the pure compounds glycine betaine ($R_t$ = 3.8 min), trigonelline ($R_t$ = 5.4 min), trigonelline methylester ($R_t$ = 5.4 min), trigonelline glucosylester ($R_t$ = 6 min) and 3-carboxy-1-(2-sulfoethyl)- pyridinium ($R_t$ = 6.2 min).

### 11.3.2 Characterization of Betaine Analogs

Structures of all isolated compounds (I–V) were elucidated by spectral studies such as UV, IR, NMR ($^1$H, $^{13}$C, COSY, HOHAHA, HMQC, HMBC) and mass-spectrometric methods.

### 11.3.3 Betaine Analysis

The ESI-MS mass spectral and NMR data of compounds I, II and III, are in accordance with the results reported in the literature for glycine betaine (I), trigonelline (II) and trigonelline methylester (III). The structure of compound IV was elucidated on the basis of its considerable similarities to compound II. The ESI-MS mass spectrum of compound IV exhibited a quasimolecular ion peak at *m/z* 301 [M + H]+, indicating the molecular formula $C_{13}H_{18}NO_7$ in accordance with $^{13}$C and $^{13}$C DEPT NMR data. The ESI-MS mass spectrum showed a fragment ion peak at *m/z* 137, indicating the loss of one hexose. The $^1$H and $^{13}$C NMR resonances of IV were superimposable on those obtained for II, with the exception of an additional sugar group signal in each of the $^1$H NMR and $^{13}$C NMR spectra. The combined analysis of the COSY, HOHAHA, and HMQC confirmed the nature of the monosaccharide as glucose. The pattern of $^{13}$C NMR resonances confirmed this assignment. This residue was linked to the carboxy group of the aglycone, as indicated by the HMBC correlation peak between the anomeric proton Glc-1 (δ 5.41) and the upfield-shifted (ppm 167.0) carboxy group carbon of the aglycone. From all the evidence mentioned above, the structure of IV was deduced to be trigonelline glucosylester (Figure 11.8).

The ESI-MS mass spectrum of compound V exhibited a quasimolecular ion peak at *m*/*z* 233 [M + H]+, indicating the molecular formula $C_8H_{10}NO_5S$, in accordance with $^{13}C$ and $^{13}C$ DEPT NMR data. The molecular formula, and the IR spectrum (KBr), which showed two absorptions at $\nu$ max 1360 and 1175 $cm^{-1}$, suggested the presence of an alkylsulfonate moiety (taurine-like) as the portion linked to the pyridinium nitrogen. The $^1H$ NMR spectrum exhibited signals at δ 9.25 (1H, s, H-2), δ 8.93 (1H, d, *J* = 7.2 Hz, H-4), δ 8.10 (1H, t, *J* = 7.2 Hz, H-5), δ 8.90 (1H, d, *J* = 7.2 Hz, H-6), δ 5.05 (2H, t, *J* = 5.2 Hz, H-7), δ 3.87 (2H, t, *J* = 5.2 Hz, H-8). The $^{13}C$ NMR spectrum exhibited signals at 147.5 (C-2), 139.3 (C-3), 146.8 (C-4), 128.5 (C-5), 146.0 (C-6), 67.0 (C-7), 53.7 (C-8), 168.8 (COOH). Comparison of the $^1H$ and $^{13}C$ NMR spectra with those of trigonelline, showed evidence for a β-substituted pyridinium nucleus with a $-CH_2CH_2-$ moiety linked to the nitrogen atom for compound V. Two coupled triplets at δ 5.05 and 3.87 in the $^1H$ NMR spectrum suggested that, in compound V, there is a dimethylene portion. In particular, the methylene resonating at δ 5.05 must be that linked to the nitrogen atom, as demonstrated by its strong NOE effect with the doublet at δ 9.25 (proton of pyridine). As a consequence, the second triplet must link the sulfonate function, and this conclusion is further supported by the chemical shift of C-8 at δ 53.7 in the $^{13}C$ NMR spectrum. The presence of a carboxylate in the structure of V was inferred by IR (KBr) absorption bands at $\nu$ max 1645 $cm^{-1}$ and HMBC correlation peaks of the carboxylate carbon (C-9) with H-2 and H-4. This agrees with the molecular formula deduced by mass spectral data and $^{13}C$ and $^{13}C$ DEPT NMR. These facts indicated that compound V was 3-carboxy-1-(2-sulfoethyl)-pyridinium (Figure 11.8).

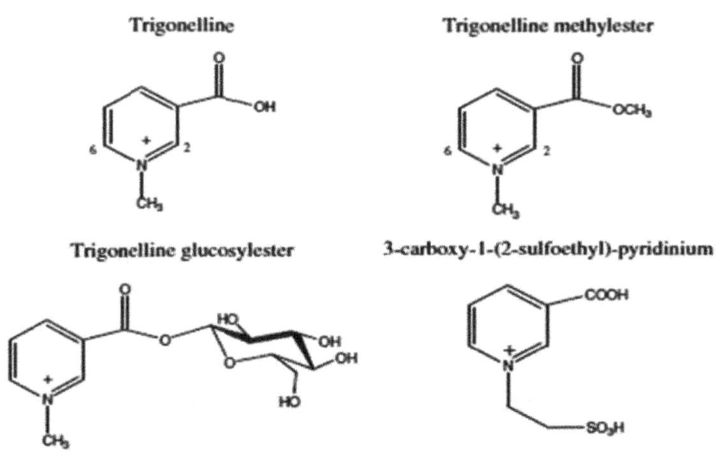


**Figure 11.8** Structures of betaine derivatives. Structures of various betaine derivatives have been confirmed and the figure appears in Dini *et al.*, *Food Chemistry* 2006. Reprinted with permission from Elsevier.

In the family Chenopodiaceae, only glycine betaine has been reported, thus the authors reported the presence of trigonelline and its derivatives and 3-carboxy-1-(2-sulfoethyl)-pyridinium in Kancolla seeds, could be used as taxonomic markers (Dini *et al.*, 2006).

## 11.4 Isolation of L-(+)-Isoleucine Betaine from *Cannabis* Seeds

*Cannabis sativa* is an annual herbaceous plant in the Cannabis genus, a species of the Cannabaceae family. Marijuana, hemp and cannabis are common names for plants of genus cannabis. Cannabis has been studied for the treatment of a number of conditions, including eczema, epilepsy, chronic pain, insomnia, and symptoms of multiple sclerosis. The most significant benefits have been found in the treatment of chronic pain and symptoms of multiple sclerosis.

### 11.4.1 Extraction and Purification of Isoleucine Betaine

A method has been described for the isolation of isoleucine betaine from Cannabis seeds Bercht *et al.* (1973). Powdered fresh hemp seeds were extracted with ethanol and the ethanol extract was acidified with glacial acetic acid. The extract was then concentrated to about half of the volume under vacuum with temperature not exceeding 40 °C. The aqueous solution was exhaustively extracted with benzene and the benzene was removed from the filtered aqueous layer. An excess of Kraut reagent* was added to the solution, the reddish ppt was filtered and suspended in water. The suspension was decomposed by hydrogen sulfide gas. After filtration, the filtrates were concentrated to one fourth of the volume and subsequently shaken with freshly prepared silver chloride. Filtration provided solutions that were concentrated under vacuum. The residue was further purified by column chromatography on cellulose, eluted with *n*-butanol, saturated with water. The fraction with $R_f$ of 0.45 confirmed the presence of alkaloid on TLC (cellulose MN 300, eluent n-butanol, saturated with water) when sprayed with Dragendorff's reagent. It was chromatographed on a basic alumina column using methanol as eluent. The compound with an $R_f$ of 0.30 (TLC, basic alumina/methanol) was further purified using a column prepared from $CaHPO_4.2H_2O$ that was washed with *n*-pentane followed by chloroform, which yielded isoleucine betaine.

### 11.4.2 Characterization of Isoleucine Betaine

Characterization of isoleucine betaine was carried out by IR, $^1H$ NMR studies. The optical rotation was measured using a polarimeter.

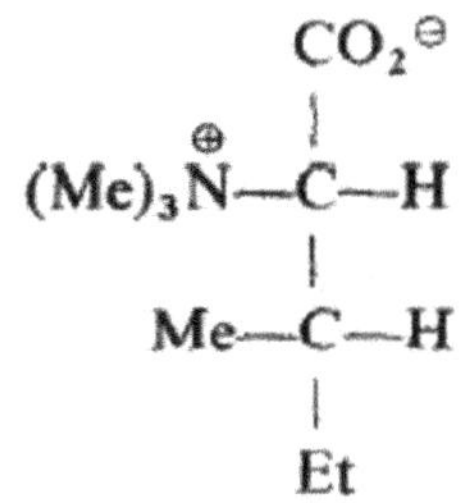

**Figure 11.9** Structure of isoleucine betaine. Structure of isoleucine betaine was confirmed and the figure appears in Bercht *et al. Phytochemistry*, 1973. Reprinted with permission from Elsevier.

### 11.4.3 Analysis of Isoleucine Betaine

The purified base showed a positive reaction towards Dragendorff's reagent. The component was optically active, $[\alpha]_D^{20} + 18.0°$ (c 1, chloroform). The IR spectrum showed main absorptions at 3340, 2960, 1625, 1490, 1450, 1390, 1030, 970, 960, 920 and 860 $cm^{-1}$, while acidification of the sample caused a remarkable shift of the 1625 $cm^{-1}$ band to 1710 $cm^{-1}$, simultaneously with the appearance of the typical broad absorption of carboxylic acids at the appropriate high wave number, and with the appearance of a strong absorption at 965 $cm^{-1}$ that could be ascribed to an antisymmetrical stretching of a trimethyl ammonium group (Bercht *et al.*, 1973).

The 100 MHz proton magnetic resonance spectrum (in $D_2O$) showed resonances at δ1.02 (t, *J* 7.3 Hz, 3H), 1.06 (δ, *J* 7-O Hz, 3H), 1.58 (m, *J* 7.0 Hz, 2H), 2.11 (m, 1H), 3.24 (s, 9H) and 3.67 (δ, *J* 1.1 Hz, 1H) ppm. Double resonance experiments indicated the presence of a secondary butyl group. The resonance at δ 3.24 ppm was ascribed to the trimethyl ammonium group, which could be confirmed by Hofmann degradation, yielding trimethylammonium hydrochloride that was identified in comparison with an authentic sample. Mass spectrometric and elemental analysis of this type of compound is notoriously difficult because of the ease of fragmentation under high vacuum and their very strong hygroscopic character. However, PMR and IR determined functional groups suggested the structure of isoleucine betaine (Figure 11.9) as being in accordance with the analytical data (Bercht *et al.*, 1973).

## 11.5 Conclusion

Betaine derivatives are important dietary constituents and have diverse biological activities. Betaine plays a vital role in health management. Further systematic research should be carried out to identify medicinal plant sources containing betaine and to isolate betaine from these sources.

## Summary Points

- The alkaloid betaine is a methyl derivative of the amino acid glycine.
- The betaine derivatives such as trimethylglycine, glycine betaine, lycine, and oxyneurine, are present in many food materials.
- Betaine is a zwitterionic quaternary ammonium compound.
- It is an alkaloid that is found in micro-organisms, plants, and animals.
- It has been reported to be present in various parts of many medicinal plants.
- A literature survey reveals that there is no adequate data available on its isolation and characterization from seeds of medicinal plants.
- This chapter includes isolation and characterization of betaine from three different sources, namely *Achyranthus aspera* (Prickly Chaff flower) seeds, *Chenopodium quinoa* (Kancolla) seeds and *Cannabis sativa* (Cannabis) seeds.
- The authors have isolated glycine betaine from the seeds of *Achyranthus aspera* by a simple preparative thin-layer chromatographic technique and the structure have been identified and characterized by chemical and spectral studies.
- Isolation and characterization of glycine betaine and its four derivatives from Kancolla seeds carried out by Dini *et al.* have been discussed.
- Isolation and characterization of isoleucine betaine from Cannabis seeds carried out by Bercht *et al.* have also been discussed.
- Betaine and its derivatives as important dietary constituents with diverse biological activities play a vital role in health management.
- Therefore, further research should be carried out to identify medicinal plant sources containing betaine and to isolate betaine from these sources.

## Key Facts

### Key Facts of $^1$H Correlation Spectroscopy (COSY)

- The one-dimensional nuclear magnetic resonance (NMR) method represents a signal to be presented as a function of a single parameter, *i.e.* chemical shift.
- The two-dimensional NMR method involves two coordinate axes, both of which represent chemical shifts and signal is presented as a function of each of these chemical shift ranges.
- $^1$H COSY is a two-dimensional NMR spectroscopic method.
- $^1$H COSY is used for determining which signals arise from neighboring proton (usually four bonds).
- Correlations appear when there is spin–spin coupling between protons, but where there is no coupling no correlation is expected.
- This method is very useful when the multiplets overlap which, complicate a one-dimensional spectrum.

## Key Facts of Distortionless Enhancement by Polarization Transfer (DEPT NMR)

- DEPT experiment is used to distinguish carbon atom according to the number of attached hydrogen atoms.
- It is a multipulse experiment that uses pulses on both the 1H and 13C channels.
- There are three types of pulse sequences – DEPT 45, DEPT 90 and DEPT 135.
- DEPT 45 only shows carbons that have attached hydrogen atoms (quarternary carbons do not appear).
- DEPT 90 only shows carbon atoms with a single attached hydrogen atom (methane carbon atoms).
- DEPT 135 distinguishes between carbon atoms based on their phasing:
- CH and $CH_3$ carbon atoms are phased in the same way (usually positive);
- $CH_2$ carbon atoms are phased the opposite way (usually negative).

## Key Facts of Electrospray Ionization-Mass Spectroscopy (ESI-MS)

- Mass spectrometry is one of the most sensitive analytical methods for the structural characterization of molecules.
- A mass spectrometer generates a characteristic qualitative and quantitative pattern of ions derived from a chemical compound, often including an ion due to the molecular mass of the compound.
- A mass spectrometer has a ionization source for generation of ions.
- Electrospray ionization is the most commonly used ionization method when dealing with liquid samples.
- It is suitable for analyses of both low-mass molecules and biomolecules.
- Other ionization techniques have a disadvantage of analyte fragmentation.
- ESI is a soft ionization technique that ionizes intact chemical species (proteins) by multiple charging without fragmentation.
- Because of multiple charging, the mass to charge ($m/z$) ratio of the resulting ions become very small and fall in mass ranges of all common mass analyzers.
- ESI is a very useful and versatile ionization technique for production of gas-phase ions from large biologically important macromolecules like proteins and nucleic acids and their mass analysis and structural characterization.
- In ESI, very weak covalent bonds are preserved in gas phase.

# Definition of Words and Terms

**Dragendorff's reagent:** Dragendorff's reagent (solution of potassium bismuth iodide in glacial acetic acid) is an alkaloidal precipitating reagent that gives an orange-colored precipitate with alkaloids.

**High-performance liquid chromatography (HPLC)**: HPLC is a separation technique that involves injection of a small volume of liquid sample through a column packed with tiny particles (3 to 5 micrometer (μm) in diameter called the stationary phase. Individual components of the sample are forced down the through the column by the high pressure of the liquid delivered by a pump. The individual components are separated from one another on the column packing by various chemical and/or physical interactions with the stationary phase particles and are detected at the exit of the column by a detector.

**Hofmann degradation**: A reaction in which a quaternary ammonium salt is reacted with a strong base (often aqueous silver oxide), resulting in elimination of a tertiary amine and formation of an alkene. When multiple alkene products are possible, the least stable (usually the least highly substituted) alkene is favored.

**Homocystinuria**: Homocystinuria is an inherited disorder of methionine metabolism, leading to an abnormal accumulation of homocysteine and its metabolites (homocystine, homocysteine–cysteine complex, and others) in blood and urine.

**Hydrophilic interaction liquid chromatography (HILIC)**: HILIC is a normal-phase chromatography performed using a polar stationary phase (for example, unmodified silica-, amino-, cyano- or diol-bonded phases). The mobile phase used is highly organic in nature (>70% solvent, typically acetonitrile) also containing a small percentage of aqueous solvent/buffer or other polar solvent. The water/polar solvent forms an aqueous-rich sublayer adsorbed to the polar surface of the stationary phase into which analytes partition.

**Kraut reagent**: Modified Dragendorff's reagent (solution of potassium bismuth iodide in nitric acid) is an alkaloidal precipitating reagent that gives a heavy brown precipitate.

**Limit of detection (LOD)**: The detection limit of an individual analytical procedure is the lowest amount of analyte in a sample that can be detected but not necessarily quantitated as an exact value.

**Limit of quantitation (LOQ)**: The quantitation limit of an individual analytical procedure is the lowest amount of analyte in a sample that can be quantitatively determined with suitable precision and accuracy.

**Optical rotation**: Optical activity is the ability of a chiral molecule to rotate the plane of plane-polarized light, measured using a polarimeter. The degree of rotation of polarized light passing through optically active material is called optical rotation.

**Standardization**: Standardization (of herbal drugs) means confirmation of its identity, determination of its quality, purity and detection of nature of adulterant by various parameters like morphological, microscopical, physical, chemical and biological evaluations.

**Thin-layer chromatography**: Thin-layer chromatography (TLC) is a chromatographic technique used to separate the components of a mixture using a thin stationary phase supported by an inert backing. A compound will have different affinities for the mobile and stationary phases, and this affects the speed at which it migrates. After a separation is complete, individual compounds appear as spots separated vertically. Each spot has a retention factor ($R_f$) which is equal to the distance migrated over the total distance covered by the solvent.

## List of Abbreviations

| | |
|---|---|
| COSY | Correlation Spectroscopy |
| DEPT NMR | Distortionless Enhancement by Polarization Transfer |
| ESI-MS | Electrospray ionization- Mass spectroscopy |
| FTIR | Fourier Transform Infrared Spectroscopy |
| HPLC | High-Performance Liquid Chromatography |
| HILIC | Hydrophilic Interaction Liquid Chromatography |
| HNMR | Proton Nuclear Magnetic Resonance |
| HOHAHA | Homonuclear Hartmann-Hahn Spectroscopy |
| HMQC | Heteronuclear Multiple–Quantum Correlation |
| HMBC | Heteronuclear Multiple Bond Correlation |
| LOD | Limit of Detection |
| LOQ | Limit of Quantitation |
| KH2PO4 | Potassium Dihydrogen Phosphate |
| NAD | Nicotinamide Adenine Dinucleotide |
| NMR | Nuclear Magnetic Resonance |
| NH4OH | Ammonium Hydroxide |
| NOE | Nuclear Overhauser effect |
| ROESY | Rotating Frame Overhauser Spectroscopy |
| SCX | Strong Cation Exchange |
| TLC | Thin Layer Chromatography |
| UV | Ultraviolet |

## References

Ahmad, U. V. and Sultan, M. A., 1980. Isolation of betaine from Lycium barbarum. *Journal of the Chemical Society of Pakistan*. 2: 113.

Bercht, L., Lousberg, J. J., Kuppers, J. E. M. and Salemink, C. A., 1973. L-( +)-isoleucine betaine in Cannabis seeds. *Phytochemistry*. 12: 2457–2459.

Bullock, S., 2010. *High Sensitivity Analysis of Betaine in Chinese Lycium by HPLC with ELSD Polymer Laboratories, Application note-SI-01239/TB1074*, pp. 1–2.

Chambers, S. T. and Kunin, C. M., 1987. Isolation of glycine betaine and proline betaine from human urine. Assessment of their role as osmoprotective agents for bacteria and the kidney. *Journal of Clinical Investigation*. 79: 731–737.

Chendrimada, T. P., Neto, M. G., Pesti, G. M., Davis, A. J. and Bakalli, R. I., 2002. Determination of the betaine content of feed ingredients using high-performance liquid chromatography. *Journal of the Science of Food and Agriculture*. 82: 1556–1563.

Chumak, D., Stonik, V. A. and Elyakov, G. V., 1978. Pacific Ocean Institute of Organic Chemistry, Far-Eastern Center of the Academy of Sciences of the USSR, Vladivostok. *Translated from Khimiya Prirodnykh Soedinenii*. 3: 421.

Desai, S., Tatke, P. and Gabhe, S., 2013. Isolation and characterization of betaine from dried seeds of *Achyranthes aspera*. *Indian Drugs*. 50: 56–59.

Dini, I., Tenore, G. C., Trimarco, E. and Dini, A., 2006. Two novel betaine derivatives from kancolla seeds (Chenopodiaceae). *Food Chemistry*. 98: 209–213.

Gorham, J., 1984. Separation of plant betaines and their sulphur analogues by cation-exchange high-performance liquid chromatography. *Journal of Chromatography A*. 281: 345–351.

Jancurová, M., Minarovicova, L. and Dandár, A., 2009. Quinoa- a review. *Czech Journal of Food Sciences*. 27: 71–79.

Khan, M. I., Aeri, V., Shahzad, N. and Khan, M., 2011. A validated HPLC method for estimation of betaine in whole plant of *Achyranthes aspera* Linn, marketed formulation. *International Journal of Pharma World Research*. 2(1): 1–14.

Kikuchi, N., Matsuno, K. and Miki, T., 1993. Separation and determination of betaine in an oriental medicine by liquid chromatography. *Analytica Chimica Acta*. 283: 338–343.

Laryea, M. D., Steinhagen, F., Pawliczek, S. and Wendel, U., 1998. Simple method for the routine determination of betaine and *N,N*-dimethylglycine in blood and urine. *Clinical Chemistry*. 44: 1937–1941.

Laryea, M. D., Zab, R., Ritgen, J. and Wendel, U., 1994. Simultaneous determination of betaine and *N,N*-dimethylglycine in urine. *Clinical Chimica Acta*. 230: 169–175.

Liscovitch, M., Freese, A., Blusztajn, J. K. and Wurtman, R. J., 1985. High-performance liquid chromatography of water-soluble choline metabolites. *Analytical Biochemistry*. 151: 182–187.

Mar, M. H., Ridky, T. W., Garner, S. C. and Zeisel, S. H., 1995. A method for the determination of betaine in tissues using high performance liquid chromatography. *Journal of Nutritional Biochemistry*. 6: 392–398.

Mehrafarin, A., Qaderi, A., Rezazadeh, S., Naghdi, B. H., Noormohammadi, G., and Z and E., 2010. Bioengineering of important secondary metabolites and metabolic pathways in fenugreek (Trigonella Foenum-Graecum L.) *Journal of Medicinal Plants*. 9: 1–18.

Murray, W. and Walter, E. D., 1945. Isolation of betaine from guayule. *Journal of American Society*. 67: 1422.

Rustembekova, G. B., Goryaev, M. I. and Gladyshev, P. P., 1973. Institute of Chemical Sciences, Academy of Sciences of the Kazakh SSR. *Translated from Khimiya Prirodnykh Soedinenii*. 4: 569.

Shin, H. D., Suh, J. H., Kim, J., Lee, H., Eom, H. Y., Kim, U., Yang, D. H., Han, S. B. and Youm, J. R., 2012. Determination of betaine in *Fructus Lycii* using HILIC-ELSD. *Bulletin of the Korean Chemical Society*. 33: 553–558.

Storer, M. K., McEntyre, C. J. and Lever M., 2006. Separation of cationic aracyl derivatives of betaines and related compounds. *Journal of Chromatography A*. 1104: 263–271.

Strange, N., Majer, J. R. and Smith, H., 1974. The isolation and identification of choline and betaine as the two major components in anthers and wheat germ that stimulate Fusarium graminearum *in vitro*. *Physiological Plant Pathology*. 4: 277–290.

Vialle, J. and Kolosky, M., 1981. Determination of betaine in sugar and wine by liquid chromatography. *Journal of Chromatography A*. 204: 429–435.

Zhang J., Blusztajn, J. K. and Zeisel, S. H., 1992. Measurement of the formation of betaine aldehyde and betaine in rat liver mitochondria by a high pressure liquid chromatography-radioenzymatic assay. *Biochimica Biophysica Acta*. 1117: 333–339.

CHAPTER 12

# Occurrence and Analysis of Betaines in Fruits

LUIGI SERVILLO*[a], ALFONSO GIOVANE[a], ROSARIO CASALE[a], NUNZIA D'ONOFRIO[a], GIOVANNA FERRARI[b], DOMENICO CAUTELA[c], DOMENICO CASTALDO[d], AND MARIA LUISA BALESTRIERI[a]

[a]Department of Biochemistry, Biophysics and General Pathology, Seconda Università di Napoli, Via Luigi De Crecchio 7, 80138 Napoli, Italy; [b]Dipartimento di Ingegneria Industriale e ProdAL scarl, Università degli Studi di Salerno, Via Ponte Don Melillo, 84084 Fisciano (SA), Italy; [c]Stazione Sperimentale per le Industrie delle Essenze e dei derivati dagli Agrumi (SSEA)-Azienda Speciale della Camera di Commercio di Reggio Calabria, Via Tommaso Campanella, 12, 89127, Reggio Calabria, Italy; [d]Ministero dello Sviluppo Economico (MiSE), Via Molise 2, 00187 Roma, Italy
*E-mail: luigi.servillo@unina2.it

## 12.1 Introduction

Betaines, fully *N*-methylated amino acid derivatives, and other related quaternary ammonium compounds (QACs) are small molecules used by plants in the physiological mechanisms of accumulation and adaptation (Rhodes and Hanson, 1993). These compounds take part to the defensive response towards biotic or abiotic stresses suffered by plants. The biotic stress is a consequence of attacks by other living organisms, such as herbivores or parasites, whereas the abiotic stress is caused by environmental factors, which

Food and Nutritional Components in Focus No. 7
Betaine: Chemistry, Analysis, Function and Effects
Edited by Victor R Preedy

Published by the Royal Society of Chemistry, www.rsc.org

include lack of nutrients, air pollutants, low temperatures, salinity, water scarcity and heavy-metal toxicity.

The water scarcity is definitely the type of stress that, on the whole, most affects the plant development and growth. At the cellular level, water stress activates specific metabolic pathways aimed to stabilize the osmotic potential whose variations can be responsible for the cell death (Kavi Kishor *et al.*, 2005). Among the mechanisms used by plants to counteract the water shortage, the osmotic adjustment is the main adaptive response. This mechanism is aimed to synthesize at least three different classes of specific low molecular weight solutes and to accumulate them in intracellular fluids. The low molecular weight solutes are betaines, polyols (mannitol, glycerol, sorbitol, ononitol, pinitol), and specific amino acids, such as proline that, unlike other more specific metabolites, is accumulated in all plant species in response to both biotic and abiotic stress conditions (Hanson *et al.*, 1994; Nolte *et al.*, 1997). Overall, all those metabolites produced through specific biosynthetic pathways contribute to modulate the cellular osmotic potential limiting, in this way, the efflux of water from cells with subsequent dehydration. For such a vital function these substances, which stabilize proteins and other important cellular components, are called "osmoprotectants" or "compatible solutes" because they can be accumulated into the cells in considerable amounts without interfering with the normal cellular enzymatic activities (Kurz, 2008).

The plant defence mechanism against the stress induced by the sudden temperature decrease is the acclimatization, which plays an important role in the cold resistance. This resistance is accomplished through the modification of the cell-membrane structure and the accumulation of the "compatible solutes", which lower the freezing point of the intracellular fluids due to their strong protective "antifreeze" characteristics. Glycine betaine and 3-dimethylsulfoniopropionate (DMSP), well-known molecules involved in the mechanisms of cryoprotection, accumulate under conditions of osmotic stress and low temperatures. Glycine betaine occurs in several angiosperm, such as beetroot, spinach, wheat flour (de Zwart *et al.*, 2003; Zeisel *et al.*, 2003; Slow *et al.*, 2005), whereas DMSP, which is biosynthesized from methionine (Gage *et al.*, 1997), has been mainly found in marine bacteria and phytoplankton. Some filter feeders and marine animals accumulate DMSP as a consequence of feeding on those organisms (Malin and Kirst, 1997). Furthermore, besides abiotic stress, some QACs are also involved in the plant defence toward biotic stress. Indeed, plants activate defence mechanisms in response to invasion of pathogens aimed at eliminating or limiting their spreading. The induction of these defence mechanisms occurs through the interaction of signaling molecules, named "elicitors", with the cell wall. The elicitors are synthesized either by the pathogens (exogenous elicitors) or released by the cell itself as a result of pathogen attacks (endogenous elicitors). The interaction with the elicitors activates a cascade of biochemical defensive responses that include: (i) the rapid production of reactive oxygen species (ROS), particularly superoxide anion and hydrogen peroxide, which directly inhibit the growth of pathogens (Jaspers and Kangasjarvi, 2010); (ii) the accumulation of phytoalexins and the production of nitric oxide (Neill *et al.*, 2008); (iii) the activation of the salicylic acid biosynthesis, which acts as a signal for the systemic expression of pathogen-related proteins (PRP) (Yoshioka *et al.*, 2003); and (iv) the production of defence metabolites mostly derived from the enzymatic decarboxylation of tryptophan, tyrosine and 3,4-dihydroxyphenylalanine (often referred to as L-DOPA). Not surprisingly, these last substances are mainly expressed in leaf epidermis and seeds. It is well known, in fact, that the reproductive structures of plants (seeds), but also leaves, especially the young ones, have a greater need of more complex and articulated physiological defences than other plant tissues. From the chemical point of view, this necessity mainly results in an increased presence of low molecular weight metabolites (secondary metabolites) or proteins that confer toxic or antinutritional properties to these vital plant organs (Howe and Jander, 2008). Tyramine is formed by the action of the enzyme tyrosine decarboxylase (TYDC). Plant TYDC belongs to the family of pyridoxal-5′-phosphate (PLP)-dependent enzymes that decarboxylate tyrosine, or even L-DOPA, to yield tyramine, or dopamine. These enzymes play an important role in response to biotic stress of plants, by producing the biosynthetic precursors (tyramine and dopamine) of toxic phytochemicals (alkaloids or protoalkaloids) against plant aggressors (herbivores and pathogens) (Facchini *et al.*, 2000). A similar fate also applies to the other aromatic amino acid, tryptophan, which is decarboxylated to tryptamine. Subsequently, a series of secondary defensive metabolites stems from tryptamine. Although the toxicology of most of these compounds does not appear sufficiently tested, studies on some QACs deriving from tryptamine, such as *N,N,N*-trimethyltryptamine and 5-hydroxy-*N,N,N*-trimethyltriptamine, suggest that their production constitutes a biosynthetic response of plants against parasites and herbivores (Barlow and Burston, 1980). These studies showed that these compounds are cholinergic agonists and that the hydroxylated form of *N,N,N*-trimethyltryptamine is an even more potent parasympathetic ganglion stimulant (about 10-fold) than *N,N,N*-trimethyltryptamine (Barlow and Burston, 1980). As suggested by Servillo *et al.* (2012a, 2013a), the high cholinergic activity of most of the QACs derived from tryptamine, but also of those deriving from tyramine, is probably the basis of the chemical defence mechanism adopted by some plant *genera* against herbivores (Deguchi *et al.*, 1963). These results substantiate and further confirm some recent data, about the important contribution of the hydroxylated indole metabolites, both in terms of deterrence and postinvasive plant defence. In fact, studies have shown that the synthesis of 5-hydroxytryptamine (5-HT, serotonin), is drastically implemented in rice leaves as a response to phytopathogen infections (Ishihara *et al.*, 2008; Kang *et al.*, 2009).

## 12.2 Biosynthesis of Betaines and QACs

Glycine betaine (betaine or *N,N,N*-trimethylglycine) represents the archetypal betaine in the plant Kingdom and the simplest representative of this class of substances. This widely distributed osmoprotectant is present at

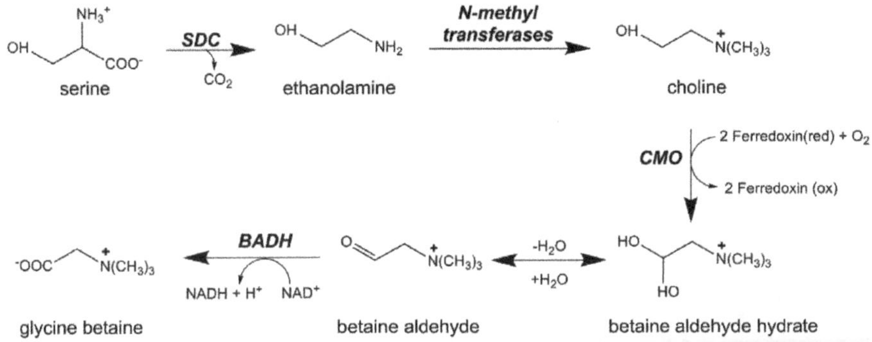


**Figure 12.1** Pathway of glycine betaine synthesis. The biosynthetic route of glycine betaine starts from serine that is decarboxylated to ethanolamine by serine decarboxylase (SDC). The methylation of ethanolamine produces choline that is oxidized by choline mono-oxygenase (CMO) into betaine aldehydes subsequently converted into glycine betaine by betaine aldehyde dehydrogenase (BADH).

significant levels in *Beta vulgaris* (sugar beet), from which the trivial name betaine derives, and in other *Chenopodiacee*. Glycine betaine is a zwitterion at neutral pH, characterized by a permanent positive charge on the quaternary nitrogen atom and a negative charge on the carboxylate group. In plants, glycine betaine is synthesized from choline that, in turn, derives from ethanolamine produced by serine decarboxylation (Rhodes and Hanson, 1993). In the biosynthetic route of glycine betaine, choline is first oxidized by choline mono-oxygenase (CMO) into betaine aldehydes and then converted into glycine betaine by betaine aldehyde dehydrogenase (BADH) (Figure 12.1).

For the biosynthesis of betaines and QACs the presence of a biosynthetic system is required for the production and transfer of the methyl groups to the precursors of these substances. In this system, the amino acid methionine play a major role through its "active" form, *S*-adenosylmethionine (SAM), a substance capable of donating methyl groups in reactions catalyzed by *N*-methyltransferase enzymes. SAM, a universal donor of methyl groups used for the methylation of many compounds such as amino acids, nucleic acids, proteins and phospholipids, is produced by reaction of methionine with ATP (Figure 12.2). The *N*-methyltransferases catalyze the progressive transfer of the methyl groups from SAM to the acceptor. As a consequence, SAM is converted into S-adenosylhomocysteine (SAHC) that is then transformed into homocysteine by the enzyme adenosylhomocysteinase (Figure 12.2). The conversion of homocysteine to methionine (remethylation process) and the subsequent conversion into SAM (the active form of methionine) is used to restart a new cycle of methylation (activated methyl cycle). Methionine is reformed through the methionine synthase action that transfers the methyl group from *N*5-methyl-tetrahydrofolate ($CH_3$–THF) to homocysteine, thus restoring the activated methyl cycle (Giovannelli *et al.*, 1980) (Figure 12.2).

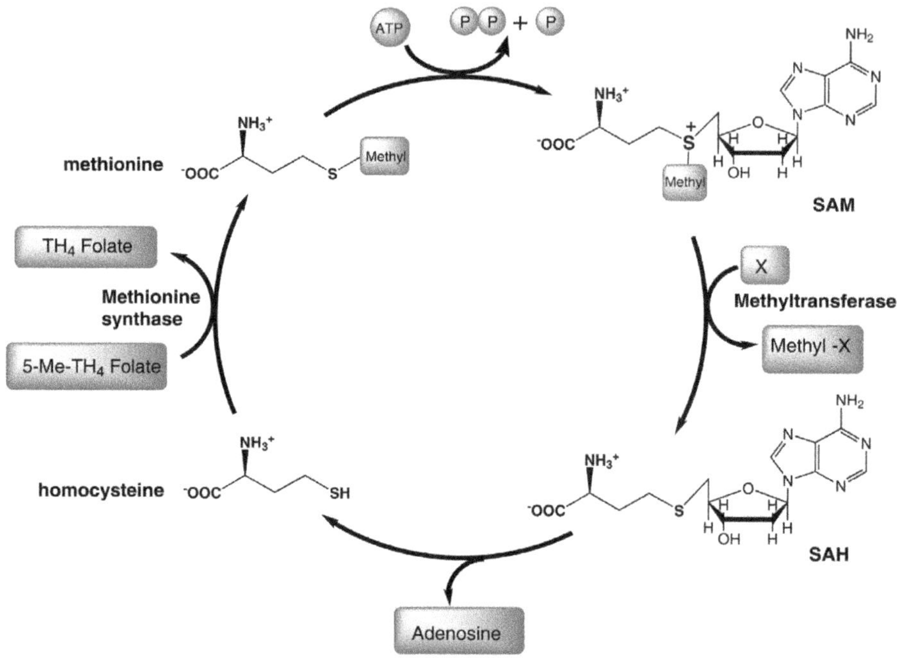


**Figure 12.2** Cyclic pathway of *S*-adenosylmethionine biosynthesis. *S*-adenosylmethionine (SAM) is produced by reaction of methionine with ATP. SAM donates the methyl group in reactions catalyzed by *N*-methyltransferase enzymes. As a consequence, SAM is converted into *S*-adenosylhomocysteine (SAHC), which is then transformed into homocysteine by the enzyme *S*-adenosylhomocysteinase. The conversion of homocysteine into methionine and the subsequent conversion into SAM restarts a new methylation cycle. Methionine is reformed through the methionine synthase action that transfers the methyl group from *N*5-methyltetrahydrofolate (5-Me-$TH_4$Folate) to homocysteine.

To date, the *N*-methyltransferases have been identified only in vegetable crude extracts and, currently, it is still unclear how the three methylation steps occur. For this reason, betaine biosynthesis is generically described as a sequential methylation process catalyzed by one or more *N*-methyltranferases. As an example, proline, a cyclic imino acid widely distributed in plants, is considered one of the most common compatible solutes. Proline accumulation has been documented in many living organisms as a consequence of several stress conditions (Kavi Kishor *et al.*, 2005). Stachydrine (*N,N*-dimethylproline) is the cyclic betaine derived from proline *N*-methylation. This compound occurs at high levels in important plant genera such as *Medicago* and *Citrus*. However, the stachydrine synthesis has been only hypothesized to be due to *N*-methylation of proline through the action of SAM-dependent methyltransferases, but the enzymes involved in this hypothetical biosynthetic process have not yet been isolated and characterized. Also, the

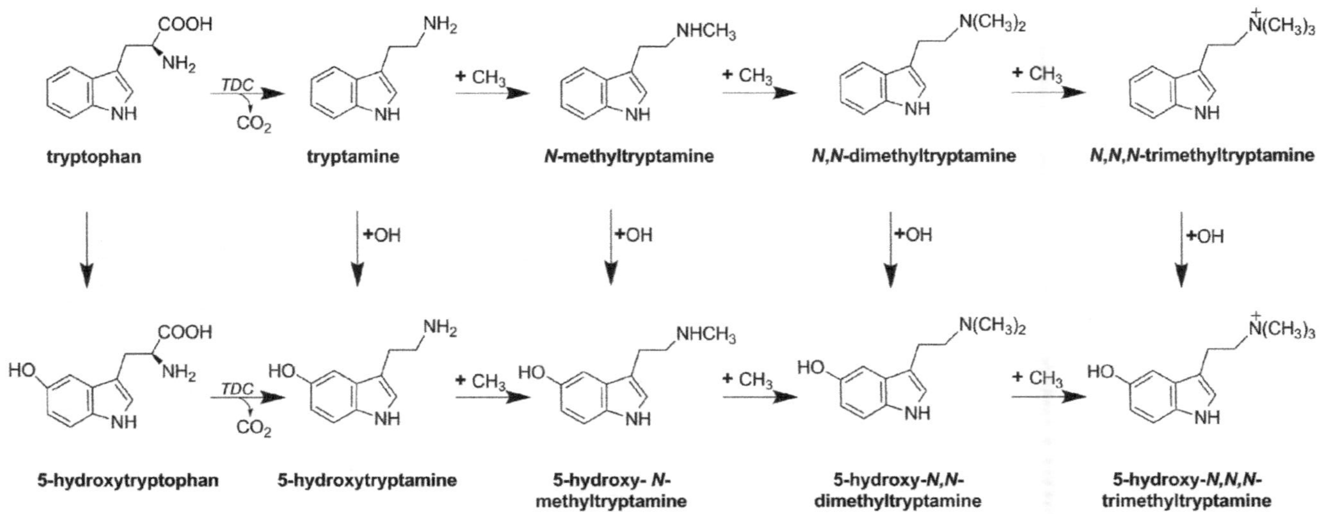


**Figure 12.3** Hypothetical steps for the tryptamine derivative biosynthesis. Tryptophan and 5-tryptophan are decarboxylated to tryptamine and 5-hydroxytryptamine, respectively, by tryptophan decarboxylase (TDC). Sequential methylation steps of tryptamine and 5-hydroxytryptamine produce the two QACs, *N,N,N*-trimethyltryptamine and 5-hydroxy-*N,N,N*-trimethyltryptamine, respectively. 5-hydroxy-*N,N,N*-trimethyltryptamine could be also produced from hydroxylation of *N,N,N*-trimethyltryptamine.

enzymes involved in the biosynthesis of betonicine, the hydroxylated form of stachydrine, presents at noticeable levels in citrus fruit juices, have not been characterized so far. Betonicine is supposed to be produced by an enzymatic hydroxylation process of stachydrine. Similarly, the enzymes involved in the biosynthesis of QACs derived from tryptamine have not yet been characterized. It has been hypothesized that their synthesis proceeds through steps of methylation and hydroxylation (Servillo *et al.*, 2013a) (Figure 12.3).

## 12.3 Analytical Methods for Determination of Betaines and QACs

The majority of the current knowledge on the betaines and QACs has been obtained through qualitative and/or semiquantitative methods, mainly thin-layer chromatography, thin-layer electrophoresis and $^{1}H$ and $^{13}C$ nuclear magnetic resonance (NMR), which has been used for the qualitative analysis of these compounds in marine algae and plant materials (Blunden *et al.*, 1986). However, besides the requirement of extensive purification steps to remove large part of the possible interfering substances, NMR is not suited for the quantitative determination of secondary metabolites, generally present in low amounts. High-performance liquid chromatography (HPLC) methods have been widely employed for quantification of betaines even if the application of this technique is strongly affected by the intrinsic characteristics of these compounds. Betaines, in fact, are substances that exhibit poor absorption in the UV range and do not possess an appreciable fluorescence. For this reason, they can only be detected in the low-UV region (190–220 nm) where there is a strong contribution of interfering substances. Due to the objective difficulty to detect these substances, the HPLC techniques have been scarcely used for a long period.

Thanks to efficient and selective derivatization methods of the carboxylic function present in all betaines, the HPLC quantification of these compounds in plants and food matrices has been achieved in a more reliable way (Happer *et al.*, 2004). The carboxyl group of betaines has been derivatized by the use of aracyl trifluoromethanesulfonates (aracyl triflates). In particular, 2'-phenanthreneacyltriflate became the most useful due to its UV absorption and fluorescence intensity (Happer *et al.*, 2004). This derivatization method has been used to quantify some major betaines and similar compounds (glycine betaine, proline betaine, trigonelline and 3-dimethylsulfoniopropionate) in beverages, dairy products, fruits, grains, meat, nuts, canned seafood and vegetables (Slow *et al.*, 2005).

Among the chromatographic methods for separation of underivatized betaines, hydrophilic interaction liquid chromatography (HILIC) has proven to be suitable for such polar compounds. This technique differs from the classic normal phase chromatography (NP-LC) in that it allows separation of polar compounds on silica-based polar stationary phases and utilizes as mobile phases eluents also used in the reverse-phase liquid chromatography

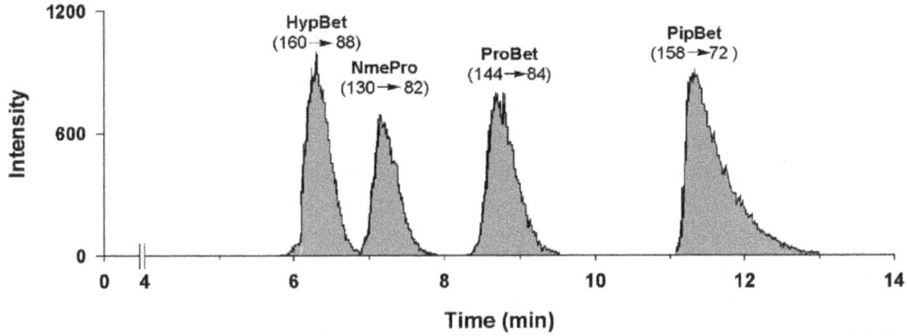


**Figure 12.4** Chromatography of betaines on C8-bonded silica column. Chromatography of a betaine standard mixture employing a C8-bonded silica column. The analysis was performed as described by Servillo *et al.* (2011a). The detection was made by tandem mass spectrometry with multiple reaction monitoring (MRM). The monitored transition are indicated on each peak. The standard mixture contained hydroxyproline betaine (HypBet), *N*-methylproline (NmePro), proline betaine (ProBet), pipecolic acid betaine (PipBet).

(RP-LC). For example, the betaine content in *Lycii fructus* has been determined by HILIC with isocratic elution of acetonitrile and 30 mM of ammonium acetate buffer (80/20, v/v) (Zhao *et al.*, 2013).

A novel analysis of betaines and some QACs was performed by employing a C8-bonded silica columns that is normally used for chromatographic separation of hydrophobic compounds using a mixture of aqueous–organic eluents. Instead, the C8-bonded silica column surprisingly showed a discrete retention even for highly polar compounds such as betaines and QACs. In this way, the separation and quantification of several betaines and QACs in citrus juices was obtained in short times using 0.1% solution of formic acid in water as the unique elution solvent (Servillo *et al.*, 2011a,b). The chromatography of a standard mixture of some betaines performed in this way is reported (Figure 12.4).

The real breakthrough in the quantitative analysis of betaine has occurred with the advent of the mass-spectrometric detection. Betaines and QACs, by possessing a permanent positive charge, are particularly suitable for this kind of detection. The techniques first employed were plasma desorption (PD), desorption chemical ionization, and fast atom bombardment (FAB) mass spectrometry (Wood *et al.*, 1991). Although quantification of the main betaines in plants was satisfactorily achieved in this way, such ionization techniques showed a rather low sensitivity. The FAB-MS sensitivity in the analysis of betaines could be improved by derivatization of the carboxyl group into n-propyl or n-butyl esters (Rhodes *et al.*, 1987) but this requires longer analysis times and adds uncertainty due to sample manipulation and the derivatization yield. The mass-spectrometric determination of betaines has been greatly improved by using HPLC-ESI-MS/MS (Wood *et al.*, 2002).

Tandem mass spectrometry is of paramount importance to obtain the structural information necessary for the unambiguous identification of unknown compounds and for highly selective quantification especially when it is necessary to discriminate isomeric ions. Furthermore, betaines and QACs, possessing the permanent positive charge on the quaternary nitrogen, are particularly suitable for ESI because their solutions produce parent ions in high yields, which greatly increases the detection sensitivity.

However, especially for complex vegetal matrices, a more or less extensive sample preparation is necessary before HPLC-MS analysis. Some guidelines for sample preparation are shown in Figure 12.5. For matrices with low lipid content, such as tissue or foods of plant origin, the analytical determination can be performed by extracting samples with 0.2% formic acid in water in the ratio 1:5, for about 15 min. After centrifuging at 12 000 × g for 10 min, samples are further diluted with 0.2% formic acid in water in the ratio 1:1 or 1:5 (10 ≤ final dilution factor ≤ 25). In some cases, especially for matrices containing high sugar levels, the samples can be purified by ion-exchange chromatography before analysis by LC-ESI MS/MS. In this case, the sample is centrifuged at 14 000 × g for 20 min, and 1 mL is applied to a 5 × 1 cm column filled with Bio-Rad AG 50WX8-(H+) resin. After loading, the column is washed with five volumes of Milli-Q water and then the amino acids and betaines are eluted with 20 mL of ammonia solution (1:1 v/v water) and 20 mL of water. The eluted solution is dried with a rotavapor and reconstituted with 0.2% formic acid in water in a volume corresponding to an initial sample dilution of 1:10 or 1:25. When analyzing matrices with high lipid content, such as sauces and lipid-rich fruits, the determinations may be carried out by extracting the betaines and QACs present in the samples with a solution of methanol/chloroform (2:1 v/v) at −20 °C overnight and recovering the betaines and similar compounds from the aqueous phase after centrifugation. The aqueous phase is then dried with a rotavapor and reconstituted with 0.2% formic acid in water in a volume corresponding to an initial sample dilution either 1:10 or 1:25 (Figure 12.5).

As for particular fruit samples, such as industrial citrus fruit juice concentrates, they are preliminarily treated by diluting samples with Milli-Q water to a level of soluble solids (expressed as Brix degrees) equal to the corresponding natural juice, that is 8° Brix for reconstituted lemon juice, 11.2° Brix for reconstituted orange juice, and 10.0° Brix for reconstituted mandarin juice. After dilution, the juices are centrifuged, diluted with 0.2% formic acid in water in the ratios (v/v) 1:10 or 1:25, and analyzed by LC-ESI-MS/MS without any further preparation. Indeed, in some case, particularly when the levels of betaines are high and the matrix is sufficiently clean, only a sample dilution and centrifugation before analysis can be sufficient. For example, stachydrine and betonicine quantification in citrus fruit juices can be accomplished in a very fast way, without needing the chromatographic analysis, by simply diluting the natural juices about 5–10 thousand times and, after centrifugation, analyzing samples by flow injection analysis, (FIA) (Servillo *et al.*, 2011a).

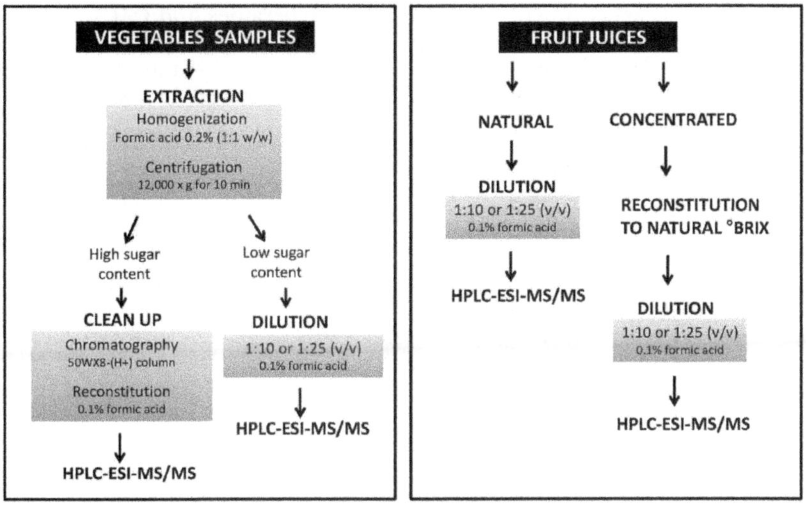


**Figure 12.5** Fruit juice and vegetable sample preparations before HPLC-ESI-MS/MS. Left panel. The analytical determination of betaines in vegetable matrices with low lipid content can be performed by extracting samples with 0.2% formic acid in water. After centrifugation, samples are further purified (clean up) by ion-exchange chromatography before analysis by LC-ESI-MS/MS, or diluted with formic acid before LC-ESI-MS/MS. Right panel. Fruit juices are first properly diluted with 0.2% formic acid in water and then analyzed by LC-ESI-MS/MS without any further preparation. Indeed, in some cases, particularly when the levels of betaines are high and the matrix is sufficiently clean, only a sample dilution and centrifugation before analysis may be sufficient.

## 12.4 Identification and Quantification of Betaines and QACs by ESI-Tandem Mass Spectrometry

Tandem mass spectrometry is a structural elucidation technique fundamentally based on the analyte ionization in the gas phase and its subsequent fragmentation into ions (fragment ions) that are discriminated on the basis of their mass/charge ratio. In the collision-induced dissociation (CID) the fragmentation is usually obtained by collision of parent ions with a gas (helium or nitrogen). Many substances having common structural features present similar features in their fragmentation patterns. In the case of acyclic betaines, the presence of the quaternary ammonium generally leads to the formation of the fragment ions at $m/z$ $[M + H–C_3H_9N]^+$, corresponding to the neutral loss of trimethylamine, although the most intense fragments usually are those retaining the quaternary ammonium group. Moreover, ions at $m/z$ 58, 59 and 60 are always present, in different reciprocal ratios according to the structural characteristics of the betaine chain to which the quaternary nitrogen is linked. Actually, the fragment at $m/z$ 60 corresponds to trimethylammonium ion ($C_3H_{10}N^+$), the fragment at

**Table 12.1** Mass spectrometric features of some betaines and QACs. The precursor amino acids of the betaines or QAC are reported in column 1. The mass/charge ratios (*m*/*z*) of the parent ions of betaines and QACs and their main MS/MS fragment ions are reported in columns 3 and 4, respectively.

| Precursor Amino acid | Betaines and QACs | $[M + H]^+$ | MS/MS Fragments |
|---|---|---|---|
| Serine | Choline | 104 | 58, 60 |
| | Glycine betaine | 118 | 58, 59 |
| γ-Aminobutyric acid | γ-Butyrobetaine | 146 | 60, 87 |
| Proline | Proline betaine (stachydrine) | 144 | 58, 84, 102 |
| | Hydroxyproline betaine (betonicine) | 160 | 58, 88, 102 |
| Pipecolic acid | Pipecolic acid betaine (homostachydrine) | 158 | 58, 102, 112 |
| Nicotinic acid | Trigonelline | 138 | 94, 110 |
| β-Alanine | β-Alanine betaine | 132 | 60, 73 |
| Lysine | $N^{\varepsilon}$-Trimethyllysine | 189 | 60, 84, 130 |
| Histidine | Hercynine | 198 | 60, 95, 154 |
| | Ergothioneine | 230 | 60, 127, 186 |
| Tryptophan | *N,N,N*-Trimethyltryptamine | 203 | 60, 144 |
| | 5-Hydroxy-trimethyltryptamine | 219 | 60, 160 |
| Tyrosine | *N,N,N*-Trimethyltyramine | 180 | 60, 121 |

*m*/*z* 59 ($C_3H_9N^+$) is the radical ion arising from the homolytic cleavage of the bond between the quaternary nitrogen and the carbon atom of the betaine chain to which the nitrogen is linked, and the fragment at *m*/*z* 58 is the dimethyl(methylene)ammonium ion ($C_3H_8N^+$) (Naresh Chary *et al.*, 2012). It has been shown that acyclic betaines, depending on the structure of their chain, present different abundances of these three fragments. For example, in the MS/MS fragmentation pattern of glycine betaine, the fragment at *m*/*z* 60 is absent and the fragments at *m*/*z* at 58 and 59 have similar high abundance. Instead, in the MS/MS fragmentation pattern of γ-butyrobetaine, the fragment at *m*/*z* 60 is well represented but the other two are practically absent (Naresh Chary *et al.*, 2012).

The cyclic betaines, proline betaine (stachydrine), hydroxyproline betaine (betonicine) and pipecolic acid betaine (homostachydrine), besides specific fragment ions, also show common aspects in their MS/MS fragmentation patterns. In fact, in each of them, the ions at *m*/*z* 58 and at *m*/*z* 102 are present. The first has been shown to be dimethyl(methylene)ammonium ion, present also in the MS/MS fragmentation pattern of acyclic betaines, the other, at *m*/*z* 102, has elemental composition $C_4H_8N_2$ (Naresh Chary *et al.*, 2012). The major MS/MS fragments of the main betaines and QACs are briefly summarized in Table 12.1, where for each compound are showed the corresponding biological precursor (generally an amino acid).

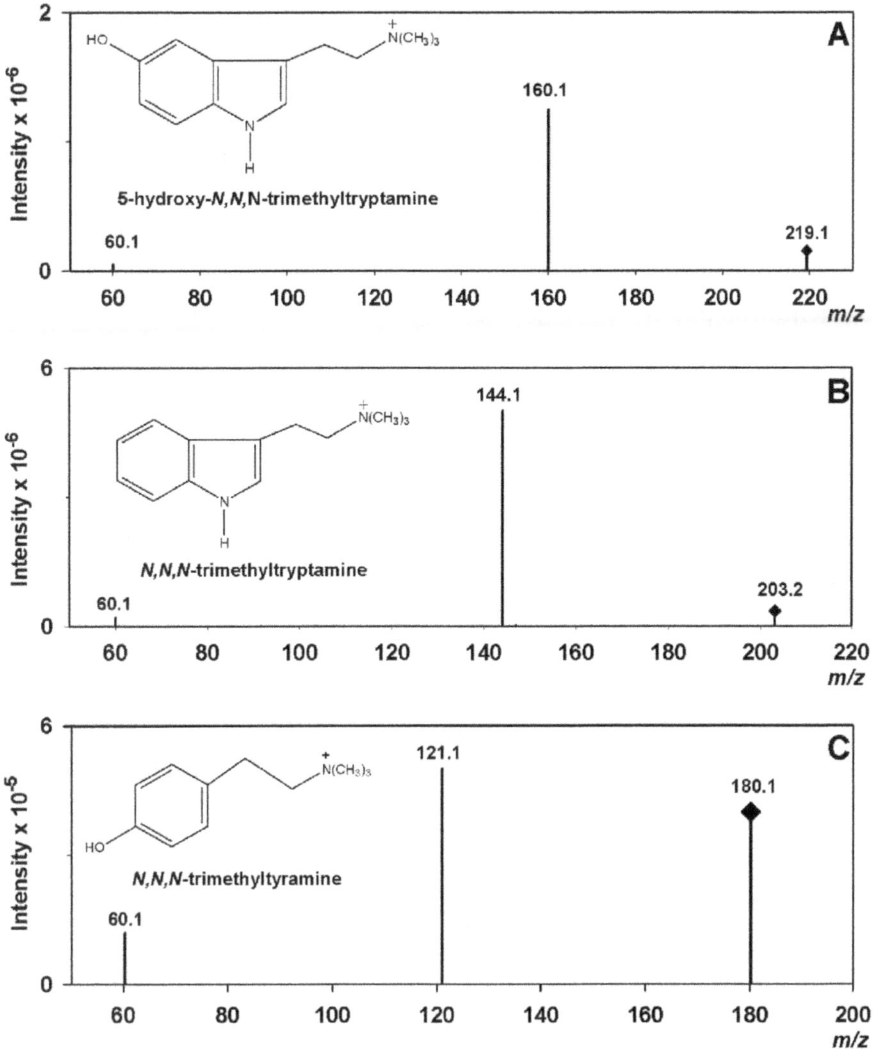


**Figure 12.6** MS/MS fragmentation patterns of QACs derived from tryptamine and tyramine. Panel A. The MS/MS fragmentation pattern of 5-hydroxy-*N,N,N*-trimethyltryptamine, in which predominate the two fragments at *m*/*z* 160, corresponding to 219 minus 59, and at *m*/*z* 60, besides the parent ion at *m*/*z* 219.1. Panel B. The MS/MS fragmentation pattern of *N,N,N*-trimethyltryptamine, in which predominate the two fragments at *m*/*z* 144, corresponding to 203 minus 59, and at *m*/*z* 60, besides the parent ion at *m*/*z* 203.2. Panel C. The MS/MS fragmentation pattern of *N,N,N*-trimethyltyramine, in which predominate the two fragments at *m*/*z* 121, corresponding to 180 minus 59, and at *m*/*z* 60, besides the parent ion at *m*/*z* 180.2. The mass diminution of 59 amu is due the trimethylamine neutral loss from the parent ions. The fragment at *m*/*z* 60 corresponds to the trimethylammonium ion.

It is important to point out that betaines, when in solution in the presence of potassium or sodium ions, easily form adducts with these ions in the gas phase. Therefore, in such case, they are detected at *m*/*z* values corresponding to $[M + Na]^+$ and $[M + K]^+$. However, the adducts are readily revealed by the presence in their CID spectra of the ions $Na^+$ and $K^+$, easily detected at *m*/*z* 23 and *m*/*z* 39, respectively. It is worth noting that some mass spectrometers, especially those based on a quadrupole ion trap, have limits that do not allow detection of ions at *m*/*z* below 40–50 amu. However, as other fragments produced from the cationized species, closely resemble those from the protonated species $[M + H]^+$ (Naresh Chary *et al.*, 2012), the analysis of the whole fragmentation pattern can lead to the unambiguous identification of the cationized betaines.

Many quaternary nitrogen compounds formally can be considered as betaines lacking the carboxyl group. These compounds usually show MS/MS fragmentation patterns simpler to that of the corresponding betaine. Two ions usually predominate in their patterns, *i.e.*, that at *m*/*z* 60 corresponding to trimethylammonium ion and the fragment ion at *m*/*z* $[M + H–C_3H_9N]$, corresponding to the neutral loss of trimethylamine. As an example (Figure 12.6), in the MS/MS fragmentation spectrum of 5-hydroxy-*N*,*N*,*N*-trimethyltryptamine, it can be seen, besides the parent ion (*m*/*z* 219), only the two fragments at *m*/*z* 60 and at *m*/*z* 160 corresponding to 219 minus 59 $[M + H–C_3H_9N]$. Similarly, in the MS/MS fragmentation pattern of *N*,*N*,*N*-trimethyltryptamine ($M + H^+$ *m*/*z* 203) predominate the ions at 144 (203 minus 59) and at *m*/*z* 60 and in that of *N*,*N*,*N*-trimethyltyramine ($M + H^+$ *m*/*z* 180) predominate the ions at *m*/*z* 121 (180 minus 59) and at *m*/*z* 60 (Servillo *et al.*, 2012a, 2013a, 2014a).

## 12.5 Betaines and QACs Distribution in Fruits

The analysis of betaines and QACs has been achieved for numerous foods, including fruits and vegetables, meat, seafood and many other products such as coffee, chocolate, honey, milk, tea, and yogurt (de Zwart *et al.*, 2003; Zeisel *et al.*, 2003; Slow *et al.*, 2005). Particularly interesting are the results concerning the changes of betaine content in some foods subjected to different cooking methods, such as boiling, steaming and microwaving (de Zwart *et al.*, 2003). Indeed, the content of betaines can undergo a significant decrease depending on the method of cooking and these changes should be taken into consideration when estimating the dietary intakes of betaines. In more detail, large losses (60–80%) of betaines occur during boiling and not during baking, microwaving, or frying (de Zwart *et al.*, 2003). In addition only small to medium losses were observed during steaming (de Zwart *et al.*, 2003). The distribution of glycine betaine and choline has been reported in 145 foods. (Zeisel *et al.*, 2003). More recently, the distribution of betaines and related QACs in natural and frozen concentrated juices of citrus fruits and, for the first time, the occurrence of *N*ε-trimethyllysine in fruits and other vegetable sources has

**Table 12.2** Distribution of some betaines in fruits. The content of the betaines is expressed in mg per kg of product. (–) indicates the absence of analytical data in the literature.

| Fruits | Glycine betaine | Proline betaine | Betonicine | Pipecolic acid betaine | Trigonelline | Choline |
|---|---|---|---|---|---|---|
| Apple | ≤ 2 | — | — | — | — | ≤ 7 |
| Apricots | Trace | Trace | — | — | Trace | — |
| Avocado | 3–35 | ≤ 8 | — | — | 3–60 | ≤ 90 |
| Banana | < 5 | Trace | — | — | 23 | ≤ 30 |
| Bergamot | — | 300–500 | 130–300 | 1–2 | — | — |
| Chinotto | — | 420–480 | 170–220 | — | 4–6 | ≤ 20 |
| Cranberry | ≤ 2 | — | — | — | — | < 4 |
| Grape | ≤ 2 | Trace | — | — | Trace | < 40 |
| Grapefruit | ≤ 3 | 190–260 | 50–70 | — | ≤3 | < 8 |
| Kiwifruit | < 5 | Trace | — | — | Trace | — |
| Lemon | — | 120–180 | 40–50 | ≤ 0.5 | 6–10 | 6–10 |
| Mandarin | — | 300–400 | 180–300 | — | 7–14 | 10–20 |
| Mango | Trace | Trace | — | — | Trace | — |
| Yellow orange | ≤ 3 | 400–500 | 100–250 | ≤ 0.5 | 5–10 | 10–15 |
| Red orange | — | 500–600 | 120–250 | — | 3–6 | 5–20 |
| Bitter orange | — | 400–500 | 100–150 | — | 5–10 | 5–10 |
| Peach | < 5 | Trace | — | — | — | < 10 |
| Pear | < 10 | < 5 | — | — | — | ≤ 20 |
| Pineapple | Trace | Trace | — | — | Trace | — |
| Prunes | ≤ 4 | Trace | — | — | Trace | < 8 |
| Strawberries | ≤ 2 | — | — | — | — | < 8 |
| Tomato | ≤ 2 | Trace | — | — | ≤ 30 | ≤ 40 |

been reported (Servillo *et al.* 2011a,b; Servillo *et al.*, 2014b). The distribution of the main betaines in fruits is briefly summarized in Table 12.2.

### 12.5.1 Glycine Betaine

Data regarding the glycine betaine content in fruits are rather scarce. Levels of this compound have been quantified in apple, avocado, banana, cranberry, grape, kiwifruit, orange, peach, pear, prunes, strawberries, and tomato. Overall, the content of glycine betaine in fruits, generally very small, ranges between 2 and 3 mg $kg^{-1}$. By contrast, some vegetables appear to be particularly rich in this betaine that is nutritionally important for its role as methyl-group donor in the bioconversion of homocysteine into methionine. These vegetables include spinach (6 g $kg^{-1}$) and sugar beet (2 g $kg^{-1}$). Other sources particularly rich in this betaine are some cereal derived foods largely consumed, such as pasta and crackers with values ranging between 1 and 2 g $kg^{-1}$ (de Zwart *et al.*, 2003; Slow *et al.*, 2005).

### 12.5.2 Betaines and QAC Deriving from Proline

This class of compounds is species specific for the *Citrus* genus. Stachydrine and betonicine are described in fruits, leaves and seeds of all *Citrus* genus plants (Nolte *et al.*, 1997; Servillo *et al.*, 2011a,b). Proline betaine is particularly abundant in citrus fruits. Red orange juice contains the highest levels (about 550 mg $L^{-1}$). The stachydrine content in yellow orange juice was found between 400 and 500 mg $L^{-1}$. Another rich source of stachydrine is bergamot juices with levels between 300 and 500 mg $L^{-1}$, while in mandarin juice, it ranged between 300 and 400 mg $L^{-1}$. Instead, significantly lower levels of stachydrine are contained in lemon juice (between 120 and 180 mg $L^{-1}$). As for betonicine, it is well represented in mandarin juice with an average value of 270 mg $L^{-1}$. Conversely, lower values of betonicine are reported for lemon and grapefruit juice with mean values of about 50 mg $L^{-1}$ (Servillo *et al.* 2011a,b).

### 12.5.3 γ-Butyrobetaine

To date, the occurrence of γ-butyrobetaine (3-carboxypropyltrimethylammonium) in fruits has never been reported. γ-Butyrobetaine was not detected even in juices of citrus plant fruits, where high levels of γ-aminobutyric acid occur (Servillo *et al.*, 2011b).

### 12.5.4 Pipecolic Acid Betaine and β-Alanine Betaine

The occurrence of pipecolic acid betaines (homostachydrine) has been recently reported for the first time in the *Citrus* genus plants by Servillo *et al.* (2012b). Levels of homostachydrine in bergamot juice ranges between 1 and 2 mg $L^{-1}$ whereas lower levels are described in lemon juice and orange juice (0.5 mg $L^{-1}$). Homostachydrine has also been found in fruits, seeds, and leaves of orange, lemon and bergamot (Citrus bergamia et Poit Risso). However, homostachydrine is present at high concentration (200–500 mg $kg^{-1}$) in *Medicago* and *Achillea genera* (Wood *et al.*, 1991). As for β-alanine betaine, it is known to be effectively accumulated by the high stress-tolerant plants of the Plumbaginaceae family (Hanson *et al.*, 1994). However, no data are reported of β-alanine betaine occurrence in fruits.

### 12.5.5 Trigonelline and Choline

Trigonelline is the nicotinic acid betaine. It is well represented in coffee with values ranging between 2 and 12 g $kg^{-1}$ (de Zwart *et al.*, 2003; Zeisel *et al.*, 2003). Other food products with discrete amounts of this metabolite are alfalfa sprouts (about 200 mg $kg^{-1}$) and pea (about 300 mg $kg^{-1}$). Low levels of trigonelline are present in fruits (about 30 mg $kg^{-1}$) (Slow *et al.*, 2005). In citrus juices trigonelline content is between 3 and 15 mg $L^{-1}$. In particular, mandarin juice shows the highest trigonelline content (between 7 and 15 mg $L^{-1}$) followed by lemon and yellow orange (between 5 and 10 mg $L^{-1}$). Red orange juice shows trigonelline values between 3 and 6 mg $L^{-1}$ (Servillo *et al.*, 2011b). Other sources of trigonelline are tomato, avocado, and bananas with values between 5 and 30 mg $kg^{-1}$ (Slow *et al.*, 2005). As for choline, foods containing this QAC have been widely reviewed by Zeisel *et al.* (2003). The richest fruit sources seem to be avocado (90 mg $kg^{-1}$), grape (50 mg $kg^{-1}$), tomato (40 mg $kg^{-1}$), banana (30 mg $kg^{-1}$), and pear (20 mg $kg^{-1}$) (Zeisel *et al.*, 2003). Apple, apricots, cranberry, peach, prunes, and strawberries shows levels of choline lower than 10 mg $kg^{-1}$. As for citrus fruits, the highest content was found in tangerine (10–20 mg $L^{-1}$) and red orange juices (5–20 mg $L^{-1}$). The choline content in citrus fruits does not exceed 20 mg $L^{-1}$ (Zeisel *et al.*, 2003).

### 12.5.6 $N^{\varepsilon}$-trimethyllysine

$N^{\varepsilon}$-trimethyllysine (TML), is a peculiar betaine inasmuch as it can also be considered an amino acid, the α-amino group still being present in its molecular structure. TML levels do not seem to be related to biotic or abiotic stress. TML is the precursor of carnitine, a metabolite essential for the fatty acid transport and utilization in mitochondria. The current opinion is that mammals utilize the breakdown of their endogenous proteins containing TML residues as the starting point for carnitine biosynthesis. TML was believed to be rather rare in vegetables. Indeed, novel findings reveal that TML is present in all the vegetables with particularly high levels in sweet pepper fruit (10–30 mg $kg^{-1}$) followed by tomato (2–7 mg $kg^{-1}$), potato tuber (1–4 mg $kg^{-1}$), leaf vegetables as spinach and chicory (0.5–2 mg $kg^{-1}$), nettle (0.5–3 mg $kg^{-1}$), and lettuce (0.1–1mg $kg^{-1}$) (Servillo *et al.*, 2014b). It is of particular interest that high levels of TML were found in leaves of alfalfa, a plant of paramount importance for cattle feeding, and in fruit juices (Servillo *et al.*, 2014b). Orange and grapefruit juices contain levels of TML in the same range of the content of some essential amino acids, such as lysine and leucine. Grape juice shows a high content of TML (0.5 and 5 mg $kg^{-1}$) compared with grapefruit juice (0.5–4 mg $kg^{-1}$), bergamot juice (0.3–2 mg $kg^{-1}$), orange juice (0.2–2 mg $kg^{-1}$), mandarin juice (0.1–2 mg $kg^{-1}$), and pineapple juice (0.5–2 mg $L^{-1}$) (Servillo *et al.*, 2014b).

### 12.5.7 Hercynine and Ergothioneine

Hercynine is the betaine derived from histidine. Its distribution in the natural world is almost completely unknown. Its importance consists in the fact that it is the biological precursor of ergothioneine (EGT), which is chemically 2-thiol-L-histidine betaine. Although EGT is biosynthesized only by a few types of fungi and prokaryotes, it is widespread in vegetables and animals, which effectively accumulate this betaine. A specific protein deputed to the EGT transport into human erythrocyte has been discovered in the membrane of these blood cells (Grundemann *et al.*, 2005). The role of EGT is still a matter of debate, but it is prevalently believed that EGT has a role as an antioxidant in eukaryotic cells (Cheah and Halliwell, 2012). The highest ergothioneine concentrations were found in the mushrooms *Boletus edulis* (400–600 mg $kg^{-1}$) and *Pleurotus ostreatus* (90–150 mg $kg^{-1}$), while in other species of mushrooms the EGT levels were much lower. In vegetables, EGT was found in black turtle bean and red kidney bean in the range 4–15 mg $kg^{-1}$, in oat bran at about 4.5 mg $kg^{-1}$, in wheat bran and wheat germ in the range 0.5–1 mg $kg^{-1}$. Values below 0.5 mg $kg^{-1}$ were found in fruits such as apple, pear, apricot, orange, lemon, grapefruit, pineapple, mango, and papaya (Ey *et al.*, 2007).

## 12.6 Conclusions

Analysis of betaine and $QAC_S$ by HPLC-ESI-MS/MS is now much more rapid, specific, and sensitive than other techniques used in the past. Thanks to the use of HPLC-ESI-MS/MS, the determination of these important phytochemical indicators can be obtained with minimal sample preparation, in some cases even without any cleanup step, and with short analysis times. This advanced technology could be helpful in the analysis of betaines as a biomarker in the evaluation of the food quality. Indeed, in the light of the fact that betaine distribution in plants has been defined as species specific (Hanson *et al.*, 1994), it seems reasonable to assume that betaine content can help to characterize vegetable sources and their derivatives, such as fruit juice. Finally, beside the extension of the analysis of betaine and QACs in a wide variety of plant sources, future directions should also focus on the effects of food processing (thermal treatment, storage, cooking) on the content of betaines.

## Summary points

- This chapter focuses on the occurrence and distribution of betaines in fruits and on the analytical techniques for their analysis.
- Betaines are a group of low molecular weight compounds, highly soluble in water, derived from amino acids in which the nitrogen atom is fully methylated.
- Betaines take part in the defensive response of plants towards abiotic stresses.
- Numerous analytical methods have been proposed for the determination of betaines in vegetables and fruits.
- Presently, high-performance liquid chromatography electrospray tandem mass spectrometry appears to be the method of choice for the determination of betaines.
- Betaines show peculiar features of tandem mass fragmentation patterns that facilitate their identification and quantification.
- Betaines distribution in vegetables and fruits appears to be species specific.

## Key Facts of Betaines

1. Betaines are substances utilized by plant cells to counteract abiotic stress.
2. Betaines accomplish their biological function at high concentrations for they are compatible solutes that do not interfere with the cell functions.
3. Betaines are characterized by the presence of a quaternary ammonium group that confers particular sensitivity and fragmentation features useful for their detection at low levels and selective quantification by mass spectrometry.
4. Betaines can be readily detected and quantified by high-performance liquid chromatography electrospray tandem mass spectrometry.
5. The content of betaines can help to characterize vegetable sources and their derivatives, such as fruit juice.
6. Beside the extension of the knowledge of levels of betaines in additional plant sources, future directions should also concern the effects of food processing (thermal treatment, storage, cooking) on their content.

## Definitions and Explanations of Key Terms

**Abiotic stress**: Adverse effects produced by environmental factors on growth, development and productivity of plants. Examples are lack of water, excess of water, high temperature, low temperatures, excessive salinity, nutritional deficiencies, and toxic substances in the soil.

**Betaines**: Zwitterionic compounds generally bearing a permanent positive charge on a quaternary nitrogen atom and negative charge on a carboxylate group. Most betaines derive from amino acid in which the amino group is fully methylated.

**Biotic stress**: Adverse effects produced on growth, development and productivity of plants by attacks of other living organisms.

**Brix degree**: Measure unit employed to express the concentration of soluble solids in fruit juices and fruit concentrates. Brix degrees indicate the grams of soluble solids contained in 100 g of solution.

**CID**: Collision-induced dissociation, a procedure to fragment ions based on their collision with atoms of an inert gas, usually helium or argon.

**Compatible solute**: Substance that can be accumulated in considerable amounts inside cells without interfering with the normal cellular enzymatic activities and helping cells to counteract osmotic stress.

**Electrospray ionization chamber**: The part of the mass spectrometer where solutions containing the analytes are sprayed to produce ions in the gas phase.

**ESI**: Electrospray ionization, a technique to produce ions from compounds to be analyzed by mass spectrometry. It essentially consists of spraying in a strong electrical field the solution containing the analyte. A stream of hot gas dries the solution droplets until analyte ions are burst in the gas phase. The ions produced are then driven in other parts of the mass spectrometer for the detection and fragmentation.

**FIA**: Flow injection analysis. Analytical mass spectrometric approach where the solution containing the analyte is not subjected to chromatography but directly injected into the eluent flowing in the mass spectrometer electrospray ionization chamber. FIA analyses are very fast but may be unreliable if the samples contain high levels of interfering substances.

**Fragmentation pattern**: The distribution according to *m*/*z* ratios and intensities of fragment ions produced from the parent ion.

**N-methyltransferases**: Enzymes catalyzing methylation reactions in which methyl groups are transferred from a donor molecule, usually SAM, to a nitrogen atom of acceptor molecules.

**Osmoprotectant**: The same as a compatible solute.

**Osmotic stress**: Abrupt variation of intracellular osmotic pressure generally caused by sudden changes of intracellular or extracellular solute concentrations.

**QACs**: Compounds bearing a permanent positive charge on a quaternary nitrogen atom. Most QACs from vegetables derive from biogenic amines whose amino group has been fully methylated.

**SAM**: *S*-Adenosylmethionine, a universal donator of methyl groups, involved in methylation reactions catalyzed by methyltransferase enzymes.

**Tandem mass spectrometry**: A sequential mass spectrometric technique where in a first step the ionized analyte of interest is selected according to its *m*/*z* ratio and in a successive step is subjected to the fragmentation procedure. The pattern of fragment ions produced is then analyzed according to their *m*/*z* ratios and intensities.

## List of Abbreviations

| | |
|---|---|
| BADH | Betaine aldehyde dehydrogenase |
| CID | Collision Induced Dissociation |
| CMO | Choline mono-oxygenase |
| DMSP | 3-Dimethylsulfoniopropionate |
| EGT | Ergothioneine |
| ESI | Electrospray ionization |
| FAB | Fast atom bombardment |
| FIA | Flow injection analysis |
| HILIC | Hydrophilic interaction liquid chromatography |
| HPLC | High-performance liquid chromatography |
| HypBet | Hydroxyproline betaine |
| 5-Me-TH4folate | *N*5-methyltetrahydrofolate |
| MS | Mass spectrometry |
| NMePro | *N*-Methylproline |
| NMR | Nuclear magnetic resonance |
| NP-LC | Normal-phase liquid chromatography |
| PipBet | Pipecolic acid betaine |
| PD | Plasma desorption |

| | |
|---|---|
| Pro | Proline |
| ProBet | Proline betaine |
| QACs | Quaternary ammonium compounds |
| ROS | Reactive oxygen species |
| RP-LC | Reverse phase liquid chromatography |
| SAH | *S*-Adenosylhomocysteine |
| SAM | *S*-Adenosylmethionine |
| SDC | Serine decarboxylase |
| TH4folate | Tetrahydrofolate |
| TYDC | Tyrosine decarboxylase |
| TML | $N^{\varepsilon}$-Trimethyllysine |

## References

Barlow, R. B. and Burston, K. N., 1980. A comparison of cinobufotenin (the quaternary derivative of 5-HT) and some related compounds with coryneine (the quaternary derivative of dopamine) on the frog rectus, guinea-pig ileum and rat fundus strip preparations. *British Journal of Pharmacology*. 69: 597–600.

Blunden, G., Gordon, S. M., Crabb, T. A., Roch, O. G., Rowan, M. G. and Wood, B., 1986. NMR spectra of betaines from marine algae. *Magnetic Resonance in Chemistry*. 24: 965–971.

Cheah, I. K. and Halliwell, B., 2012. Ergothioneine; antioxidant potential, physiological function and role in disease. *Biochimica et Biophysica Acta*. 1822: 784–793.

Deguchi, T., Urakawa, N., Hayama, T. and Ohkubo., Y., 1963. Ganglion stimulating action of candicine. *Japanese Journal of Pharmacology*. 13: 143–159.

de Zwart, F. J., Slow, S., Payne, R. J., Lever, M., George, P. M., Gerrard, J. A. and Chambers, S. T., 2003. Glycine betaine and glycine betaine analogues in common foods. *Food Chemistry*. 83: 197–204.

Ey, J., Schoming, E. and Taubert, D., 2007. Dietary sources and antioxidant effects of ergothioneine. *Journal of Agricultural and Food Chemistry*. 55: 6466–6474.

Facchini, P. J., Huber-Allanach, K. L. and Tari, L. W., 2000. Plant aromatic L-amino acid decarboxylase: Evolution, biochemistry, regulation, and metabolic engineering applications. *Phytochemistry*. 54: 121–138.

Gage, D. A., Rhodes, D., Nolte, K. D., Hicks, W. A., Leustek, T., Cooper, A. J. L. and Hanson, A. D., 1997. A new route for synthesis of dimethylsulphoniopropionate in marine algae. *Nature*. 387: 891–894.

Giovanelli, J., Mudd, S. H. and Datkp, A. H., 1980. Sulfur amino acids in plants. In: Miflin, B. J. (ed.) *The Biochemistry of Plants*. Academic Press, New York, pp. 453–505.

Grundemann, D., Harlfinger, S., Golz, S., Geerts, A., Lazar, A., Berkels, R., Jung, N., Rubbert, A. and Shomig, E., 2005. Discovery of the ergothioneine transporter. *Proceedings of the National Academy of Sciences of the United States of America*. 102: 5256–5261.

Hanson, A. D., Rathinasabapathi, B., Rivoal, J., Burnet, M., Dillon, M. O. and Gage, D. A., 1994. Osmoproctetive compounds in the Plumbaginaceae: a natural experiment in metabolic engineering of stress tolerance. *Proceedings of the National Academy of Sciences of the United States of America*. 91: 306–310.

Happer, D. A. R., Hayman, C. M., Storer, M. K. and Lever, M., 2004. Aracyl triflates as derivatizing agents for biological betaine. *Australian Journal of Chemistry*. 57: 467–472.

Howe, G. A. and Jander, G., 2008. Plant immunity to insect herbivores. *Annual Review of Plant Biology*. 59: 41–66.

Ishihara, A., Hashimoto, Y., Miyagawa, H. and Wakasa, K., 2008. Induction of serotonin accumulation by feeding of rice striped stem borer in rice leaves. *Plant Signaling and Behavior*. 3: 714–716.

Jaspers, P. and Kangasjarvi, J., 2010. Reactive oxygen species in abiotic stress signaling. *Physiologia Plantarum*. 138: 405–413.

Kang, K., Kim, Y. S., Park, S. and Back, K., 2009. Senescence-induced serotonin biosynthesis and its role in delaying senescence in rice leaves. *Plant Physiology*. 150: 1380–1393.

Kavi Kishor, P. B., Sangam, S., Amrutha, R. N., Sri Laxmi, P., Naidu, K. R., Rao, K. R. S. S., Rao, S., Reddy, K. J., Theriappan, P. and Sreenivasulu, N., 2005. Regulation of proline biosynthesis, degradation, uptake and transport in higher plants: Its implications in plant growth and abiotic stress tolerance. *Current Science*. 88: 424–438.

Kurz, M., 2008. Review. Compatible solute influence on nucleic acids: many questions but few answers. *Saline Systems*. 4: 6. DOI: 10.1 1186/ 1746-1448-4-6.

Malin, G. and Kirst, G. O., 1997. Algal production of dimethyl sulphide and its atmosheric role. *Journal of Phycology*. 33: 889–896.

Naresh Chary, V., Dinesh Kumar, C., Vairamani, M. and Prabhakar, S., 2012. Characterization of amino acid-derived betaines by electrospray ionization tandem mass spectrometry. *Journal of Mass Spectrometry*. 47: 79–88.

Neill, S., Barros, R., Bright, J., Desikan, R., Hancock, J., Harrison, J., Morris, P., Ribeiro, D. and Wilson, I., 2008. Nitric oxide, stomatal closure, and abiotic stress. *Journal of Experimental Botany*. 59: 165–176.

Nolte, K. D., Hanson, A. D. and Gage, D. A., 1997. Proline accumulation and methylation to proline betaine in Citrus : implications for genetic engineering of stress resistance. *Journal of the American Society for Horticultural Science*. 122: 8–13.

Rhodes, D. and Hanson, A. D., 1993. Quaternary ammonium and tertiary sulphonium compounds in higher plants. *Annual Review of Plant Physiology and Plant Molecular Biology*. 44: 357–384.

Rhodes, D., Rich, P. J., Myers, A. C., Reuter, C. C. and Jamieson, G. C., 1987. Determination of betaines by fast atom bombardment mass spectrometry. *Plant Physiology*. 84: 781–788.

Servillo, L., Giovane, A., Balestrieri, M. L., Cautela, D. and Castaldo, D., 2011a. Proline derivatives in fruits of bergamot (Citrus bergamia Risso et Poit):

presence of *N*-methylproline and 4-hydroxyproline betaine. *Journal of Agricultural and Food Chemistry*. 59: 274–281.

Servillo, L., Giovane, A., Balestrieri, M. L., Bata-Csere, A., Cautela, D. and Castaldo, D., 2011b. Betaines in fruits of Citrus genus plants. *Journal of Agricultural and Food Chemistry*. 59: 9410–9416.

Servillo, L., Giovane, A., Balestrieri, M. L., Cautela, D. and Castaldo, D., 2012a. *N*-methylated tryptamine derivatives in citrus genus plants: identification of *N,N,N*-trimethyltryptamine in bergamot. *Journal of Agricultural and Food Chemistry*. 60: 9512–9518.

Servillo, L., Giovane, A., Balestrieri, M. L., Ferrari, G., Cautela, D. and Castaldo, D., 2012b. Occurrence of pipecolic acid and pipecolic acid betaine (homostachydrine) in Citrus genus plants. *Journal of Agricultural and Food Chemistry*. 60: 315–321.

Servillo, L., Giovane, A., Balestrieri, M. L., Casale, R., Cautela, D. and Castaldo, D., 2013a. Citrus genus plants contain *N*-methylated tryptamine derivatives and their 5-hydroxylated forms. *Journal of Agricultural and Food Chemistry*. 61: 5156–5162.

Servillo, L., Giovane, A., D'Onofrio, N., Casale, R., Cautela, D., Ferrari, G., Balestrieri, M. L. and Castaldo, D., 2014a. *N*-Methylated derivatives of tyramine in citrus genus plants: Identification of *N,N,N*-trimethyltyramine (candicine). *Journal of Agricultural and Food Chemistry*. DOI: 10.1021/jf5001698.

Servillo, L., Giovane, A., Cautela, D., Castaldo, D. and Balestrieri, M. L., 2014b. Where does $N^{\varepsilon}$- trimethyllysine for the carnitine biosynthesis in mammals come from? *PLos One*. 9: e84589.

Slow, S., Donaggio, M., Cressey, P. J., Lever, M., George, P. M. and Chambers, S. T., 2005. The betaine content of New Zealand foods and estimated intake in the New Zealand diet. *Journal of Food Composition and Analysis*. 18: 473–485.

Wood, K. V., Stringham, K. J., Smith, D. L., Volenec, J. J., Hendershot, K. L., Jackson, K. A., Rich, P. J., Yang, W.-J. and Rhodes, D., 1991. Betaines of Alfalfa. Characterization by fast atom bombardment and desorption chemical ionization mass spectrometry. *Plant Physiology*. 96: 892–897.

Wood, K. V., Bonham, C. C., Miles, D., Rothwell, A. P., Peel, G., Wood, B. C. and Rhodes, D., 2002. Characterization of betaines using electrospray MS/MS. *Phytochemistry*. 59: 759–765.

Yoshioka, H., Numata, N., Nakajima, K., Katou, S., Kawakita, K., Rowland, O., Jones J. D. G. and Doke, N., 2003. Nicotiana benthamiana gp91$^{phox}$ homologs NbrbohA and NbrbohB participate in $H_2O_2$ accumulation and resistance to *Phytophtora infestans*. *Plant Cell*. 15:706–718.

Zeisel, S. H., Mar, M. H., Howe, J. C. and Holden, J. M., 2003. Concentrations of choline-containing compounds and betaine common foods. *Journal of Nutrition*. 133: 1302–1307.

Zhao, B. T., Tian, B., Jeong, S. Y., Hwangbo, K., Moon, D. C., Seo, E., Lee, D., Lee, J., Min, B., Ma, F., Son, J. K. and Woo, M. H., 2013. Quantitative analysis of betaine in Lycii Fructus by HILIC-ELSD. *Archives of Pharmacal Research*. 36: 1231–1237.

# Section IV
# Function and Effects

CHAPTER 13

# *Betaine and Metabolic Stress in Different Tissues*

KAISA OLLI*[a] AND KIRSTI TIIHONEN[a]

[a]DuPont Nutrition and Health, Kantvik Active Nutrition, Sokeritehtaantie 20, FI-02460, Kantvik, Finland
*E-mail: kaisa.olli@dupont.com

## 13.1 Introduction

Betaine (glycine betaine, trimethylglycine) is one of the most powerful organic osmolytes in living organisms. Plants, in particular, produce betaine as a response to environmental stresses such as drought, salinity, extreme temperature and UV radiation (Ashraf and Foolad, 2007). The plant that is best known for accumulating betaine is sugar beet (*Beta vulgaris*), however, betaine can be found in a wide variety of other plant, animal and microbial species.

The aim of this chapter is to evaluate the role of dietary betaine in nutrition with a focus on its role in animal tissues under metabolic and osmotic stress. The bioavailability of the betaine molecule to various tissues is essential when evaluating the role of betaine in stress. The bioavailability of betaine involves its digestion, absorption, distribution, and metabolism as well as elimination from the body.

### 13.1.1 Physicochemical Properties of Betaine

Betaine is a naturally occurring, nontoxic, stable and highly water soluble (160 g per 100 g) *N*-methylated amino acid with three chemically reactive methyl groups attached to the nitrogen atom of a glycine molecule. This

Food and Nutritional Components in Focus No. 7
Betaine: Chemistry, Analysis, Function and Effects
Edited by Victor R Preedy

Published by the Royal Society of Chemistry, www.rsc.org

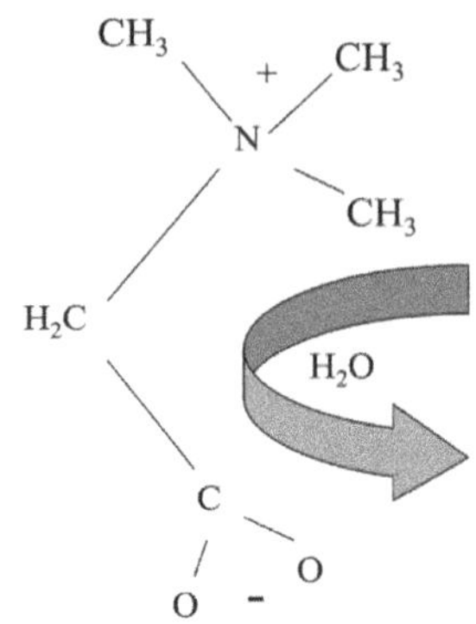


**Figure 13.1** The chemical structure of betaine. A water molecule can temporarily occupy the space between the $COO^-$ and methyl groups when in contact with the betaine molecule.

small molecule (117.2 Da) is very hygroscopic, retaining several water molecules in its vicinity. Opposite hydrophopic (trimethyl) and hydrophilic (glycine) sides of the betaine molecule increase the mobility of water around it. The zwitterionic characteristic of betaine increases its hydrophilicity and allows it to adapt to a large pH range (Mathlouthi, 1997). The molecular structure and charge distribution of betaine are shown in Figure 13.1.

The unique interaction between betaine and water molecules is important for water regulation in various cell types. The negative hydration number ($h \sim -0.797$) and the high Huggins constant ($k' \sim 1.16$–$1.19$) of betaine indicates that a water molecule attracted by a betaine molecule is rapidly substituted by another water molecule (Mathlouthi, 1997). Thus, betaine carries water molecules without immobilizing them, leaving them readily available for biological reactions.

### 13.1.2 Digestion, Absorption and Distribution

Most of the body's betaine originates from dietary sources containing betaine or choline. The synthesis of betaine from choline involves a two-step oxidation process with choline dehydrogenase and betaine aldehyde dehydrogenase in the liver mitochondria (Craig, 2004). However, only a small proportion of dietary choline is converted to betaine (Siljander-Rasi *et al.*, 2003).

The basal betaine concentration in human plasma is about 50 $\mu mol\ l^{-1}$, but can be increased up to 1 $mmol\ l^{-1}$ after a 6 g dose (Schwab *et al.*, 2006). In humans, free dietary betaine is absorbed quickly into the blood circulation. Peak betaine concentrations are observed at 60–80 min after ingestion (Schwab *et al.*, 2006). In chickens, the absorption kinetics of betaine occurs quickly (peak concentration levels at 30 min after ingestion) when compared to the related nutrients choline and methionine, as presented in Figure 13.2 (Kettunen *et al.*, 2001b).

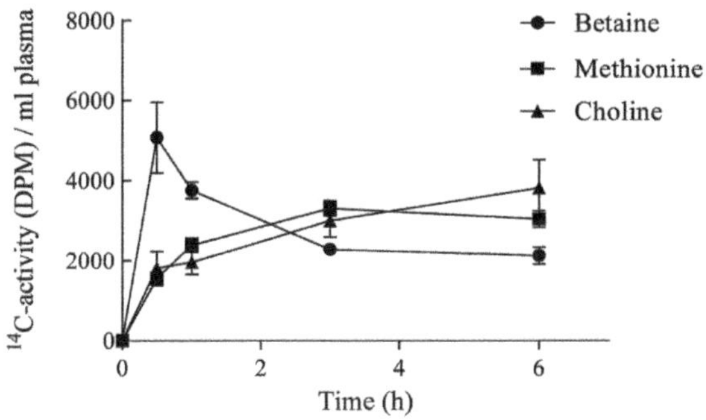


**Figure 13.2** Absorption kinetics of betaine, methionine and choline. The appearance of $^{14}$C-activity from [methyl-$^{14}$C]-labeled betaine, methionine and choline in chicken plasma after feeding. The activity is presented as DPM per ml plasma. Values are presented as Mean ± SEM ($n$ = 3–4). This figure has been reproduced from Kettunen *et al.* (2001b) with kind permission from Elsevier. DPM, disintegration per minute.

Ingested betaine is absorbed in the brush-border membrane of the small intestinal epithelium, mainly by the imino acid transporter (SLC36A1). However, betaine is also known to be carried by other transporters with different substrate selectivities (Thwaites and Anderson, 2007). A significant proportion of betaine is absorbed in the chicken duodenum using $Na^+$-dependent transporters and, in addition, the absorption capacity can be induced by the exposure to dietary betaine (Kettunen *et al.*, 2001b). In comparison to betaine supplements, the digestion of food products containing endogenous betaine takes more time and may prolong the absorption of betaine.

In animals, the highest concentrations of betaine are found in the kidney and liver, however betaine can also be detected in all measured tissues (Kettunen, 2001; Slow *et al.*, 2009). The distribution of dietary betaine throughout a chicken, as measured by [methyl-$^{14}$C]-labeled betaine, has shown that there is a considerable accumulation in the gut epithelium, kidney and especially the liver 24 h after ingestion (Figure 13.3). In the plasma and muscle, betaine was located predominantly in the water phase, however, in the liver and intestine it was distributed more evenly within the water, lipid and protein phases. In comparison, dietary choline was located predominantly in the lipid phase of these tissues (Kettunen *et al.*, 2001b).

Betaine is carried from the plasma to tissues mainly by the $Na^+$-coupled betaine/gamma-amino-butyric acid (GABA) transporter 1 (BGT1). Different BGT1 types are regulated and expressed in a tissue-specific manner (Takenaka *et al.*, 1995). Zhou *et al.* (2012) have shown that hepatocytes have a high capacity to accumulate betaine *via* BGT1 that is expressed predominantly in plasma membranes facing the portal circulation. BGT1 is also expressed in the kidney medulla, mainly in the basolateral membrane of the medullary thick ascending tubules and collecting ducts, where it extracts betaine from the vascular system (Zhou *et al.*, 2012). The accumulation of betaine in renal

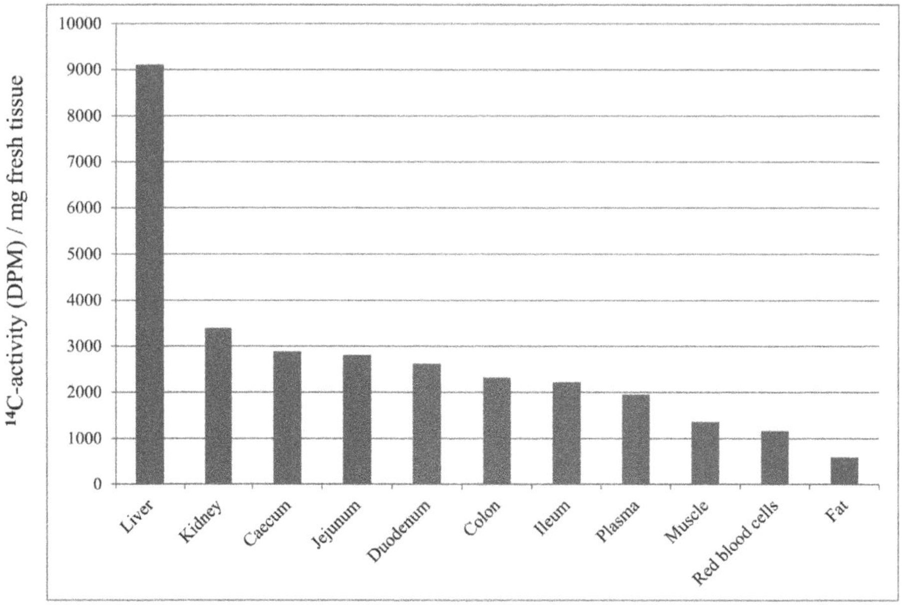


**Figure 13.3** The distribution of dietary betaine in chicken tissues. The distribution of $^{14}$C-activity from [methyl-$^{14}$C]-labeled betaine in different chicken tissues ($n$ = 2) 24 h after feeding (K. Tiihonen & H. Kettunen, unpublished observations). DPM, disintegration per minute.

cells is regulated by hypertonic stress (Kempson and Montrose, 2004). BGT1 is also found in other tissues such as the brain, Kupffer cells and keratinocytes (Takenaka *et al.*, 1995; Warskulat *et al.*, 1996, 2004).

## 13.2 Betaine and Methylation in the Liver

### 13.2.1 Methylation Functions

Methyl groups are required for many metabolic processes, such as the synthesis of phospholipids, adrenal hormones, RNA and DNA. Methyl groups cannot be synthesized in the body and must be provided in the diet. Choline, methionine and betaine can donate methyl groups for these reactions using different biochemical routes (for a review see Craig, 2004). The primary methyl donor in most metabolic reactions in the body is activated methionione, known as *S*-adenosylmethionine (*S*-AM). When *S*-AM donates its methyl group, it forms *S*-adenosylhomocysteine, which is converted to homocysteine. Defects in homocysteine metabolism, either genetic or nutritional, lead to elevated plasma homocysteine that is associated with cardiovascular disease risk (Craig, 2004).

Betaine can remethylate homocysteine to methionine *via* the activity of betaine-homocysteine *S*-methyltransferase (BHMT), thus reducing plasma homocysteine (Olthof *et al.*, 2003; Schwab *et al.*, 2002). BHMT is the only

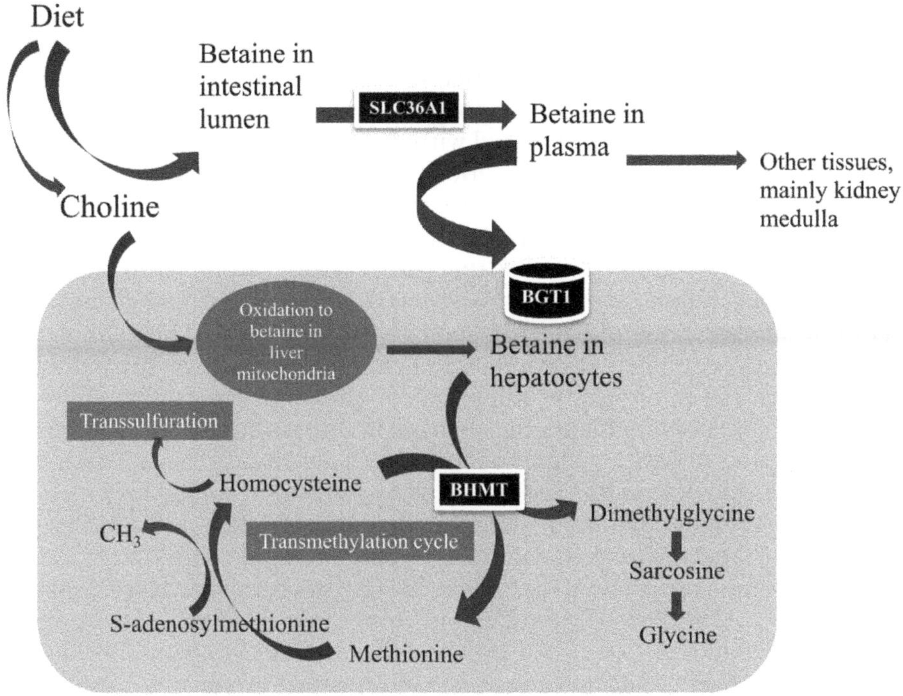


**Figure 13.4** The metabolism of dietary betaine. Imino acid transporter (SLC36A1) is expressed in the small intestine. The betaine/gamma-aminobutyric acid transporter 1 (BGT1) is expressed mainly in liver. Betaine is metabolized in liver hepatocytes by the betaine–homocysteine *S*-methyltransferase (BHMT). A proportion of dietary choline is converted to betaine in the liver mitochondria.

known betaine-metabolizing enzyme and it is expressed in hepatocytes and the kidney cortex (Figure 13.4) (Delgado-Reyes *et al.*, 2001; Sunden *et al.*, 1997). Betaine contains three methyl groups, all of which can be utilized in the transmethylation cycle. During methyl donation, betaine is metabolized to glycine *via* dimethylglycine and sarcosine. Only a small amount of betaine is excreted in the urine.

## 13.2.2 Betaine and Liver Fat Metabolism

Dietary betaine has shown to increase liver betaine concentrations in animals (Kettunen *et al.*, 2001c; Siljander-Rasi *et al.*, 2003). Evidence suggests that the liver is the body's primary site for storing betaine reserves. A recent study in mice (Pekkinen *et al.*, 2013) using nontargeted metabolite profiling demonstrated that betaine levels increased in the liver when drinking water was supplemented with 1% (wt/vol) betaine over a four-week period. In addition, betaine supplementation increased methionine and dimethylglycine levels and decreased choline levels in the liver of mice. Metabolic profiling of mice liver also found that betaine supplementation increased the number of short-chain acylcarnitine species, which are recognized carnitine metabolites. Thus, supplemented dietary betaine seems to affect hepatic carnitine metabolism *via* the transmethylation cycle. Pekkinen and coworkers (2013) also noted that a diet supplemented with dietary betaine decreased the concentration of triglycerides and a variety of other lipid species in mice liver tissue. These data support the evidence that betaine can reduce hepatic lipid accumulation induced by a high-fat diet.

There are several animal studies that demonstrate the ability of betaine to alleviate nonalcoholic fatty liver disease (NAFLD) by correcting the adverse or harmful changes in methionine metabolism that occur with a high-fat diet. In a recent study with mice having NAFLD, betaine alleviated hepatic pathological changes and improved insulin sensitivity in adipose tissue (Wang *et al.*, 2010). The effect of betaine on the changes in transsulfuration and oxidative reactions associated with a high-fat intake has been studied using a rat model (Kwon *et al.*, 2009). Results showed that disrupted sulfur-amino acid metabolism can be re-maintained by betaine supplementation. Supplementation can also protect the liver from nonalcoholic steatosis and oxidative stress. However, the mechanisms behind the beneficial effects of betaine in NAFLD may be dependent on several factors. A further mice study revealed that betaine might affect the hepatic AMP-activated protein kinase (AMPK) system and thereby serve as a therapeutic agent against hepatic steatosis (Song *et al.*, 2007). Evidence from a recent rat trial (Zhang *et al.*, 2013) also suggests that the mechanism by which betaine improves liver function and offers protection against high-fat-induced NAFLD may be related to the inhibition of signaling pathways involving nuclear proteins and pattern-recognition receptors, *i.e.* toll-like receptors (TLR). In addition to high-fat diet-induced liver stress, evidence from mice and rats demonstrates that betaine supplementation could protect against hepatotoxicity induced by a toxic dose of chloroform (0.5 ml $kg^{-1}$) (Kim and Kim, 2005) or alcohol (Kharbanda *et al.*, 2005). Although hepatoprotective effects of dietary betaine have been reported in several other animal studies, proving similar results in humans with NAFLD is challenging.

### 13.2.3 Liver Macrophages and the Immunological Effects of Betaine

Liver macrophages, *i.e.* Kupffer cells, are specialized immune cells involved with the mononuclear phagocyte system. Their function is heavily influenced by changes in cell–volume homeostasis. Betaine is used as an organic osmolyte to protect against changing osmolarity in liver macrophages (Warskulat *et al.*, 1996). The betaine transporter, BGT1, is shown to be upregulated in these cells when exposed to hyperosmotic incubation.

Macrophages play a central role in immune defence by phagocytosing pathogens and producing immune molecules. Phagocytosis by liver macrophages is inhibited by hyperosmotic exposure, but this inhibition can be prevented in the presence of 1–10 mM betaine (Warskulat *et al.*, 1996). Changes in osmotic conditions can affect the expression of various cytokines such as tumor necrosis factor-alpha (TNF-α) and proinflammatory enzymes such as cyclo-oxygenase-2 (COX-2) (Zhang *et al.*, 1996a). TNF-α is produced by Kupffer cells in response to an invasion of pathogens and it is known to mediate endotoxin effects *in vivo*. Supplemented betaine has been shown to attenuate the stimulation of inducible COX-2 and TNF-α in hyperosmotically stimulated Kupffer cells *in vitro* (Zhang *et al.* 1996a,b). Thus, osmolytes like betaine have the potential to modulate the cellular immune response *in vivo*.

## 13.3 Betaine and Osmoregulation in the Kidney

### 13.3.1 Osmoregulation

Water balance is important for the optimal function of intracellular macromolecules. Cells maintain their volume through changes in their concentrations of intracellular organic or inorganic osmolytes. Even small fluctuations in cell hydration act as signals for cellular metabolism and gene expression (Häussinger, 1996). Cells preserve water levels by pumping ions into the cell, which causes water to follow by diffusion. In addition to inorganic ions, low molecular weight organic osmolytes, such as betaine, attract water. These small organic osmolyte molecules function to preserve intracellular solution concentrations that are optimal for metabolic activities even in the face of osmotic stress.

In an aqueous solution, betaine can change the arrangement of water molecules. It retains a certain number of neighboring water molecules, but allows them to be constantly exchanged with the surrounding water (Mathlouthi, 1997). Thus, when assimilated in to a cell, betaine retains water, while at the same time allowing it to be made available for metabolic processes within that cell. Betaine, as a compatible osmolyte, also inhibits the accumulation of inorganic ions, therefore protecting enzymes and membranes from osmotic inactivation. Betaine acts as an organic osmolyte in a variety of cells, but especially in those that are faced with hypertonic conditions such as the intestinal epithelium, keratinocyte cells, kidney and liver macrophages (Kettunen *et al.*, 2001a; Warskulat *et al.*, 2004; Yancey and Burg 1990; Zhang *et al.* 1996b). Although endothelial and muscle cells are not normally faced with hyperosmotic conditions, it has been shown that they can also adapt to and survive extreme conditions (*e.g.* dehydration) by accumulating organic osmolytes such as betaine (Alfieri *et al.*, 2006).

### 13.3.2 The Role of Betaine in the Kidney

The kidney is constantly exposed to hyperosmotic stress under normal physiological conditions. Urea is an osmolyte but its properties disrupt protein structure and function. Organic osmolytes such as betaine, sarcosine, inositol

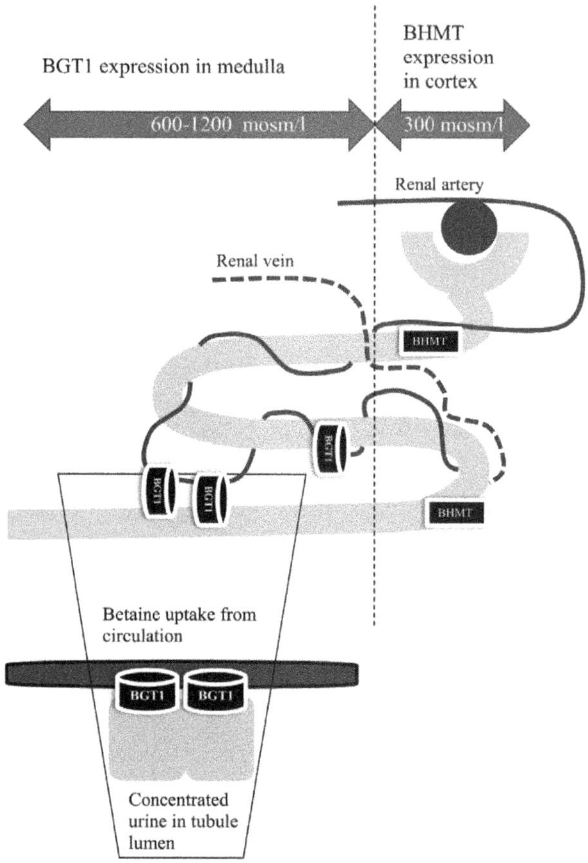


**Figure 13.5** Betaine as an osmolyte in kidney cells. Betaine is absorbed from circulation to kidney epithelial cells by the betaine/gamma-amino-butyric acid transporter 1 (BGT1). It is expressed in the kidney medulla where osmolarity can be as high as 1200 mosm $l^{-1}$. Betaine can be metabolized by betaine-homocysteine *S*-methyltransferase (BHMT) in the kidney cortex where the osmolarity is close to blood osmolarily. The figure has been reproduced from Zhou *et al.* (2012) with kind permission from the American Physiological Society.

and glycerophosphorylcholine are found in the kidney medulla where they counteract the harmful effects of urea on protein function (Yancey and Burg, 1990). BGT1 is found in the renal medulla, mainly in the basolateral membrane of the medullary thick ascending tubules and collecting ducts, where it transports betaine from the vascular system. However, BGT1 is virtually absent in the renal cortex that is not normally exposed to hypertonic stress (Figure 13.5) (Zhou *et al.*, 2012). The brush-border of the cortical tubules has additional transporters that can recover betaine that has been lost in urine (Broer, 2008). Cortical cells contain BHMT and can metabolize betaine, whereas the medulla is almost lacking in BHMT, which shows the different roles of betaine in the kidney cortex and medulla (Zhou *et al.*, 2012).

# 13.4 Betaine in Epithelial Cells

## 13.4.1 Betaine in the Gut Epithelium

Betaine appears to have an important protective role within the intestinal tissue. In chickens, high levels of betaine are found in different sections of the small intestine and to a lesser extent in the colon (Kettunen, 2001). The protective effect of betaine can be explained partly by improved phagocytosis by resident intestinal macrophages, reduction of energy expenditure when animals are exposed to energy-limited diets and improved osmoregulation (Klasing *et al.*, 2002; Schrama *et al.*, 2003). *In vitro* studies with chicken tissues have shown that the duodenal epithelium is particularly capable of using betaine as an osmolyte when exposed to a hyperosmotic environment (Kettunen *et al.*, 2001a). Dietary betaine has also been shown to stabilize the intestinal morphology of chickens (Kettunen *et al.*, 2001c).

## 13.4.2 Betaine in Skin and Oral Mucosa

Human skin cells, particularly keratinocytes located in the outer epidermis (stratum granulosum), are typically exposed to UV radiation and low water concentrations. Both of these stressors may cause disturbances in cell hydration and macromolecular function. To maintain cell–water balance, epidermal keratinocytes are able to import and efflux organic osmolytes, such as betaine. Betaine can be found in the epidermis of human skin (H. Putaala, personal communication) and its transporter, BGT1, is also known to be expressed in skin keratinocytes. Warskulat *et al.* (2004) have shown that when normal human keratinocytes are subject to hyperosmotic exposure (405 mosmol $l^{-1}$), the expression and transporter activity of BGT1 will increase. Moreover, they demonstrated that UV radiation, which is the most significant source of oxidative stress in the skin, increases BGT1 expression and betaine uptake in keratinocytes. Betaine can also increase skin–barrier function and thus water balance by improving tight junction integrity in skin keratinocytes (Putaala *et al.*, 2010).

Dermal fibroblasts of the lower skin layer produce collagen fibers that are responsible for the structural integrity of the skin. Various stress factors, such as toxins and UV radiation, speed up collagen degradation in dermal fibroblasts. Osmolytes including betaine are known to have a protective role within fibroblasts, as these cells express and transport betaine in response to hyperosmotic and UV exposure (Warskulat *et al.*, 2008).

In addition to the skin, betaine is used to protect sensitive oral mucosa. Betaine-containing toothpaste is known to alleviate symptoms of dry mouth syndrome and to protect oral mucosa against commonly used detergents in toothpastes (Rantanen *et al.*, 2002, 2003).

# 13.5 Betaine in Muscle Cells

## 13.5.1 Betaine and Lean Weight Gain

In animal nutrition betaine is known to have a positive impact on lean weight gain (for a review see Eklund *et al.*, 2005). According to animal studies, dietary betaine is shown to be taken up by muscles where it contributes to metabolism (Kettunen *et al.*, 2001b; Pekkinen *et al.*, 2013). Betaine supplementation increases lean muscle mass but the effect is also dependent on various other factors, such as the availability of energy and protein in the diet (Huang *et al.*, 2006). In humans, betaine is used to improve muscle performance. In trained human males, supplemental betaine appeared to increase anabolic signaling (Apicella *et al.*, 2013) suggesting an increase in protein synthesis in muscle. Moreover, an *in vitro* study showed that betaine can improve muscle fiber differentiation in myotube hypertrophy in myoblast culture (Senesi *et al.*, 2013). Betaine supplementation can also improve muscle cell survival in hypertonic incubations (Alfieri *et al.*, 2006).

## 13.5.2 Betaine and Carnitine Metabolism in Muscle

The effect of betaine supplementation on murine muscle has been studied using nontargeted metabolite profiling (Pekkinen *et al.*, 2013). In addition to an increased betaine concentration in muscle, the concentration of metabolites in the L-carnitine biosynthesis pathway was also increased. As L-carnitine improves dietary fat utilization in mitochondria, the increased rate of L-carnitine biosynthesis after betaine supplementation may explain the favorable effects of betaine on muscle energy metabolism. A high-fat diet decreases the level of muscle carnitine, which leads to serious disruptions with mitochondrial metabolism (Noland *et al.*, 2009). However, a recent animal study (Pekkinen *et al.*, 2013) revealed that the amount of free L-carnitine increased in the muscles of mice being given a betaine supplement, despite being fed a high-fat diet at the same time. Both carnitine and acylcarnitines are end-products in a chain of biochemical reactions involving betaine that acts as a methyl donor. Additional evidence from a human study (Lever *et al.*, 2012) shows an association between plasma betaine and carnitine levels. Maintaining adequate carnitine levels in the tissue is essential for muscle metabolism. A diet supplemented with betaine could provide a useful way to increase endogenous carnitine production.

# 13.6 Betaine and Fat Cells

## 13.6.1 Betaine and Hypoxic Stress in Adipose Tissue

Despite the small accumulations of dietary betaine in adipose tissue (Figure 13.2), here betaine may still present significant physiological effects, especially during metabolic stress. White adipose tissue is an important endocrine organ that produces a vast variety of signaling hormones called

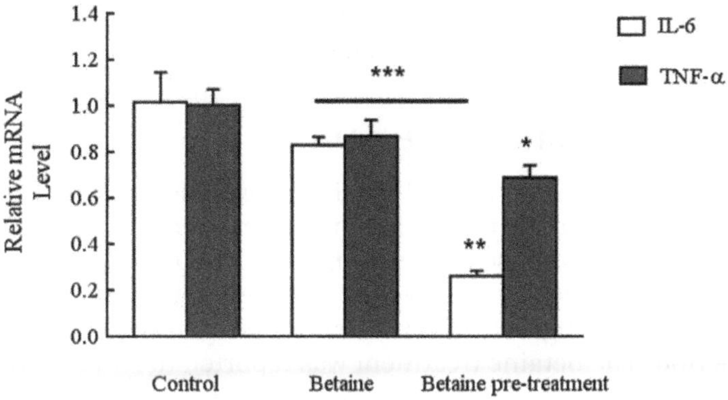


**Figure 13.6** The effect of betaine on adipokine mRNA expression in human adipocytes in hypoxia. The relative mRNA levels of interleukin-6 (IL-6) and tumor necrosis factor-alpha (TNF-α) are measured by qPCR after 8 h in hypoxia. Betaine (250 μmol $l^{-1}$) was administered to the cells either simultaneously with or 16 h before (=pretreatment) the hypoxia induction. Values are mean ± SEM ($n = 3$). *$P < 0.05$, and **$P < 0.01$, compared to the medium control, and ***$P < 0.001$ between the indicated treatments (Student's *t*-test). The figure is copyright-protected and reproduced from Olli *et al.* (2013) with kind permission from Cambridge University Press.

adipokines. In obesity, the production of proinflammatory adipokines, such as interleukin-6 (IL-6), leptin and TNF-α is increased. One reason for this phenomenon is the relative hypoxia affecting the adipose tissue cells in advance of angiogenesis (Trayhurn and Wood, 2004). In obese subjects the expanding fat mass induces hypoxic conditions within the tissue. As a consequence of the high levels of proinflammatory adipokines and other acute-phase proteins secreted from the adipocytes into the blood circulation, a chronic inflammatory state may develop (Trayhurn and Wood, 2004). There is evidence to indicate that a betaine supplement reduces the inflammation-related changes in the expression of adipokines. The physiologically relevant concentration (250 μM) of natural betaine has been shown to decrease the level of mRNA expression of IL-6 and leptin in a human *in vitro* adipocyte model (Olli *et al.*, 2013). When visceral adipocytes were cultivated under hypoxia (1% $O_2$), mimicking the obese state of adipose tissue, supplemented betaine managed to decrease the level of mRNA expression of IL-6 and TNF-α. The effect of betaine was also tested as a pretreatment, *i.e.* betaine was applied to the cells before the hypoxia induction. This additional application significantly fortified the inflammation-reducing effect of betaine. The effect of betaine on hypoxia-related adipokine expression is demonstrated in Figure 13.6. Thus, by reducing visceral adipocyte inflammation, betaine may also help to reduce obesity-related systemic inflammation, especially inflammation in adipose tissue overlaying the abdominal organs within the peritoneal cavity. An excessive amount of visceral fat is typically linked to type 2 diabetes, insulin resistance, and other obesity-related diseases. Therefore, soothing inflamed conditions inside the peritoneal cavity is important for overall health.

### 13.6.2 Betaine and a High-Fat Diet

In addition to *in vitro* trials, animal studies have demonstrated the protective and potentially therapeutic role of betaine that can reduce the harmful effects of a high-fat diet on adipose tissue. Wang *et al.* (2010) studied the effects of betaine in mice fed with a high-fat diet. They fed mice a high-fat diet simultaneously with drinking water supplemented with 1% (wt/vol) betaine over a 12 week period. The betaine treatment was reported to enhance the high-fat diet-induced changes in the insulin signaling pathway by enhancing insulin sensitivity in adipose tissue whilst lowering the levels of abnormal adipokine (adiponectin, resistin and leptin) production in isolated adipocytes. This effect of betaine is important for the overall health of the subjects, especially because insulin resistance and altered adipokine production in adipose tissue are related to the initiation and progression of several chronic liver diseases identified as part of the NAFLD (Delgado, 2008).

### 13.6.3 Betaine and Fat Metabolism

Endoplasmic reticulum (ER) stress has been shown to contribute to adipose tissue dysfunction and insulin resistance in animal models subject to a high-fat diet (Gregor and Hotamisligil, 2007). In a study by Wang *et al.* (2010) betaine supplementation was shown to prevent a high-fat diet-induced ER stress response in isolated epididymal adipose tissue. These results imply that by improving the adipose tissue function, *i.e.* reducing adipose tissue inflammation and ER stress, betaine may prevent the early stages of metabolic syndrome and related diseases. In pigs, dietary betaine is also known to improve lean mass gain without fat accumulation (Siljander-Rasi *et al.*, 2003). In other studies involving pigs, betaine supplementation resulted in reduced fat deposition and a lower percentage of overall fat compared to nonsupplemented diets (Fernandez-Figares *et al.*, 2002). The current understanding of the mechanism suggests that lipogenic gene expression is suppressed in the adipose tissue of animals consuming a betaine supplemented diet, which may result in less fat accumulation (Xing *et al.*, 2011). Thus, the role of betaine in enhancing fat catabolism *via* fatty acid oxidation may have minor importance in adipose tissue.

## 13.7 Concluding Remarks

Supplemented dietary betaine affects various metabolic pathways in different tissues. In addition to the stress conditions and tissues discussed in this chapter, betaine has also been shown to affect neural and cognitive functions. Furthermore, recent studies indicate that it offers myocardial protection and antistress properties, including the prevention of oxidative stress.

Although natural betaine is found in various food products and is a part of a normal daily diet, this chapter summarizes the potential benefits of betaine supplementation, especially when cells and tissues are tested under different stress conditions. It should be noted that both *in vivo* and *in vitro* trials have shown that betaine functions even more efficiently when administered as a pretreatment, *i.e.* before exposing tissues or cells to stress factors. Current studies suggest that further evaluation of the effects of betaine is required within a broader context of biological systems.

## Summary Points

- This chapter evaluates the physiological role of dietary betaine focusing on its role in animal tissues under metabolic and osmotic stress.
- The bioavailability of the betaine molecule to various tissues is essential when evaluating its role in stress.
- Dietary betaine is readily absorbed into the circulatory system and then distributed further in a tissue-specific manner.
- Dietary betaine has a role in reducing high-fat, diet-induced hepatic lipid accumulation.
- According to animal studies betaine has an ability to alleviate nonalcoholic fatty liver disease (NAFLD) by correcting abnormalities in hepatic methionine metabolism.
- Betaine supplements can reduce the inflammation-related changes in visceral adipocytes.
- Dietary betaine stabilizes the intestinal morphology of parasite-infected chickens.
- Betaine provides a protective role in skin keratinocytes subject to hyperosmotic and UV exposure.
- Dietary betaine contributes significantly to muscle metabolism.

## Key Facts

Key Facts of the Physiological Role of Betaine:

1. Most of the body's betaine originates from dietary sources containing betaine or choline.
2. Ingested betaine is absorbed in the brush-border membrane of the small intestinal epithelium.
3. The highest betaine concentrations in animals are found in the kidney and liver.
4. Betaine functions as an organic osmolyte to protect cells against high osmolarity.
5. Betaine can remethylate homocysteine to methionine, thus reducing a high plasma homocysteine level, which is considered a risk factor for cardiovascular diseases.

Key Facts of Metabolic Diseases:

1. Nonalcoholic fatty liver disease (NAFLD) is a term that encompasses a spectrum of metabolic diseases, such as hepatic steatosis, nonalcoholic steatohepatitis, fibrosis and cirrhosis, manifested by excessive accumulation of fat in the liver.
2. Long-term feeding with high fat and/or high sugar containing feed can cause animals to develop NAFLD.
3. Insulin resistance is a condition in which the cells have a decreased sensitivity to insulin.
4. Hypoxia is a condition in which the normal oxygen level has decreased and cells or tissues are left without a sufficient oxygen supply.
5. Nontargeted metabolite profiling involves the identification of small-molecule biomarkers that are characteristic of a particular physiological condition.

## Definitions of Words and Terms

**Adipocyte**: An adipose tissue cell capable of changing its size by accumulating fat.
**Adipokine**: An endocrine-signaling hormone secreted by white adipose tissue.
**Basolateral**: The layer of cell membrane that is adjacent to the basement membrane and facing away from the lumen.
**Epidermis**: The outermost layer of vertebrate skin.
**Dermal fibroblast**: A skin cell producing proteins, *e.g.* collagen, for extracellular matrix.
**Hepatocyte**: A specific type of liver cell originating from the parenchymal layer of the liver.
**Insulin**: A hormone excreted by the pancreas that regulates glucose homeostasis.
**Keratinocyte**: A skin cell originating from the outermost layer of the vertebrate skin.
**Kupffer cell**: A specialized immunological liver cell type originating from the walls of liver sinusoids.
**Macrophage**: A large, active immune cell, *e.g.* a phagocyte, which participates in the immune responses.
**Methylation**: An enzyme-catalyzed process in which a methyl group is added to a substrate molecule.
**mRNA**: A single-stranded group of RNA molecules formed during genetic transcription and used as a template for protein synthesis.
**Mucosa**: A mucous-containing membrane covering the inner organs of, *e.g.* respiratory and gastrointestinal tract.
**Organic osmolyte**: A nonperturbing small solute that functions to preserve optimal intracellular solution concentrations for metabolic activities.

**Osmoregulation:** The regulation of intracellular concentrations of organic or inorganic osmolytes to maintain the volume of a cell or tissue.
**Phagocytosis:** A process during which phagocytes engulf micro-organisms, other cells or foreign particles.
**Quantitative polymerase chain reaction (qPCR):** A quantitative mechanism that measures the changes in gene function in different cells, tissues, *etc.*
**Receptor:** A molecular structure that selectively binds specific compounds in a lock–key manner.
**Tight junction:** An intracellular protein-containing structure through which the adjacent cells are in contact.

## List of Abbreviations

| | |
|---|---|
| AMP | Adenosine monophosphate |
| ATP | Adenosine triphosphate |
| AMPK | AMP-activated protein kinase |
| BGT1 | Betaine/gamma-amino-butyric acid transporter 1 |
| BHMT | Betaine-homocysteine *S*-methyltransferase |
| COX-2 | Cyclo-oxygenase-2 |
| DNA | Deoxyribonucleic acid |
| DPM | Disintegration per minute |
| ER | Endoplasmic reticulum |
| GABA | Gamma-amino-butyric acid |
| IL-6 | Interleukin-6 |
| mRNA | Messenger RNA |
| NAFLD | Nonalcoholic fatty liver disease |
| qPCR | Quantitative polymerase chain reaction |
| RNA | Ribonucleic acid |
| *S*-AM | *S*-adenosylmethionine |
| SEM | Standard error of the mean |
| SLC36A1 | Imino acid transporter |
| TNF-α | Tumor necrosis factor-alpha |
| TLR | Toll-like receptors |
| UV | Ultraviolet |

## Acknowledgements

This work was sponsored by DuPont Nutrition and Health. The authors kindly thank Elaine Krul, Stuart Craig and David Bishop (DuPont Nutrition and Health) for proofreading the text and Kirsti Jutila (DuPont Industrial Biosciences), Juha Nurmi (DuPont Manufacturing Process Technology) and Mohamed Mathlouthi (University of Reims) for their scientific contributions.

## References

Alfieri, R. R., Bonelli, M. A., Cavazzoni, A., Brigotti, M., Fumarola, C., Sestili, P., Mozzoni, P., De Palma, G., Mutti, A., Carnicelli, D., Vacondio, F., Silva, C., Borghetti, A. F., Wheeler, K. P. and Petronini, P. G., 2006. Creatine as a compatible osmolyte in muscle cells exposed to hypertonic stress. *Journal of Physiology*. 576 (Pt 2): 391–401.

Apicella, J., Lee, E., Bailey, B., Saenz, C., Anderson, J., Craig, S., Kraemer, W., Volek, J. and Maresh, C., 2013. Betaine supplementation enhances anabolic endocrine and Akt signaling in response to acute bouts of exercise. *European Journal of Applied Physiology*. 113: 793–802.

Ashraf, M. and Foolad, M. R., 2007. Roles of glycine betaine and proline in improving plant abiotic stress resistance. *Environmental and Experimental Botany*. 59(2): 206–216.

Broer, S., 2008. Amino acid transport across mammalian intestinal and renal epithelia. *Physiological Reviews*. 88(1): 249–286.

Craig, S., 2004. Betaine in human nutrition. *American Journal of Clinical Nutrition*. 80(3): 539–549.

Delgado-Reyes, C. V., Wallig, M. A. and Garrow, T. A., 2001. Immunohistochemical detection of betaine–homocysteine *S*-methyltransferase in human, pig, and rat liver and kidney. *Archives of Biochemistry and Biophysics*. 393(1): 184–186.

Delgado, J.-S., 2008. Evolving trends in nonalcoholic fatty liver disease. *European Journal of Internal Medicine*. 19(2): 75–82.

Eklund, M., Bauer, E., Wamatu, J. and Mosenthin, R., 2005. Potential nutritional and physiological functions of betaine in livestock. *Nutrition Research Reviews*. 18(1): 31–48.

Fernandez-Figares, I., Wray-Cahen, D., Steele, N. C., Campbell, R. G., Hall, D. D., Virtanen, E. and Caperna, T. J., 2002. Effect of dietary betaine on nutrient utilization and partitioning in the young growing feed-restricted pig. *Journal of Animal Science*. 80(2): 421–428.

Gregor, M. F. and Hotamisligil, G. S., 2007. Thematic review series: Adipocyte Biology. Adipocyte stress: the endoplasmic reticulum and metabolic disease. *Journal of Lipid Research*. 48(9): 1905–1914.

Häussinger, D., 1996. The role of cellular hydration in the regulation of cell function. *Biochemical Journal*. 313 (Pt 3): 697–710.

Huang, Q., Xu, Z., Han, X. and Li, W., 2006. Changes in hormones, growth factor and lipid metabolism in finishing pigs fed betaine. *Livestock Science*. 105(1): 78–85.

Kempson, S. A. and Montrose, M. H., 2004. Osmotic regulation of renal betaine transport: transcription and beyond. *Pflugers Archiv*. 449(3): 227–234.

Kettunen, H., May 5, 2001. Betaine in the nutrition of broiler chiks – absorption, methyl group metabolism, and intestinal osmoregulation. In: *PhD thesis*. University of Helsinki, Helsinki, Finland, pp. 1–32.

Kettunen, H., Peuranen, S. and Tiihonen, K., 2001a. Betaine aids in the osmoregulation of duodenal epithelium of broiler chicks, and affects the movement of water across the small intestinal epithelium in vitro. *Comparative Biochemistry and Physiology, Part A: Molecular and Integrative Physiology*. 129 (2–3): 595–603.

Kettunen, H., Peuranen, S., Tiihonen, K. and Saarinen, M., 2001b. Intestinal uptake of betaine in vitro and the distribution of methyl groups from betaine, choline, and methionine in the body of broiler chicks. *Comparative Biochemistry and Physiology, Part A: Molecular and Integrative Physiology*. 128(2): 269–278.

Kettunen, H., Tiihonen, K., Peuranen, S., Saarinen, M. T. and Remus, J. C., 2001c. Dietary betaine accumulates in the liver and intestinal tissue and stabilizes the intestinal epithelial structure in healthy and coccidia-infected broiler chicks. *Comparative Biochemistry and Physiology, Part A: Molecular and Integrative Physiology*. 130(4): 759–769.

Kharbanda, K. K., Rogers, D. D., 2nd, Mailliard, M. E., Siford, G. L., Barak, A. J., Beckenhauer, H. C., Sorrell, M. F. and Tuma, D. J., 2005. Role of elevated S-adenosylhomocysteine in rat hepatocyte apoptosis: protection by betaine. *Biochemical Pharmacology*. 70(12): 1883–1890.

Kim, S. K. and Kim, Y. C., 2005. Effects of betaine supplementation on hepatic metabolism of sulfur-containing amino acids in mice. *Journal of Hepatology*. 42(6): 907–913.

Klasing, K. C., Adler, K. L., Remus, J. C. and Calvert, C. C., 2002. Dietary betaine increases intraepithelial lymphocytes in the duodenum of coccidia-infected chicks and increases functional properties of phagocytes. *Journal of Nutrition*. 132(8): 2274–2282.

Kwon, D. Y., Jung, Y. S., Kim, S. J., Park, H. K., Park, J. H. and Kim, Y. C., 2009. Impaired sulfur-amino acid metabolism and oxidative stress in nonalcoholic fatty liver are alleviated by betaine supplementation in rats. *Journal of Nutrition*. 139(1): 63–68.

Lever, M., Storer, M. K., Lewis, J. G., George, P. M. and Chambers, S. T., 2012. Plasma betaine concentrations correlate with plasma cortisol but not with C-reactive protein in an elderly population. *Clinical Chemistry and Laboratory Medicine*. 50(9): 1635–1640.

Mathlouthi, M., 1997. Physicochemical Properties of Betaine-Water Solutions and Their Possible Applications to Cosmetic Formulations. In: Final report of contract with Finnsugar Bioproducts. University of Reims, Reims, France.

Noland, R. C., Koves, T. R., Seiler, S. E., Lum, H., Lust, R. M., Ilkayeva, O., Stevens, R. D., Hegardt, F. G. and Muoio, D. M., 2009. Carnitine insufficiency caused by aging and overnutrition compromises mitochondrial performance and metabolic control. *Journal of Biological Chemistry*. 284(34): 22840–22852.

Olli, K., Lahtinen, S., Rautonen, N. and Tiihonen, K., 2013. Betaine reduces the expression of inflammatory adipokines caused by hypoxia in human adipocytes. *British Journal of Nutrition*. 109(1): 43–49.

Olthof, M. R., van Vliet, T., Boelsma, E. and Verhoef, P., 2003. Low dose betaine supplementation leads to immediate and long term lowering of plasma homocysteine in healthy men and women. *Journal of Nutrition*. 133(12): 4135–4138.

Pekkinen, J., Olli, K., Huotari, A., Tiihonen, K., Keski-Rahkonen, P., Lehtonen, M., Auriola, S., Kolehmainen, M., Mykkänen, H., Poutanen, K. and Hanhineva, K., 2013. Betaine supplementation causes increase in carnitine metabolites in the muscle and liver of mice fed a high-fat diet as studied by nontargeted LC-MS metabolomics approach. *Molecular Nutrition and Food Research*. 57(11): 1959–1968.

Putaala, H., Tiihonen, K. and Rautonen, N., 2010. Betaine increases tight junction integrity in epidermal keratinocytes. In: 40th Annual Meeting of the European Society for Dermatological Research. September 8-11, 2010 Helsinki, Finland. p. S27.

Rantanen, I., Nicander, I., Jutila, K., Ollmar, S., Tenovuo, J. and Soderling, E., 2002. Betaine reduces the irritating effect of sodium lauryl sulfate on human oral mucosa in vivo. *Acta Odontologica Scandinavica*. 60(5): 306–310.

Rantanen, I., Tenovuo, J., Pienihakkinen, K. and Soderling, E., 2003. Effects of a betaine-containing toothpaste on subjective symptoms of dry mouth: a randomized clinical trial. *Journal of Contemporary Dental Practice*. 4(2): 11–23.

Schrama, J. W., Heetkamp, M. J. W., Simmins, P. H. and Gerrits, W. J. J., 2003. Dietary betaine supplementation affects energy metabolism of pigs. *Journal of Animal Science*. 81(5): 1202–1209.

Schwab, U., Törrönen, A., Meririnne, E., Saarinen, M., Alfthan, G., Aro, A. and Uusitupa, M., 2006. Orally administered betaine has an acute and dose-dependent effect on serum betaine and plasma homocysteine concentrations in healthy humans. *Journal of Nutrition*. 136(1): 34–38.

Schwab, U., Törrönen, A., Toppinen, L., Alfthan, G., Saarinen, M., Aro, A. and Uusitupa, M., 2002. Betaine supplementation decreases plasma homocysteine concentrations but does not affect body weight, body composition, or resting energy expenditure in human subjects. *American Journal of Clinical Nutrition*. 76(5): 961–967.

Senesi, P., Luzi, L., Montesano, A., Mazzocchi, N. and Terruzzi, I., 2013. Betaine supplement enhances skeletal muscle differentiation in murine myoblasts via IGF-1 signaling activation. *Journal of Translational Medicine*. 11(1): 174.

Siljander-Rasi, H., Peuranen, S., Tiihonen, K., Virtanen, E., Kettunen, H., Alaviuhkola, T. and Simmins, P. H., 2003. Effect of equi-molar dietary betaine and choline addition on performance, carcass quality and physiological parameters of pigs. *Animal Science*. 76(1): 55–62.

Slow, S., Lever, M., Chambers, S. T. and George, P. M., 2009. Plasma dependent and independent accumulation of betaine in male and female rat tissues. *Physiological Research*. 58(3): 403–410.

Song, Z., Deaciuc, I., Zhou, Z., Song, M., Chen, T., Hill, D. and McClain, C. J., 2007. Involvement of AMP-activated protein kinase in beneficial effects of betaine on high-sucrose diet-induced hepatic steatosis. *American Journal of Physiology: Gastrointestinal and Liver Physiology*. 293(4): G894–G902.

Sunden, S. L., Renduchintala, M. S., Park, E. I., Miklasz, S. D. and Garrow, T. A., 1997. Betaine–homocysteine methyltransferase expression in porcine and human tissues and chromosomal localization of the human gene. *Archives of Biochemistry and Biophysics*. 345(1): 171–174.

Takenaka, M., Bagnasco, S. M., Preston, A. S., Uchida, S., Yamauchi, A., Kwon, H. M. and Handler, J. S., 1995. The canine betaine gamma-amino-n-butyric acid transporter gene: diverse mRNA isoforms are regulated by hypertonicity and are expressed in a tissue-specific manner. *Proceedings of the National Academy of Sciences of the United States of America*. 92(4): 1072–1076.

Thwaites, D. T. and Anderson, C. M. H., 2007. Deciphering the mechanisms of intestinal imino (and amino) acid transport: the redemption of SLC36A1. *Biochimica et Biophysica Acta*. 1768(2): 179–197.

Trayhurn, P. and Wood, I. S., 2004. Adipokines: inflammation and the pleiotropic role of white adipose tissue. *British Journal of Nutrition*. 92(3): 347–355.

Wang, Z., Yao, T., Pini, M., Zhou, Z., Fantuzzi, G. and Song, Z., 2010. Betaine improved adipose tissue function in mice fed a high-fat diet: a mechanism for hepatoprotective effect of betaine in nonalcoholic fatty liver disease. *American Journal of Physiology Gastrointestinal and Liver Physiology*. 298(5): G634–G642.

Warskulat, U., Brookmann, S., Felsner, I., Brenden, H., Grether-Beck, S. and Häussinger, D., 2008. Ultraviolet A induces transport of compatible organic osmolytes in human dermal fibroblasts. *Experimental Dermatology*. 17(12): 1031–1036.

Warskulat, U., Reinen, A., Grether-Beck, S., Krutmann, J. and Häussinger, D., 2004. The osmolyte strategy of normal human keratinocytes in maintaining cell homeostasis. *Journal of Investigative Dermatology*. 123(3): 516–521.

Warskulat, U., Zhang, F. and Häussinger, D., 1996. Modulation of phagocytosis by anisoosmolarity and betaine in rat liver macrophages (Kupffer cells) and RAW 264.7 mouse macrophages. *FEBS Letters*. 391(3): 287–292.

Xing, J., Kang, L. and Jiang, Y., 2011. Effect of dietary betaine supplementation on lipogenesis gene expression and CpG methylation of lipoprotein lipase gene in broilers. *Molecular Biology Reports*. 38(3): 1975–1981.

Yancey, P. H. and Burg, M. B., 1990. Counteracting effects of urea and betaine in mammalian cells in culture. *American Journal of Physiology*. 258 (1 Pt 2): R198–R204.

Zhang, F., Warskulat, U. and Häussinger, D., 1996a. Modulation of tumor necrosis factor-α release by anisoosmolarity and betaine in rat liver macrophages (Kupffer cells). *FEBS Letters*. 391(3): 293–296.

Zhang, F., Warskulat, U., Wettstein, M. and Häussinger, D., 1996b. Identification of betaine as an osmolyte in rat liver macrophages (Kupffer cells). *Gastroenterology*. 110(5): 1543–1552.

Zhang, W., Wang, L.-W., Wang, L.-K., Li, X., Zhang, H., Luo, L.-P., Song, J.-C. and Gong, Z.-J., 2013. Betaine protects against high-fat-diet-induced liver injury by inhibition of high-mobility group box 1 and toll-like receptor 4 expression in rats. *Digestive Diseases and Sciences*. 58(11): 3198–3206.

Zhou, Y., Holmseth, S., Hua, R., Lehre, A. C., Olofsson, A. M., Poblete-Naredo, I., Kempson, S. A. and Danbolt, N. C., 2012. The betaine-GABA transporter (BGT1, slc6a12) is predominantly expressed in the liver and at lower levels in the kidneys and at the brain surface. *American Journal of Physiology: Renal Physiology*. 302(3): F316–F328.

CHAPTER 14

# *Beneficial Effects of Betaine on Water-Immersion Restraint Stress-Induced Memory Impairment in Mice*

MASAYUKI HIRAMATSU*[a,b], YUKARI TAKAHASHI[a], ANNA AMANO[b], KIWAMU KIDO[b], TAKUYA MATSUKURA[b], TORU YAMANASHI[b], YUSUKE KAKEHI[b], TOSHITAKA NABESHIMA[b], AND MASAYA MIWA[a]

[a]Laboratory of Neuropsychopharmacology, Graduate School of Environmental and Human Sciences, Meijo University, 150 Yagotoyama, Tenpaku-ku, Nagoya 468-8503, Japan; [b]Department of Chemical Pharmacology, Faculty of Pharmacy, Meijo University, 150 Yagotoyama, Tenpaku-ku, Nagoya 468-8503, Japan
*E-mail: mhiramt@meijo-u.ac.jp

## 14.1 Introduction

Betaine (glycine betaine or trimethylglycine) is widely distributed in plants and micro-organisms, as well as in various dietary sources (Zeisel *et al.*, 2003; Craig, 2004). Some plants accumulate high levels of betaine in response to abiotic stress, and both the exogenous application of betaine and introduction of the betaine-biosynthetic pathway into plants that do not naturally accumulate betaine *via* transgenes increase the tolerance of these plants to

Food and Nutritional Components in Focus No. 7
Betaine: Chemistry, Analysis, Function and Effects
Edited by Victor R Preedy

Published by the Royal Society of Chemistry, www.rsc.org

various types of abiotic stress, such as drought, high salinity, and temperature stress (Chen and Murata, 2008).

In humans, betaine is obtained from the diet (Zeisel *et al.*, 2003) or from its metabolic precursor choline (Zeisel *et al.*, 2003; Craig, 2004). Betaine is utilized as a methyl donor in the reaction that converts homocysteine into methionine *via* betaine-homocysteine methyltransferase. Betaine also plays a role in osmotic regulation in the kidneys, which are routinely exposed to high extracellular osmolarity during normal operation of the urine-concentrating mechanism (Kempson and Montrose, 2004; Ueland, 2011). Furthermore, dietary betaine suppresses the activation of nuclear factor-κB (NF-κB) with oxidative stress, and the protein expression of proinflammatory molecules, such as cyclooxygenase-2 (COX-2), inducible nitric oxide synthase (iNOS), and tumor necrosis factor (TNF)-α, in aged rat kidneys (Go *et al.*, 2005, 2007). Betaine has beneficial effects that may be due to inhibition of the expression of proinflammatory molecules that increase oxidative stress in aged rat kidneys (Go *et al.*, 2005) and liver tissue (Çoban *et al.*, 2013).

In everyday life, stress is ubiquitous; we are always exposed to social, psychological, and environmental stress. Moderate stress may have a favorable influence on the body; however, prolonged and/or excessive stress generally has negative effects on learning and memory processes (Vanitallie, 2002), and has differential effects on the memory function depending on a number of factors (such as stress duration, intensity, timing and the source of the stress, as well as the learning type under study) (Bisaz *et al.*, 2009). For example, Stillman *et al.* (1998) demonstrated that restraint and cold stress impair performance in a memory task, and the extent of impairment is related to the stress severity. One of the mechanisms responsible for the observed behavioral deficits under cold stress may involve an altered cholinergic function. Furthermore, environmental stress has adverse affects on memory in humans and rodents.

Since the hippocampus and frontal cortex play an important role in learning and memory, it is thought that these brain areas develop dysfunction induced by stress (Chauveau *et al.*, 2009; Goodman *et al.*, 2012). There are several reasons why learning and memory are impaired by stress. For example, when the hypothalamic – anterior pituitary gland – adrenal cortex (HPA) system is activated, glucocorticoid, which is an adrenal cortical hormone, is secreted excessively (Roozendaal *et al.*, 2004; Oosthuizen *et al.*, 2005), and excessive stress also releases excessive glutamic acid and nitric oxide (NO), so hippocampal neurons are disordered (Lowy *et al.*, 1993; Oosthuizen *et al.*, 2005). Furthermore, increased glucocorticoid also enhances the effects of glutamic acid in the extracellular hippocampus (Danilczuk *et al.*, 2005). Repeated stress has been reported to cause reversible impairment of the central nervous system (CNS). It was proposed that alterations in glutamatergic, cholinergic, and monoamine neurotransmitter systems after exposure to stress are initial CNS events contributing to this impairment (Pung *et al.*, 2006).

In this study, each mouse was restrained in a restriction cage and immersed in water up to the clavicle for a long period. We used this water-immersion restraint stress (WIRS) model to examine the effect of stress on learning and memory function. WIRS is a method that causes acute, severe stress and is used as a model of post-traumatic stress disorder (PTSD) (Tamaki *et al.*, 2008) and stress-induced gastric ulcer (Konturek *et al.*, 2001). After chronic WIRS, the number of neurons in the hippocampus was decreased (Mizoguchi *et al.*, 1992), acute WIRS raised serum lipid peroxide and NOx levels (Ohta *et al.*, 2007), and microglia in the brain were activated (Sugama *et al.*, 2007). It is thought that oxidative stress is associated with the onset of many mental disorders including PTSD, and that stress disorders are caused by reactive oxygen species (Brown and Morey, 2012; Wilson *et al.*, 2013).

Therefore, the aim of the present work was to ameliorate the neuronal dysfunction after acute stress, and examine the effect of betaine after WIRS. We recently showed that betaine has protective effects against lipopolysaccharide (LPS)-induced memory impairment, and that the prevention of LPS-induced changes in GAT2 mRNA expression is crucial for this ameliorating effect (Miwa *et al.*, 2011). Since betaine may have antioxidative effects and may protect against neurodegenerative diseases, such as Alzheimer's disease, catechin, which is an ingredient in green tea with antioxidant effects, and scavenges reactive oxygen species, is also used as a positive control drug. Catechin has been reported to protect against Alzheimer's disease (Rezai-Zadeh *et al.*, 2008); therefore, we used catechin to investigate whether reactive oxygen species were involved in the learning and memory impairment induced by WIRS.

## 14.2 Methods

### 14.2.1 Animals

Male ddY strain mice (7–9 weeks old; Japan SLC., Hamamatsu, Japan) were used. The mice were kept in a regulated environment (24 ± 1 °C, 55 ± 5% humidity) under a 12 h light/dark cycle (lights on 7:45 a.m.) and given food and tap water *ad libitum*. The experimental protocols concerning the use of laboratory animals were approved by the ethical committee of Meijo University and followed the guidelines of the Japanese Pharmacological Society (Folia pharmacol. japon, 1992, 99: 35A); the Interministerial Decree of May 25th, 1987 (Ministry of Education, Japan); and the National Institutes of Health Guide for the Care and Use of Laboratory animals (NIH Publications No. 8023, revised 1978). All efforts were made to minimize animal suffering and the number of animals used.

### 14.2.2 Drugs

Betaine hydrochloride (betaine; Sigma, St. Louis, MO, USA) was dissolved in 0.9% saline and injected subcutaneously (s.c.). Catechin hydrate (Funakoshi, Tokyo) was also dissolved in 0.9% saline and injected s.c. at a volume of 0.1 mL/10 g body weight. The control mice were administered the vehicle (s.c.) instead of one of the drug solutions.

### 14.2.3 Experimental Schedules

Behavioral experimental schedules are shown in each figure. In schedule #1, we investigated whether memory impairment was induced after WIRS using the Y-maze and passive avoidance tests. The spontaneous alternation performance was conducted at 1, 3, 5, 7, and 14 days and the passive avoidance test was conducted at 4, 6, 8, and 10 days after WIRS. In the previous experiment, betaine showed a bell-shaped dose–response effect, and the dose of 0.163 mmol $kg^{-1}$ was significantly effective. Therefore, in schedule #2, betaine (0.163 mmol $kg^{-1}$, s.c.) was administered repeatedly for 13 days from 4 days before to 9 days after WIRS. Spontaneous alternation performance and the passive avoidance test were conducted from 7 and 9–10 days after WIRS. On the day of WIRS, betaine or catechin was administered 1 h before and/or 1 h after WIRS. In schedule #3 for subacute and acute experiments, mice received only betaine or catechin 1 h before and/or 1 h after WIRS. In acute experiments, the spontaneous alternation performance was completed on the 5th day, and the training trial of the passive avoidance test was conducted on the 7th day after WIRS, after mice received betaine 1 h before or 1 h after WIRS.

Proinflammatory molecules, glial activation, and the GABA transporter (GAT) may be important for the pathogenesis of WIRS-induced memory impairment, so we measured whether the mRNA expressions of proinflammatory molecules, oxidative stress, and GATs were changed. The expressions of each mRNA were measured 2 to 6 h after WIRS. Furthermore, the plasma corticosterone level was measured 0 to 3 h after WIRS, and lipid peroxidation was measured 9 days after WIRS.

### 14.2.4 Water-Immersion Restraint Stress (WIRS)

Mice starved overnight were restrained in a metallic cage (Natsume Seisakusho Co., Ltd., Tokyo, Japan) and immersed in water at 23–25 °C up to the clavicle for 3 h in a container (60 × 30 × 19 cm high) with a stirrer (STR-1; Tokyo Rikakikai Co., Ltd., Tokyo, Japan) and then kept on a hot-plate (KN-205D; Natsume Seisakusho Co., Ltd.) at 39 °C for 1 h to keep the body temperature in the normal range.

### 14.2.5 Spontaneous Alternation Behavior in Y-Maze Test

Immediate working memory was assessed by recording spontaneous alternation behavior during a single session in the Y-maze (Hiramatsu and Inoue, 1999), with a minor modification from the original report on rats. Each mouse, new to the maze, was placed at the end of one arm and allowed to move freely through the maze during an 8 min session, and arm entries were counted. Each series of arm entries was recorded visually, and an arm entry was defined as when the hind paws of the mouse were completely within the arm. Alternation was defined if mice entered different arms three times in succession based on the results of consecutive arm entering. The number of overlapping entrance sequences (*e.g.*, ABC, BCA) was defined as the number of alternations. The effect was calculated as a percentage alternation according to the following formula:

Percent alternation = (number of alternations/total number of arm entries 2) × 100

In this test, mice have to memorize the arm that they entered immediately before the current arm. Mice have a tendency to enter a newer environment, which, in this case, would be an arm not entered previously; therefore, the percent alternation is used as an index of short-term memory. The total number of arm entries is measured as an index of locomotor activity, although it is not direct evidence.

### 14.2.6 Step-Down-Type Passive Avoidance Test

A step-down-type passive avoidance test was performed as described previously (Hiramatsu and Inoue, 2000) with minor modifications. The apparatus consisted of a transparent acrylic rectangular cage (30 × 30 × 40 cm high) comprising a grid floor with a wooden platform (4 × 4 × 4 cm) in the center, set in a semisoundproof wooden outer box (35 × 35 × 90 cm high). Illumination was provided by a 15 W lamp above the apparatus. An electric current (1 Hz, 0.5 s, 60 V, DC) was delivered to the grid floor by an isolated stimulator (SEN-3201; Nihon Koden, Tokyo, Japan).

In the training test, each mouse was placed on the wooden platform. When the mouse stepped down from the platform onto the grid floor, an electric shock was delivered for 15 s. Almost all mice received an electric shock for 15 s. When the step-down latency in the training session was less than 3 s or more than 30 s, data were omitted from the results. A retention test was carried out 24 h after the training session, similar to the training, except that no electric shock was delivered to the grid floor. Each mouse was placed on the platform and the step-down latency was recorded. An upper cut-off time of 300 s was set.

### 14.2.7 Responses to Electric Shock

The responses to electric shock during the training session were recorded. The following scores were given based on the responses to the electric shock: 3 = jumping, 2 = vocalization, 1 = flinching, 0 = no response. Shock sensitivity is shown as the total score, which is the sum of each score for 15 s.

### 14.2.8 Corticosterone Assay

Plasma corticosterone was measured using the Corticosterone EIA kit (Immunodiagnostic System Ltd., Fountain, USA) according to the manufacturer's protocol. Briefly, plasma samples were diluted 50 times, and 100 μL of this sample was mixed with 100 μL of enzyme–conjugate solution. After being incubated overnight at 4 °C, 200 μL of tetramethylbenzidine was added as a substrate, and the mixture was incubated for 30 min at room temperature. After adding 100 μL of stop solution, the absorbance was determined by a plate reader (iMark Microplate Reader, Bio-Rad, USA) at 450 nm. Corticosterone is expressed as ng $mg^{-1}$ protein.

### 14.2.9 Real-Time RT-PCR

The mice were sacrificed after WIRS and/or betaine. Immediately after decapitation, their hippocampus and frontal cortex were rapidly dissected and immersed in liquid nitrogen. The frozen hippocampus was stored at −80 °C until use. Total RNA was extracted using RNA-Bee Reagent (Tel-Test, Inc., Friendswood, TX, USA) according to the manufacturer's instructions. Reverse transcription was performed with the PrimeScript RT reagent Kit (Perfect Real Time) (Takara Bio Inc., Otsu, Japan) under the conditions recommended by the manufacturer. Real-time PCR analysis was undertaken using SYBR Premix Ex Taq II (Takara Bio Inc.). Data collection involved using a Chromo4 real-time PCR detector and analysis with an Opticon Monitor 3 (Bio-Rad Laboratories Inc., Hercules, CA, USA). The real-time PCR primers used in this study are listed in Table 14.1. All primers were purchased from Takara Bio Inc. The real-time PCR conditions were as follows: initial denaturation at 95 °C for 10 s followed by 40 cycles of 95 °C for 5 s and 60 °C for 20 s. The expression levels of the genes analyzed by real-time PCR were normalized relative to the levels of β-actin.

**Table 14.1** Gene-specific real time RT-PCR primer sequences.

| Gene | Sequence (5′-3′) |
|---|---|
| β-actin | Forward TGACAGGATGCAGAAGGAGA<br>Reverse GCTGGAAGGTGGACAGTGAG |
| GAT2 | Forward CCATCTTGGGCTTCATGTCTCA<br>Reverse CAGCTGGGACAAAGGCATCA |
| GAT1 | Forward TAACAACAACAGCCCATCCA<br>Reverse GGAGTAACCCTGCTCCATGA |
| GAT3 | Forward AACTGCTCCTGCGACACCGATGA<br>Reverse TTGGATTTAATGACATGGAAGGAA |
| IL-1β | Forward TCCAGGATGAGGACATGAGCAC<br>Reverse GAACGTCACACACCAGCAGGTTA |
| iNOS | Forward GGAATGGAGACTGTCCCAGCA<br>Reverse GTCATGAGCAAAGGCGCAGA |
| TNF-α | Forward AAGCCTGTAGCCCACGTCGTA<br>Reverse GGCACCACTAGTTGGTTGTCTTTG |
| COX-2 | Forward GTGTGCGACATACTCAAGCAGGA<br>Reverse TGAAGTGGTAACCGCTCAGGTG |
| Heme oxygenase-1 | Forward TGCAGGTGATGCTGACAGAGG<br>Reverse TGTCTGGGATGAGCTAGTGCTGA |

### 14.2.10 Determination of Lipid Peroxidation Level

Mice were decapitated, and the hippocampus and frontal cortex were removed. Brain samples were frozen immediately in liquid $N_2$ and stored at −80 °C in a deep freezer until the assay.

Malondialdehyde (MDA) was measured with a thiobarbituric acid-reactive substance (TBARS) assay kit (Cayman, Ann Arbor, MI, USA). Briefly, the isolated brain samples were homogenized in cooled RIPA buffer (50 mM Tris–HCl buffer pH 7.4, 150 mM NaCl, 1% NP-40, 1% sodium deoxycholate, 0.1% SDS, 1 mM PMSF, 1 mM $Na_3VO_4$, 1 mM NaF as Protease Inhibitor Cocktail) with an ultrasonic homogenizer (Microson Ultrasonic Cell Disruptor, Misonix., N.Y., USA) for 15–20 s and centrifuged at 4500 g for 10 min at 4 °C. The homogenate of brain samples was incubated with 8.1% sodium dodecylsulfate for 10 min followed by the addition of 20% acetic acid (pH 3.5). The reaction mixture was incubated with 0.6% TBA in boiling water for 2 h. After a 10 min cooling period in an ice bath, the mixture was centrifuged at 1600 g for 10 min at 4 °C. The absorbance was determined by a plate reader (Wallac 1420 ARVOsx, USA) at 550 nm. The MDA content is expressed in μmol $mg^{-1}$ protein.

### 14.2.11 Protein Assay

The protein content in diluted samples (×50 and ×100) was measured based on a standard of BSA using the Bio-Rad DC protein assay (Bio-Rad). Samples were read on a photometer (iMark Microplate Reader) set at 595 nm.

### 14.2.12 Data Analysis

The behavioral data are expressed as medians and interquartile ranges, with first to third quartiles. In general, since data for these memory tests did not always show a Gaussian distribution, they were analyzed using nonparametric statistical methods. Thus, the significance of differences was evaluated using the Mann–Whitney *U*-test for two-group comparison or the Kruskal–Wallis test followed by Bonferroni's multiple comparison tests (Dunn's test) where appropriate.

The expression levels of each mRNA and MDA levels are expressed as the means ±SEM. An unpaired t-test was used to compare two groups, and one-way ANOVA followed by Dunnett's *post hoc* test was used for multiple comparisons.

All statistical tests were performed with Prism 4 or 5 software (GraphPad Software, Inc., San Diego, CA, USA). The criterion for significance was $p < 0.05$ in all statistical evaluations.

# 14.3 Results

## 14.3.1 Onset of Memory Impairment Induced by Water-Immersion Restraint Stress

After exposure to WIRS, mice were tested for spontaneous alternation behavior using the Y-maze task on days 1, 3, 5, 7, and 14 after stress exposure. The percent alternation significantly decreased 1, 3, 5, and 7 days after WIRS as compared with the control group (Figure 14.1A). On the 14th day after WIRS, the percent alternation returned to the control level. There was no significant effect on the total number of arm entries after WIRS (data not shown).

To test long-term memory, the step-down-type passive avoidance test was performed 4, 6, 8, and 10 days after WIRS. The median step-down latency was about 290 s in control mice, and it shortened the latency to half at 4, 6, and 10 days after WIRS, although this was not significant. On the 8th day, the latency was shortened significantly (Figure 14.1B). There was no significant difference in the total score after WIRS (data not shown).

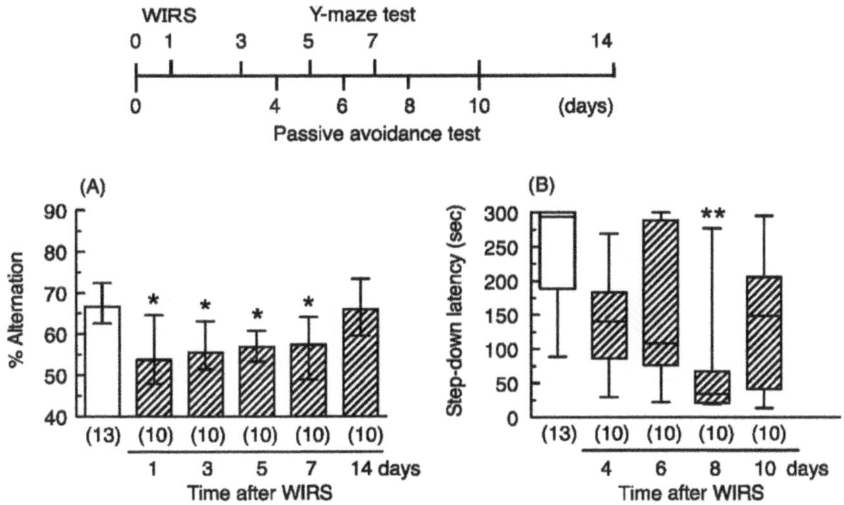


**Figure 14.1** Onset of memory impairment induced by water-immersion restraint stress (WIRS) on spontaneous alternation (A) and step-down-type passive avoidance test (B). The Y-maze test was carried out 1, 3, 5, 7, and 14 days after WIRS and the passive avoidance test was carried out 4, 6, 8, and 10 days after WIRS. Data are shown as the median (vertical column) and first and third quartiles (vertical line) for the Y-maze test and median (horizontal bar), first and third quartiles (vertical column), and 10th and 90th percentiles (vertical line). The number of mice used is shown in parentheses. Significant levels; $*p < 0.05$, $**p < 0.01$ *vs.* control (Bonferroni's test).

### 14.3.2 Effects of Repeated Administration of Betaine on Learning and/or Memory Impairment after Water-Immersion Restraint Stress

After WIRS, many physiological events occur in different time courses, and betaine may protect against those events. For example, WIRS leads to significant alterations in endogenous levels of both GABA and glutamate (Yoneda *et al.*, 1983), and other stresses induce increases in cytokines, corticosterone, and oxidative stress in different time courses (Ohta *et al.*, 2013). In our previous study, we found that the repeated administration of betaine showed a bell-shaped dose–response relationship, and a dose of 0.163 mmol kg$^{-1}$ (s.c.) significantly reversed the LPS-induced impairment of spontaneous alternations in the Y-maze test and the novel object recognition test (Miwa *et al.*, 2011). Therefore, betaine was repeatedly administered from 4 days before the WIRS to the end of the experiment. Y-maze and passive avoidance tests were performed 7 and 9–10 days after WIRS, respectively.

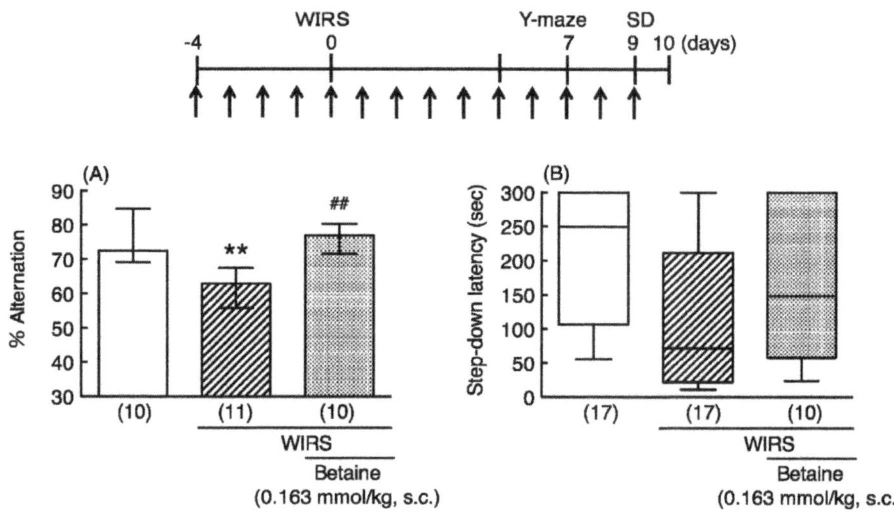


**Figure 14.2** Effects of the repeated administration of betaine on water-immersion restraint stress (WIRS)-induced memory impairment in percent alternation (A) and passive avoidance (B). The Y-maze and passive avoidance tests were carried out 7 and 11–12 days after WIRS, respectively. The mice were continuously treated with betaine (0.163 mmol kg$^{-1}$, s.c.) for 13 days starting 4 day before WIRS. On the day of the Y-maze and step-down tests, betaine was administered after the tests. Data are shown as the median (vertical column) and first and third quartiles (vertical line) for the Y-maze test and median (horizontal bar), first and third quartiles (vertical column), and 10th and 90th percentiles (vertical line). The number of mice used is shown in parentheses. Significant levels; $^{**}p < 0.01$ *vs.* control (Mann–Whitney's *U*-test). $^{\#\#}p < 0.01$ *vs.* WIRS alone (Mann–Whitney's *U*-test).

In the Y-maze test, the percent alternation 7 days after WIRS was significantly lower compared with control mice. When betaine (0.163 mmol $kg^{-1}$, s.c.) was injected repeatedly, the delayed impairment of spontaneous alternation after WIRS was significantly inhibited (Figure 14.2A). In the passive avoidance test, the step-down latency 10 days after WIRS tended to decrease, and the median latency was much shorter than in control mice. Betaine tended to suppress the impairment of long-term memory.

### 14.3.3 Effects of Subacute Administration of Betaine or Catechin on Learning and/or Memory Impairment after Water-Immersion Restraint Stress

When betaine was administered both 1 h before and 1 h after WIRS, the lowered percent alternation and step-down latency were significantly inhibited (Figure 14.3). Interestingly, catechin (0.324 mmol $kg^{-1}$, s.c.) also significantly inhibited impairment of the percent alternation; however, it did not prevent the impairment of step-down latency.

### 14.3.4 Effects of Acute Administration of Betaine on Learning and/or Memory Impairment after Water-Immersion Restraint Stress

It has been reported that severe stress leads to significant alterations in endogenous levels of both glutamate and GABA (Gao *et al.*, 2014), and corticosterone is also released during WIRS (Ohta *et al.*, 2013). To assess whether betaine protects against behavioral impairments mediated by these events, betaine was administered once either 1 h before or 1 h after WIRS. Betaine prevented the lowered percent alternation both 1 h before and 1 h after WIRS (Figure 14.4A). On the passive avoidance test, the step-down latency was significantly prolonged when betaine was administered 1 h after WIRS, but not 1 h before WIRS (Figure 14.4B).

### 14.3.5 Effects of Acute Administration of Betaine on Plasma Corticosterone Level after Water-Immersion Restraint Stress

After WIRS, plasma corticosterone levels were measured immediately, 0.5, 1, and 2 h after WIRS. Plasma corticosterone levels were significantly higher immediately and 0.5 h after WIRS and tended to be high at 1 h after and returned to the control level at 2 h after WIRS (Figure 14.5A). In the next experiment, betaine (0.163 mmol $kg^{-1}$) was administered 0.5 h after and mice were sacrificed 1 h after WIRS. Betaine could not suppress the increase in plasma corticosterone levels (Figure 14.5B).

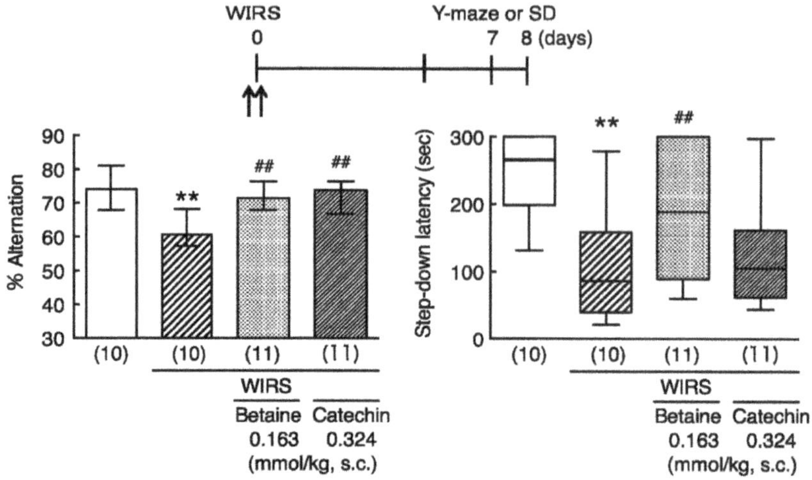


**Figure 14.3** Effects of subacute administration of betaine or catechin on learning and/or memory impairment after water-immersion restraint stress (WIRS). The Y-maze and passive avoidance tests were carried out 7 and 7–8 days after WIRS, respectively. The mice were treated with betaine (0.163 mmol $kg^{-1}$, s.c.) both 1 h before and 1 h after WIRS. Data are shown as the median (vertical column) and first and third quartiles (vertical line) for the Y-maze test and median (horizontal bar), first and third quartiles (vertical column), and $10^{th}$ and $90^{th}$ percentiles (vertical line). The number of mice used is shown in parentheses. Significant levels; $^{**}p < 0.01$ *vs.* control (Mann–Whitney's *U*-test). $^{\#\#}p < 0.01$ *vs.* WIRS alone (Mann–Whitney's *U*-test).

### 14.3.6 Effects of Acute Administration of Betaine on Interleukin-1β mRNA Level in the Hippocampus and Frontal Cortex after Water-Immersion Restraint Stress

After WIRS, the mRNA expression of IL-1β significantly increased in the hippocampus and frontal cortex. The IL-1β mRNA level increased 2 and 3 h after WIRS (Figure 14.6A), and this increase was significantly inhibited by betaine 1 h after WIRS in the frontal cortex (Figure 14.6C), while it was not significant in the hippocampus (Figure 14.6B). Since IL-1β changed significantly, other cytokines and proinflammatory molecules may change. We therefore investigated whether or not WIRS induces these parameters such as TNF-α, iNOS, and COX-2. However, mRNA expression levels of these inflammatory molecules did not change 2 h after WIRS (data not shown). The mRNA expression level of heme oxygenase-1, a known marker of oxidative stress, also did not change (data not shown).

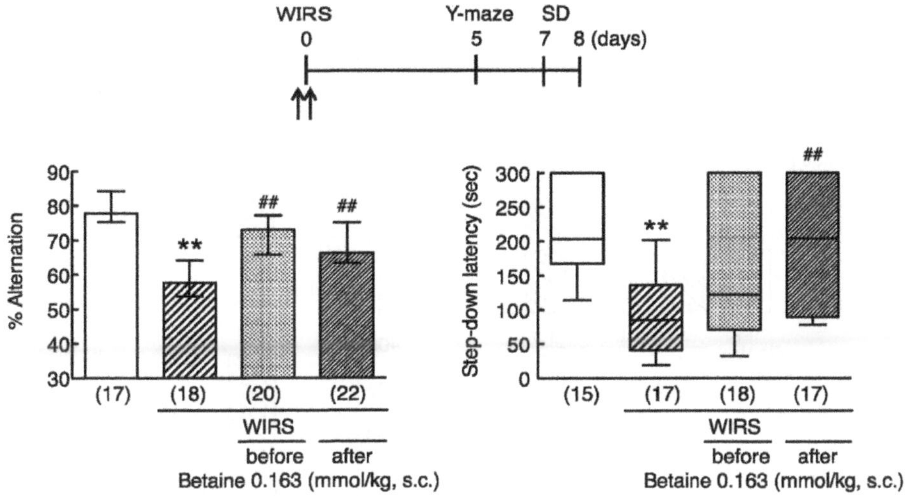


**Figure 14.4** Effects of acute administration of betaine on learning and/or memory impairment after water-immersion restraint stress (WIRS). The Y-maze and passive avoidance tests were carried out 5 and 7–8 days after WIRS, respectively. The mice were treated with betaine (0.163 mmol $kg^{-1}$, s.c.) either 1 h before or 1 h after WIRS. Data are shown as the median (vertical column) and first and third quartiles (vertical line) for the Y-maze test and median (horizontal bar), first and third quartiles (vertical column), and 10th and 90th percentiles (vertical line). The number of mice used is shown in parentheses. Significant levels; $**p < 0.01$ *vs.* control (Mann–Whitney's *U*-test). $\#\#p < 0.01$ *vs.* WIRS alone (Mann–Whitney's *U*-test).

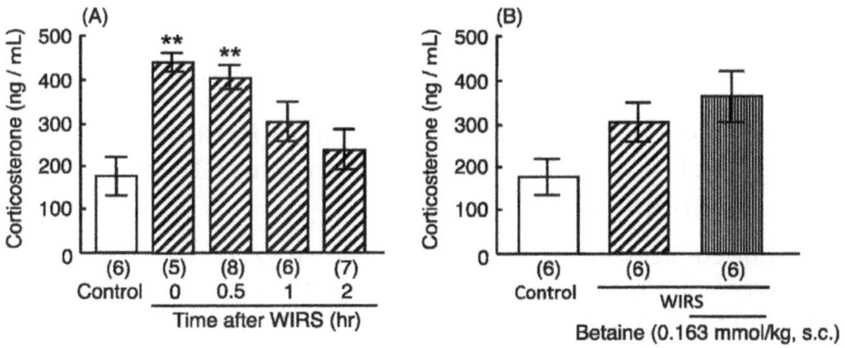


**Figure 14.5** Plasma corticosterone levels after water-immersion restraint stress (WIRS) and effects of acute administration of betaine. Plasma blood samples were collected 0, 0.5, 1, and 2 h after WIRS (A), and 1 h after WIRS (B). Betaine (0.163 mmol $kg^{-1}$, s.c.) was administered 0.5 h after WIRS. The plasma corticosterone levels were measured using ELISA. The number of mice used is shown in parentheses. Significant levels; $**p < 0.01$ *vs.* control (Dunnett's test).

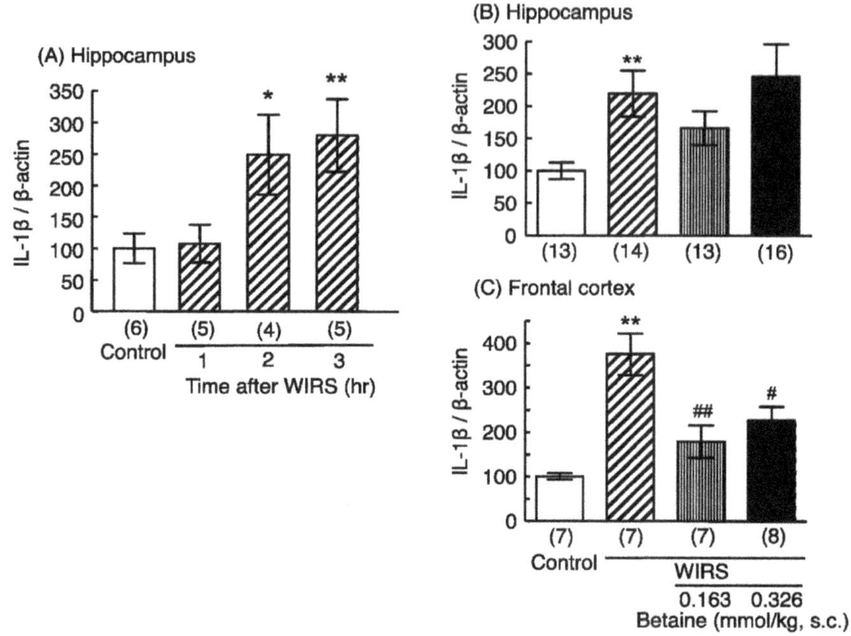


**Figure 14.6** Effects of acute administration of betaine on interleukin-1β mRNA expression level in the hippocampus and frontal cortex after water-immersion restraint stress (WIRS). Mice were sacrificed at 1, 2, or 3 h after WIRS (A). Mice were treated with betaine (0.163 mmol kg$^{-1}$, s.c.) 1 h after WIRS, and sacrificed at 2 h after WIRS (B, C). Their mRNA levels in the hippocampus (A, B) or frontal cortex (C) were assessed by real-time RT-PCR. The level of each mRNA was normalized to the mRNA level of β-actin as an endogenous control. Values are shown as the mean ± SEM, as shown in parentheses. Significance levels: $*p < 0.05$, $**p < 0.01$ *vs.* control (unpaired *t*-test), $\#p < 0.05$, $\#\#p < 0.01$ *vs.* WIRS alone (Bonferroni's test).

### 14.3.7 Effects of Acute Administration of Betaine on MDA Level in the Hippocampus after Water-Immersion Restraint Stress

It was reported that betaine prevented the generation of reactive oxygen species (Lee *et al.*, 2013). Furthermore, it has anti-inflammatory and antioxidant actions and helps to maintain liver and heart health (Ueland, 2011). Therefore, we assessed the MDA level 9 days after WIRS. The MDA level did not change after WIRS (data not shown).

### 14.3.8 Effects of Acute Administration of Betaine on the Expression of GABA Transporter mRNA Levels in the Hippocampus after Water-Immersion Restraint Stress

In our previous report, LPS treatment significantly increased the mRNA expression of GAT2 at 24 h after its injection. Interestingly, betaine (0.163 mmol kg$^{-1}$) prevented the LPS-induced increase in the GAT2 mRNA

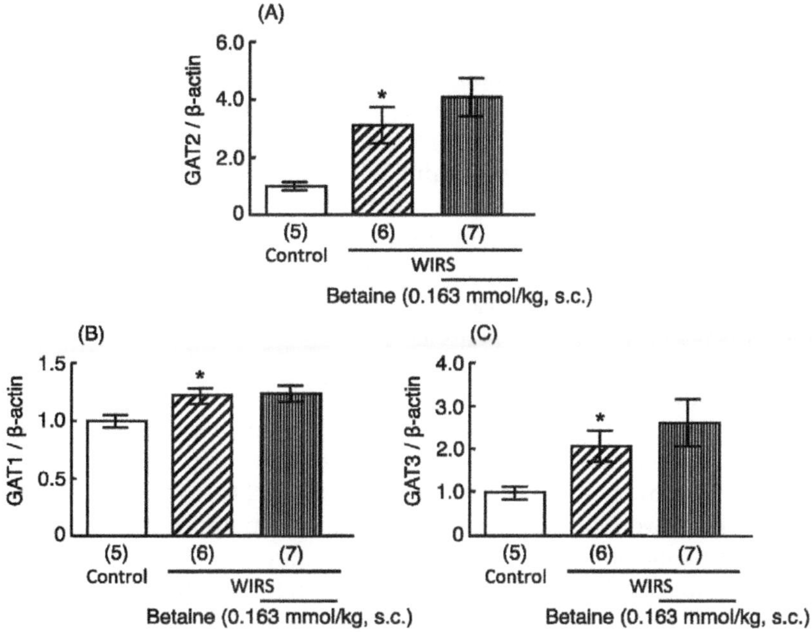


**Figure 14.7** Effects of acute administration of betaine on the expression of GABA transporters mRNA levels in the hippocampus after water-immersion restraint stress (WIRS). Mice were treated with betaine (0.163 mmol $kg^{-1}$, s.c.) 1 h after WIRS and sacrificed at 2 h after WIRS. The mRNA levels of GAT2 (A), GAT1 (B), and GAT3 (C) in the hippocampus were assessed by real-time RT-PCR. The level of each mRNA was normalized to the mRNA level of β-actin. Values are shown as the mean ± SEM of 5–7 mice, in parentheses. Significance levels: $*p < 0.05$ *vs.* control (unpaired *t*-test).

level (Miwa *et al.*, 2011). Therefore, we suggested that the suppression of LPS-induced memory impairment by betaine might act on GAT2/BGT-1 expressed on neurons and/or glial cells (Miwa *et al.*, 2011); therefore, we examined the effects of WIRS and betaine on the mRNA expression of GAT2. The mRNA expression of GAT2 increased significantly 2 h after WIRS (Figure 14.7A). Furthermore, the mRNA expression of other subtypes of GAT, GAT1 and GAT3, was also increased (Figure 14.7B and C). Under the present experimental conditions, none of the increased GAT mRNA expression was prevented by betaine (0.163 mmol $kg^{-1}$) 2 h after WIRS (Figure 14.7).

## 14.4 Discussion

Betaine suppressed the expression of proinflammatory molecules, such as COX-2, iNOS, and TNF-α, in aged rat kidneys (Go *et al.*, 2005, 2007). Furthermore, increased oxidative stress is suppressed by betaine in aged rat kidneys (Go *et al.*, 2007) and liver tissue (Çoban *et al.*, 2013). Betaine also prevented chronic ethanol consumption-induced oxidative stress in brain synaptosomes (Alirezaei *et al.*, 2011). As neuroinflammation and oxidative stress are critical components of the pathogeneses of some neurodegenerative disorders, including Alzheimer's disease (Amor *et al.*, 2013), and they induce learning and memory impairment in rats (Tanaka *et al.*, 2006), it is important to elucidate whether betaine is a useful compound for preventing neurodegenerative disorders and/or other diseases involving oxidative stress and inflammatory processes; however, the effects of betaine on memory impairment involving oxidative stress and/or neuroinflammatory processes are not well known.

In our previous study, the repeated administration of betaine (0.163 mmol $kg^{-1}$) improved LPS-induced memory impairment in the Y-maze and novel object recognition tests in a bell-shaped dose–response manner (Miwa *et al.*, 2011). Furthermore, the subacute (1 h before, 1 and 24 h after LPS injection) or single (1 h after LPS injection) administration of betaine also prevented LPS-induced memory impairment (Miwa *et al.*, 2011). In the present study, similar findings were obtained on WIRS-induced memory impairment after repeated and subacute administrations of betaine. It is known that the biochemical changes after WIRS, like LPS treatment, occur during the early phase, and induce delayed neuronal impairment such as memory impairment. To elucidate the mechanisms underlying the effects of betaine, it was administered once in the early phase after WIRS, and we assessed which factor is necessary to prevent WIRS-induced memory impairment. Consistent with the LPS experiment, a single administration of betaine 1 h after WIRS also prevented the delayed memory impairment induced by WIRS. On the other hand, although betaine prevented WIRS-induced long-term memory impairment, catechin did not prevent it. Catechin displays strong antioxidant and anti-inflammatory properties (Rodrigeus *et al.*, 2013). Moreover, inflammatory mediators like TNF-α, IL-1β levels, and the expression of iNOS were significantly attenuated by catechin pretreatment. Therefore, catechin is effective for preventing memory loss, ameliorating oxidative stress, and might be beneficial for the treatment of sporadic dementia of the Alzheimer's type (SDAT) (Ejaz Ahmed *et al.*, 2013). It has been reported that, since catechin is effective for preventing memory loss, ameliorating oxidative stress (Ejaz Ahmed *et al.*, 2013), the mechanism of action of betaine may not be related to oxidative stress.

Choi *et al.* (2008) reported that the i.c.v. injection of LPS induced neuronal damage and the activation of microglia and astrocytes in the hippocampus at 24 h after LPS injection. As reported in our previous report, LPS induced the expression of proinflammatory molecules and glial activation within several days of LPS injection in the mouse hippocampus and cortex (Miwa *et al.*, 2011), consistent with a previous report (Szczepanik and Ringheim, 2003), and these increases in the expression levels of proinflammatory cytokines peaked at about 6–9 h after LPS injection. Therefore, we investigated whether WIRS increases the mRNA expression levels of proinflammatory molecules and glial markers in the hippocampus and/or frontal cortex, and betaine affects the changes using real-time RT-PCR. WIRS induced a transient increase in the mRNA expression levels of IL-1β, while the expressions of TNF-α, iNOS, and COX-2 mRNA did not change. As inflammatory mediators like TNF-α, IL-1β, and the expression of iNOS were significantly attenuated by catechin pretreatment (Ejaz Ahmed *et al.*, 2013), betaine (0.163 and 0.326 mmol $kg^{-1}$) significantly antagonized the WIRS-induced increase in IL-1β mRNA expression in the frontal cortex, but not in the hippocampus. These results indicate that betaine did not suppress the mRNA expression of proinflammatory molecules or glial markers, and the preventive effect on IL-1β was limited in the specific brain area.

Furthermore, since the expression of heme oxygenase-1 and SOD2 mRNA and MDA formation in the hippocampus also did not change after WIRS, oxidative stress may not be important. From a different point of view, the present findings are very interesting, because in aged rat kidneys, betaine suppressed the expression of proinflammatory molecules, increased oxidative stress, and showed cytoprotective effects (Go *et al.*, 2005, 2007), while the mechanism underlying the ameliorating effects of betaine on memory impairment is different and has different physiological roles in the CNS. It is necessary to elucidate further whether the protective effect of betaine on WIRS-induced memory impairment is only controlled *via* IL-1β-mediated responses.

Four different subtypes of GAT have been cloned and are termed GAT1, GAT2, GAT3, and GAT4 in mice (GAT-1, BGT-1, GAT-2, and GAT-3, respectively, in rats and humans) (Madsen *et al.*, 2009). GAT2/BGT-1 transports both GABA and betaine (Borden, 1996). In renal epithelial cells, GAT2/BGT-1 is a basolateral membrane protein that protects cells in the hypertonic inner medulla by mediating betaine uptake and accumulation (Kempson and Montrose, 2004). In the CNS, it was reported that the betaine content and BGT-1 mRNA level were increased in rat brains with hyperosmotic serum (Olsen *et al.*, 2005). In addition, the protein and mRNA expressions of GAT2/BGT-1 were upregulated in mouse and rat astrocyte primary cultures under hyperosmotic conditions (Bitoun and Tappaz, 2000). These results suggest that betaine and GAT2/BGT-1 play important roles in osmotic regulation in the CNS. Moreover, the expression of BGT-1 was increased in astrocytes after kainate-induced neuronal injury in the rat hippocampus (Zhu and Ong, 2004). Therefore, betaine and GAT2/BGT-1 may be involved in neuronal dysfunction caused by neurodegeneration or neuronal injury. In the present study, we examined the mRNA expression of GAT2 after WIRS with/without betaine in the mouse hippocampus. The mRNA expression of GAT2 significantly increased at 2 h after WIRS. Interestingly, the expression of other GAT subtypes, GAT1 and GAT3, also significantly increased at 2 h after WIRS. However, betaine (0.163 mmol $kg^{-1}$) did not block these WIRS-induced increases in mRNA expressions of GAT1 and GAT3. In our preliminary experiments, although a nonselective GAT inhibitor, nipecotic acid, could not prevent WIRS-induced memory impairment, a selective GAT2 inhibitor, NNC 05-2090, significantly inhibited delayed memory impairment. In addition, the mRNA expression of GAT2 transiently increased after LPS injection, and the increase in the GAT2 mRNA level in the hippocampus was significantly inhibited by betaine (Miwa *et al.*, 2011). In future experiments, therefore, it will be necessary to elucidate whether GAT2/BGT-1 plays important roles in neuronal dysfunction caused by neuronal injury.

## 14.5 Conclusions

Betaine inhibited WIRS-induced memory impairment and blocked the increase in the mRNA expression of IL-1β induced by WIRS; however, it did not prevent the WIRS-induced increase in the mRNA expression of proinflammatory molecules, glial markers, and GATs. These results suggest that betaine has protective effects against WIRS-induced memory impairment that are mediated through unique mechanisms. Betaine may act on GAT2, which is involved in the development of memory impairment, without affecting proinflammatory molecules or glial markers.

## Competing Interests

The authors declare that they have no competing interests.

## Authors' Contributions

M. H. conceived the study, participated in its design and coordination, and wrote the manuscript. Y. T., A. A., K. K., T. M., T. Y., and Y. K. carried out the experiments. M. M. helped to carry out the experiments. T. N. discussed some of the results. All of the authors have read and approved the final manuscript.

## Summary Points

- This chapter focuses on the beneficial effects of betaine on WIRS-induced delayed memory impairment.
- The repeated administration of betaine (0.163 mmol $kg^{-1}$) prevented WIRS-induced delayed memory impairment.
- The subacute and acute administrations of betaine (0.163 mmol $kg^{-1}$) also prevented WIRS-induced memory impairment in the Y-maze test.
- The interleukin 1β (IL-1β) mRNA level was significantly increased after WIRS, and the administration of betaine 1 h after WIRS blocked this increase in the frontal cortex.
- Although GABA transporter (GAT)-1, 2, and 3 mRNA expressions were significantly increased at 2 h after WIRS, betaine could not inhibit these increases.
- Betaine has protective effects against WIRS-induced memory impairment, and the prevention of WIRS-induced changes in IL-1β mRNA expression may be one of the crucial factors in this ameliorating effect.

# Key Facts

## Key Facts of Delayed Neurodegeneration

1. Brain dysfunction after prolonged and/or excessive stress is a significant contributor to memory impairment and other psychiatric disorders.
2. Selectively vulnerable neuron populations undergo delayed neurodegeneration in the hours to days following stress.
3. Multiple factors influence neuronal death after stress, including excitatory synaptic input, mitochondrial dysfunction, oxidative stress, inflammation, and disruption of intracellular $Ca^{2+}$ homeostasis.
4. The hippocampal CA1 pyramidal neurons are vulnerable to these factors.

# Definitions of Words and Terms

**Betaine (glycine betaine or trimethylglycine)**: widely distributed in plants and micro-organisms, as well as in various dietary sources. Some plants accumulate high levels of betaine in response to abiotic stress, and both the exogenous application of betaine and introduction of the betaine-biosynthetic pathway into plants that do not naturally accumulate betaine *via* transgenes increase the tolerance of these plants to various types of abiotic stress, such as drought, high salinity, and temperature stress.

**Catechin**: One of the antioxidants found in the leaves of the tea plant, red wine, and chocolate. Catechin polyphenols are part of a molecular family called flavonoids, and are able to prevent the activity of free radicals, the molecules that cause cellular damage that can lead to cancer and neurodegeneration.

**Water-immersion restraint stress (WIRS)**: A method that causes acute, severe stress and is used as a model of post-traumatic stress disorder (PTSD) and stress-induced gastric ulcers.

**Post-traumatic stress disorder (PTSD)**: A severe condition that may develop after a person experienced or witnessed a traumatic or terrifying event in which serious physical harm occurred or was threatened. The diagnosis may be given when a group of symptoms such as disturbing recurring flashbacks, avoidance or numbing of memories of the event, and hyperarousal continue for more than a month after the traumatic event.

**GABA transporter (GAT)**: γ-Aminobutyric acid (GABA) transporters (GATs) are members of the $Na^{+}/Cl^{-}$-dependent neurotransmitter transporter gene family (SLC6 gene family) with a homology to the GATs, GAT-1, BGT-1, GAT-2, and GAT-3 in human and rat, and in the mouse referred to as mGAT1, mGAT2, mGAT3, and mGAT4, respectively.

**BGT-1**: Betaine/GABA transporter-1 (mGAT2) transports both GABA and betaine. In renal epithelial cells, mGAT2/BGT-1 is a basolateral membrane protein that protects cells in the hypertonic inner medulla by mediating betaine uptake and accumulation.

**Alzheimer's disease**: A neurological disorder in which the death of brain cells causes memory loss and cognitive decline. A neurodegenerative type of dementia, the disease starts mild and gets progressively worse. The abnormal production of proinflammatory molecules, as well as reactive oxygen and nitrogen species, can disrupt nerve terminals activity causing dysfunction and loss of synapses, which correlates with memory decline; these are phenomena preceding the neuronal damage associated with Alzheimer's disease.

**Lipopolysaccharide (LPS)**: A component of the cell wall of Gram-negative bacteria, is used to experimentally induce memory impairment, neuroinflammatory responses, and oxidative stress such as increases in mRNA levels of interleukin (IL)-1β, IL-6, heme oxygenase-1, TNF-α, microglial activation, and iNOS activity.

**Proinflammatory molecules**: Some of the molecules that accelerate inflammation and regulate inflammatory reactions. The major proinflammatory molecules such as IL-1β, TNF-α, iNOS, and COX-2 are responsible for early responses of inflammation.

**Oxidative stress**: is caused by an imbalance between production of reactive oxygen and ability to detoxify the reactive intermediates. Disturbances in normal redox system can cause toxic effects through the production of peroxides and free radicals that damage all components of the cell, including proteins, lipids, and DNA.

**Malondialdehyde (MDA)**: One of the most prevalent byproducts of enzymatic and oxygen radical-induced lipid peroxidation during oxidative stress, which can be quantified colorimetrically following its controlled reaction with a thiobarbituric acid-reactive substance (TBARS).

**Real-time RT-PCR**: The real-time reverse transcription polymerase chain reaction (RT-PCR) uses fluorescent reporter molecules to monitor the production of amplification products during each cycle of the PCR reaction.

## Abbreviations

| | |
|---|---|
| BGT-1 | Betaine/GABA transporter-1 |
| BSA | Bovine serum albumin |
| CNS | Central nervous system |
| COX-2 | Cyclooxygenase-2 |
| EIA | Enzyme immunoassay |
| GABA | γ-Aminobutyric acid |
| GAT | GABA transporter |
| IL-1β | Interleukin 1β |
| iNOS | Inducible nitric oxide synthase |
| LPS | Lipopolysaccharide |
| MDA | Malondialdehyde |
| NF-κB | Nuclear factor-κB |
| NO | Nitric oxide |
| PTSD | Post-traumatic stress disorder |

| | |
|---|---|
| RT-PCT | Reverse transcription polymerase chain reaction |
| SDAT | Sporadic dementia of the Alzheimer's type |
| TBARS | Thiobarbituric acid-reactive substance |
| TNF-α | Tumor necrosis factor-α |
| WIRS | Water-immersion restraint stress |

## Acknowledgements

This study was supported in part by collaboration with the Local Communities Project from MEXT (Ministry of Education, Culture, Sports, Science, and Technology) and the Academic Frontier Project for Private Universities, which matched the subsidy provided by MEXT and the International Center for Green Biotechnology of Meijo University.

## References

Alirezaei, M., Jelodar, G., Niknam, P., Ghayemi, Z. and Nazifi, S., 2011. Betaine prevents ethanol-induced oxidative stress and reduces total homocysteine in the rat cerebellum. *Journal of Physiology and Biochemistry*. 67: 605–612.

Amor, S., Peferoen, L. A., Vogel, D. Y., Breur, M., van der Valk, P., Baker, D. and van Noort, J. M., 2013. Inflammation in neurodegenerative diseases – an update. *Immunology*. (in press).

Bisaz, R., Conboy, L. and Sandi, C., 2009. Learning under stress: a role for the neural cell adhesion molecule NCAM. *Neurobiology of Learning and Memory*. 91: 333–342.

Bitoun, M. and Tappaz, M., 2000. Gene expression of the transporters and biosynthetic enzymes of the osmolytes in astrocyte primary cultures exposed to hyperosmotic conditions. *Glia*. 32: 165–176.

Borden, L. A., 1996. GABA transporter heterogeneity: pharmacology and cellular localization. *Neurochemistry International*. 29: 335–356.

Brown, V. M. and Morey, R. A., 2012. Neural systems for cognitive and emotional processing in posttraumatic stress disorder. *Frontiers in Psychology*. 3: 449.

Chauveau, F., Piérard, C., Tronche, C., Coutan, M., Drouet, I., Liscia, P. and Béracochéa, D., 2009. The hippocampus and prefrontal cortex are differentially involved in serial memory retrieval in non-stress and stress conditions. *Neurobiology of Learning and Memory*. 91: 447–455.

Chen, T. H. and Murata, N., 2008. Glycinebetaine: an effective protectant against abiotic stress in plants. *Trends in Plant Sciences*. 13: 499–505.

Choi, S. H., Langenbach, R. and Bosetti, F., 2008. Genetic deletion or pharmacological inhibition of cyclooxygenase-1 attenuate lipopolysaccharide-induced inflammatory response and brain injury. *FASEB Journal*. 22: 1491–1501.

Çoban, J., Bingül, I., Yesil-Mizrak, K., Dogru-Abbasoglu, S., Oztezcan, S. and Uysal, M., 2013. Effects of carnosine plus vitamin E and betaine treatments on oxidative stress in some tissues of aged rats. *Current Aging Science*. 6: 199–205.

Craig, S. A., 2004. Betaine in human nutrition. *American Journal of Clinical Nutrition*. 80: 539–549.

Danilczuk, Z., Ossowska, G., Lupina, T., Cielik, K. and Zebrowska-upina, I., 2005. Effect of NMDA receptor antagonists on behavioral impairment induced by chronic treatment with dexamethasone. *Pharmacological Reports*. 57: 47–54.

Ejaz Ahmed, M., Khan, M. M., Javed, H., Vaibhav, K., Khan, A., Tabassum, R., Ashafaq, M., Islam, F., Safhi, M. M. and Islam, F., 2013. Amelioration of cognitive impairment and neurodegeneration by catechin hydrate in rat model of streptozotocin-induced experimental dementia of Alzheimer's type. *Neurochemistry International*. 62: 492–501.

Gao, J., Wang, H., Liu, Y., Li, Y. Y., Chen, C., Liu, L. M., Wu, Y. M., Li, S., Yang, C., 2014. Glutamate and GABA imbalance promotes neuronal apoptosis in hippocampus after stress. *Medical Science Monitor*. 20: 499–512.

Go, E. K., Jung, K. J., Kim, J. M., Lim, H., Lim, H. K., Yu, B. P. and Chung, H. Y, 2007. Betaine modulates age-related NF-κB by thiol-enhancing action. *Biological and Pharmaceutical Bulletin*. 30: 2244–2249.

Go, E. K., Jung, K. J., Kim, J. Y., Yu, B. P. and Chung, H. Y., 2005. Betaine suppresses proinflammatory signaling during aging: the involvement of nuclear factor-κB via nuclear factor-inducing kinase/IκB kinase and mitogen-activated protein kinases. *The Journals of Gerontology Series A: Biological Sciences and Medical Sciences*. 60: 1252–1264.

Goodman, J., Leong, K. C. and Packard, M. G., 2012. Emotional modulation of multiple memory systems: implications for the neurobiology of post-traumatic stress disorder. *Reviews of Neuroscience*. 23: 627–643.

Hiramatsu, M. and Inoue, K., 2000. Improvement by low doses of nociceptin/orphanin FQ on scopolamine-induced impairment of learning and/or memory in mice. *European Journal of Pharmacology*. 395: 149–156.

Hiramatsu, M. and Inoue, K., 1999. Nociceptin/orphanin FQ and nocistatin on learning and memory impairment induced by scopolamine in mice. *British Journal of Pharmacology*. 127: 655–660.

Kempson, S. A. and Montrose, M. H., 2004. Osmotic regulation of renal betaine transport: transcription and beyond. *Pflügers Archiv*. 449: 227–234.

Konturek, P. C., Brzozowski, T., Duda, A., Kwiecien, S., Löber, S., Dembinski, A., Hahn, E. G. and Konturek, S. J., 2001. Epidermal growth factor and prostaglandin $E_2$ accelerate mucosal recovery from stress-induced gastric lesions via inhibition of apoptosis. *Journal of Physiology*. 95: 361–367.

Lee, E. K., Jang, E. J., Jung, K. J., Kim, D. H., Yu, B. P. and Chung, H. Y., 2013. Betaine attenuates lysophosphatidylcholine-mediated adhesion molecules in aged rat aorta: modulation of the nuclear factor-κB pathway. *Experimental Gerontology*. 48: 517–524.

Lowy, M. T., Gault, L. and Yamamoto, B. K., 1993. Adrenalectomy attenuates stress-induced elevations in extracellular glutamate concentrations in the hippocampus. *Journal of Neurochemistry*. 61: 1957–1960.

Madsen, K. K., Clausen, R. P., Larsson, O. M., Krogsgaard-Larsen, P., Schousboe, A. and White, H. S., 2009. Synaptic and extrasynaptic GABA transporters as targets for anti-epileptic drugs. *Journal of Neurochemistry*. 109(suppl. 1): 139–144.

Miwa, M., Tsuboi, M., Noguchi, Y., Enokishima, A., Nabeshima, T. and Hiramatsu, M., 2011. Effects of betaine on lipopolysaccharide-induced memory impairment in mice and the involvement of GABA transporter 2. *Journal of Neuroinflammation*. 8: 153.

Mizoguchi, K., Kunishita, T., Chui, D. H. and Tabira, T., 1992. Stress induces neuronal death in the hippocampus of castrated rats. *Neuroscience Letters*. 138: 157–160.

Ohta, Y., Chiba, S., Tada, M., Imai, Y. and Kitagawa, A., 2007. Development of oxidative stress and cell damage in the liver of rats with water-immersion restraint stress. *Redox Report*. 12: 139–147.

Ohta, Y., Yashiro, K., Kaida, S., Imai, Y., Ohashi, K. and Kitagawa, A., 2013. Water-immersion restraint stress disrupts nonenzymatic antioxidant defense systems through rapid and continuous ascorbic acid depletion in the adrenal gland of rats. *Cell Biochemistry and Function*. 31: 254–262.

Olsen, M., Sarup, A., Larsson, O. M. and Schousboe, A., 2005. Effect of hyperosmotic conditions on the expression of the betaine-GABA-transporter (BGT-1) in cultured mouse astrocytes. *Neurochemical Research*. 30: 855–865.

Oosthuizen, F., Wegener, G. and Harvey, B. H., 2005. Nitric oxide as inflammatory mediator in post-traumatic stress disorder (PTSD): evidence from an animal model. *Neuropsychiatric Disease and Treatment*. 1: 109–123.

Pung, T., Klein, B., Blodgett, D., Jortner, B. and Ehrich, M., 2006. Examination of concurrent exposure to repeated stress and chlorpyrifos on cholinergic, glutamatergic, and monoamine neurotransmitter systems in rat forebrain regions. *International Journal of Toxicology*. 25: 65–80.

Rezai-Zadeh, K., Arendash, G. W., Hou, H., Fernandez, F., Jensen, M., Runfeldt, M., Shytle, R. D. and Tan, J., 2008. Green tea epigallocatechin-3-gallate (EGCG) reduces beta-amyloid mediated cognitive impairment and modulates tau pathology in Alzheimer transgenic mice. *Brain Research*. 1214: 177–187.

Rodrigues, J., Assunção, M., Lukoyanov, N., Cardoso, A., Carvalho, F. and Andrade, J. P., 2013. Protective effects of a catechin-rich extract on the hippocampal formation and spatial memory in aging rats. *Behavioural Brain Research*. 246: 94–102.

Roozendaal, B., Hahn, E. L., Nathan, S. V., de Quervain, D. J. and McGaugh, J. L., 2004. Glucocorticoid effects on memory retrieval require concurrent noradrenergic activity in the hippocampus and basolateral amygdala. *The Journal of Neuroscience*. 24: 8161–8169.

Stillman, M. J., Shukitt-Hale, B., Levy, A. and Lieberman, H. R., 1998. Spatial memory under acute cold and restraint stress, *Physiology & Behavior*. 64: 605–659.

Sugama, S., Fujita, M., Hashimoto, M. and Conti, B., 2007. Stress induced morphological microglial activation in the rodent brain: involvement of interleukin-18. *Neuroscience*. 146: 1388–1399.

Szczepanik, A. M. and Ringheim, G. E., 2003. IL-10 and glucocorticoids inhibit Aβ(1-42)- and lipopolysaccharide-induced pro-inflammatory cytokine and chemokine induction in the central nervous system. *Journal of Alzheimer's Disease*. 5: 105–117.

Tamaki, K., Yamada, K., Nakamichi, N. and Taniura, H., Yoneda, Y., 2008. Transient suppression of progenitor cell proliferation through NMDA receptors in hippocampal dentate gyrus of mice with traumatic stress experience. *Journal of Neurochemistry*. 105: 1642–1655.

Tanaka, S., Ide, M., Shibutani, T., Ohtaki, H., Numazawa, S., Shioda, S. and Yoshida, T., 2006. Lipopolysaccharide-induced microglial activation induces learning and memory deficits without neuronal cell death in rats. *Journal of Neuroscience Research*. 83: 557–566.

Ueland, P. M., 2011. Choline and betaine in health and disease. *Journal of Inherited Metabolic Disease*. 34: 3–15.

Vanitallie, T. B., 2002. Stress a risk factor for serious illness. *Metabolism*. 51 (6 suppl. 1): 40–45.

Wilson, C. B., McLaughlin, L. D., Nair, A., Ebenezer, P. J., Dange, R. and Francis, J., 2013. Inflammation and oxidative stress are elevated in the brain, blood, and adrenal glands during the progression of post-traumatic stress disorder in a predator exposure animal model. *PLoS One*. 8: e76146.

Yoneda, Y., Kanmori, K., Ida, S. and Kuriyama, K., 1983. Stress-induced alterations in metabolism of gamma-aminobutyric acid in rat brain. *Journal of Neurochemistry*. 40: 350–356.

Zeisel, S. H., Mar, M. H., Howe, J. C. and Holden, J. M., 2003. Concentrations of choline-containing compounds and betaine in common foods. *Journal of Nutrition*. 133: 1302–1307.

Zhu, X. M. and Ong, W. Y., 2004. Changes in GABA transporters in the rat hippocampus after kainate-induced neuronal injury: decrease in GAT-1 and GAT-3 but upregulation of betaine/GABA transporter BGT-1. *Journal of Neuroscience Research*. 77: 402–409.

CHAPTER 15

# Serum Betaine Concentrations and the Effects of Bariatric Surgery

JULIAN SWIERCZYNSKI*[a], ARETA HEBANOWSKA[a], AND TOMASZ SLEDZINSKI[b]

[a]Department of Biochemistry, Medical University of Gdansk, ul. Debinki 1, 80-211, Gdansk, Poland; [b]Department of Pharmaceutical Biochemistry, Medical University of Gdansk, ul. Debinki 1, 80-211, Gdansk, Poland
*E-mail: juls@gumed.edu.pl

## 15.1 Introduction

Betaine is present in a variety of foods, rapidly absorbed from the intestine and accumulated in many organs, particularly in liver and kidneys (Craig, 2004; Slow *et al.*, 2009; Zeisel *et al.*, 2013). Choline, present in consumed food, is absorbed from the intestine and transported to the liver. In the liver, choline can be incorporated into phosphatidylcholine (the most abundant phospholipid in mammalian tissues and a substrate for sphingomyelin synthesis) or converted into betaine *via* betaine aldehyde (Figure 15.1). Betaine is an osmolyte protecting cells under stress, and a source of methyl groups in reactions catalyzed by betaine-homocysteine methyltransferase (BHMT). The latter reaction contributes to the reduction of the elevated levels of serum homocysteine, thus diminishing the risk of cardiovascular disease,

Food and Nutritional Components in Focus No. 7
Betaine: Chemistry, Analysis, Function and Effects
Edited by Victor R Preedy

Published by the Royal Society of Chemistry, www.rsc.org

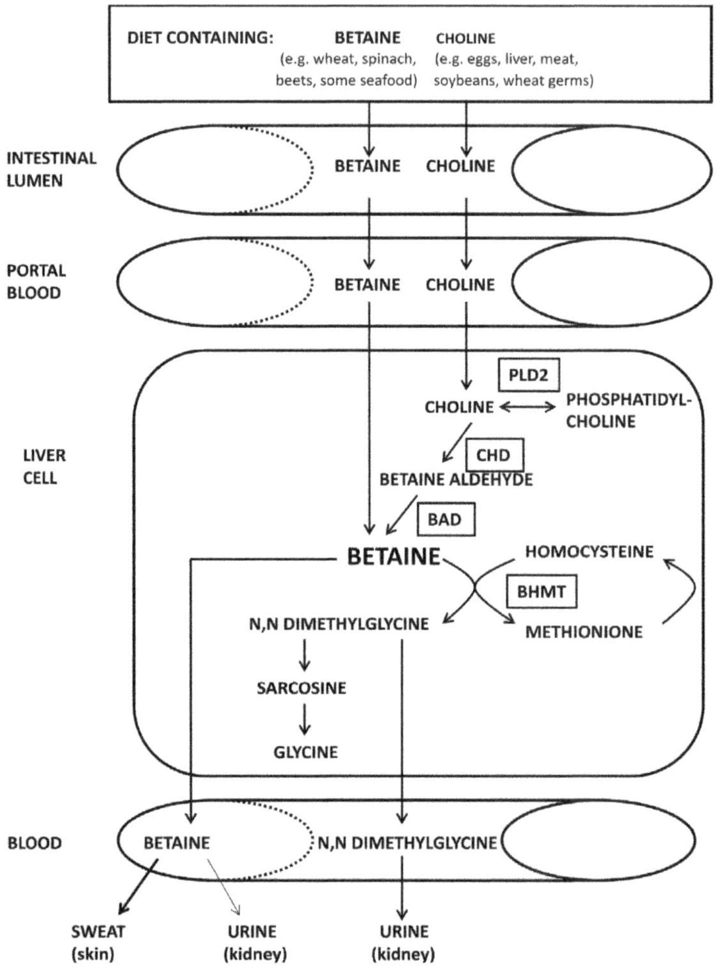


**Figure 15.1** Sources and metabolism of betaine in mammals. Betaine and choline are absorbed from the alimentary tract and then transported to the liver. In the liver, choline can be incorporated into phosphatidylcholine or converted into betaine *via* betaine aldehyde in reactions catalyzed by the mitochondrial enzymes: choline dehydrogenase (CHD; EC 1.1.99.1) and betaine aldehyde dehydrogenase (BAD; EC 1.2.1.8). Choline may also derive from hydrolysis of membrane phosphatidylcholine by phospholipase D2 (PLD2). Betaine, in reaction catalyzed by betaine-homocysteine methyltransferase (BHMT) (EC 2.1.1.5), is metabolized to N,N-dimethylglycine (DMG), that can be in turn metabolized to sarcosine (N-methylglycine) and further to glycine, or excreted in urine. Most of the DMG is converted to sarcosine, rather than eliminated with urine.

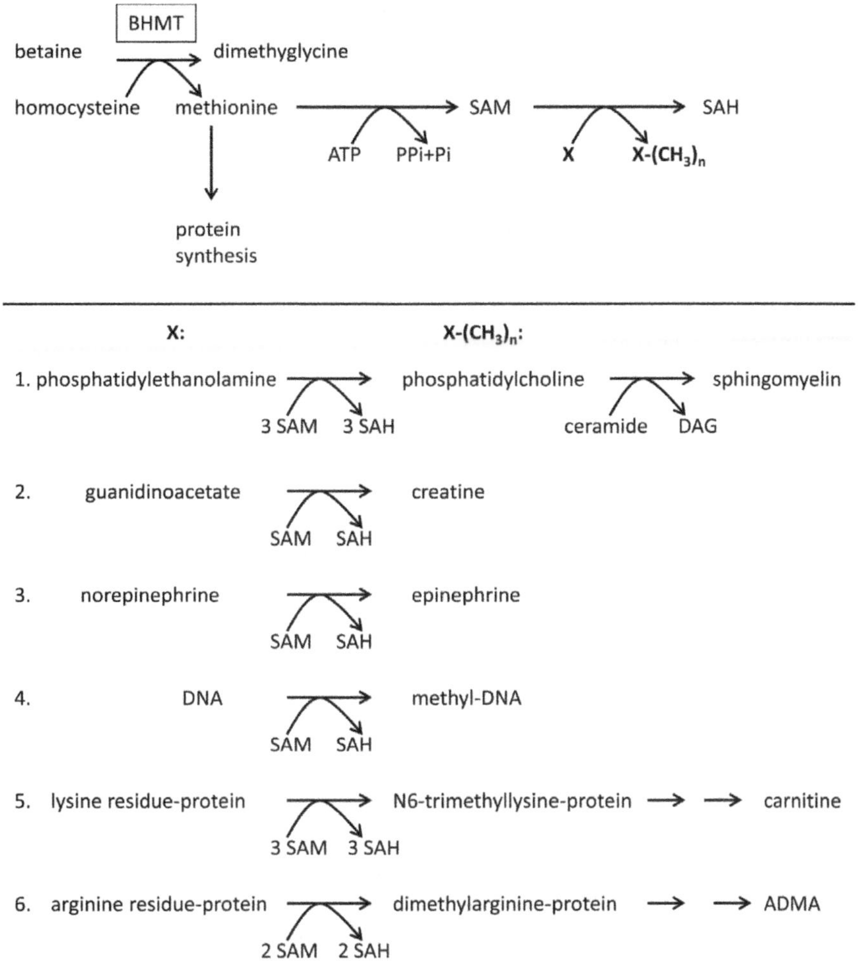


**Figure 15.2** The role of betaine in the supply of *S*-adenosylmethionine for methylation reactions. Betaine is used for conversion of homocysteine to methionine catalyzed by betaine-homocysteine methyltransferase (BHMT). In turn, methionine is incorporated into protein or converted to *S*-adenosylmethionine (SAM). SAM is a methyl donor for presented methylation reactions. SAH – *S*-adenosylhomocysteine; ADMA – asymmetric dimethylarginine.

stroke, and other disorders, such as Alzheimer's disease, dementia, and neural tube defects (Craig, 2004). Methionine formed in reaction catalyzed by BHMT can be used for S-adenosylmethionine (SAM) or protein biosynthesis (Figure 15.2). SAM is an important methyl-group donor in many reactions presented in Figure 15.2. Betaine also plays a regulatory role (Song *et al.*, 2007; Jung *et al.*, 2013; Wang *et al.*, 2013; Zeisel *et al.*, 2013) and is required to maintain normal structure and function of mitochondria (Zeisel *et al.*, 2013). Potential physiological roles of betaine are summarized in Figure 15.3.

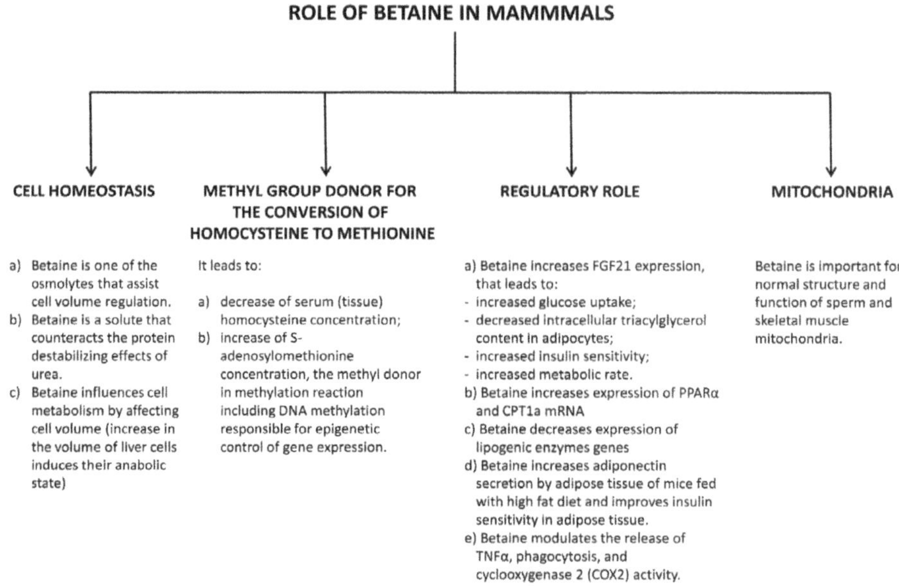


**Figure 15.3** The summary of potential physiological roles of betaine in mammals. Based on the data reported in: Craig (2004); Eklund *et al.* (2005); Song *et al.* (2007); Wang *et al.* (2010); Jung *et al.* (2013); Wang *et al.* (2013); Zeisel *et al.* (2013).

The data presented above suggest that betaine plays an important role in physiology of humans and other mammals, and can be involved in pathogenesis of some diseases. Therefore, it is important to identify patients with betaine deficiency and correct their betaine levels by means of supplementation. The focus of this review is to briefly describe the pathophysiological role of betaine, and characterize tissue and circulating betaine concentrations in healthy subjects and in patients with various diseases. In view of the global obesity epidemic and its treatment by bariatric surgery, special attention is paid to circulating betaine concentrations in morbidly obese patients before and after bariatric surgery. Bariatric surgery is designed to promote body weight loss, mainly *via* changes in anatomical structure of alimentary tract and eating habits. Since some region of the alimentary tract is bypassed, bariatric surgery leads to weight loss, but unfortunately, patients develop some complications including deficiencies in macronutrients and micronutrients, such as folate and vitamin $B_{12}$, which, like betaine, are related to the intracellular levels of methyl groups. Considering that betaine and its precursor, choline, are obtained from food, and absorbed from the intestinal lumen, obese patients may develop betaine deficiency after bariatric surgery. This may lead to low levels of SAM in tissues and disturbance in methylation reactions, and consequently, to side effects of bariatric surgery, such as hyperhomocysteinemia and related disorders. Betaine deficiency after bariatric surgery can be treated similarly as deficiency of most micronutrients. Thus, attention should be paid to serum betaine concentrations before and after bariatric surgery to identify patients with low concentrations of circulating betaine. Data regarding serum betaine concentration in obese subjects are sparse and the number of papers reporting the effects of bariatric surgery on serum betaine concentrations is limited (Sledzinski *et al.*, 2011). Moreover, no data regarding betaine supplementation after bariatric surgery have been reported. Therefore, this research area is in its initial stages, and we need new data for the development of strategies that would prevent betaine deficiency and its consequences after bariatric surgery.

## 15.2 The Role of Betaine in Pathological Conditions

Betaine-dependent conversion of homocysteine to methionine can be responsible for the decrease in circulating homocysteine concentrations. Perturbations in this process may contribute to hyperhomocysteinemia and, consequently, to cardiovascular disease, because increased plasma homocysteine concentration has been identified as a risk factor for atherosclerosis-related conditions (Craig, 2004). Concentrations of betaine in liver and kidneys are affected mainly by BHMT activity. Reduced BHMT activity was documented in experimental cirrhotic rat liver (Craig, 2004), which is associated with the increase in plasma homocysteine concentration in cirrhosis. Betaine can protect the structure and function of rat liver from certain toxins (*i.e.* chloroform and carbon tetrachloride), ethanol and drugs (methotrexate), and prevents cirrhosis (Craig, 2004). Betaine exerts hepatoprotective effect in an animal model of nonalcoholic fatty liver disease (NAFLD), often associated with obesity, diabetes and insulin resistance. Betaine also exerts an antiatherosclerotic effect in ApoE$^{-/-}$ mice. Reduction of aortic inflammation and decrease in Apo-B containing lipoprotein could be one possible mechanism behind the antiatherogenic action of betaine (Ueland, 2011). Some data suggest that betaine can be used to treat NAFLD and nonalcoholic steatohepatitis (NASH) in humans (Ueland, 2011). Betaine, alone or combined with other metabolites (*i.e.* guanidinoacetate – a precursor of creatine) can alleviate certain symptoms in patients with heart, and renal disorders and in individuals with poliomyelitis (Craig, 2004). High doses of betaine (up to 20 g per day) are already given to subjects with inherited metabolic disorders associated with hyperhomocysteinemia. Betaine intake is inversely correlated with breast cancer risk (Zhang *et al.*, 2013). However, other data suggest that betaine intake has little or no influence on colorectal cancer risk in a population of well-nourished men (Lee *et al.*, 2010). Beneficial effects of betaine supplementation on power performance and fatigue were also reported (Hoffman *et al.*, 2009). The data discussed above suggest that betaine plays an important role in physiology and can be involved in pathogenesis of some diseases. Thus, betaine deficiency could be responsible for metabolic disturbances and consequently lead to development of health problems. Therefore, it is important to identify subjects with betaine deficiency and to correct their circulating and tissues levels of this compound by means of supplementation.

## 15.3 Serum Betaine and Dimethylglycine Concentrations in Health and Disease

The concentrations of betaine in most of rat tissues (except of brain and skeletal muscle) are significantly higher than in plasma. Ineffective transport of betaine across the blood–brain barrier could be the one possible reason behind lower betaine concentrations in the brain than in circulation. Most of choline transported to the brain from intravascular compartment across the blood–brain barrier (note that mammalian brain has limited capacity to synthesize choline) is converted to acetylcholine or to phospholipids instead of betaine. This could be the second reason behind very low levels of betaine observed in the brain. The concentrations of betaine in skeletal muscles were comparable with plasma betaine level (Slow *et al.*, 2009). Similarly as in plasma, the betaine concentrations in some tissues (organs) of males (except for liver, kidneys and skin) were higher than in females (Slow *et al.*, 2009).

Determination of metabolites, vitamins, and microelements in serum or plasma is the best way to diagnose nutritional deficiencies. Whether serum (plasma) betaine concentrations can be good markers of betaine status in humans is still an open question. Animal studies suggest that betaine is stored in the cells, and constitutes a ready-to-use methyl-group donor (Obeid, 2013). Therefore, plasma betaine concentration is a rather poor predictor of intracellular betaine content under physiological conditions, but may reflect changes in the levels of the methyl group under conditions associated with high demand for the latter (Obeid, 2013).

Plasma betaine concentration is undoubtedly an important determinant of plasma homocysteine level. The changes in circulating betaine concentrations observed under pathophysiological conditions may result from: (a) varied ingestion of foods containing different amounts of betaine and choline, (b) varied intestinal absorption of betaine (and choline) associated with alimentary tract disease or gut surgery, (c) stimulation or inhibition of choline conversion to betaine catalyzed by mitochondrial choline dehydrogenase and aldehyde betaine dehydrogenase, (d) changes in BHMT activity, and (e) betaine excretion. Normal dietary variation has no significant effect on plasma betaine concentrations and betaine excretion, whereas high doses of betaine can be reflected by increased plasma betaine concentrations and elevated betaine excretion. Folate supplementation increases plasma betaine concentrations, which suggests that it causes a lowering of betaine utilization (Obeid, 2013). Circulating betaine concentrations in healthy human range between 16 and 60 μmol $L^{-1}$ and are slightly higher in adult males than in females (Table 15.1). The use of different methods for serum (plasma) betaine determination could be reflected by some discrepancies between results from different

**Table 15.1** Betaine and dimethylglycine concentrations in plasma and serum of healthy subjects. *-statistically significant *vs.* fasted group. DMG – N, N-dimethylglycine; SEM – standard error of the mean; SD – standard deviation.

| | | Betaine (μmol $L^{-1}$) | DMG (μmol $L^{-1}$) | DMG/Betaine | References |
|---|---|---|---|---|---|
| Serum | women and men (av. 43 years) | | | | Holm (2003) |
| | fasted | 32.8 (28.3–42.1) | 1.88 (1.39–2.19) | 0.057 | |
| | non-fasted | 41.5 (37.4–47.9)* | 1.93 (1.64–2.52) | 0.046* | |
| Plasma | women (<21 years) | 25 ± 10 (SD) | 2.3 ± 1.8 (SD) | 0.092 | Yan (2012a) |
| Plasma | women and men (5–17 years) | 28.9 (16.4–44.1) | 3.5 (1.6–5.5) | 0.121 | Obeid (2012) |
| Plasma | women (49–71 years) | 29.8 ± 24.4 (SD) | 3.86 ± 3.13(SD) | 0.129 | Wallace (2012) |
| Plasma | men (19–40 years) | 36.1 ± 2.7 (SEM) | 0.9 ± 0.3 (SEM) | 0.024 | Atkinson (2008) |
| Plasma | women and men (4–12 years) | 37.0 ± 7 (SD) | | | Schoendorfer (2012) |
| Plasma | men (av.56 years) | 37.30 (36.9–37.7) | | | de Vogel (2011) |
| Plasma | women (20–50 years) | 32.3 ± 4.4 (SEM) | | | Ross (2011) |
| | men (20–50 years) | 59.4 ± 10.6 (SEM) | | | |
| Plasma | women (47–49 years) | 33.7 ± 10.8 (SD) | | | Konstantinova (2008a) |
| | women (71–74 years) | 37.3 ± 10.7 (SD) | | | |
| | men (47–49 years) | 44.1 ± 12.6 (SD) | | | |
| | men (71–74 years) | 45.4 ± 12.5 (SD) | | | |
| Plasma | men (av. 21 years) | 40.1 (24.4–58.2) | | | Yan (2012b) |
| Plasma | women and men (20–65 years) | 41.7 (35.9–46.8) | 2.07 (1.60–2.52) | 0.049 | Holm 2003 |
| | Serum women and men (20–65 years) | 41.5 (37.4–47.9) | 1.93 (1.64–2.52) | 0.47 | |
| Plasma | women (av.5 years) | 47.1 ± 10.7 (SD) | 4.7 ± 2.6 (SD) | 0.1 | Strain (2013) |
| | men (av.5 years) | 44.4 ± 10.8 (SD) | 4.9 ± 2.7 (SD) | 0.11 | |
| Plasma | women (av. 30 years) | 50.9 ± 18.1 (SD) | 4.1 ± 2.1 (SD) | 0.08 | Alsayed (2013) |
| | men (av.16 months) | 59.8 ± 25.3 (SD) | 6.0 ± 2.4 (SD) | 0.1 | |
| Serum | women and men (20–50 years) | 22.6 ± 10.0 (SD) | 16.2 ± 4.4 (SD) | 0.72 | Schwab (2011) |

studies. Fasting significantly decreases, whereas feeding increases serum betaine concentrations (Holm *et al.*, 2003). Strong positive correlations between dietary intake and plasma betaine concentrations were observed in autistic children (Hamlin *et al.*, 2013). Betaine concentrations were positively correlated with intake of high-fiber diet, whereas negatively associated with diet containing excess of meat, fat, and sugar (Konstantinova *et al.*, 2008b). Although serum betaine concentrations seem to be independent of age, this issue requires further study (Table 15.1). Plasma DMG concentrations corresponded to approximately 10% of betaine levels in healthy subjects (Table 15.1). Consequently, the ratio of plasma DMG/betaine concentrations, which reflects the flux through BHMT is close to 0.1, with only few reported exception, where the ratio was either significantly lower (0.024), or higher (0.7) (Table 15.1). Perhaps, these discrepancies resulted from either underestimation (in the case of lower serum DMG/betaine ratio), or overestimation (in the case of higher serum DMG/betaine ratio) of DMG concentrations. However, it cannot be excluded that the flux through BHMT was lower in subjects displaying DMG/betaine = 0.024 or higher in subjects with DMG/betaine = 0.7. Plasma betaine concentrations correlate significantly with plasma levels of choline (Holm *et al.*, 2003). Table 15.2 presents plasma and serum betaine concentrations in certain pathological conditions. In the case of various lipid disorders, plasma betaine and DMG concentrations in men were higher than in women, however, the DMG/betaine ratio remained close to 0.1 (Lever *et al.*, 2007). Plasma DMG concentrations increased significantly with renal function decline in the course of chronic renal failure, whereas betaine concentrations remained unchanged (McGregor *et al.*, 2001). Also, plasma betaine concentration of diabetic patients turned out to be diminished and inversely correlated with plasma homocysteine level (Chen *et al.*, 2013). Lower plasma betaine concentrations are also associated with increased risk of microangiopathy, a serious complication of diabetes. Serum betaine concentrations of patients with diabetic retinopathy and nephropathy are lower than in other patients with diabetes (Chen *et al.*, 2013). Significantly enhanced excretion of betaine observed in many patients with diabetes and vascular disease can constitute a potential reason for a decrease in circulating betaine concentration (Lever *et al.*, 2005). Despite elevated plasma betaine concentrations in pregnant women with pre-eclampsia, higher plasma homocysteine concentrations were observed in this group. While the plasma concentrations of betaine in infants with congenital heart defects are lower than in controls, the concentrations of DMG are higher. This can result from increased activity of BHMT, leading to enhanced biosynthesis of DMG, or from decreased DMG metabolism through enzymes, DMG dehydrogenase or sarcosine dehydrogenase. Interestingly, increased levels of DMG were also observed in the plasma of mothers, suggesting a genetic background of this phenomenon (Alsayed *et al.*, 2013). Cystic fibrosis is another disease associated with abnormal concentrations of circulating betaine. Plasma concentrations of betaine and DMG are significantly decreased, probably due to disturbances of digestion and absorption of nutrients observed in cystic fibrosis. Disturbances of

**Table 15.2** Betaine plasma concentration in pathological conditions. *-statistically significant *vs.* control group, ⁺-statistically significant *vs.* woman. DMG – N, N-dimethylglycine; SEM – standard error of the mean; SD – standard deviation.

| | | Betaine ($\mu mol\ L^{-1}$) | DMG ($\mu mol\ L^{-1}$) | DMG/Betaine | References |
|---|---|---|---|---|---|
| Serum | control women and men (59 ± 13 years) | 28.06 (22.6–36.0) | | | Chen (2013) |
| | Diabetes type 2 (60 ± 14 years) | 30.79 (24.3–39.5) | | | |
| | Diabetes type 2/microangiopathy (60 ± 13 years) | 25.11 (17.8–31.3)* | | | |
| Plasma | women with lipid disorders (48–63 years) | 20.0 (15.0–25.6) | 1.8 (1.4–2.2) | 0.09 | Lever (2007) |
| | Men with lipid disorders (43–59 years) | 24.9 (19.2–31.9)⁺ | 2.1 (1.5–2.7)⁺ | 0.08 | |
| Plasma | men, normal weight (17–41 years) | 40.1 (24.4–58.2) | | | Yan (2012b) |
| | Obese (18–56 years) | 37.3 (21.2–79.4)* | | | |
| Plasma | women and men | | | | de Vogel (2011) |
| | No neoplasia (56.1 ± 8 years) | 37.3 (36.9–37.8) | | | |
| | Low risk colorectal adenoma (56.7 ± 3.7 years) | 36.8 (36.1–37.5) | | | |
| | High risk colorectal adenoma (57.2 ± 3.6 years) | 35.8 (34.6–37.0) | | | |
| Plasma | women and men | | | | Alsayed (2013) |
| | control (16.9 ± 9.4 months) | 59.8 ± 25.3 (SD) | 6.0 ± 2.4 (SD) | 0.1 | |
| | Congenital heart defects (19.9 ± 12.3 months) | 56.1 ± 31.4 (SD) | 8.7 ± 5.3* (SD) | 0.16 | |
| Plasma | women and men | | | | Obeid (2012) |
| | control (5.7–17 years) | 28.9 (16.4–44.1) | 3.5 (1.6–5.5) | 0.12 | |
| | Down's syndrome (1.9–27 years) | 34.0 (24.0–52.6)* | 5.9 (4–10.5)* | 0.17 | |
| Plasma | women and men | | | | Schoendorfer (2012) |
| | control (7.5 ± 2 years) | 37 ± 7 (SD) | | | |
| | Cerebral palsy (7 ± 2 years) | 31 ± 12 (SD)* | | | |
| Plasma | women and men | | | | McGregor (2001) |
| | control (53–75 years) | 33.6 (23.9–42.1) | 2.6 (1.8–3.7) | 0.077 | |
| | Chronic renal failure (54–74 years) | 32.5 (23.9–42.5) | 4.8 (3.6–6.2)* | 0.15 | |
| Plasma | women and men | | | | Sledzinski (2011) |
| | control | 16 ± 0.84 (SEM) | | | |
| | Morbidly obese | 20 ± 1.8 (SEM)* | | | |
| | Morbidly obese six month after bariatric surgery | 14 ± 0.86 (SEM) | | | |

one-carbon metabolism (including betaine and folate metabolism dysregulation) observed in patients with cystic fibrosis lead to impaired growth and development (Schoendorfer *et al.*, 2012). A higher dietary intake of betaine, or its higher plasma levels, are correlated with lower risk of colorectal cancer. This probably results from reduction of DNA instability through efficient methylation processes (de Vogel *et al.*, 2011). Methylation disturbances are also supposed to induce neurodegeneration. Increased cystathionine biosynthesis resulting from overexpression of cystathionine synthase (associated with trisomy of 21 chromosome), leads to a decrease in plasma homocysteine concentrations. Simultaneously, a higher plasma concentration of betaine, DMG and methionine, point to increased BHMT activity. Those differences of methyl cycle metabolites observed in Down's syndrome patients, suggest the changes in methylation processes, increasing risk of dementia (Obeid *et al.*, 2012).

## 15.4 Human Plasma Betaine Concentration after Betaine Supplementation

Average daily intake of betaine in humans varies between populations, and depends mainly on the kind of consumed food. Usually it does not exceed approximately 0.5 g per day (most humans intake between 100 mg and 300 mg of betaine per day), although the recommended daily intake has not been established. Betaine is absorbed from the intestinal lumen by nonspecific amino acid transporter and then transported to the liver and other organs. Betaine intake in humans displayed rapid absorption and redistribution, with a peak increase in serum betaine concentration at approximately 1.5 h (1–2 h) (Schwahn *et al.*, 2003).

The main purpose of betaine supplementation is to decrease the circulating homocysteine concentrations, which may potentially reduce the risk of cardiovascular disease and stroke. One meta-analysis proved that betaine supplementation, at least 4 g per day for 6 weeks, decreased plasma homocysteine concentrations in healthy adults by approximately 12% of baseline values (McRae, 2013). Few studies concerning association between betaine supplementation and cardiovascular risk have been published thus far, and no reported evidence has indicated that betaine supplementation reduces significantly the risk of heart disease and stroke. The data presented in a recent review suggest that long-term supplementation of betaine may prevent cardiovascular disease and mortality by decreasing inflammation and other risk factors (Rajaie and Esmaillzadeh, 2011). Administration of betaine (6–20 g per day) in classical homocystinuria resulting from cystationine β-synthase deficiency, significantly (up to 90%) decreased circulating homocysteine concentrations, however, no clinical benefit of betaine administration (6 g per day) was reported (Olthof *et al.*, 2005). A pharmacokinetic study showed the maximum concentrations (148–258 μmol $L^{-1}$) of circulating betaine were found between 1 and

2 h after its oral administration (100 mg per kg body weight) to healthy adults, with apparent elimination half-life of one hour (Olthof *et al.*, 2005). Another study revealed an increase in plasma betaine concentration (up to 940 μmol $L^{-1}$) within one hour after administering a single dose of betaine (50 mg per kg body weight) to healthy male subjects (Olthof *et al.*, 2005). Plasma homocysteine concentrations in patients with methylenetetrahydrofolate reductase (MTHFR) deficiency were affected by betaine administration in an amount that caused an increase in betaine concentration above 400 μmol $L^{-1}$. The effect of circulating betaine on plasma homocysteine concentration in patients with MTHFR is negligible, when serum betaine concentration is below 400 μmol $L^{-1}$ (*i.e.* approximately 10 times higher than in the controls) (Olthof *et al.*, 2005). The results of pharmacokinetic and pharmacodymic studies suggest that 150 mg per kg of betaine administered twice a day is an optimal dose to treat homocystynuria in patients with cystationine β-synthase deficiency. In these patients the mean 24 h plasma betaine concentrations are approximately 450 μmol $L^{-1}$ (Olthof *et al.*, 2005). Therefore, increasing the daily dose of betaine from 150 mg to 250 per kg body weight results in negligible health benefits. However, more data are needed to resolve this problem. The results presented above suggest that betaine should be administered twice a day at 150 mg per kg body weight to significantly reduce plasma homocysteine concentrations. Noticeably high doses of betaine (4–6 g per day) were revealed to increase plasma levels of total cholesterol, LDL-cholesterol, and triacylglycerol in humans (Olthof *et al.*, 2005).

## 15.5 Betaine and Body Composition

Betaine is commercially available in purified form as an animal feed additive (mainly for pigs and chickens). Feeding animals with diet supplemented with betaine leads to a decrease in fat tissue mass and an improved carcass lean percentage of the animals (Eklund *et al.*, 2005). Betaine seems to alter nutrient partitioning in young animals under metabolic or nutritional stress, leading to enhancement of protein deposition at the expense of carcass fat, and in part, at the expense of visceral adipose tissue. Thus, betaine is often dubbed a “carcass modifier”. In general, betaine is considered as lipotropic and growth-promoting factor in animal nutrition. Some authors suggest that the betaine-induced increase in the percentage of lean body weight can result from higher availability of methionine (formed from homocysteine in a reaction catalyzed by BHMT) for protein biosynthesis in animals fed with betaine supplemented diet. Increased protein biosynthesis from dietary amino acids results in decreased availability of the latter as substrate for fatty acid synthesis and triacylglycerol synthesis (Eklund *et al.*, 2005), leading to a decrease in fat mass. Diet rich in betaine has also been shown to significantly increase carnitine levels in liver and muscles of some animals. Carnitine and CTP1a are important factors regulating long-chain fatty acid oxidation, because the entry of long-chain fatty acids into mitochondria determines the supply of substrate for fatty acid β-oxidation in the mitochondrial matrix. This supports the hypothesis according to which betaine-dependent synthesis of carnitine (namely, carnitine synthesis dependent on SAM formed from methionine, which in turn is formed in betaine-dependent process of homocysteine remethylation) could be linked, at least in part, with the body composition of animals. Feeding obese mice (*db*/*db* mice) with betaine-rich diet led to induction of PPARα and CPT1a mRNA levels (Jung *et al.*, 2013). Increase in PPARα level and its activation are associated with upregulation of genes involved in long-chain fatty acid transport across the inner mitochondrial membrane (due to upregulation of CPT1a) and in mitochondrial fatty acid β-oxidation. Moreover, *db*/*db* mice fed with betaine-rich diet displayed reduced acetyl-CoA carboxylase activity, a key lipogenic enzyme, which may lead to a decrease in fatty acid synthesis. Taken together, these data suggest that betaine can increase fatty acid oxidation and decrease fatty acid synthesis. Also, a decrease in body weight associated with upregulation of PPARα and CPT1, as well as downregulation of fatty acids synthase (another key lipogenic enzyme) gene expression, were observed in $ApoE^{-/-}$ mice fed with betaine-supplemented diet (Wang *et al.*, 2013). Thus, stimulation of fatty acid oxidation and inhibition of fatty acid synthesis after high betaine intake could constitute another mechanism underlying the reduction of body weight and fat mass in animals (Figure 15.4). Contrary to the results discussed above, no effect of betaine-rich diet on body weight was observed in control mice (C57BL/6 mice) and in mice on a high fat diet (Wang *et al.*,

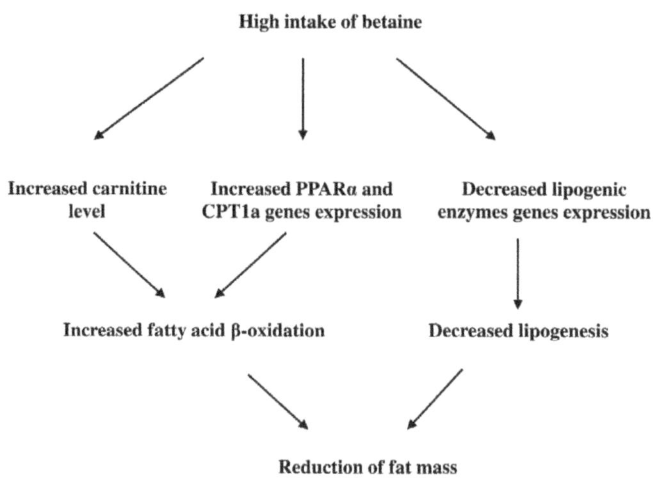


**Figure 15.4** The possible mechanism of betaine influence on fat mass. High betaine intake can increase fatty acids oxidation and decrease fatty acids synthesis, leading to reduction of body weight and fat mass. PPARα – peroxisome proliferator activated receptor alpha; CPT1a – carnitine palmitoyltransferase.

2010). Despite the lack of significant effect of betaine on body mass, this compound improved adipose tissue function in mice on a high fat diet and prevented NAFLD induced by this type of diet (Wang *et al.*, 2010). According to some authors, betaine can reduce protein turnover rate, as a "carcass modifier", which results in higher nitrogen retention and increases, (probably indirectly – *via* glycine, plasma concentrations of growth affecting factors: growth hormone (GH), insulin-like growth factor 1 (IGF-1) and thyroid hormones ($T_3$ and $T_4$) (Eklund *et al.*, 2005). Recently published data indicate that BHMT deficiency reduces body fat mass *via* increasing energy expenditure and impairing lipid synthesis and glucose oxidation in white adipose tissue of mice (Teng *et al.*, 2012). That points to a potential relation between serum/tissue betaine concentrations and the composition of an animal body.

## 15.6 Serum Betaine Concentrations in Obese Patients after Bariatric Surgery

The results of large population-based study (including approximately 7100 men and women) suggest that serum betaine concentration is inversely correlated with BMI, waist circumference, and body fatness (Konstantinova *et al.*, 2008a). Moreover, this study revealed an inverse correlation between serum betaine levels and serum concentrations of triacylglycerol and non-HDL-cholesterol, blood pressure, and smoking. Furthermore, a positive correlation between serum betaine levels and serum HDL-cholesterol concentrations and physical activity was documented (Konstantinova *et al.*, 2008a). Surprisingly, plasma choline concentrations correlated inversely with plasma betaine levels (Konstantinova *et al.*, 2008a), contrary to the results presented by Holm *et al.* for serum and plasma from healthy individuals (Holm *et al.*, 2003). Plasma betaine concentration in young overweight Mexican American men with median BMI 28 kg $m^{-2}$, were slightly lower than in those with normal BMI (22 kg $m^{-2}$) (Yan *et al.*, 2012b). One possible explanation for lower plasma betaine concentrations in overweight men could be the increased use of betaine as a methyl donor for methionine and then for phosphatidylcholine biosynthesis (Yan *et al.*, 2012b). This suggestion is based on the results of murine studies indicating that high-fat diet leads to upregulation of phosphatidylcholine synthesis, obesity, and a decrease in liver betaine concentrations (Jacobs *et al.*, 2010; Rubio-Aliaga *et al.*, 2011). Plasma betaine concentrations in overweight subjects correlated inversely with the levels of serum glucose, total cholesterol, LDL-cholesterol, and total cholesterol/HDL-cholesterol ratio. Moreover, a positive correlation was documented between serum betaine and HDL-cholesterol (Yan *et al.*, 2012b). However, some authors suggested that betaine supplementation exerts no effect on body weight and body composition (Schwab *et al.*, 2011).

Serum betaine concentrations in morbidly obese subjects were significantly higher than in the controls. The fact that plasma betaine concentrations correlated positively with a high-fiber diet (Konstantinova *et al.*, 2008b), and obese subjects consumed a fiber-rich diet prior to surgery (to lose body fat mass), likely contributed to elevated serum betaine concentrations. A decrease in serum betaine concentrations, to values lower than those observed in the controls, was documented six months after bariatric surgery, when body weight and fat mass were reduced considerably (Sledzinski *et al.*, 2011). Positive correlations between: BMI and serum betaine concentrations ($r = 0.38, p < 0.05$), between body mass and serum betaine concentrations ($r = 0.36, p < 0.05$) and between serum leptin and serum betaine concentrations ($r = 0.54, p < 0.01$) were also documented. Moreover, morbidly obese patients had higher serum homocysteine concentrations as compared to healthy subjects (Sledzinski *et al.*, 2011). This suggests that serum betaine concentrations in morbidly obese patients increase proportionally to serum homocysteine levels and body mass. However, in contrast to betaine, serum homocysteine concentrations did not decrease after bariatric surgery. Although based on a small number of patients, our observations emphasize the importance of serum betaine and homocysteine determination in the prevention of betaine deficiency and hyperhomocysteinemia in patients after bariatric surgery. The decrease in serum betaine concentrations after bariatric surgery is probably associated with reduced food intake. However, other causative mechanisms cannot be excluded. Exclusion of an important fragment of the alimentary tract from the passage of nutrients, where absorption of betaine may occur, is crucial and may lead to malabsorption of betaine. A decrease in serum betaine concentrations after bariatric surgery could be reflected by elevated serum concentrations of homocysteine. Betaine acts on homocysteine into methionine conversion *via* different mechanism from those involved by folic acid and vitamin $B_{12}$. Nonbetaine-dependent process of homocysteine remethylation, catalyzed by methionine synthase, requires 5-methyltetrahydrofolate (folate) as methyl donor and cobalamin (vitamin $B_{12}$) as cofactor (Figure 15.5A). In the case of folate and vitamin $B_{12}$ deprivation (often observed in patients after bariatric surgery), more betaine is used for homocysteine remethylation, thereby increasing the demand for betaine (Figure 15.5B). This phenomenon can, at least in part, also contribute to the decrease in serum betaine concentration observed after bariatric surgery. In the case of betaine deficiency, methionine synthase plays a key role in homocysteine remethylation, thereby increasing demand for folate and vitamin $B_{12}$. Thus, supplementation of betaine together with vitamin $B_{12}$ and folate, should be considered in patients after bariatric surgery, in order to decrease serum homocysteine concentrations. Patients with homocystynuria are treated with high doses of betaine (6 g per day) combined with folic acid and vitamin $B_{12}$. Such treatment reduces plasma homocysteine concentrations significantly, and improves the clinical condition of patients.

Due to an inverse relation between serum betaine concentrations and BMI, waist circumference, and body fatness (Yan *et al.*, 2012b), it is tempting to speculate that the bariatric surgery-induced decrease in BMI, and

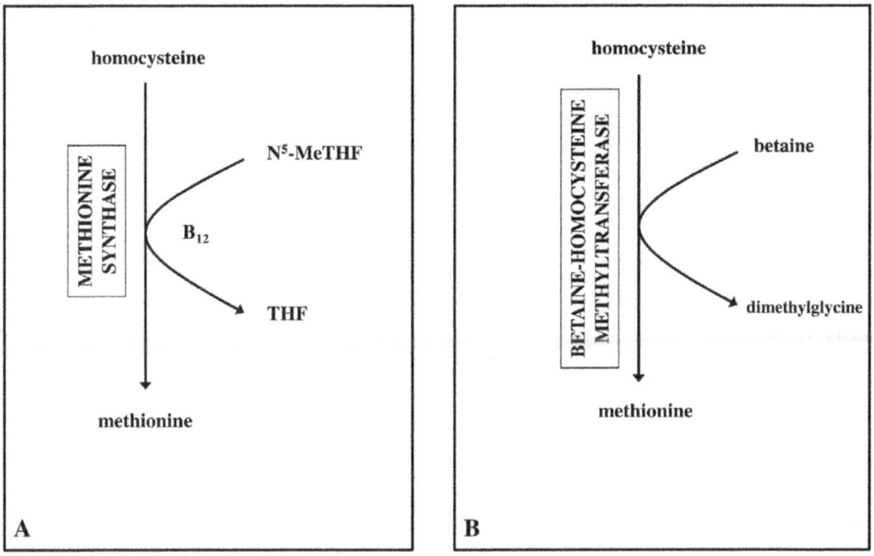


**Figure 15.5** Conversion of homocysteine into methionine catalyzed by methionine synthase (A) and betaine-homocysteine methyltransferase (B). Reactions catalyzed by methionine synthase and betaine-homocysteine methyltransferase are alternative pathways of conversion of homocysteine into methionine. $N^5$-MeTHF – 5-methyltetrahydrofolate; THF – tetrahydrofolate.

especially in body fat mass could be greater if serum betaine concentrations were higher. This could be another argument for betaine supplementation after bariatric surgery.

## 15.7 Directions of Future Research

The facts that: a) betaine plays an important role in methyl group metabolism; b) serum betaine concentrations are inversely correlated with hyperhomocysteinemia and potentially associated with some diseases; c) betaine is derived from food (directly, or indirectly as a choline), and d) betaine is absorbed from intestine, indicate that special attention should be paid to serum betaine concentrations after bariatric surgery in order to avoid metabolic and clinical complications. Bariatric surgery (especially the malabsorptive type of bariatric surgery) usually leads to macronutrient and micronutrient deficiency, including hyperhomocysteinemia-associated deficiency of micronutrients, such as folate, vitamin $B_{12}$ and $B_6$ in some patients, with potential deprivation of methyl groups. This can impact not only on plasma homocysteine concentration, but also on methionine availability for the synthesis of protein, phospholipids, and other important metabolites. These abnormalities can be easily corrected. The data discussed in this review suggest that bariatric surgery may lead to a decrease in circulating betaine concentrations (it should be noted that betaine seems to be more effective than folate at reducing hyperhomocysteinemia (Bertolo and McBreairty, 2013), and that the metabolisms of betaine, folate and vitamin $B_{12}$ are strictly interrelated. Moreover, the deficiency of one of them may cause metabolic disturbances, and can be at least in part compensated by the others. This suggests that supplementation with betaine, together with vitamin $B_{12}$ and folate should be considered in patients after bariatric surgery in order to decrease serum homocysteine concentration and improve other metabolic abnormalities. However, additional research is required to fully clarify: (a) the effect of bariatric surgery, especially different types of this procedure, on plasma betaine concentrations and other related compounds, such as DMG, choline, methionine, folate and vitamin $B_{12}$, (b) the effect of betaine supplementation after bariatric surgery on plasma homocysteine concentrations and other metabolic and functional disturbances including plasma vitamin $B_{12}$, folate, methionine, choline, and (c) whether folate and $B_{12}$ supplementation after bariatric surgery is able to maintain serum/tissue betaine concentrations as a methyl donor, or whether betaine supplementation can maintain the folate level.

## Summary Points

1. Betaine is an important osmolyte, methyl-group donor, and regulatory molecule.
2. Betaine is obtained from food, can also be synthesized from choline, and is stored in tissues.
3. Blood concentration of betaine (16–60 mmol $L^{-1}$ in healthy subjects) can be altered in some pathological conditions.
4. Bariatric surgery leads to a decrease in serum betaine concentration in morbidly obese patients.
5. The decrease in serum betaine concentrations after bariatric surgery is probably associated with exclusion of an important fragment of the alimentary tract.
6. A decrease in serum betaine concentrations after bariatric surgery could be reflected by elevated serum concentrations of homocysteine.
7. Control of blood betaine concentration, and its supplementation in the case of deficiency, seems vital for avoiding the complications associated with bariatric surgery.

## Definitions of Words and Terms

1. Bariatric surgery (metabolic surgery) – Surgical intervention on alimentary tract to promote body weight loss.
2. Osmolyte – A compound that regulates cell volume by countering changes in extracellular tonicity.
3. Hyperhomocysteinemia – Elevated serum homocysteine level resulting from altered homocysteine metabolism.

4. Lipogenic enzymes – Enzymes of fatty acid synthesis.
5. Phosphatidylcholine – The most abundant phospholipid in mammalian tissues.
6. Betaine-homocysteine methyltransferase (BHMT) – An enzyme converting homocysteine to methionine with the use of methyl group from betaine.
7. *N*,*N*-Dimethylglycine – A product of betaine catabolism (*i.e.* by BHMT).
8. Sarcosine (*N*-methylglycine) – A metabolite of DMG.
9. *S*-Adenosylmethionine – An important methyl-group donor in many reactions.
10. Carcass modifier – A compound influencing body fat content.
11. Peroxisome proliferator activated receptor alpha (PPARα) – Nuclear factor promoting fatty acid catabolism.
12. Carnitine palmitoyltransferase (CPT1) – An enzyme involved in fatty acid transport into mitochondria.
13. Carnitine – A compound necessary for fatty acid transport by CPT1.
14. Fatty acid β-oxidation – Degradation of fatty acids into acetyl-CoA with simultaneous production of $NADH + H^+$ and $FADH_2$.
15. Tetrahydrofolate (THF) – An enzymatic reaction cofactor providing one carbon groups.
16. $N^5$ –MeTHF – Methylated THF.
17. Epinephrine – Catecholamine acting as hormone and neurotransmitter.
18. Norepinephrine – Catecholamine acting as hormone and neurotransmitter.

## Key Facts of Bariatric Surgery

- Obesity is an epidemic disease.
- Patients having BMI > 35 kg $m^{-2}$ with obesity associated comorbidities are defined as morbidly obese.
- BMI – Body mass index (weight/squared height (kg $m^{-2}$)). Normal BMI should be in between 18.5 and 25.
- Bariatric surgery (metabolic surgery) – Surgery leading to weight loss.
- Bariatric surgery is the only effective treatment for morbidly obese patients.
- Malabsorptive bariatric surgery techniques lead to reduction of absorption of nutrients.
- Restrictive bariatric surgery techniques lead to a decrease of food intake and promote satiety.
- Bariatric surgery can result in half of the weight loss in one year after the surgery.
- Bariatric surgery leads to improvement of many metabolic parameters, *e.g.* serum lipids, insulin sensitivity.
- Bariatric surgery reduces the risk of obesity associated comorbidities.
- Bariatric surgery can result in deficiencies of some nutrients.

## List of Abbreviations

| | |
|---|---|
| ADMA | Asymmetric dimethylarginine; |
| ApoB | Apolipoprotein B |
| ApoE | Apolipoprotein E |
| BAD | Betaine aldehyde dehydrogenase |
| BHMT | Betaine–homocysteine methyltransferase |
| BMI | Body mass index |
| CHD | Choline dehydrogenase |
| CoA | Coenzyme A |
| COX2 | Cyclooxygenase 2 |
| CPT1a | Carnitine palmitoyltransferase 1a |
| DMG | *N,N*-Dimethylglycine |
| DNA | Deoxyribonucleic acid |
| $FADH_2$ | Flavin adenine dinucleotide |
| FGF 21 | Flbroblast growth factor 21 |
| GH | Growth hormone |
| HDL | High-density lipoprotein |
| IGF–1 | Insulin-like growth factor 1 |
| LDL | Low-density lipoprotein |
| MTHFR | Methylenetetrahydrofolate reductase |
| NADH | Nicotinamide adenine dinucleotide |
| NAFLD | Nonalcoholic fatty liver disease |
| NASH | Nonalcoholic steatohepatitis |
| $N^5$–MeTHF | 5–Methyltetrahydrofolate |
| PLD2 | Phospholipase D2 |
| PPARα | Peroxisome proliferator activated receptor alpha |
| SAM | *S*–Adenosylmethionine |
| SAH | *S*–Adenosylhomocysteine |
| $T_3$ | Triiodothyronine |
| $T_4$ | Thyroxine |
| THF | Tetrahydrofolate |
| TNFα | Tumor necrosis factor alpha |

## Acknowledgements

This work was supported by the Medical University of Gdansk grants no. ST-41 and ST-40.

## References

Alsayed, R., Al Quobaili, F., Srour, S., Geisel, J. and Obeid, R., 2013. Elevated dimethylglycine in blood of children with congenital heart defects and their mothers. *Metabolism*. 62: 1074–1080.

Atkinson, W., Elmslie, J., Lever, M., Chambers, S. T. and George, P. M., 2008. Dietary and supplementary betaine: acute effects on plasma betaine and homocysteine concentrations under standard and postmethionine load conditions in healthy male subjects. *American Journal of Clinical Nutrition*. 87: 577–585.

Bertolo, R. F. and McBreairty, L. E., 2013. The nutritional burden of methylation reactions. *Current Opinion in Clinical Nutrition and Metabolic Care*. 16: 102–108.

Chen, L., Chen, Y. M., Wang, L. J., Wei, J., Tan, Y. Z., Zhou, J. Y., Yang, Y., Chen, Y. M., Ling, W. H. and Zhu, H. L., 2013. Higher homocysteine and lower betaine increase the risk of microangiopathy in patients with diabetes mellitus carrying the GG genotype of PEMT G774C. *Diabetes/Metabolism Research and Reviews*. 29: 607–617.

Craig, S. A., 2004. Betaine in human nutrition. *American Journal of Clinical Nutrition*. 80: 539–549.

de Vogel, S., Schneede, J., Ueland, P. M., Vollset, S. E., Meyer, K., Fredriksen, A., Midttun, O., Bjorge, T., Kampman, E., Bretthauer, M. and Hoff, G., 2011. Biomarkers related to one-carbon metabolism as potential risk factors for distal colorectal adenomas. *Cancer Epidemiology, Biomarkers & Prevention*. 20: 1726–1735.

Eklund, M., Bauer, E., Wamatu, J. and Mosenthin, R., 2005. Potential nutritional and physiological functions of betaine in livestock. *Nutrition Research Reviews*. 18: 31–48.

Hamlin, J. C., Pauly, M., Melnyk, S., Pavliv, O., Starrett, W., Crook, T. A. and James S. J., 2013. Dietary intake and plasma levels of choline and betaine in children with autism spectrum disorders. *Autism Research and Treatment*. 2013: 578429.

Hoffman, J. R., Kang, J., Ratamess, N. A., Hoffman, M. W., Tranchina, C. P. and Faigenbaum, A. D., 2009. Examination of a pre-exercise, high energy supplement on exercise performance. *Journal of the International Society of Sports Nutrition*. 6: 2.

Holm, P. I., Ueland, P. M., Kvalheim, G. and Lien, E. A., 2003. Determination of choline, betaine, and dimethylglycine in plasma by a high-throughput method based on normal-phase chromatography-tandem mass spectrometry. *Clinical Chemistry*. 49: 286–294.

Jacobs, R. L., Zhao, Y., Koonen, D. P., Sletten, T., Su, B., Lingrell, S., Cao, G., Peake, D. A., Kuo, M. S., Proctor, S. D., Kennedy, B. P., Dyck, J. R. and Vance, D. E., 2010. Impaired de novo choline synthesis explains why phosphatidylethanolamine *N*-methyltransferase-deficient mice are protected from diet-induced obesity. *Journal of Biological Chemistry*. 285: 22403–22413.

Jung, G. Y., Won, S. B., Kim, J., Jeon, S., Han, A. and Kwon, Y. H., 2013. Betaine alleviates hypertriglycemia and Tau hyperphosphorylation in db/db mice. *Toxicology Research*. 29: 7–14.

Konstantinova, S. V., Tell, G. S., Vollset, S. E., Nygard, O., Bleie, O. and Ueland, P. M., 2008a. Divergent associations of plasma choline and betaine with components of metabolic syndrome in middle age and elderly men and women. *Journal of Nutrition*. 138: 914–920.

Konstantinova, S. V., Tell, G. S., Vollset, S. E., Ulvik, A., Drevon, C. A. and Ueland, P. M., 2008b. Dietary patterns, food groups, and nutrients as predictors of plasma choline and betaine in middle-aged and elderly men and women. *American Journal of Clinical Nutrition*. 88: 1663–1669.

Lee, J. E., Giovannucci, E., Fuchs, C. S., Willett, W. C., Zeisel, S. H. and Cho, E., 2010. Choline and betaine intake and the risk of colorectal cancer in men. *Cancer Epidemiology, Biomarkers & Prevention*. 19: 884–887.

Lever, M., Atkinson, W., George, P. M. and Chambers, S. T., 2007. Sex differences in the control of plasma concentrations and urinary excretion of glycine betaine in patients attending a lipid disorders clinic. *Clinical Biochemistry*. 40: 1225–1231.

Lever, M., George, P. M., Dellow, W. J., Scott, R. S. and Chambers, S. T., 2005. Homocysteine, glycine betaine, and *N,N*-dimethylglycine in patients attending a lipid clinic. *Metabolism*. 54: 1–14.

McGregor, D. O., Dellow, W. J., Lever, M., George, P. M., Robson, R. A. and Chambers, S. T., 2001. Dimethylglycine accumulates in uremia and predicts elevated plasma homocysteine concentrations. *Kidney International*. 59: 2267–2272.

McRae, M. P., 2013. Betaine supplementation decreases plasma homocysteine in healthy adult participants: a meta-analysis. *Journal of Chiropractic Medicine*. 12: 20–25.

Obeid, R., 2013. The metabolic burden of methyl donor deficiency with focus on the betaine homocysteine methyltransferase pathway. *Nutrients*. 5: 3481–3495.

Obeid, R., Hartmuth, K., Herrmann, W., Gortner, L., Rohrer, T. R., Geisel, J., Reed, M. C. and Nijhout, H. F., 2012. Blood biomarkers of methylation in Down syndrome and metabolic simulations using a mathematical model. *Molecular Nutrition and Food Research*. 56: 1582–1589.

Olthof, M. R., van Vliet, T., Verhoef, P., Zock, P. L. and Katan, M. B., 2005. Effect of homocysteine-lowering nutrients on blood lipids: results from four randomised, placebo-controlled studies in healthy humans. *PLoS Medicine*. 2: e135.

Rajaie, S. and Esmaillzadeh, A., 2011. Dietary choline and betaine intakes and risk of cardiovascular diseases: review of epidemiological evidence. *ARYA Atherosclerosis*. 7: 78–86.

Ross, A. B., Bruce, S. J., Blondel-Lubrano, A., Oguey-Araymon, S., Beaumont, M., Bourgeois, A., Nielsen-Moennoz, C., Vigo, M., Fay, L. B., Kochhar, S., Bibiloni, R., Pittet, A. C., Emady-Azar, S., Grathwohl, D. and Rezzi, S., 2011. A whole-grain cereal-rich diet increases plasma betaine, and tends to decrease total and LDL-cholesterol compared with a refined-grain diet in healthy subjects. *British Journal of Nutrition*. 105: 1492–1502.

Rubio-Aliaga, I., Roos, B., Sailer, M., McLoughlin, G. A., Boekschoten, M. V., van Erk, M., Bachmair, E. M., van Schothorst, E. M., Keijer, J., Coort, S. L., Evelo, C., Gibney, M. J., Daniel, H., Muller, M., Kleemann, R. and Brennan, L., 2011. Alterations in hepatic one-carbon metabolism and related pathways following a high-fat dietary intervention. *Physiological Genomics*. 43: 408–416.

Schoendorfer, N. C., Obeid, R., Moxon-Lester, L., Sharp, N., Vitetta, L., Boyd, R. N. and Davies, P. S., 2012. Methylation capacity in children with severe cerebral palsy. *European Journal of Clinical Investigation*. 42: 768–776.

Schwab, U., Alfthan, G., Aro, A. and Uusitupa, M., 2011. Long-term effect of betaine on risk factors associated with the metabolic syndrome in healthy subjects. *European Journal of Clinical Nutrition*. 65: 70–76.

Schwahn, B. C., Hafner, D., Hohlfeld, T., Balkenhol, N., Laryea, M. D. and Wendel, U., 2003. Pharmacokinetics of oral betaine in healthy subjects and patients with homocystinuria. *British Journal of Clinical Pharmacology*. 55: 6–13.

Sledzinski, T., Goyke, E., Smolenski, R. T., Sledzinski, Z. and Swierczynski, J., 2011. Decreased serum betaine concentrations in patients after bariatric surgery. *Obesity Surgery*. 21: 1634–1639.

Slow, S., Lever, M., Chambers, S. T. and George, P. M., 2009. Plasma dependent and independent accumulation of betaine in male and female rat tissues. *Physiological Research*. 58: 403–410.

Song, Z., Deaciuc, I., Zhou, Z., Song, M., Chen, T., Hill, D. and McClain, C. J., 2007. Involvement of AMP-activated protein kinase in beneficial effects of betaine on high-sucrose diet-induced hepatic steatosis. *American Journal of Physiology: Gastrointestinal and Liver Physiology*. 293: G894–G902.

Strain, J. J., McSorley, E. M., van Wijngaarden, E., Kobrosly, R. W., Bonham, M. P., Mulhern, M. S., McAfee, A. J., Davidson, P. W., Shamlaye, C. F., Henderson, J., Watson, G. E., Thurston, S. W., Wallace, J. M., Ueland, P. M. and Myers, G. J., 2013. Choline status and neurodevelopmental outcomes at 5 years of age in the Seychelles Child Development Nutrition Study. *British Journal of Nutrition*. 110: 330–336.

Teng, Y. W., Ellis, J. M., Coleman, R. A. and Zeisel, S. H., 2012. Mouse betaine-homocysteine S-methyltransferase deficiency reduces body fat via increasing energy expenditure and impairing lipid synthesis and enhancing glucose oxidation in white adipose tissue. *Journal of Biological Chemistry*. 287: 16187–16198.

Ueland, P. M., 2011. Choline and betaine in health and disease. *Journal of Inherited Metabolic Disease*. 34: 3–15.

Wallace, J. M., McCormack, J. M., McNulty, H., Walsh, P. M., Robson, P. J., Bonham, M. P., Duffy, M. E., Ward, M., Molloy, A. M., Scott, J. M., Ueland, P. M. and Strain, J. J., 2012. Choline supplementation and measures of choline and betaine status: a randomised, controlled trial in postmenopausal women. *British Journal of Nutrition*. 108: 1264–1271.

Wang, L., Chen, L., Tan, Y., Wei, J., Chang, Y., Jin, T. and Zhu, H., 2013. Betaine supplement alleviates hepatic triglyceride accumulation of apolipoprotein E deficient mice via reducing methylation of peroxisomal proliferator-activated receptor alpha promoter. *Lipids in Health and Disease*. 12: 34.

Wang, Z., Yao, T., Pini, M., Zhou, Z., Fantuzzi, G. and Song, Z., 2010. Betaine improved adipose tissue function in mice fed a high-fat diet: a mechanism for hepatoprotective effect of betaine in nonalcoholic fatty liver disease. *American Journal of Physiology: Gastrointestinal and Liver Physiology*. 298: G634–G642.

Yan, J., Jiang, X., West, A. A., Perry, C. A., Malysheva, O. V., Devapatla, S., Pressman, E., Vermeylen, F., Stabler, S. P., Allen, R. H. and Caudill, M. A., 2012a. Maternal choline intake modulates maternal and fetal biomarkers of choline metabolism in humans. *American Journal of Clinical Nutrition*. 95: 1060–1071.

Yan, J., Winter, L. B., Burns-Whitmore, B., Vermeylen, F. and Caudill, M. A., 2012b. Plasma choline metabolites associate with metabolic stress among young overweight men in a genotype-specific manner. *Nutrition and Diabetes*. 2: e49.

Zeisel, S. H., 2013. Metabolic crosstalk between choline/1-carbon metabolism and energy homeostasis. *Clinical Chemistry and Laboratory Medicine*. 51: 467–475.

Zhang, C. X., Pan, M. X., Li, B., Wang, L., Mo, X. F., Chen, Y. M., Lin, F. Y. and Ho, S. C., 2013. Choline and betaine intake is inversely associated with breast cancer risk: a two-stage case-control study in China. *Cancer Science*. 104: 250–258.

CHAPTER 16

# Folate and Betaine Interactions during Pregnancy

MICHELLE M. MURPHY*[a]

[a]Area of Preventive Medicine and Public Health, Department of Basic Medical Sciences, Faculty of Medicine and Health Sciences, IISPV, CIBERobn, Universitat Rovira i Virgili, C/ Sant Llorenç, 21, 43201 Reus, Spain
*E-mail: michelle.murphy@urv.cat

## 16.1 One-Carbon Metabolism

### 16.1.1 Folate and Betaine and One-Carbon Metabolism

Folate and betaine each participate in the one-carbon (1-C) metabolic network. This complex system provides 1-C units for DNA synthesis or repair and methyl groups that participate in regulatory epigenetic processes such as DNA methylation. It clearly has a direct effect on early development and anomalies in this system have been extensively associated with developmental defects or adverse reproductive or pregnancy outcomes in animal and human studies. Its key components are methionine, homocysteine, folate, cobalamin, choline and betaine as well as other B vitamins such as riboflavin and pyridoxine that act as enzyme cofactors. Its regulation and function through an intricate gene – enzyme backbone, depends on a balanced supply of each of these micronutrients and their interaction through positive or negative feedback from intra- or interdependent pathways. Methyl groups

Food and Nutritional Components in Focus No. 7
Betaine: Chemistry, Analysis, Function and Effects
Edited by Victor R Preedy

Published by the Royal Society of Chemistry, www.rsc.org

from folate or betaine are used to remethylate homocysteine to methionine in separate reactions that converge at the same point of the methionine cycle. Folate supply determines the flux through methylenetetrahydrofolate reductase (MTHFR) and newly formed 1-C units that are directed towards nucleotide synthesis or towards the *de novo* synthesis of transferable methyl groups. Methionine synthase (MS), of which cobalamin is a cofactor, is responsible for the transfer of methyl groups from the folate cycle to homocysteine forming tetrahydrofolate and methionine as a result. Methionine, betaine and choline are dietary methyl donors and direct sources of transferable methyl groups. Endogenous betaine is produced from choline oxidation, in a pathway that culminates in dimethylglycine (DMG) and methionine production following the remethylation of homocysteine by betaine, catalyzed by betaine:homocysteine methyltransferase (BHMT). Homocysteine remethylation to methionine is central to determining the fate of methyl groups supplied either by folate or betaine. Subsequently *S*-adenosylmethionine (SAM) is produced and it provides methyl groups that are used by DNA methyltransferases to methylate CpG dinucleotides, thus affecting gene transcription and chromosomal stability.

Given their influence on this process and ultimately on SAM production, folate and betaine status affect methyl group availability, gene expression, and fetal programming. The importance of folate status in preventing neural tube defects (NTDs) during very early development has led to the mandatory fortification of flour with folic acid in the USA since 1998 and subsequently in other countries as a preventive measure. This has led to widespread improvement in folate status in affected populations including women of child-bearing age. The effects have been beneficial in reducing the incidence of NTDs and folate as a single nutrient has been the focus of much research. However, less is known regarding the role of betaine in early development and pregnancy outcome.

#### *16.1.1.1 Crosstalk Between MTHFR/MS and BHMT Pathways*

The dominance of either remethylation pathway in methionine synthesis can shift in response to external stresses such as nutrient deficiencies or imbalance, or the physiological changes and demands of pregnancy on the system. Given the complexity of the effects of different stages of pregnancy on this system, it is relevant to recap first, the evidence for crosstalk between the pathways in the nonpregnant state.

MTHFR and MS are ubiquitous and catalyze the reactions in which folate and cobalamin participate, respectively. The BHMT pathway is important in methionine generation and maintenance in the liver (Finkelstein and Martin, 1984) and is also present in the kidney and eye. Globally, remethylation of homocysteine to methionine by this pathway is less than by MS. However, it is enhanced in situations of excess methionine, as observed following a methionine load test (Holm *et al.*, 2004).

There is also evidence to suggest upregulation of the BHMT pathway when remethylation *via* MS is insufficient due to reduced folate, cobalamin or B vitamin status. Normal serum betaine but elevated DMG, indicative of homocysteine remethylation by the BHMT pathway, were reported in folate-deficient or in some cobalamin-deficient patients (Allen *et al.*, 1993). Data collected in the sixth examination of the Framingham study showed evidence for an interaction between folate status and the role of choline and betaine in homocysteine remethylation (Lee *et al.*, 2010). An inverse correlation between choline or betaine and homocysteine was confined to participants with low folate or cobalamin status and was no longer observed after mandatory fortification of flour with folic acid had been introduced in the USA. Various European studies have reported an inverse association between plasma betaine and homocysteine in adults (Holm *et al.*, 2004, 2005; Melse-Boonstra *et al.*, 2005). A placebo-controlled randomized control trial tested the association between betaine and homocysteine after a methionine load (PML) before and after an intervention with different B vitamin supplement regimes or placebo. At baseline, PML betaine and DMG increased indicating enhanced flux through the BHMT pathway. An inverse association between betaine and PML homocysteine was also observed. After 3 months of intervention, these observations were not repeated following B vitamin treatment, suggesting that the MS pathway was upregulated and the BHMT pathway downregulated when B vitamin status was improved (Holm *et al.*, 2004). In a separate study, betaine was shown to be a strong predictor of PML homocysteine, but the association between betaine and fasting plasma total homocysteine (tHcy) was weaker (Holm *et al.*, 2005). The association between betaine and tHcy or PML homocysteine was weaker with increasing quartiles of plasma folate status and multivariate linear regression analysis of predictors of tHcy revealed a significant interaction between plasma folate and betaine. Another placebo-controlled randomized control trial investigated the effect of a 12 week intervention with different folic acid supplement regimes, ranging from 50 to 800 μg per day, on the association between betaine and tHcy and on plasma betaine concentrations (Melse-Boonstra *et al.*, 2005). An inverse correlation between betaine and tHcy both before and after the intervention was observed. In contrast to the previously described studies, the intervention regime in this one was limited to folic acid rather than a B vitamin mix and PML homocysteine was not investigated. However, plasma betaine increased in a dose-dependent manner with the different folic acid supplement regimes. Another randomized placebo-controlled trial showed that plasma betaine increased following use of a combined supplement of 400 μg of folic acid and 1000 μg of cobalamin (Eussen *et al.*, 2007). tHcy was reduced and DMG was unchanged following the supplement regime, suggesting that betaine is spared as a methyl source for remethylating homocysteine in situations of replete folate and/or cobalamin status in adults.

## 16.2 Early Development

### 16.2.1 Betaine, BHMT, Embryo and Fetal Development

*BHMT* gene expression and enzyme activity in the embryo or foetus appears to vary between species and according to time of development. However, available evidence suggests that betaine is essential during embryogenesis and early development. Betaine addition to culture medium protects preimplantation mouse embryos from the deleterious effect of NaCl (Biggers *et al.*, 1993) or from the effects of raised osmolarity (Dawson *et al.*, 1997). It has been suggested that it may act as an osmolyte in the developing embryo (Anas *et al.*, 2007). Endogenous betaine synthesis (Anas *et al.*, 2008), high betaine concentrations and *BHMT* gene expression and enzyme activity in the mouse blastocyst suggest its importance in methylation reactions at this early stage of development (Lee *et al.*, 2012). Inhibition of BHMT activity during embryogenesis results in developmental abnormalities. Prenatal supplementation of rat dams with folic acid led to reduced liver *BHMT* gene expression in the progeny, suggesting an effect of folate status on the BHMT pathway during development (Chmurzynska and Malinowska, 2011) and that its role is secondary to that of MS when folate/folic acid supply is plentiful. BHMT knockout mice were susceptible to developing fatty liver and hepatic tumors (Teng *et al.*, 2011). Human studies have mainly focussed on the effect of maternal compartment betaine status on homocysteine metabolism or pregnancy outcome. However, it has been reported that human fetal liver *BHMT* gene expression and enzyme activity is considerably lower than in adult liver but increases with gestational age (Feng *et al.*, 2011). Further work is required to elucidate the role of BHMT in the human embryo and foetus but based on the evidence from studies to date, it might be expected to vary depending on time of pregnancy and with maternal folate status or folic acid supplementation regime.

The National Birth Defects Prevention Study in the USA reported that betaine intake above the median (≥75.3 mg per day), in a study of mothers of 954 anencephaly or spina bifida cases and 6268 controls, reduced by about half the risk of a pregnancy affected by anencephaly in Hispanic women (Chandler *et al.*, 2012). There have been some studies on the association between the 742 G > A polymorphism of the *BHMT* gene and different types of NTDs in the USA. In a study of father/mother/child triads it was reported to be associated with myelomeningocele or anencephaly and the association was stronger when mothers had taken folic acid supplements preconceptionally (Boyles *et al.*, 2006). The polymorphism in the children was associated with an 80% increase in risk of spina bifida (Shaw *et al.*, 2009). It was associated with increased risk of omphalocele in a large population-based case control study of newborns in New York between 1998 and 2005 (Mills *et al.*, 2012). The effect was specific to African-American and Asian ethnic groups. The heterozygote variant doubled the risk, whereas the homozygote mutant form increased the risk almost fivefold in African Americans. The effect size was larger in Asians but only significant for heterozygotes, probably due to the small size of this ethnic group in the study. The same SNP was associated with increased NTD risk when prenatal folic acid supplementation was absent in a Chinese study (Liu *et al.*, 2014). Other reports, have found a protective tendency of the 742 G > A *BHMT* polymorphism against orofacial clefts in a Polish study of the effect of the polymorphism in the child (Mostowska, 2010), spina bifida in a Canadian study of the effect of the polymorphism in mothers or children (Morin *et al.*, 2003) or Down syndrome in a Brazilian study of the effect of the polymorphism in mothers (Amorim *et al.*, 2013). However, these studies vary in design and small sample size is an important limitation of the latter two, so their results should be confirmed in adequately powered studies.

## 16.3 Pregnancy and Fetal Development

### 16.3.1 One-Carbon Metabolism, Pregnancy and Fetal Development

The relevance of the 1-C metabolic system in achieving an optimal pregnancy outcome is evident given the association of inadequate status or imbalance of the nutrients involved with increased risk of pregnancy complications.

Deficiencies or imbalance in any of the nutrients can enhance or inhibit the different cycles or pathways involved in 1-C metabolism. As a result, shifts in the destiny of 1-C units may occur and thus affect cellular proliferation and DNA stability or epigenetic processes such as DNA methylation.

Deficiencies in different components of the 1C metabolic network, polymorphisms that affect it, impaired folate transport or elevated tHcy that results from the global imbalance in the network have all been associated with reproductive or pregnancy complications such as infertility (Berrocal-Zaragoza *et al.*, 2009), miscarriage (Dodds *et al.*, 2008), pre-eclampsia (Dodds *et al.*, 2008), preterm birth or other adverse pregnancy outcomes such as congenital malformations (neural tube (Hibbard and Smithells, 1965) or heart defects (Shaw *et al.*, 2010)), intrauterine growth retardation (Onalan *et al.*, 2006) or reduced birth weight (Murphy *et al.*, 2004).

tHcy is elevated when its remethylation to methionine is reduced due to folate or B12 deficiency or polymorphisms that affect the activity of enzymes involved in methyl availability such as methylenetetrahydrofolate reductase. tHcy has proven an interesting biomarker of global 1-C metabolism status. Elevated tHcy has been linked with pathological conditions that stem from vascular abnormalities such as stroke or dementia and pregnancy complications such as pre-eclampsia or intrauterine growth retardation/suboptimal fetal growth due to insufficient placental perfusion. Given that betaine can regulate tHcy by providing the methyl group required to convert homocysteine to methionine, the potential of betaine to prevent elevated tHcy and maintain methionine synthesis during pregnancy is of interest.

#### 16.3.1.1 Folate and Betaine Interactions during Pregnancy

Few pregnancy studies to date have focused on whether folate and betaine interact. Given the proven risk of folate deficiency for grave NTDs and the high prevalence of deficiency/low status in this vitamin, many studies during the last 20–25 years have focused on folate status/intervention with folic acid alone. Folate status has improved considerably in many countries as a result of mandatory fortification of flour with folic acid in countries such as the USA or due to increased exposure to voluntarily fortified foods with folic acid. However, differences in laws regulating the availability of fortified foods on the market or differences in consumption of industrially prepared foods leads to substantial variation in exposure to added folic acid between countries. Prenatal care protocols now widely recommend folic acid supplement use during the first trimester of pregnancy to provide sufficient vitamin during this critical period of early development.

Choline, folate and betaine status, and the associations of folate or betaine with tHcy, have been tracked from early pregnancy in longitudinal studies in Curaçao (Velzing-Aarts *et al.*, 2005) and in the Reus-Tarragona Birth Cohort in Spain (clinicaltrials.gov, NCT:01778205, Fernandez-Roig *et al.*, 2013). Other studies have also investigated some or all of these components at different stages of pregnancy and are summarized in Table 16.1.

The Curaçao study followed 50 women from a mean of 9 gestational weeks (GW) until approximately 14 weeks postpartum. 86% of the participants reported taking supplements containing 800–1000 µg of folic acid, albeit on an irregular basis, during pregnancy. The Reus-Tarragona Birth Cohort included 522 women that were recruited following ultrasound scan confirmation of a viable foetus of less than 12 GW. Fasting blood samples were collected at <12, 15, 24–27 and 34 GW. Further blood samples were collected on admission to hospital with confirmed labor and from the cord after delivery. Plasma folate, choline, betaine, DMG and tHcy, among other determinants of 1-C metabolism, were measured throughout pregnancy and in the cord. Detailed information regarding folic acid and/or multivitamin supplement use, in the 6 months prior to and during pregnancy, was collected from the participants at 20 and 32 GW. Participants were recommended to take 400 µg per day of folic acid throughout the first trimester of pregnancy. 95% of them took some form of folic acid supplement during this time. 28% took supplements containing >400 µg of folic acid and 52% continued to take supplements ≥400 µg for the duration of pregnancy. Each of these studies reported a progressive increase in plasma choline during pregnancy, a decrease in betaine during the first half of pregnancy and an increase in DMG during the second half of pregnancy.

Physiological effects of pregnancy such as haemodilution, albumin decline or increased glomerular filtration rate are known to be reflected in reduced plasma concentrations of nutrients and metabolites. Plasma choline and DMG increased despite these effects, illustrating the importance of both choline synthesis and choline oxidation during pregnancy. The Reus-Tarragona Birth Cohort study set out to investigate whether choline oxidation and

**Table 16.1** Plasma choline, betaine and dimethylglycine concentrations during pregnancy and in the cord, ranked by midpregnancy maternal folate status.

| | | ***Gestational weeks*** | | | | | |
|---|---|---|---|---|---|---|---|
| | | <12 | 15–16 | 24–27 | 34–36 | Labor | Cord |
| **Folate, nmol $L^{-1}$** | | | | | | | |
| Fernandez-Roig *et al.* (2013)[a] | Low[b] | 14.4 | 17.7 | 7.48 | 6.40 | 6.31 | 18.8 |
| Wallace *et al.* (2008)[a] | | | | 14.1 | | 9.0 | 40.2 |
| Velzing-Aarts *et al.* (2005)[c] | | 16.4 | 19.6 | 17.3 | 20.5 | | |
| Fernandez-Roig *et al.* (2013)[a] | High[d] | 41.4 | 34.6 | 21.0 | 17.1 | 18.0 | 30.4 |
| Molloy *et al.* (2005)[c] | | — | — | — | — | 24.8 | 44.8 |
| Wu *et al.* (2013)[a] | | — | 35.9 | — | 35.8 | — | — |
| **Choline, µmol $L^{-1}$** | | | | | | | |
| Fernandez-Roig *et al.* (2013)[a] | Low[b] | 7.5 | 7.9 | 9.4 | 10.2 | 11.4 | 27.4 |
| Velzing-Aarts *et al.* (2005)[c] | | 6.6 | 7.3 | 8.8 | 10.7 | | |
| Fernandez-Roig *et al.* (2013)[a] | High[d] | 7.8 | 7.5 | 9.3 | 10.4 | 11.7 | 27.8 |
| Molloy *et al.* (2005)[c] | | — | — | — | — | 12.3 | 36.6 |
| Wu *et al.* (2013)[a] | | — | 6.9 | — | 9.4 | — | — |
| **Betaine, µmol $L^{-1}$** | | | | | | | |
| Fernandez-Roig *et al.* (2013)[a] | Low[b] | 20.5 | 14.6 | 12.3 | 12.7 | 12.8 | 24.2 |
| Wallace *et al.* (2008)[a] | | | 14.2 | 10.6 | | 11.2 | 21.0 |
| Velzing-Aarts *et al.* (2005)[c] | | 16.3 | 11.5 | 10.3 | 10.9 | | |
| Fernandez-Roig *et al.* (2013)[a] | High[d] | 21.7 | 15.0 | 13.5 | 13.7 | 13.8 | 25.3 |
| Molloy *et al.* (2005)[c] | | — | — | — | — | 9.5 | 21.1 |
| Wu *et al.* (2013)[a] | | — | 13.0 | — | 13.1 | — | — |
| **DMG, µmol $L^{-1}$** | | | | | | | |
| Fernandez-Roig *et al.* (2013)[a] | Low[b] | 2.48 | 2.14 | 2.16 | 2.50 | 2.86 | 3.67 |
| Velzing-Aarts *et al.* (2005)[c] | | 1.96 | 1.54 | 1.61 | 1.76 | | |
| Fernandez-Roig *et al.* (2013)[a] | High[d] | 2.37 | 2.08 | 2.00 | 2.09 | 2.55 | 3.33 |
| Molloy *et al.* (2005)[c] | | — | — | — | — | 1.8 | 2.5 |
| Wu *et al.* (2013)[a] | | — | 1.0 | — | 1.3 | — | — |
| **DMG/betaine** | | | | | | | |
| Fernandez-Roig *et al.* (2013)[a] | Low[b] | 0.13 | 0.17 | 0.21 | 0.23 | 0.28 | 0.16 |
| Velzing-Aarts *et al.* (2005)[c] | | 0.12 | 0.13 | 0.16 | 0.16 | | |
| Fernandez-Roig *et al.* (2013)[a] | High[d] | 0.13 | 0.15 | 0.16 | 0.18 | 0.21 | 0.13 |
| **tHcy, µmol $L^{-1}$** | | | | | | | |
| Fernandez-Roig *et al.* (2013)[a] | Low[b] | 5.5 | 4.6 | 5.0 | 5.7 | 6.7 | 5.3 |
| Wallace *et al.* (2008)[a] | | | 5.8 | 6.8 | | 12.4 | 10.2 |
| Velzing-Aarts *et al.* (2005)[c] | | 9.4 | 7.3 | 7.3 | 7.6 | | |
| Fernandez-Roig *et al.* (2013)[a] | High[d] | 5.0 | 4.4 | 4.2 | 4.8 | 5.6 | 4.5 |
| Molloy *et al.* (2005)[c] | | — | — | — | — | 7.9 | 7.4 |
| Wu *et al.* (2013)[a] | | — | 4.1 | — | 4.8 | — | — |

[a]Medians.
[b]plasma folate below study median (27.6 nmol $L^{-1}$) at <12 GW.
[c]geometric means.
[d]plasma folate above study median (11.4 nmol $L^{-1}$) at 24–2 GW.

the BHMT pathway were affected by folate status and this is illustrated in Figure 16.1.

Multiple linear regression analysis was used to determine the contribution of the MS and BHMT pathways to homocysteine remethylation at different stages of pregnancy. During the first trimester of pregnancy when folic acid

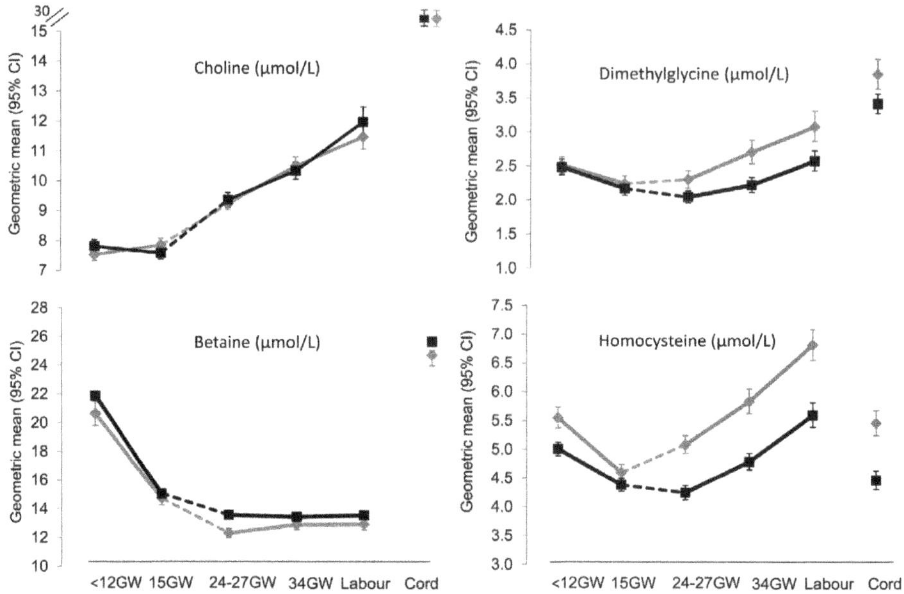


**Figure 16.1** Pregnancy changes in plasma choline, betaine, dimethylglycine and homocysteine according to folate status. Figure adapted from Fernandez-Roig *et al.*, (*American Journal of Clinical Nutrition*. 2013. 97: 1252–1259) shows fluctuations in plasma choline, betaine, dimethylglycine and homocysteine according to pregnancy plasma folate status (above median; below median). GW: gestational weeks.

use was high, betaine was not significantly associated with tHcy. After cessation of first-trimester folic acid use, differences in components of the choline oxidation pathway were observed according to folate status. Betaine was lower in women with plasma folate below the median (11.4 nmol $L^{-1}$ at 24–27 GW) compared to those with plasma folate above it. DMG was higher and so was the DMG : betaine ratio. This is indicative of greater BHMT pathway activity in the women with lower folate status. Further confirmation of this was seen in the fact that the contribution of betaine to homocysteine remethylation was evident by midpregnancy (after cessation of folic acid supplement use). In the women with plasma folate above the median, the relationship between betaine and homocysteine was not evident until late pregnancy, when folate status was at its lowest. These results are illustrated in Figure 16.2.

Despite the enhancement of the BHMT pathway and the increased contribution of betaine to homocysteine remethylation, tHcy remained persistently higher in the low folate status group throughout pregnancy (Figure 16.1). This suggests that increased betaine utilization did not compensate for the deficit in folate. Arguably, if the women had continued to use folic acid regularly throughout pregnancy remethylation by the MS pathway would have been sufficient and by BHMT would not have been necessary. Mandatory fortification of flour with folic acid is absent in Spain and when first-trimester folic acid

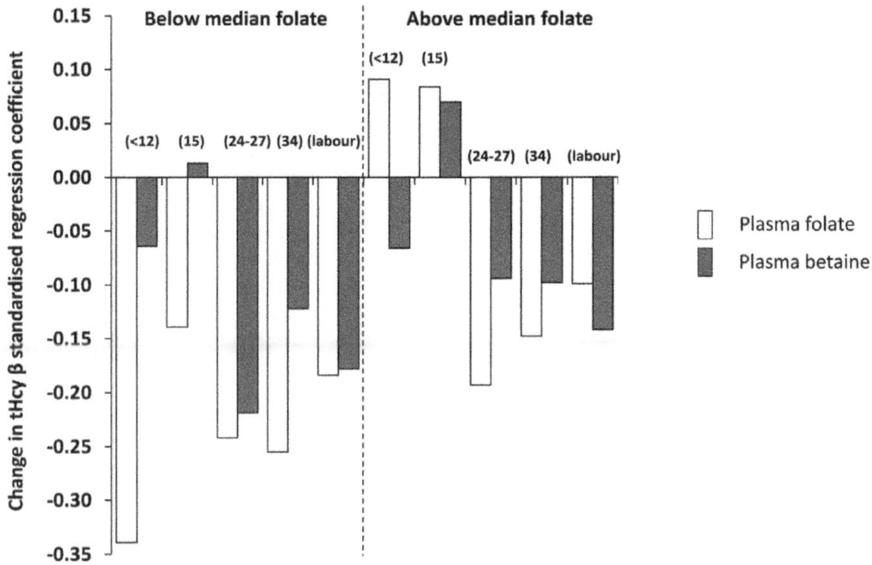


**Figure 16.2** Changes in multiple linear regression β standardized regression coefficients in models of the associations between folate or betaine and homocysteine at different stages of pregnancy and according to plasma folate status. Figure adapted from Fernandez-Roig *et al.*, (*American Journal of Clinical Nutrition*. 2013. 97: 1252–1259) shows how the associations between folate or betaine and homocysteine during pregnancy vary with plasma folate status and time of pregnancy. Numbers in brackets refer to gestational weeks. The models were adjusted for confounding factors such as maternal age, smoking, $B_{12}$ status, serum creatinine, *MTHFR* 677C > T genotype.

supplement use stopped there was a significant drop in folate status for the remainder of pregnancy. Two studies that reported higher folate status than in this study also reported lower DMG concentrations to those observed at most of the comparable time points of pregnancy that were studied (Velzing-Aarts *et al.*, 2005, Wu *et al.*, 2013). These results are summarized in Table 16.2.

Another study carried out in The Seychelles reported that betaine was a significant predictor of tHcy during pregnancy, only in situations of low methionine status (Wallace *et al.*, 2008).

**16.3.1.1.1 Choline and Betaine Requirements during Pregnancy.** Methyl-rich choline is essential for fetal development given its roles in cell signaling, membrane structure, lipid transport and in neurotransmission. The demands of pregnancy on 1-C metabolism are high. Methionine and choline are especially important for fetal growth and development. Under normal circumstances, cord folate, cobalamin, choline, betaine and DMG concentrations are higher than in maternal blood (Molloy *et al.*, 2005, Velzing Aarts *et al.*, 2005, Fernandez-Roig *et al.*, 2013). This suggests that fetal

**Table 16.2** Pregnancy folate status and plasma dimethylglycine (DMG).

| Study | Location | N | Gestational weeks | Plasma folate (nmoL $L^{-1}$) | Plasma DMG (μmol $L^{-1}$) |
|---|---|---|---|---|---|
| Velzing-Aarts *et al.* (2005) | Curaçao | 50 | 16 | 19.61[a] | 1.54[a] |
| | | | 24 | 17.31 | 1.61 |
| | | | 36 | 20.46 | 1.76 |
| Fernandez-Roig *et al.* (2013) | Spain | 514 | 15 | 17.0[a] | 2.3[a] |
| | | | | 34.8 | 2.3 |
| | | | 24–27 | 7.2 | 2.4 |
| | | | | 22.4 | 2.0 |
| | | | 34 | 6.4 | 2.7 |
| | | | | 18.0 | 2.2 |
| Wu *et al.* (2013) | Canada | 270 | 16 | 35.5[b] | 1.12[b] |
| | | | 36 | 35.2 | 1.41 |

[a]Geometric means.
[b]Arithmetic means.

requirements are high for these nutrients or reflect transport across the placenta. The DMG : betaine ratio is lower in the cord than in maternal plasma at labor (Fernandez-Roig *et al.*, 2013), which is consistent with the previous observation that BHMT activity is lower in fetal than in adult liver (Feng *et al.*, 2011).

It is believed that the pregnancy associated increase in maternal plasma choline is due to the enhancement of the endogenous synthesis of choline to meet elevated maternal and fetal requirements (Molloy *et al.*, 2005). Oestrogen was shown to enhance phosphatidylethanolamine *N*-methyltransferase (*PEMT*) gene expression in human and in rat hepatocytes (Resseguie *et al.*, 2007). Oestrogen production also increases with advancing pregnancy and this may stimulate endogenous choline synthesis.

A study that compared the effect of giving women in their third trimester of pregnancy and nonpregnant women 480 or 930 mg per day of choline for 12 weeks, found that pregnancy has a profound effect on choline metabolism (Yan *et al.*, 2012). Plasma choline was 30% higher in pregnant than in nonpregnant women, regardless of the assigned choline feeding regime. Although plasma betaine and DMG were lower in pregnant women, concentrations of both derivatives of the choline oxidation pathway were higher in those that received the higher choline dose. This finding suggested that greater choline intake during pregnancy was destined to enhanced use of choline as a methyl donor, *via* its conversion to betaine in the choline oxidation pathway. Interestingly, this study was carried out in the USA in 2009, over a decade after the introduction of the mandatory folic acid fortification policy, so folate status as determined by plasma and red cell folate concentrations was good. Both folate status markers were higher in the pregnant than in the nonpregnant women and the authors attributed this to the use of supplements containing folic acid by the pregnant women. However, despite this, it was evident that choline-derived betaine synthesis was enhanced when choline intake was higher. This could be interpreted as responding to the need to use betaine for homocysteine remethylation to methionine in late pregnancy, even when folate status is good. This was not observed in the women on the lower dietary choline regime and in this situation, dietary choline may have been preferentially used to maintain phosphatidylcholine synthesis. This suggests that the choline oxidation pathway is necessary for methionine synthesis during pregnancy but that choline is spared if dietary intake is not sufficient to meet pregnancy demands. If maternal choline requirements are not adequately met by dietary choline and endogenous synthesis alone (Yan *et al.*, 2012), the irreversible oxidation of choline to betaine to maintain methionine synthesis from homocysteine would strain choline supply further. Table 16.1 also shows that there is relatively little difference in choline at corresponding time points of pregnancy between different studies. This is despite important differences in folate status. In fact, a tracer study of the fate of labeled choline during human pregnancy confirmed that dietary choline was preferentially destined to the PEMT pathway for phosphatidylcholine synthesis than to the choline oxidation pathway (Yan *et al.*, 2013). A recent Canadian study reported that low cobalamin status is associated with lower choline and betaine status in late pregnancy (Wu *et al.*, 2013). However, there was no difference in DMG among the different cobalamin status groups. This, and the fact that there was no difference in free choline : betaine, betaine : DMG or methionine : homocysteine ratios between the different cobalamin status groups, led the authors to suggest that the association between low cobalamin and low choline was due to decreased transmethylation rather than increased choline oxidation. Data from the Reus-Tarragona Birth Cohort (Fernandez Roig *et al.*, 2013), however, supports the hypothesis that choline oxidation is enhanced when the MS pathway is downregulated when folate supply is low. Across multiple time points of pregnancy, plasma choline and betaine concentrations were lower, while DMG and the betaine : DMG ratio were higher in women with plasma folate below the median compared to above it. Furthermore, the contribution of betaine to homocysteine remethylation was also greater in the women with lower folate status. These data provide strong evidence for upregulation of the BHMT pathway in response to downregulation of the MS pathway during pregnancy. An important difference between the two studies is that folate status was considerably higher in the Canadian study. High folate status combined with low cobalamin status is known to produce adverse metabolic effects associated with the metabolic trap of folate in its 5-methyltetrahydrofolate form and with circulating unmetabolized folic acid (Troen *et al.*, 2006; Miller *et al.*, 2009). One of the consequences of the methylfolate trap is reduced methionine synthesis that leads to reduced SAM production. These events would lead to reduced transmethylation of methionine as suggested by the authors of the Canadian study. Alternatively, the choline oxidation pathway could be affected by the cellular accumulation of 5-methyltetrahydrofolate. The complete choline oxidation pathway leads to glycine and 5,10-methylenetetrahydrofolate formation. 5-methyltetrahydrofolate inhibits glycine *N*-methyltransferase activity, thus reducing sarcosine formation, unless dietary methionine is inadequate.

**Table 16.3** Summary of animal and human studies that have investigated how folate or other 1-C micronutrients interact with betaine. The table summarizes the study type, source and main findings.

| | Study, year | Effect |
|---|---|---|
| **Folic acid fortification** *(Population)* | Lee *et al.*, 2010 (Framingham, USA) | Choline or betaine inversely correlated with tHcy in participants with low folate or cobalamin status. This was no longer observed after mandatory fortification with folic acid was introduced. |
| **B Vitamin supplements** *(Population)* | Holm *et al.*, 2005 (Europe) | Following treatment with B vitamins, PML choline oxidation no longer observed. |
| **Folic acid supplements** *(Pregnancy-rats)* | Chmurzynska and Malinowska, 2011 | BHMT gene expression and activity not observed in offspring born to rat dams treated with folic acid during pregnancy |
| **Folic acid supplement use** *(Population)* | Melse-Boonstra, 2005 (Europe) | Plasma betaine increased in dose-dependent manner with folic acid supplement use |
| **Folic acid and $B_{12}$ supplements** *(population)* | Eussen *et al.*, 2007 (Europe) | Plasma betaine increased in dose-dependent manner with folic acid supplement use and BHMT downregulated |
| **Folate status** *(Population)* | Allen *et al.*, 1993 (USA) | BHMT pathway upregulated in folate or B12-deficient patients |
| **Folate status** *(pregnancy)* | Fernàndez-Roig, 2013 (Europe) | Pregnancy BHMT pathway activity upregulated with low folate status |
| **$B_{12}$ status** *(pregnancy)* | Wu *et al.*, 2013 (Canada) | Low cobalamin status associated with low choline and betaine status but BHMT pathway unaffected. |
| **Methionine status** *(pregnancy)* | Wallace *et al.*, 2005 (Seychelles) | Pregnancy BHMT pathway activity upregulated with low methionine status |

# Summary Points

- Folate status is associated with betaine status and interacts with the betaine-homocysteine methyltransferase (BHMT) pathway of homocysteine remethylation. The evidence from animal and human studies is summarized in Table 16.3.
- The role of the choline oxidation pathway or betaine in pregnancy varies with time of pregnancy and with methionine synthase (MS) pathway activity.
- Betaine-homocysteine methyltransferase (BHMT) activity is upregulated in situations of a reduction in or low folate status or low methionine.
- Choline and betaine are spared during pregnancy in situations of replete folate or methionine status.

## Key Facts

1. Betaine status affects early development and pregnancy outcome
   - Betaine is essential in the preimplantation embryo in which it may play a role as an osmolyte and for correct neural tube formation.
   - Polymorphisms of the betaine-homocysteine methyltransferase (*BHMT*) gene are associated with increased risk of different neural tube or other congenital defects.
   - Betaine-homocysteine methyltransferase (BHMT) expression and activity are lower in the fetal than in the adult liver.
   - Betaine-homocysteine methyltransferase (BHMT) expression and activity in the fetal liver increase as pregnancy progresses.
2. Betaine derived from choline oxidation contributes to methionine synthesis during pregnancy
   - Pregnancy choline requirements are high and choline is preferentially used by the phosphatidylethanolamine *N*-methyltransferase (PEMT) pathway to generate phosphatidylcholine over the choline oxidation pathway during pregnancy.
   - Betaine, produced by the choline oxidation pathway is a methyl source used to remethylate homocysteine to methionine during pregnancy.
3. Folate status affects betaine utilization for methionine synthesis during pregnancy
   - Betaine-homocysteine methyltransferase (*BHMT*) expression is not detected in the offspring of rat dams supplemented with folic acid during pregnancy.
   - The use of betaine derived from choline oxidation as a methyl source for methionine synthesis is enhanced when prenatal folic acid supplementation is ceased.
   - The betaine-homocysteine methyltransferase (BHMT) pathway is upregulated in situations of low folate status during pregnancy and this is more widespread in late pregnancy.
   - Despite upregulation of the betaine-homocysteine methyltransferase (BHMT) pathway, elevated tHcy persists in the presence of low compared to high folate status during pregnancy.

## Definitions of Words and Terms

**Betaine**: Methyl-rich nutrient that can be obtained from dietary sources or synthesized endogenously by choline oxidation.
**BHMT pathway**: Betaine-homocysteine methyltransferase (BHMT) pathway in which methyl groups provided by betaine are used to remethylate homocysteine to methionine in a reaction catalyzed by BHMT.
**Choline oxidation pathway**: One of choline's metabolic pathways in which it is converted to betaine and dimethylglycine and eventually to glycine and sarcosine.

**Fasting plasma total homocysteine (tHcy):** Total homocysteine concentration in fasting plasma samples. It is elevated in the presence of suboptimal status in any of the micronutrients that participate in one-carbon metabolism.

**Folate cycle:** Metabolic cycle of the different folate forms that leads to the recycling of tetrahydrofolate and to the release of a methyl group to methionine synthase (MS) when 5-methyltetrahydrofolate is converted to tetrahydrofolate.

**Homocysteine:** Sulfur-containing amino acid intermediate of methionine cycle that is responsible for binding methyl groups from the folate cycle or from betaine metabolism to form methionine.

**MS pathway:** Methionine synthase (MS) pathway in which methyl groups provided by the folate cycle are used to remethylate homocysteine to methionine in a reaction catalyzed by the cobalamin-dependent MS.

**One carbon metabolism:** Metabolic network fuelled by different B vitamins, choline, betaine and methionine that is responsible for the production and transfer of 1-C units and methyl groups used for DNA synthesis and repair and for methylation reactions.

**Post methionine load test:** Test of changes in plasma homocysteine concentration in the hours following the administration of an oral methionine load.

**Remethylation:** Binding of methyl group to homocysteine to convert it into methionine at a point in the methionine cycle where the MS and BHMT pathways converge.

# Abbreviations

| | |
|---|---|
| 1-C | One carbon |
| BHMT | Betaine-homocysteine methyltransferase |
| DMG | Dimethylglycine |
| GW | Gestational weeks |
| MS | Methionine synthase |
| MTHFR | Methylenetetrahydrofolate reductase |
| NTD | Neural tube defect |
| PEMT | Phosphatidylethanolamine *N*-methyltransferase |
| PML | Postmethionine load |
| SAM | *S*-Adenosylmethionine |
| tHcy | Fasting plasma total homocysteine |

# References

Allen, R. H., Stabler, S. P. and Lindenbaum, J., 1993. Serum betaine, *N,N*-dimethylglycine and *N*-methylglycine levels in patients with cobalamin and folate deficiency and related inborn errors of metabolism. *Metabolism.* 42: 1448–1460.

Amorim, M. R., Moura, C. M., Gomes, A. D., Barboza, H. N., Lopes, R. B. and Ribeiro, M. G., Costa Lima M. A., 2013. Betaine-homocysteine methyltransferase 742G > A polymorphism and risk of down syndrome offspring in a Brazilian population. *Molecular Biology Reports*. 40: 4685–4689.

Anas, M. K., Hammer, M. A., Lever, M., Stanton, J. A. and Baltz, J. M., 2007. The organic osmolytes betaine and proline are transported by a shared system in early preimplantation mouse embryos. *Journal of Cellular Physiology*. 210: 266–277.

Anas, M. K., Lee, M. B., Zhou, C., Hammer, M. A., Slow, S., Karmouch, J., Liu, X. J., Bröer, S., Lever, M. and Baltz, J. M., 2008. SIT1 is a betaine/proline transporter that is activated in mouse eggs after fertilization and functions until the 2-cell stage. *Development*. 135: 4123–4130.

Berrocal-Zaragoza, M. I., Fernandez-Ballart, J. D., Murphy, M. M., Cavallé-Busquets, P., Sequeira, J. M. and Quadros, E. V., 2009. Association between blocking folate receptor autoantibodies and subfertility. *Fertility and Sterility*. 91: 1518–1521.

Biggers, J. D., Lawitts, J. A. and Lechene, C. P., 1993. The protective action of betaine on the deleterious effects of NaCl on preimplantation mouse embryos in vitro. *Molecular Reproduction and Development*. 34: 380–390.

Boyles, A. L., Billups, A. V., Deak, K. L., Siegel, D. G., Mehltretter, L., Slifer, S. H., Bassuk, A. G., Kessler, J. A., Reed, M. C., Nijhout, H. F., George, T. M., Enterline, D. S., Gilbert, J. R. and Speer, M. C., NTD Collaborative Group, 2006. Neural tube defects and folate pathway genes: family-based association tests of gene-gene and gene-environment interactions. *Environmental Health Perspectives*. 114: 1547–1552.

Chandler, A. L., Hobbs, C. A., Mosley, B. S., Berry, R. J., Canfield, M. A., Qi, Y. P., Siega-Riz, A. M. and Shaw, G. M., National Birth Defects Prevention Study, 2012. Neural tube defects and maternal intake of micronutrients related to one-carbon metabolism or antioxidant activity. *Birth Defects Research, Part A: Clinical and Molecular Teratology*. 94: 864–874.

Chmurzynska, A. and Malinowska, A. M., 2011. Homocysteine homeostasis in the rat is maintained by compensatory changes in cystathionine β-synthase, betaine-homocysteine methyltransferase, and phosphatidylethanolamine *N*-methyltransferase gene transcription occurring in response to maternal protein and folic acid intake during pregnancy and fat intake after weaning. *Nutrition Research*. 31: 572–578.

Dawson, K. M. and Baltz, J. M., 1997. Organic osmolytes and embryos: substrates of the Gly and beta transport systems protect mouse zygotes against the effects of raised osmolarity. *Biology of Reproduction*. 56: 1550–1558.

Dodds, L., Fell, D. B., Dooley, K. C., Armson, B. A., Allen, A. C., Nassar, B. A., Perkins, S. and Joseph, K. S., 2008. Effect of homocysteine concentration in early pregnancy on gestational hypertensive disorders and other pregnancy outcomes. *Clinical Chemistry*. 54: 326–334.

Eussen, S. J., Ueland, P. M., Clarke, R., Blom, H. J., Hoefnagels, W. H., vanStaveren, W. A. and de Groot, L. C., 2007. The association of betaine, homocysteine and related metabolites with cognitive function in Dutch elderly people. *British Journal of Nutrition*. 98: 960–968.

Feng, Q., Kalari, K., Fridley, B. L., Jenkins, G., Ji, Y., Abo, R., Hebbring, S., Zhang, J., Nye, M. D., Leeder, J. S. and Weinshilboum, R. M., 2011. Betaine-homocysteine methyltransferase: human liver genotype-phenotype correlation. *Molecular Genetics and Metabolism*. 102: 126–133.

Fernandez-Roig, S., Cavallé-Busquets, P., Fernandez-Ballart, J. D., Ballesteros, M., Berrocal-Zaragoza, M. I., Salat-Batlle, J., Ueland, P. M. and Murphy, M. M., 2013. Low folate status enhances pregnancy changes in plasma betaine and dimethylglycine concentrations and the association between betaine and homocysteine. *American Journal of Clinical Nutrition*. 97: 1252–1259.

Finkelstein, J. D. and Martin, J. J., 1984. Methionine metabolism in mammals. Distribution of homocysteine between competing pathways. *Journal of Biological Chemistry*. 259: 9508–9513.

Hibbard, E. and Smithells, R. W., 1965. Folic acid metabolism and embryopathy. *Lancet*. 1: 1254–1256.

Holm, P. I., Bleie, Ø, Ueland, P. M., Lien, E. A., Refsum, H., Nordrehaug, J. E. and Nygård, O., 2004. Betaine as a determinant of postmethionine load total plasma homocysteine before and after B-vitamin supplementation. *Arteriosclerosis, Thrombosis, and Vascular Biology*. 24: 301–307.

Holm, P. I., Ueland, P. M., Vollset, S. E., Midttun, Ø, Blom, H. J., Keijzer, M. B. and den Heijer, M., 2005. Betaine and folate status as cooperative determinants of plasma homocysteine in humans. *Arteriosclerosis, Thrombosis, and Vascular Biology*. 25: 379–385.

Lee, J. E., Jacques, P. F., Dougherty, L., Selhub, J., Giovannucci, E., Zeisel, S. H. and Cho, E., 2010. Are dietary choline and betaine intakes determinants of total homocysteine concentration? *American Journal of Clinical Nutrition*. 91: 1303–1310.

Lee, M. B., Kooistra, M., Zhang, B., Slow, S., Fortier, A. L., Garrow, T. A., Lever, M., Trasler, J. M. and Baltz, J. M., 2012. Betaine homocysteine methyltransferase is active in the mouse blastocyst and promotes inner cell mass development. *Journal of Biological Chemistry*. 287: 33094–33103.

Liu, J., Qi, J., Yu, X., Zhu, J., Zhang, L., Ning, Q. and Luo, X. J., 2014. Investigations of single nucleotide polymorphisms in folate pathway genes in Chinese families with neural tube defects. *Neurological Sciences*. 337: 61–66.

Melse-Boonstra, A., Holm, P. I., Ueland, P. M., Olthof, M., Clarke, R. and Verhoef, P., 2005. Betaine concentration as a determinant of fasting total homocysteine concentrations and the effect of folic acid supplementation on betaine concentrations. *American Journal of Clinical Nutrition*. 81: 1378–1382.

Miller, J. W., Garrod, M. G., Allen, L. H., Haan, M. N. and Green, R., 2009. Metabolic evidence of vitamin B-12 deficiency, including high homocysteine and methylmalonic acid and low holotranscobalamin, is more pronounced in older adults with elevated plasma folate. *American Journal of Clinical Nutrition*. 90: 1586–1592.

Mills, J. L., Carter, T. C., Kay, D. M., Browne, M. L., Brody, L. C., Liu, A., Romitti, P. A., Caggana, M. and Druschel, C. M., 2012. Folate and vitamin B12-related genes and risk for omphalocele. *Human Genetics*. 131: 739–746.

Molloy, A. M., Mills, J. L., Cox, C., Daly, S. F., Conley, M., Brody, L. C., Kirke, P. N., Scott, J. M. and Ueland, P. M., 2005. Choline and homocysteine interrelations in umbilical cord and maternal plasma at delivery. *American Journal of Clinical Nutrition*. 82: 836–842.

Morin, I., Devlin, A. M., Leclerc, D., Sabbaghian, N., Halsted, C. H., Finnell, R. and Rozen, R., 2003. Evaluation of genetic variants in the reduced folate carrier and in glutamate carboxypeptidase II for spina bifida risk. *Molecular Genetics and Metabolism*. 79: 197–200.

Mostowska, A., Hozyasz, K. K., Wojcicki, P., Dziegelewska, M. and Jagodzinski, P. P., 2010. Associations of folate and choline metabolism gene polymorphisms with orofacial clefts. *Journal of Medical Genetics*. 47: 809–815.

Murphy, M. M., Scott, J. M., Arija, V., Molloy, A. M. and Fernandez-Ballart, J. D., 2004. Maternal homocysteine before conception and throughout pregnancy predicts fetal homocysteine and birth weight. *Clinical Chemistry*. 50: 1406–1412.

Onalan, R., Onalan, G., Gunenc, Z. and Karabulut, E., 2006. Combining 2nd-trimester maternal serum homocysteine levels and uterine artery Doppler for prediction of preeclampsia and isolated intrauterine growth restriction. *Gynecologic and Obstetric Investigation*. 61: 142–148.

Resseguie, M., Song, J., Niculescu, M. D., da Costa, K. A., Randall, T. A. and Zeisel, S. H., 2007. Phosphatidylethanolamine *N*-methyltransferase (PEMT) gene expression is induced by estrogen in human and mouse primary hepatocytes. *FASEB Journal*. 21: 2622–2632.

Shaw, G. M., Lu, W., Zhu, H., Yang, W., Briggs, F. B., Carmichael, S. L., Barcellos, L. F., Lammer, E. J. and Finnell, R. H., 2009. 118 SNPs of folate-related genes and risks of spina bifida and conotruncal heart defects. *BMC Medical Genetics*. 10: 49. DOI: 10.1186/1471-2350-10-49.

Shaw, G. M., Carmichael, S. L., Yang, W. and Lammer, E. J., 2010. Periconceptional nutrient intakes and risks of conotruncal heart defects. *Birth Defects Research, Part A: Clinical and Molecular Teratology*. 88: 144–151.

Teng, Y. W., Mehedint, M. G., Garrow, T. A. and Zeisel, S. H., 2011. Deletion of betaine-homocysteine *S*-methyltransferase in mice perturbs choline and 1-carbon metabolism, resulting in fatty liver and hepatocellular carcinomas. *Journal of Biological Chemistry*. 286: 36258–36267.

Troen, A. M., Mitchell, B., Sorensen, B., Wener, M. H., Johnston, A., Wood, B., Selhub, J., McTiernan, A., Yasui, Y., Oral, E., Potter, J. D. and Ulrich, C. M., 2006. Unmetabolized folic acid in plasma is associated with reduced natural killer cell cytotoxicity among postmenopausal women. *Journal of Nutrition*. 136: 189–194.

Velzing-Aarts, F. V., Holm, P. I., Fokkema, M. R., van der Dijs, F. P., Ueland, P. M. and Muskiet, F. A., 2005. Plasma choline and betaine and their relation to plasma homocysteine in normal pregnancy. *American Journal of Clinical Nutrition*. 81: 1383–1389.

Wallace, J. M., Bonham, M. P., Strain, J., Duffy, E. M., Robson, P. J., Ward, M., McNulty, H., Davidson, P. W., Myers, G. J., Shamlaye, C. F., *et al.*, 2008. Homocysteine concentration, related B vitamins, and betaine in pregnant women recruited to the Seychelles Child Development Study. *American Journal of Clinical Nutrition*. 87: 391–397.

Wu, B. T., Innis, S. M., Mulder, K. A., Dyer, R. A. and King, D. J., 2013. Low plasma vitamin B-12 is associated with a lower pregnancy-associated rise in plasma free choline in Canadian pregnant women and lower postnatal growth rates in their male infants. *American Journal of Clinical Nutrition*. 98: 1209–1217.

Yan, J., Jiang, X., West, A. A., Perry, C. A., Malysheva, O. V., Devapatla, S., Pressman, E., Vermeylen, F., Stabler, S. P., Allen, R. H., *et al.*, 2012. Maternal choline intake modulates maternal and fetal biomarkers of choline metabolism in humans. *American Journal of Clinical Nutrition*. 95: 1060–1071.

Yan, J., Jiang, X., West, A. A., Perry, C. A., Malysheva, O. V., Brenna, J. T., Stabler, S. P., Allen, R. H., Gregory 3rd, J. F. and Caudill, M. A., 2013. Pregnancy alters choline dynamics: results of a randomized trial using stable isotope methodology in pregnant and nonpregnant women. *American Journal of Clinical Nutrition*. 98: 1459–1467.

CHAPTER 17

# Multiple Roles of Betaine in Protecting Against Alcohol-Induced Liver Injury

PAUL G. THOMES[†a,b], SARAH M. BLIGH[†b], AND KUSUM K. KHARBANDA*[a,b,c]

[a]Research Service, Veterans Affairs Nebraska-Western Iowa Health Care System, Omaha, Nebraska, 68105, USA; [b]Department of Internal Medicine and Biochemistry & Molecular Biology, University of Nebraska Medical Center, Omaha, Nebraska, 68198, USA; [c]Department of Biochemistry & Molecular Biology, University of Nebraska Medical Center, Omaha, Nebraska, 68198, USA

*E-mail: kkharbanda@unmc.edu

## 17.1 Introduction

Ethanol consumption leads to many adverse functional and structural changes in humans and accounts for 2.5 million deaths globally each year (World Health Organization, 2011). Liver is one of the tissues most susceptible to the toxic effects, as it is the main organ responsible for metabolizing ethanol (Zakhari, 2006; Cederbaum, 2012). Ethanol consumption produces a wide spectrum of hepatic lesions, the most characteristic being steatosis, hepatitis and fibrosis/cirrhosis (Ishak *et al.*, 1991). Fatty liver, an early response to heavy

---

[†]Both authors contributed equally to the manuscript.

---

Food and Nutritional Components in Focus No. 7
Betaine: Chemistry, Analysis, Function and Effects
Edited by Victor R Preedy

Published by the Royal Society of Chemistry, www.rsc.org

alcohol consumption, develops in more than 90% of problem drinkers, with early, mild steatosis, initially developing in zone 3 (perivenular) hepatocytes; progressing to zone 2 and even zone 1 (periportal) hepatocytes, as the severity of liver injury increases. Interestingly, only about 30% of heavy drinkers develop more severe forms of alcoholic liver disease (ALD), such as advanced fibrosis and cirrhosis. This suggests that other environmental (smoking, obesity, viral infection or drug abuse) or genetic (gender or polymorphisms in key genes) factors contribute to the overall risk in heavy drinkers for developing more severe liver damage (Beier and McClain, 2010).

Approximately 51.5% of US citizens use alcohol regularly; nearly 18% of these individuals meet the criteria for alcohol abuse (Schiller *et al.*, 2012). These statistics highlight the need for both prevention of alcohol-induced liver disease by a reduction in consumption, as well as the need for adequate treatments that ameliorate or reverse existing liver damage. Betaine is one such potential treatment, given its role in promoting methylation reactions, in preventing homocysteine accumulation and in its ability to act as an antioxidant. There is a growing body of evidence supporting the beneficial effects of betaine in these roles and in reducing alcohol-induced liver damage, which we will review here.

## 17.2 Mechanisms of Alcohol-Induced Liver Injury

Research conducted over the past several decades has demonstrated that chronic alcohol exposure causes liver damage by several mechanisms, as recently reviewed (Altamirano and Bataller, 2011; Gao and Bataller, 2011; Orman *et al.*, 2013). It is well established that the pathogenesis of alcoholic liver disease involves not only multiple factors but also involves crosstalk among multiple organs/tissues, the most notable being the adipose–liver (Song *et al.*, 2008; Tang *et al.*, 2012) and the gut–liver (Zhong *et al.*, 2010; Hartmann *et al.*, 2012; Bull-Otterson *et al.*, 2013) axes.

Above all, the effects of ethanol administration on the liver methionine metabolic pathway have mainly been studied based on the observations made as early as 1948 that dietary methionine or choline deficiency results in early pathology of steatosis similar to that seen following ethanol consumption (Handler and Dubin, 1946; Sidransky and Farber, 1958). Therefore, many investigations have focused on ethanol-induced perturbations in methionine metabolism in this organ in order to better understand the mechanism of associated pathologies and to devise appropriate therapeutic interventions.

## 17.3 Methionine Metabolism

The methionine cycle, which intersects folate, methyl group transfers and homocysteine metabolism, is found in all tissues. In this metabolic cycle (Figure 17.1), the conversion of homocysteine to methionine is an essential reaction that maintains sufficient levels of methionine to form

*S*-adenosylmethionine (SAM) and to detoxify homocysteine and *S*-adenosylhomocysteine (SAH). The biochemical step between homocysteine and methionine utilizes folate, and through the action of methionine synthase (MS) a methyl group is transferred from $N^5$-methyltetrahydrofolate (MTHF) *via* vitamin $B_{12}$ to homocysteine to produce methionine. There is an additional pathway for homocysteine remethylation that is folate independent and tissue specific, existing primarily in the liver (Delgado-Reyes *et al.*, 2001). This pathway is as important as the folate-dependent pathway and utilizes betaine as a source of methyl groups in a reaction catalyzed by betaine homocysteine methyltransferase (BHMT) to form methionine (Finkelstein and Martin, 1984). The other constituent enzymes of the methionine cycle are methionine adenosyltransferases (MAT), numerous SAM-dependent methyltransferases and *S*-adenosylhomocysteine hydrolase (SAHH). Some specifics

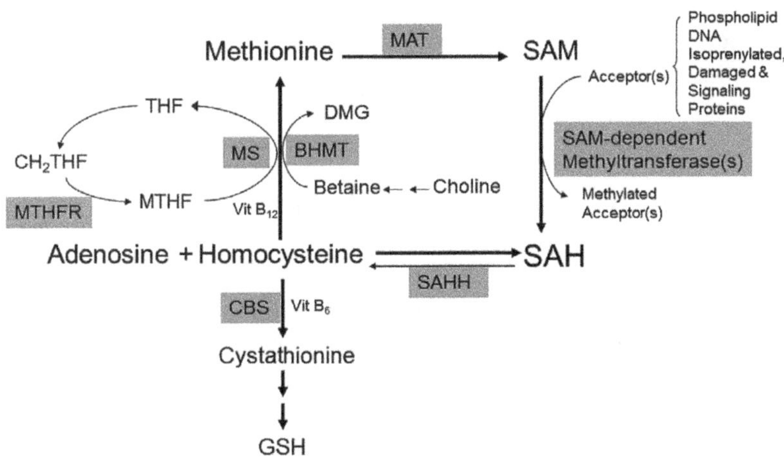


**Figure 17.1** Reactions of methionine metabolism in the liver. There are four main participants of this pathway, which are methionine, *S*-adenosylmethionine (SAM), *S*-adenosylhomocysteine (SAH) and homocysteine. Methionine adenosyltransferase (MAT) converts methionine to SAM, which then serves as a methyl-group donor substrate for methyltransferase-catalyzed reactions. The other product of these reactions is SAH, which is hydrolyzed to adenosine and homocysteine by *S*-adenosylhomocysteine hydrolase (SAHH). The thermodynamics of this reaction favors the formation of SAH if the products are not removed. The methionine cycle is completed when homocysteine is remethylated back to methionine by $B_{12}$-dependent methionine synthase (MS) using 5-methyltetrahydrofolate (MTHF) as a substrate. MTHF is derived from dietary folate and from endogenous 5,10, methylenetetrahydrofolate ($CH_2$-THF) by way of its reductase, 5,10 methylenetetrahydrofolate reductase (MTHFR). Homocysteine can also be remethylated to methionine by an alternate pathway in the liver *via* betaine homocysteine methyltransferase (BHMT)-catalyzed reactions with the formation of dimethylglycine (DMG). Homocysteine can also be catabolized through the transulfuration pathway initiated by $B_6$-dependent cystathionine β-synthase (CBS) to generate glutathione (GSH).

concerning these enzymes are as follows: MAT utilizes adenosine-5'-triphosphate (ATP) to convert methionine to SAM, the principal methylating agent. Under normal conditions, most of the 6–8 g of SAM generated per day is utilized in SAM-dependent transmethylation reactions (Finkelstein, 1990). At least 50 such reactions catalyzed by diverse methyltransferases have been identified (Schubert *et al.*, 2003) that results in the production of methylated biomolecules that serve vital biological functions. The other product of the methylation reaction is SAH, which is hydrolyzed to adenosine and homocysteine by SAHH. Although the equilibrium dynamic of SAHH reactions strongly favors SAH synthesis over homocysteine synthesis, the efficient removal of homocysteine and adenosine by multiple pathways (indicated in Figure 17.1) allows homocysteine synthesis to proceed (Finkelstein *et al.*, 1974).

Homocysteine is also irreversibly catabolized through the transsulfuration pathway and utilized for the synthesis of cysteine and its derivative, glutathione (GSH) that plays a critical role in protecting cells from metabolic oxidants.

Given the critical roles of transmethylation reactions and GSH levels in determining various cellular processes and antioxidant status, respectively, changes in methionine metabolism have far-reaching effects.

## 17.4 Effect of Ethanol Consumption on Methionine Metabolism

Using different animal models and human studies over the years, alcohol researchers have established that ethanol consumption affects multiple steps in the hepatic methionine metabolic cycle (Figure 17.2) as recently reviewed (Kharbanda, 2007, 2009, 2013). Briefly, ethanol exposure predominantly inhibits the activities of two vital cellular enzymes, MS and MAT, involved in remethylating homocysteine and generating SAM, respectively. In some species, to compensate for these losses, ethanol feeding increases the activity of BHMT that utilizes betaine to remethylate homocysteine and maintain normal levels of SAM. However, during extended periods of ethanol feeding, the alternate homocysteine remethylation pathway cannot be maintained because of a depletion of endogenous hepatic betaine stores. This results in decreasing the hepatic SAM levels, while increasing the levels of two toxic metabolites, homocysteine and SAH.

## 17.5 Functional Consequences of Ethanol-Induced Alteration in Methionine Metabolism

As a consequence of the aforementioned alterations in methionine metabolic pathways by chronic ethanol consumption, two major changes occur in the liver:

a. reduction in hepatocellular SAM : SAH ratio; and
b. increase in homocysteine levels.

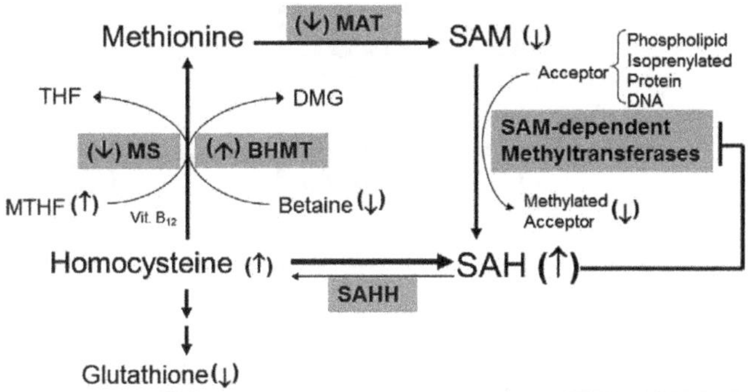


**Figure 17.2** Ethanol-induced alterations in methionine metabolic pathway in the liver. Ethanol consumption predominantly inhibits the activity of two vital enzymes, methionine synthase (MS) and methyl adenosyltransferase (MAT), involved in remethylating homocysteine and generating *S*-adenosylmethionine (SAM), respectively. In some species, to compensate for these losses, ethanol feeding increases the activity of the enzyme, betaine homocysteine methyltransferase that utilizes betaine to remethylate homocysteine and attempt to maintain normal levels of SAM. However, during extended periods of ethanol feeding, the alternate homocysteine remethylation pathway cannot be maintained because of a depletion of endogenous hepatic betaine stores. This results in decreasing the hepatic *S*-adenosylmethionine and glutathione (GSH) level while increasing the levels of two toxic metabolites, homocysteine and *S*-adenosylhomocysteine (SAH).

Both of these alcohol-induced changes have profound functional consequences in the liver.

### 17.5.1 Defects in Crucial Methylation Reactions

The ratio of SAM to SAH is an important metabolic indicator for cellular methylation status and is termed methylation potential. A decrease in the SAM : SAH ratio in the presence of increased SAH is associated with impaired activity of many members of SAM-dependent methyltransferases (Figure 17.3) that ultimately results in compromised liver function and progressive liver injury (Kharbanda, 2009).

### 17.5.2 Homocysteine Toxicity

Torres *et al.* suggested a role of homocysteine in liver fibrosis based on their observations that cultured hepatic stellate cells exposed to homocysteine produce increased collagen and tissue inhibitor of metalloproteinases (Torres *et al.*, 1999). More than a decade later, many studies unequivocally demonstrated that ethanol-induced hyperhomocysteinemia activates the endoplasmic reticulum (ER) stress response that contributes to the pathogenesis of alcoholic liver injury (Ji and Kaplowitz, 2003; Esfandiari *et al.*, 2005).

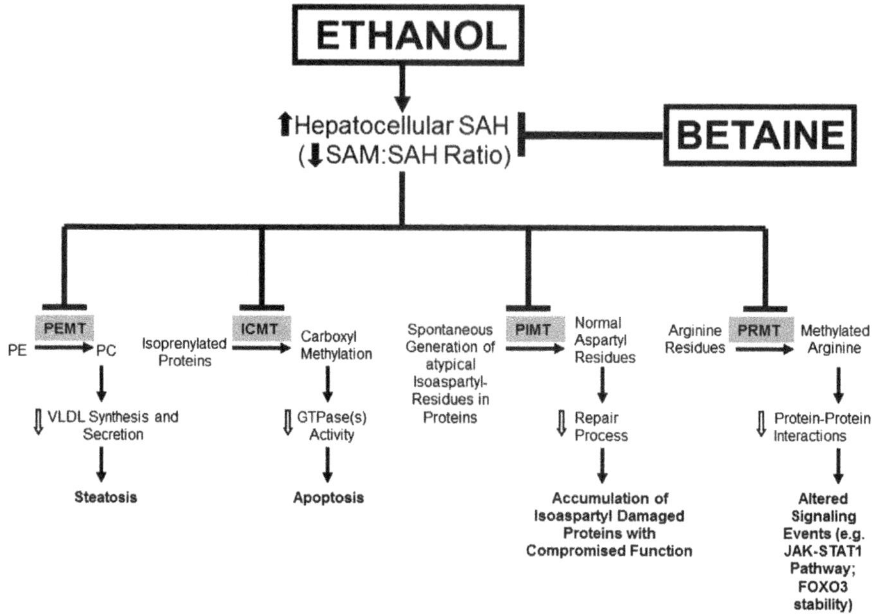


**Figure 17.3** Detrimental consequences of reduced hepatocellular *S*-adenosylmethione: *S*-adenosylhomocysteine (SAM : SAH) ratio and protection by betaine administration. Increased hepatocellular *S*-adenosylhomocysteine (SAH) levels generated by chronic ethanol exposure can negatively affect the activities of the four investigated methyltransferases, phosphatidylethanolamine methyltransferase (PEMT), isoprenylcysteine carboxyl methyltransferase (ICMT), protein-isoaspartate methyltransferases (PIMT) and protein arginine methyltransferase (PRMT). These methylation defects, in turn, cause decreased synthesis and secretion of very low density lipoproteins (VLDL), impaired activation of GTPases, diminished protein repair process and decreased protein–protein interactions. These deficiencies ultimately contribute to the induction of steatosis, increased apoptosis, accumulation of isoaspartyl damaged dysfunctional proteins and altered signaling events Betaine administration can prevent the increase in SAH and thereby prevent these ethanol-induced pathologies. The consequences of the alterations in DNA and histone methylation that can affect gene expression and epigenetic memory as well as the effects of elevated SAH on many other crucial methylation reactions are not depicted in this figure.

## 17.6 Protective Role of Betaine in Alcoholic Liver Injury

A decrease in hepatic SAM concurrent with increases in SAH and homocysteine levels is commonly observed in many experimental models of alcoholic liver injury. Thus, it is important to remove SAH and homocysteine, to regenerate SAM and maintain a normal hepatic SAM : SAH ratio to prevent alcohol-induced liver damage. Betaine cotreatment ameliorates a number of hallmark features of alcoholic liver injury in many experimental animal models (Ji and Kaplowitz, 2003; Ji *et al.*, 2004; Kharbanda *et al.*, 2007a; Kharbanda, 2007, 2009). The protective effects of betaine are primarily due to its ability to provide a methyl group to homocysteine to form methionine *via* the BHMT-catalyzed reaction (Barak *et al.*, 2001). This helps to regulate levels of homocysteine (Barak *et al.*, 2001), SAH (Barak *et al.*, 2003; Kharbanda *et al.*, 2007a), maintain an adequate supply of methionine for the synthesis of SAM (Barak *et al.*, 1993; Ji *et al.*, 2004; Kharbanda *et al.*, 2007a) as well as maintain hepatocellular SAM : SAH ratio (Kharbanda *et al.*, 2007a). Indeed, studies have revealed that betaine administration doubles the hepatic levels of SAM in control animals and increases its level four-fold in ethanol-fed rats (Kharbanda *et al.*, 2007a). Similar increases in hepatic SAM levels are also reported in mice fed intragastrically a betaine-supplemented ethanol diet (Ji *et al.*, 2004). It has also been demonstrated that the therapeutic effect of betaine is aided by an induction in hepatic BHMT at the gene and protein level by betaine treatment (Kharbanda *et al.*, 2007a). Here, we discuss in detail the multiple roles of betaine in protecting against alcohol-induced liver injury.

### 17.6.1 Betaine Attenuates Alcohol-Induced Steatosis

Chronic alcohol exposure exerts adverse effects at multiple levels in the process of fat synthesis and disposal by the liver (Figure 17.4). Accumulating evidence supports the notion that alcohol-induced adipose tissue dysfunction also play an important role in the pathogenesis of hepatic steatosis (Song *et al.*, 2008; Tang *et al.*, 2012). In particular, changes in methionine metabolic pathway, decrease in adiponectin release and increases in triglyceride mobilization causing elevated plasma nonesterified fatty acids have been documented in adipose tissue following chronic ethanol consumption (Kang *et al.*, 2007; Song *et al.*, 2008; Tang *et al.*, 2012). Betaine also exerts its protective effects at multiple levels by attenuating ethanol-induced hepatic fat synthesis and promoting fat disposal as well as by improving adipose tissue function (Song *et al.*, 2008).

#### *17.6.1.1 Betaine Corrects Alteration in Methylation Reaction Catalyzed by Phosphatidylethanolamine Methyltransferase (PEMT) Pathway and Normalizes Defective Very Low-Density Lipoprotein (VLDL) Secretion*

Excessive neutral fat accumulation in the hepatocytes is the most common and earliest response to alcohol consumption. The liver exports triglycerides and cholesterol only as constituents of VLDL particles, any impairment in the processes of either the synthesis or export of VLDL particles leads to fat accumulation within the hepatocyte. Indeed, altered hepatic VLDL secretion is implicated to play an important role in the development of alcoholic steatosis (Venkatesan

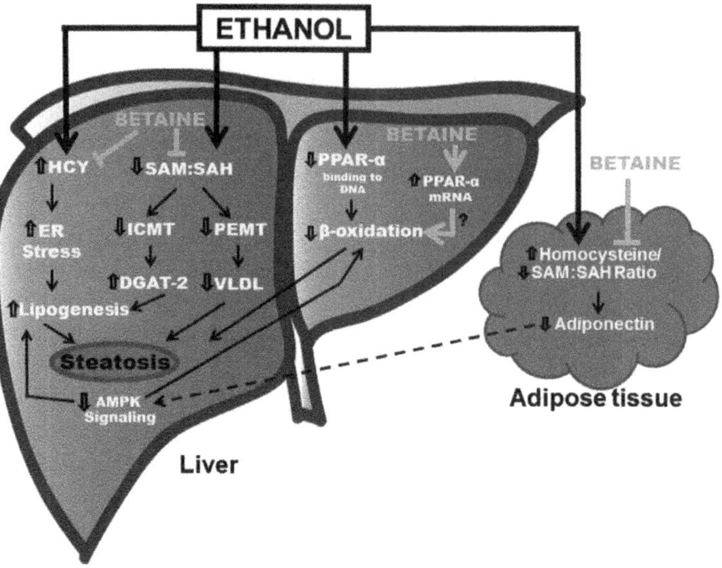


**Figure 17.4** Proposed mechanism(s) by which chronic ethanol consumption increases hepatic steatosis and the protective effects of betaine administration. Alcohol consumption adversely affects the methionine metabolic cycle in the liver and adipose tissue. This leads to increases in homocysteine (HCY) levels and decreases in SAM : SAH ratios in both tissues, ultimately causing accumulation of fat in the liver. High levels of HCY in the liver results in endoplamic reticulum (ER) stress that activates lipogenesis. In the adipose tissue, a high HCY level inhibits adiponectin production that, in turn, downregulates AMP-activated protein kinase (AMPK) activity in the liver. The latter event, concurrently triggers lipogenesis and decreases β-oxidation, causing hepatic steatosis. Concomitantly, decreased hepatocellular SAM : SAH ratio disrupts the activity of methyltransferases, ICMT and PEMT. ICMT suppression leads to activation of DGAT-2 that induces lipogenesis. PEMT suppression on the other hand, decreases VLDL formation, contributing to hepatic fat accumulation. Alcohol exposure also decreases PPAR-α binding to DNA, which greatly affects transcription of genes that encode enzymes necessary for β-oxidation. Betaine treatment reverses alcohol-induced increased HCY levels and decreased SAM : SAH ratio in both liver and adipose tissue. Betaine also increases hepatic PPAR-α mRNA level, which could likely restore hepatic β-oxidation. These effects of betaine in the liver and adipose tissue ultimately attenuate alcoholic steatosis.

*et al.*, 1988). VLDL is largely composed of phosphatidylcholine (PC) derived *via* the PEMT pathway (Nishimaki-Mogami *et al.*, 2002). In this pathway, PEMT catalyzes three successive methyl group transfers from SAM to phosphatidylethanolamine (PE) to generate PC. Many studies have demonstrated the importance of the PC synthesis *via* the PE methylation pathway for normal VLDL synthesis and secretion and have shown that any disruption of PEMT pathway leads to reduced VLDL secretion and accumulation of fat in the liver (Nishimaki-Mogami *et al.*, 2002; Noga *et al.*, 2002; Zhu *et al.*, 2003). However, the relationship between PEMT activity, SAM:SAH ratios and alcoholic steatosis became clear only after seminal findings that the SAM:SAH ratios in livers and hepatocytes of ethanol-fed rats is much lower (~2.5) than the ratios in livers and isolated hepatocytes of control diet fed rats where a ratio close to 5 was observed (Barak *et al.*, 2003; Kharbanda *et al.*, 2007a). A further study conducted using an *in vitro* system revealed that a lower SAM:SAH ratio, as seen in the liver of ethanol-fed rats, impairs PEMT activity (Barak *et al.*, 2003). Conversely, increasing the SAM:SAH ratio, as seen in control or betaine-supplemented hepatocytes, restores normal PEMT enzyme activity (Barak *et al.*, 2003). The physiological relevance of the *in vitro* study was affirmed when the metabolic conversion of PE to PC was examined in intact hepatocytes isolated from rats that were fed an ethanol diet supplemented with or without betaine (Kharbanda *et al.*, 2007a). Indeed, hepatocytes from ethanol-fed rats showed a 30–40% reduced PC synthesis *via* the PEMT pathway and this reduction correlated with lower SAM:SAH ratios (~2.5) seen in livers of these rats (Kharbanda *et al.*, 2007a). Further, *in vitro* betaine treatment of hepatocytes isolated from ethanol-fed rats (Barak *et al.*, 2003) or *in vivo* feeding of betaine-supplemented ethanol diet to rats corrected the ethanol-induced decrease in SAM:SAH ratios (Barak *et al.*, 2003; Kharbanda *et al.*, 2007a), maintained PC synthesis (Kharbanda *et al.*, 2007a), normalized VLDL secretion (Kharbanda *et al.*, 2009) and prevented alcoholic steatosis (Kharbanda *et al.*, 2007a). The protective effect of betaine was due to its ability to maintain hepatocellular SAM:SAH ratio (Kharbanda *et al.*, 2007a) as well as to induce BHMT at both the gene and protein levels (Kharbanda *et al.*, 2007a). The benefits of BHMT induction were further validated in BHMT transgenic mice fed an ethanol diet (Ji *et al.*, 2008). These mice were less susceptible to developing hyperhomocysteinemia, alterations in SAM:SAH ratio or liver steatosis and, in the presence of betaine, were completely resistant to the alcohol-induced hyperhomocysteinemia (Ji *et al.*, 2008). Collectively, these findings underscore the importance of maintaining BHMT activity, either by modulating gene expression or by feeding betaine, to prevent ethanol-induced fat accumulation in the liver (Kharbanda *et al.*, 2005b; Kharbanda *et al.*, 2007a).

#### *17.6.1.2 Betaine Restores Circulating Adiponectin Levels, Hepatic AMP-Activated Protein Kinase (AMPK) Signaling and Fat Disposal*

Adiponectin is an adipokine predominantly secreted by white adipose tissue. Alcohol exposure reduces adiponectin gene and protein expression in the adipose tissue (Song *et al.*, 2008; Tang *et al.*, 2012) The consequent reduction in the circulating adiponectin level observed reportedly associates with alcohol-induced liver injury in experimental models of ALD (Xu *et al.*, 2003; You *et al.*, 2005; Chen *et al.*, 2009). Mechanistic studies reveal that ethanol-induced hyperhomocysteinemia inhibits the extracellular signal regulated kinase (ERK) pathway and such inhibition partly accounts for the lower adiponectin production (Song *et al.*, 2008). Betaine supplementation, by decreasing homocysteine levels in adipose tissue, restores circulating adiponectin levels (Song *et al.*, 2008). This, in turn, alleviates the alcohol-induced reduced activation of AMPK, a central component in the adiponectin signaling pathway in the liver (You *et al.*, 2004), to ultimately decrease hepatic injury (Song *et al.*, 2008).

It has also been reported that ethanol feeding impairs hepatic fatty acid catabolism, in part *via* blocking peroxisome proliferator-activated receptor (PPAR)α-mediated responses that regulate transcription of genes encoding enzymes in mitochondrial and extramitochondrial fatty acid oxidation (Fischer *et al.*, 2003). While ethanol feeding has no effect on PPARα mRNA expression, it decreases the DNA binding ability of PPARα/retinoid X receptor α, resulting in reduced gene expression of several PPAR-regulated genes (Fischer *et al.*, 2003). To date, there are no studies on betaine's ability to regulate fatty acid oxidation pathways in the liver. However, betaine supplementation upregulates mRNA expression of PPARα and its target genes in apolipoprotein E deficient mice (Wang *et al.*, 2013). Thus, it is possible that betaine treatment could reverse alcohol-induced defects in the expression of the PPARα-regulated genes and thereby improve fatty acid oxidation in alcoholic liver injury models.

#### *17.6.1.3 Betaine Attenuates ER Stress*

Mice subjected to 6 weeks of intragastric ethanol feeding show a striking increase in plasma homocysteine concentrations (Ji and Kaplowitz, 2003), which induces ER stress response and triggers lipogenesis through enhanced activation of sterol regulatory element-binding proteins (SREBPs) (You *et al.*, 2002; Ji and Kaplowitz, 2003). Dietary supplementation of betaine to ethanol-fed mice lowers homocysteine levels, prevents ER stress, attenuates SREBPs, decreases the accumulation of hepatic triglycerides and cholesterol, thereby ameliorating the pathologic features of alcohol liver injury (Ji and Kaplowitz, 2003).

#### *17.6.1.4 Betaine Regulates Acyl-Coa : Diacylglycerol Acyltransferase (DGAT)*

DGAT catalyzes the final rate-limiting reaction in triglyceride synthesis in which a fatty acyl-CoA and a diacylglycerol molecule are covalently joined to form triglycerides. In mammals, DGAT occurs in two isoforms, DGAT1 and DGAT2, from distinct gene families (Cases *et al.*, 1998; Cases *et al.*, 2001). Evidence suggests that DGAT1 participates in VLDL synthesis and DGAT2 is involved in hepatic steatosis (Monetti *et al.*, 2007). Chronic ethanol administration increases DGAT2 mRNA and protein expression (Wang *et al.*, 2010). Mechanistic investigations reveal that *N*-acetyl-*S*-farnesyl-L-cysteine, a potent inhibitor of isoprenylcysteine carboxyl methyltransferase (ICMT), suppresses ERK1/2 activation, followed by enhanced DGAT2 expression and elevated triglyceride content (Wang *et al.*, 2010). Since ICMT activity is regulated by altered SAM : SAH ratio (Kharbanda *et al.*, 2005c), the study provided a mechanistic link between abnormal methionine metabolism and DGAT2 upregulation in the liver (Wang *et al.*, 2010). Betaine supplementation improved methionine metabolism, alleviated hepatic ERK1/2 suppression, attenuated hepatic DGAT2 overexpression (Wang *et al.*, 2010) and prevented fat accumulation, indicating that betaine partly regulates hepatic steatosis through modulation of DGAT2 expression.

### 17.6.2 Betaine Attenuates Alcohol-Induced Apoptosis

Ethanol administration is linked to increased hepatocyte apoptosis in both clinical and experimental alcohol-induced liver injury in a variety of animal species including rats, mice and minipigs (Halsted *et al.*, 1996; Higuchi *et al.*, 1996; Nanji, 1998; Benedetti and Marucci, 1999; Casey *et al.*, 2001; Deaciuc *et al.*, 2001; Ji and Kaplowitz, 2003; Ji *et al.*, 2004). Compelling evidence implicates elevated hepatocellular SAH and decreased SAM : SAH ratio to be responsible for alcoholic apoptosis. Mechanistic studies carried out using biochemical approaches (that were achieved by using adenosine that specifically increase SAH levels without a change in homocysteine levels) eventually confirmed that indeed it is the increased intracellular SAH levels that cause increased apoptosis of hepatocytes (Kharbanda *et al.*, 2005c). The adenosine-induced increases in apoptosis and increased SAH levels are significantly attenuated (40–50%) by simultaneous betaine exposure (Kharbanda *et al.*, 2005c). These results were further corroborated by using tubercidin, a potent inhibitor of SAHH, which increases intracellular SAH concentrations (Kharbanda *et al.*, 2005a). Betaine also prevented tubercidin-induced hepatocyte apoptosis. Mechanistic insights into how an elevated SAH or a lower SAM : SAH ratio induces apoptosis (Kharbanda *et al.*, 2005a, 2005c) indicate an involvement of a critical methyltransferase, ICMT whose activity is inhibited by elevated intracellular SAH concentrations (Kharbanda *et al.*, 2005c).

In addition to impairing methyltransferase activity, elevated SAH levels have been shown to sensitize hepatocytes to tumor necrosis factor (TNF)α-induced hepatotoxicity (Song *et al.*, 2007) as well as directly increase caspase-8 expression/activity, resulting in enhanced apoptosis (Attwood *et al.*, 2002; Song *et al.*, 2004). These presumably could also be other sites on which betaine, *via* its ability to prevent SAH accumulation, acts to protect against alcohol-induced hepatotoxicity (Figure 17.5).

Studies using the mouse model of intragastric ethanol feeding show that hyperhomocysteinemia-induced ER stress upregulates proapoptotic genes to result in increased apoptosis (Ji and Kaplowitz, 2003; Ji *et al.*, 2004). Feeding a betaine-supplemented ethanol diet prevents alcoholic apoptosis by attenuating alcohol-induced hyperhomocysteinemia, ER stress and the downstream sequelae of events (Ji and Kaplowitz, 2003; Ji *et al.*, 2004).

### 17.6.3 Betaine Prevents Accumulation of Ethanol-Induced Damaged Proteins

Proteins undergo spontaneous, thermodynamically driven changes over time that can result in decreased functionality (Clarke, 2003). A common form of such damage is the spontaneous and nonenzymatic modification of aspartate

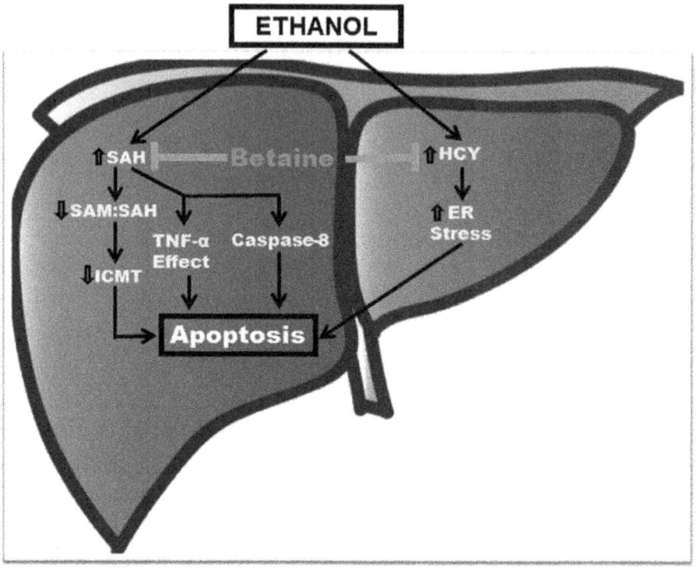


**Figure 17.5** Proposed mechanism by which betaine mitigates alcoholic apoptosis Ethanol consumption causes a concurrent increase in *S*-adenosylhomocysteine (SAH) and homocysteine (HCY) levels in the liver. Such alterations in SAH and HCY levels result in decreased SAM : SAH ratio. Both high SAH levels and low SAM : SAH ratio downregulates ICMT activity, which regulates normal apoptosis pathway. High SAH levels also activate TNF-α and caspase-8, signaling events that contribute to increased hepatocyte apoptosis. Elevated HCY levels cause ER stress, which also leads to apoptosis of the liver cells. Betaine treatment reverses alcohol-induced increases in SAH, HCY levels, and normalizes SAM : SAH ratio to protect the liver cells from alcohol-induced apoptosis.

and asparagine (which are among the most unstable residues in proteins and peptides) to form atypical isoaspartyl residues (Clarke, 2003). Isoaspartate residues have been identified in a variety of cellular structural and functional proteins including tubulin, histone 2B and collagen type-1 (Johnson *et al.*, 1987; Young *et al.*, 2001; Lanthier and Desrosiers, 2004). Such modified proteins exhibit lower biological activity that can be substantially restored *via* methylation by PIMT (Johnson *et al.*, 1987; Lanthier and Desrosiers, 2004). PIMT catalyzes the transfer of the methyl group from SAM to isoaspartyl-modified sites, allowing reisomerization and restoration of the original alpha peptide linkage after repeated rounds of methylation (Shimizu *et al.*, 2005). This physiological repair reaction prevents the accumulation of these modified protein molecules.

Environmental factors, including chronic ethanol exposure causes enhanced accumulation of damaged proteins with atypical isoaspartyl residues (Ingrosso *et al.*, 2000; D'Angelo *et al.*, 2001; Kharbanda *et al.*, 2007b). The accumulation correlates inversely with lower SAM : SAH ratios and consequently lower PIMT activity in livers of ethanol-fed rats (Kharbanda *et al.*, 2007b). Dietary supplementation of betaine to ethanol-fed rats prevents accumulation of damaged proteins by normalizing the SAM : SAH ratio and the PIMT-catalyzed protein repair (Kharbanda *et al.*, 2007b) to ultimately protect the liver from the detrimental effects of damaged protein accumulation.

### 17.6.4 Betaine Prevents Impairments in PRMT-Catalyzed Reactions

Protein arginine methyltransferase (PRMT) catalyzes the addition of monomethyl or dimethyl groups to the guanidine nitrogen atom of arginine and these methylated arginine proteins regulate protein–protein interactions and signaling events (Boisvert *et al.*, 2003).

Arginine methylation, in particular regulates the stability and the nuclear localization of forkhead box O3 (FOXO3) (Yamagata *et al.*, 2008), a transcription factor that modulates hepatic growth and many diverse metabolic pathways (Leyendecker *et al.*, 2011; Haeusler *et al.*, 2012; Tao *et al.*, 2012). Ethanol and Hepatitis C Virus (HCV) suppresses the activity of FOXO3 by inhibiting its PRMT-mediated arginine methylation (Tikhanovich *et al.*, 2013). In contrast, betaine treatment by its ability to promote methylation reactions stabilizes FOXO3, promotes its nuclear translocation and restores mRNA expression of its target genes (Tikhanovich *et al.*, 2013). In addition, betaine treatment could also normalize JAK- signal transducers and activators of transcription (STAT) signaling by restoring STAT1 methylation and thereby improve interferon (IFN)$\alpha$ signaling (Duong *et al.*, 2006).

PRMT also methylates residues on proteasome subunits that appear to regulate the proteasome activity that is vital for the removal of damaged proteins (Osna *et al.*, 2010). Studies indicate that impaired methylation of proteasome subunits suppresses its activity in liver cells that can be potentially restored by betaine treatment (Osna *et al.*, 2010). These findings clearly suggest that betaine treatment likely prevents the manifestation of hallmark features of alcoholic liver injury, in part, through restoring PRMT activity.

### 17.6.5 Betaine Affects Histone Methylation and Mallory–Denk Body Formation

Mallory–Denk bodies are inflammatory protein aggregates commonly detected in liver biopsies of patients with liver disease including those with ALD. Mechanistically, Mallory–Denk body formation results from epigenetic phenomena, in which changes in methionine metabolism alter global gene expression in hepatocytes (Bardag-Gorce *et al.*, 2008; Oliva *et al.*, 2009). In particular, decreases in MAT expression, BHMT levels and histone methylation in conjunction with increases in histone acetylation and ubiquination have been reported in association with the generation of Mallory–Denk bodies (Bardag-Gorce *et al.*, 2008; Oliva *et al.*, 2009). Betaine treatment prevents Mallory–Denk body formation as well as the changes in the methionine metabolic pathway (Oliva *et al.*, 2009).

### 17.6.6 Betaine Prevents Blood-Alcohol Cycling

Despite continuous infusion of ethanol into rodents using the intragastric feeding model, there is a cyclic increase (peak) followed by a decrease (trough) in blood-alcohol level every 6–7 days (Tsukamoto *et al.*, 1985). The blood-alcohol cycle plays an important role in the pathogenesis of alcoholic liver injury (Bardag-Gorce *et al.*, 2002; French, 2005). Increased liver hypoxia, decreased ATP and increased hepatic pathology scores are seen at the peak of blood-alcohol levels (Bardag-Gorce *et al.*, 2002; French, 2005). The cycling phenomenon correlates inversely with alternating increases (troughs) and decreases (peaks) in the ethanol metabolism and elimination (Bardag-Gorce *et al.*, 2002). Feeding a betaine-supplemented ethanol diet lowers the blood-alcohol levels and prevents the cyclic fluctuations of ethanol levels by accelerating its metabolic rate ultimately attenuating alcohol-induced liver injury (Li *et al.*, 2011). Mechanistically, betaine promotes phenylethanolamine methyltransferase-catalyzed conversion of norepinephrine to epinephrine, which is 5- to 10-fold more potent than norepinephrine in increasing the metabolic rate (Li *et al.*, 2011).

### 17.6.7 Betaine Treatment Attenuates Alcohol-Induced Alterations to the Mitochondrial Respiratory Chain Proteome

Ethanol exposure decreases both nuclear and mitochondrial encoded gene products of the oxidative phosphorylation system that decreases respiratory capacity and impairs oxidative phosphorylation; events critical to the development of alcoholic liver injury (Cunningham and Bailey, 2001; Bailey *et al.*, 2005). Dietary supplementation with betaine prevents losses in respiratory chain complexes, particularly subunits of complexes I and IV (Kharbanda *et al.*, 2012). Betaine treatment also prevents the ethanol-induced inducible nitric oxide synthase (NOS2) mRNA and protein expression and nitric oxide generation (Kharbanda *et al.*, 2012) thereby preserving the functioning of the mitochondrial electron-transport chain.

### 17.6.8 Betaine Treatment Attenuates Oxidative Stress and Inflammation

Chronic ethanol treatment increases the levels of reactive oxygen species by lowering cellular antioxidant levels and enhancing oxidative stress in the liver (Wu and Cederbaum, 2009; Gao and Bataller, 2011). The importance of antioxidant enzymes in preventing alcohol-induced liver injury is well demonstrated in glutathione peroxidase, catalase and superoxide dismutase deficient mice that exhibit greater liver damage than wild-type mice after ethanol exposure (Kessova *et al.*, 2003; Kim *et al.*, 2009). Another potent antioxidant is GSH produced by the transsulfuration pathway. Reduced GSH is the main nonprotein thiol in the cell that plays multiple roles in cell physiology and protection against a vast variety of toxic intermediates. Results of experimental models of ethanol-induced liver injury in rodents and baboons as well as in patients chronically dependent on alcohol have shown decreases in hepatic GSH content (Jewell *et al.*, 1986; Lieber *et al.*, 1990; Lu *et al.*, 1999; Kanbak *et al.*, 2001; Jung *et al.*, 2013).

As mentioned earlier, betaine treatment can dramatically increase hepatic SAM (Barak *et al.*, 1993; Kharbanda *et al.*, 2007a), which is an allosteric activator of cystathionine β-synthase (CBS) (Finkelstein *et al.*, 1975), an enzyme that catalyzes the first reaction of the transsulfuration pathway for ultimate GSH synthesis (Figure 17.1). Indeed, betaine-supplemented ethanol diet restores hepatic GSH in a variety of models of alcoholic liver injury (Kanbak *et al.*, 2001; Balkan *et al.*, 2004; Varatharajalu *et al.*, 2010; Jung *et al.*, 2013). Additionally, betaine supplementation increases the total oxyradical scavenging capability in livers of rats chronically fed alcohol (Jung *et al.*, 2013).

Betaine can act directly as an antioxidant by increasing glutathione peroxidase and superoxide dismutase activity, preventing CYP2E1 induction and decreasing the generation of hepatotoxic lipid peroxidation products (Kanbak *et al.*, 2001; Kim *et al.*, 2008; Oliva *et al.*, 2011; Alirezaei *et al.*, 2012).

Additionally, it has been shown that the level of paraoxonase, a multifunctional antioxidant enzyme that prevents low-density lipoproteins oxidation and homocysteine thiolactone accumulation, is decreased after ethanol consumption (Varatharajalu *et al.*, 2010). Betaine feeding restores expression and activity of liver paraoxonase and homocysteine thiolactonase activity (Varatharajalu *et al.*, 2010).

Betaine administration also prevents the ethanol-induced increases in serum endotoxin, alanine aminotransferase (AST), aspartate aminotransferase levels (ALT) and liver inflammatory factors, including TNFα, IFNγ, interleukin (IL)-18, cyclo-oxygenase-2 (COX-2), NOS2, toll-like receptor 4 (TLR4), cluster of differentiation 14 (CD14) (Shi *et al.*, 2010; Jung *et al.*, 2013). However, an earlier study using the intragastric ethanol-feeding model failed to show any change by betaine feeding in the ethanol-induced upregulation of CD14 and TNFα expression (Ji *et al.*, 2004). Using the same model, however, other laboratories have shown that betaine feeding prevented the alcohol-induced upregulation of a large number of genes including TLR2/4, Janus kinase 3, α-2-macroglobulin (Li *et al.*, 2011), which implies that there are multiple sites that betaine could act on to protect against the development of alcoholic liver injury.

## 17.7 Future Directions

Despite the great strides in understanding the protective role of betaine in alcohol-induced liver damage, more work is needed to fully understand the mechanism and site(s) of betaine protection. Furthermore, carefully designed clinical trials of betaine administration to patients with ALD are warranted. There have been no such clinical trials conducted to date. As a naturally acting agent, minimal side effects with betaine are expected (Craig, 2004); however, in reality this would depend on whether it is administered as betaine anhydrous oral solution or capsules (Mukherjee, 2011).

Additional knowledge is required to understand whether the severity of liver injury in patients with ALD correlates with any polymorphism in the enzymes that are involved in betaine synthesis and degradation or in the different enzymes of the methionine pathway including BHMT. This could help explain why only a subset of individuals' chronically consuming alcohol develops more serious liver damage.

## Summary Points

- Alcoholic liver disease is a major health-care problem worldwide.
- Ethanol feeding impairs several of the multiple steps in methionine metabolism resulting in decreased hepatic levels of the key methylating agent, *S*-adenosylmethionine, while increasing the levels of two toxic metabolites, homocysteine and *S*-adenosylhomocysteine.
- This ultimately causes reductions in essential methylation reactions and upregulation of the endoplasmic reticulum-stress response leading to progressive liver damage.
- Betaine, by virtue of promoting homocysteine remethylation, reduces the levels of toxic metabolites, restores *S*-adenosylmethionine level, normalizes methylation reactions and thereby attenuates alcohol-induced liver injury.
- Betainc is a promising therapeutic agent in relieving defects in hepatic methylation and other defects associated with chronic alcohol abuse.

## Key Facts of Alcohol and Liver Injury

- Alcohol is the most socially accepted addictive drug that disrupts many metabolic processes in almost all critical organ systems of the body, including liver, lung, brain, heart, pancreas and the immune system.
- It is one of the top leading causes of preventable deaths and globally accounts for up to 4% of total mortality.
- Liver is the one of the organs of the body most susceptible to the detrimental effects of alcohol abuse. This is because the two major ethanol metabolizing enzyme systems, alcohol dehydrogenase and the P-450-dependent ethanol-oxidizing system, are present mainly in the liver.
- Alcohol toxicity is linked to changes produced in the liver by ethanol metabolism that include redox state changes in the NAD+/NADH ratio, acetaldehyde formation, oxidative stress, changes in methionine metabolic pathway and mitochondrial function.
- Steatosis or accumulation of fat mainly as triglycerides within the hepatocytes is one the earliest manifestations of drinking large amounts of alcohol. This stage is called alcoholic fatty liver disease.
- It is postulated that excessive fat accumulation makes hepatocytes vulnerable to a second "hit" (virus, drug, toxin, proinflammatory cytokines or oxidative stress), leading to the progression to steatohepatitis, which is inflammation of the liver with concurrent fat accumulation.

- The ratio of SAM to SAH is an important metabolic indicator for cellular methylation status and is termed methylation potential.
- A decrease in the SAM:SAH ratio in the presence of increased SAH is associated with impaired activity of many members of SAM-dependent methyltransferases. This occurs because SAH has a high affinity binding to the catalytic region of many SAM-dependent methyltransferases enabling it to act as a potent product inhibitor; especially if the $K_i$ value for SAH is less than the $K_m$ value for SAM, as is the case for many of the liver methyltransferases.

## Definition and Explanations of Key Terms Used

**Alcoholic liver disease**: Damage to the liver attributable to alcohol misuse. It covers a range of conditions of the liver such as fat accumulation (steatosis), inflammation (hepatitis) and cirrhosis (scarring of the liver) and associated symptoms.

**Hepatocytes and liver nonparenchymal cells**: Hepatocytes are specialized epithelial cells of the liver and perform many metabolic, endocrine and secretory functions. Hepatocytes, also called the liver parenchymal cells, may constitute up to roughly 75–80% of the liver mass. The rest, 20–25%, of the liver is made up of the nonparenchymal cells comprised mainly of Kupffer cells (liver-resident macrophages), stellate cells (special liver fibroblasts) and liver endothelial cells.

**Methyltransferases**: Enzymes that catalyze the transfer of methyl group(s) to acceptor molecule(s).

**Methylation reactions**: Chemical reactions in the body where methyl groups are transferred to acceptor molecule(s) to generate methylated acceptor(s) that have a vital biological function.

**S-adenosylmethionine (SAM)**: A universal methyl-group donor.

**S-adenosylhomocysteine (SAH)**: A molecule that is formed after SAM donates its methyl group. High accumulation of SAH is harmful for the liver. SAH is also a precursor of homocysteine.

**Triglycerides**: A type of fat derived from glycerol and three fatty acids.

**Apoptosis**: Programmed cell death or sequence of steps in which cell signals self-destruct.

**Trafficking defects**: Occurs when molecules in the cell are routed inappropriately.

**Mallory-Denk bodies**: Structures made of aggregated proteins (keratin, ubiquitin, p62) seen in liver cells in the setting of alcohol-induced liver disease.

## List of Abbreviations

| | |
|---|---|
| ALD | Alcoholic liver disease |
| AMPK | AMP-activated protein kinase |
| ATP | Adenosine-5'-triphosphate |

| | |
|---|---|
| BHMT | Betaine-homocysteine-methyltransferase |
| CBS | Cystathionine-β-synthase |
| CD | Cluster of differentiation |
| COX-2 | Cyclooxygenase-2 |
| DMG | Dimethylglycine |
| ER | Endoplasmic reticulum |
| ERK | Extracellular signal regulated kinase |
| FOXO3 | Forkhead box O3 |
| GSH | Glutathione |
| HCY | Homocysteine |
| ICMT | Isoprenylcysteine carboxyl methyltransferase |
| IFN | Interferon |
| IL | Interleukin |
| MAT | Methionine adenosyltransferase |
| MS | Methionine synthase |
| MTHF | $N^5$-Methyltetrahydrofolate |
| NOS2 | Inducible nitric oxide synthase |
| PEMT | Phosphatidylethanolamine methyltransferase |
| PPAR | Peroxisome proliferator-activated receptor |
| PRMT | Protein arginine methyltransferase |
| SAH | *S*-Adenosylhomocysteine |
| SAHH | *S*-Adenosylhomocysteine hydrolase |
| SAM | *S*-Adenosylmethionine |
| SREBP | Sterol regulatory element binding protein |
| STAT | Signal transducers and activators of transcription |
| TNF | Tumor necrosis factor |
| TLR | Toll-like receptor |
| VLDL | Very low density lipoprotein. |

## Acknowledgements

This material is the result of work supported with resources and the use of facilities through the VA Nebraska-Western Iowa Health Care System. Research by the corresponding author cited in this paper has been supported by a Merit Review grant BX001155 (to Kusum K. Kharbanda, Ph.D.) from the Department of Veterans Affairs, Office of Research and Development (Biomedical Laboratory Research and Development).

## References

Alirezaei, M., Jelodar, G., Ghayemi, Z. and Mehr, M. K., 2012. Antioxidant and methyl donor effects of betaine versus ethanol-induced oxidative stress in the rat liver. *Comparative Clinical Pathology*. 1–8.

Altamirano, J. and Bataller, R., 2011. Alcoholic liver disease: pathogenesis and new targets for therapy. *Nature Reviews Gastroenterology & Hepatology*. 8: 491–501.

Attwood, J. T., Yung, R. L. and Richardson, B. C., 2002. DNA methylation and the regulation of gene transcription. *Cellular and Molecular Life Sciences*. 59: 241–257.

Bailey, S. M., Landar, A. and Darley-Usmar, V., 2005. Mitochondrial proteomics in free radical research. *Free Radical Biology & Medicine*. 38: 175–188.

Balkan, J., Oztezcan, S., Kucuk, M., Cevikbas, U., Kocak-Toker, N. and Uysal, M., 2004. The effect of betaine treatment on triglyceride levels and oxidative stress in the liver of ethanol-treated guinea pigs. *Experimental and Toxicologic Pathology*. 55: 505–509.

Barak, A. J., Beckenhauer, H. C., Junnila, M. and Tuma, D. J., 1993. Dietary betaine promotes generation of hepatic S-adenosylmethionine and protects the liver from ethanol-induced fatty infiltration. *Alcoholism: Clinical and Experimental Research*. 17: 552–555.

Barak, A. J., Beckenhauer, H. C., Kharbanda, K. K. and Tuma, D. J., 2001. Chronic ethanol consumption increases homocysteine accumulation in hepatocytes. *Alcohol*. 25: 77–81.

Barak, A. J., Beckenhauer, H. C., Mailliard, M. E., Kharbanda, K. K. and Tuma, D. J., 2003. Betaine lowers elevated S-adenosylhomocysteine levels in hepatocytes from ethanol-fed rats. *Journal of Nutrition*. 133: 2845–2848.

Bardag-Gorce, F., French, B. A., Li, J., Riley, N. E., Yuan, Q. X., Valinluck, V., Fu, P., Ingelman-Sundberg, M., Yoon, S. and French, S. W., 2002. The importance of cycling of blood alcohol levels in the pathogenesis of experimental alcoholic liver disease in rats. *Gastroenterology*. 123: 325–335.

Bardag-Gorce, F., Oliva, J., Villegas, J., Fraley, S., Amidi, F., Li, J., Dedes, J., French, B. and French, S. W., 2008. Epigenetic mechanisms regulate Mallory Denk body formation in the livers of drug-primed mice. *Experimental and Molecular Pathology*. 84: 113–121.

Beier, J. I. and McClain, C. J., 2010. Mechanisms and cell signaling in alcoholic liver disease. *Biological Chemistry*. 391: 1249–1264.

Benedetti, A. and Marucci, L., 1999. The significance of apoptosis in the liver. *Liver*. 19: 453–463.

Boisvert, F. M., Cote, J., Boulanger, M. C. and Richard, S., 2003. A proteomic analysis of arginine-methylated protein complexes. *Molecular & Cellular Proteomics*. 2: 1319–1330.

Bull-Otterson, L., Feng, W., Kirpich, I., Wang, Y., Qin, X., Liu, Y., Gobejishvili, L., Joshi-Barve, S., Ayvaz, T., Petrosino, J., Kong, M., Barker, D., McClain, C. and Barve, S., 2013. Metagenomic analyses of alcohol induced pathogenic alterations in the intestinal microbiome and the effect of *lactobacillus rhamnosus* GG treatment. *PLoS One*. 8: e53028.

Cases, S., Smith, S. J., Zheng, Y. W., Myers, H. M., Lear, S. R., Sande, E., Novak, S., Collins, C., Welch, C. B., Lusis, A. J., Erickson, S. K. and Farese, R. V., Jr., 1998. Identification of a gene encoding an acyl CoA:diacylglycerol acyltransferase, a key enzyme in triacylglycerol synthesis. *Proceedings of the National Academy of Sciences of the United States of America*. 95: 13018–13023.

Cases, S., Stone, S. J., Zhou, P., Yen, E., Tow, B., Lardizabal, K. D., Voelker, T. and Farese, R. V., Jr., 2001. Cloning of DGAT2, a second mammalian diacylglycerol acyltransferase, and related family members. *Journal of Biological Chemistry*. 276: 38870–38876.

Casey, C. A., Nanji, A., Cederbaum, A. I., Adachi, M. and Takahashi, T., 2001. Alcoholic liver disease and apoptosis. *Alcoholism: Clinical and Experimental Research*. 25: 49S–53S.

Cederbaum, A. I., 2012. Alcohol metabolism. *Clinical Liver Disease*. 16: 667–685.

Chen, X., Sebastian, B. M., Tang, H., McMullen, M. M., Axhemi, A., Jacobsen, D. W. and Nagy, L. E., 2009. Taurine supplementation prevents ethanol-induced decrease in serum adiponectin and reduces hepatic steatosis in rats. *Hepatology*. 49: 1554–1562.

Clarke, S., 2003. Aging as war between chemical and biochemical processes: protein methylation and the recognition of age-damaged proteins for repair. *Ageing Research Reviews*. 2: 263–285.

Craig, S. A., 2004. Betaine in human nutrition. *American Journal of Clinical Nutrition*. 80: 539–549.

Cunningham, C. C. and Bailey, S. M., 2001. Ethanol consumption and liver mitochondria function. *Biological Signals and Receptors*. 10: 271–282.

D'Angelo, S., Ingrosso, D., Perfetto, B., Baroni, A., Zappia, M., Lobianco, L. L., Tufano, M. A. and Galletti, P., 2001. UVA irradiation induces L-isoaspartyl formation in melanoma cell proteins. *Free Radical Biology & Medicine*. 31: 1–9.

Deaciuc, I. V., D'Souza, N. B., de Villiers, W. J., Burikhanov, R., Sarphie, T. G., Hill, D. B. and McClain, C. J., 2001. Inhibition of caspases in vivo protects the rat liver against alcohol-induced sensitization to bacterial lipopolysaccharide. *Alcoholism: Clinical and Experimental Research*. 25: 935–943.

Delgado-Reyes, C. V., Wallig, M. A. and Garrow, T. A., 2001. Immunohistochemical detection of betaine–homocysteine S-methyltransferase in human, pig, and rat liver and kidney. *Archives of Biochemistry and Biophysics*. 393: 184–186.

Duong, F. H., Christen, V., Filipowicz, M. and Heim, M. H., 2006. S-Adenosylmethionine and betaine correct hepatitis C virus induced inhibition of interferon signaling *in vitro*. *Hepatology*. 43: 796–806.

Esfandiari, F., Villanueva, J. A., Wong, D. H., French, S. W. and Halsted, C. H., 2005. Chronic ethanol feeding and folate deficiency activate hepatic endoplasmic reticulum stress pathway in micropigs. *American Journal of Physiology–Gastrointestinal and Liver Physiology*. 289: G54–G63.

Finkelstein, J. D., 1990. Methionine metabolism in mammals. *Journal of Nutritional Biochemistry*. 1: 228–237.

Finkelstein, J. D., Kyle, W. E. and Harris, B. J., 1974. Methionine metabolism in mammals: regulatory effects of S-adenosylhomocysteine. *Archives of Biochemistry and Biophysics*. 165: 774–779.

Finkelstein, J. D., Kyle, W. E., Martin, J. L. and Pick, A. M., 1975. Activation of cystathionine synthase by adenosylmethionine and adenosylethionine. *Biochemical and Biophysical Research Communications*. 66: 81–87.

Finkelstein, J. D. and Martin, J. J., 1984. Methionine metabolism in mammals. Distribution of homocysteine between competing pathways. *Journal of Biological Chemistry*. 259: 9508–9513.

Fischer, M., You, M., Matsumoto, M. and Crabb, D. W., 2003. Peroxisome proliferator-activated receptor alpha (PPARalpha) agonist treatment reverses PPARalpha dysfunction and abnormalities in hepatic lipid metabolism in ethanol-fed mice. *Journal of Biological Chemistry*. 278: 27997–28004.

French, S. W., 2005. The pathogenesis and significance of the urinary alcohol cycle in rats fed ethanol intragastrically. *Alcoholism: Clinical and Experimental Research*. 29: 158S–161S.

Gao, B. and Bataller, R., 2011. Alcoholic liver disease: pathogenesis and new therapeutic targets. *Gastroenterology*. 141: 1572–1585.

Haeusler, R. A., Kaestner, K. H. and Accili, D., 2012. FoxOs function synergistically to promote glucose production. *Journal of Biological Chemistry*. 285: 35245–35248.

Halsted, C. H., Villanueva, J., Chandler, C. J., Stabler, S. P., Allen, R. H., Muskhelishvili, L., James, S. J. and Poirier, L., 1996. Ethanol feeding of micropigs alters methionine metabolism and increases hepatocellular apoptosis and proliferation. *Hepatology*. 23: 497–505.

Handler, P. and Dubin, I. N., 1946. The significance of fatty infiltration in the development of hepatic cirrhosis due to choline deficiency. *Journal of Nutrition*. 31: 141–159.

Hartmann, P., Chen, W. C. and Schnabl, B., 2012. The intestinal microbiome and the leaky gut as therapeutic targets in alcoholic liver disease. *Frontiers in Physiology*. 3: 402.

Higuchi, H., Kurose, I., Kato, S., Miura, S. and Ishii, H., 1996. Ethanol-induced apoptosis and oxidative stress in hepatocytes. *Alcoholism: Clinical and Experimental Research*. 20: 340A–346A.

Ingrosso, D., D'Angelo, S., di Carlo, E., Perna, A. F., Zappia, V. and Galletti, P., 2000. Increased methyl esterification of altered aspartyl residues in erythrocyte membrane proteins in response to oxidative stress. *European Journal of Biochemistry*. 267: 4397–4405.

Ishak, K. G., Zimmerman, H. J. and Ray, M. B., 1991. Alcoholic liver disease: pathologic, pathogenetic, and clinical aspects. *Alcoholism: Clinical and Experimental Research*. 15: 45–66.

Jewell, S. A., Di Monte, D., Gentile, A., Guglielmi, A., Altomare, E. and Albano, O., 1986. Decreased hepatic glutathione in chronic alcoholic patients. *Journal of Hepatology*. 3: 1–6.

Ji, C., Deng, Q. and Kaplowitz, N., 2004. Role of TNF-alpha in ethanol-induced hyperhomocysteinemia and murine alcoholic liver injury. *Hepatology*. 40: 442–451.

Ji, C. and Kaplowitz, N., 2003. Betaine decreases hyperhomocysteinemia, endoplasmic reticulum stress, and liver injury in alcohol-fed mice. *Gastroenterology*. 124: 1488–1499.

Ji, C., Shinohara, M., Vance, D., Than, T. A., Ookhtens, M., Chan, C. and Kaplowitz, N., 2008. Effect of transgenic extrahepatic expression of betaine-homocysteine methyltransferase on alcohol or homocysteine-induced fatty liver. *Alcoholism: Clinical and Experimental Research*. 32: 1049–1058.

Johnson, B. A., Langmack, E. L. and Aswad, D. W., 1987. Partial repair of deamidation-damaged calmodulin by protein carboxyl methyltransferase. *Journal of Biological Chemistry*. 262: 12283–12287.

Jung, Y. S., Kim, S. J., Kwon do, Y., Ahn, C. W., Kim, Y. S., Choi, D. W. and Kim, Y. C., 2013. Alleviation of alcoholic liver injury by betaine involves an enhancement of antioxidant defense via regulation of sulfur amino acid metabolism. *Food and Chemical Toxicology*. 62: 292–298.

Kanbak, G., Inal, M. and Baycu, C., 2001. Ethanol-induced hepatotoxicity and protective effect of betaine. *Cell Biochemistry and Function*. 19: 281–285.

Kang, L., Chen, X., Sebastian, B. M., Pratt, B. T., Bederman, I. R., Alexander, J. C., Previs, S. F. and Nagy, L. E., 2007. Chronic ethanol and triglyceride turnover in white adipose tissue in rats: inhibition of the anti-lipolytic action of insulin after chronic ethanol contributes to increased triglyceride degradation. *Journal of Biological Chemistry*. 282: 28465–28473.

Kessova, I. G., Ho, Y. S., Thung, S. and Cederbaum, A. I., 2003. Alcohol-induced liver injury in mice lacking Cu, Zn-superoxide dismutase. *Hepatology*. 38: 1136–1145.

Kharbanda, K. K., 2007. Role of transmethylation reactions in alcoholic liver disease. *World Journal of Gastroenterology*. 13: 4947–4954.

Kharbanda, K. K., 2009. Alcoholic liver disease and methionine metabolism. *Seminars in Liver Disease*. 29: 155–165.

Kharbanda, K. K., 2013. Methionine metabolic pathway in alcoholic liver injury. *Current Opinion in Clinical Nutrition and Metabolic Care*. 16: 89–95.

Kharbanda, K. K., Mailliard, M. E., Baldwin, C. R., Beckenhauer, H. C., Sorrell, M. F. and Tuma, D. J., 2007a. Betaine attenuates alcoholic steatosis by restoring phosphatidylcholine generation via the phosphatidylethanolamine methyltransferase pathway. *Journal of Hepatology*. 46: 314–321.

Kharbanda, K. K., Mailliard, M. E., Baldwin, C. R., Sorrell, M. F. and Tuma, D. J., 2007b. Accumulation of proteins bearing atypical isoaspartyl residues in livers of alcohol-fed rats is prevented by betaine administration: Effects on protein-l-isoaspartyl methyltransferase activity. *Journal of Hepatology*. 46: 1119–1125.

Kharbanda, K. K. and Rogers, D. D., II, Beckenhauer, H. C., Siford, G. L., Barak, A. J., Mailliard, M. E., Sorrell, M. F. and Tuma, D. J., 2005a. Tubercidin-induced apoptosis via increased hepatocellular levels of S-adenosylhomocysteine is attenuated by betaine administration. *Alcoholism: Clinical and Experimental Research*. 29: 182A.

Kharbanda, K. K., Rogers, D. D., II, Mailliard, M. E., Siford, G. L., Barak, A. J., Beckenhauer, H. C., Sorrell, M. F. and Tuma, D. J., 2005b. A comparison of the effects of betaine and S-adenosylmethionine on ethanol-induced changes in methionine metabolism and steatosis in rat hepatocytes. *Journal of Nutrition*. 135: 519–524.

Kharbanda, K. K., Rogers, D. D., 2nd, Mailliard, M. E., Siford, G. L., Barak, A. J., Beckenhauer, H. C., Sorrell, M. F. and Tuma, D. J., 2005c. Role of elevated S-adenosylhomocysteine in rat hepatocyte apoptosis: protection by betaine. *Biochemical Pharmacology*. 70: 1883–1890.

Kharbanda, K. K., Todero, S. L., King, A. L., Osna, N. A., McVicker, B. L., Tuma, D. J., Wisecarver, J. L. and Bailey, S. M., 2012. Betaine treatment attenuates chronic ethanol-induced hepatic steatosis and alterations to the mitochondrial respiratory chain proteome. *International Journal of Hepatology*. 2012: 962183.

Kharbanda, K. K., Todero, S. L., Ward, B. W. and Cannella, J. J., III, Tuma, D. J., 2009. Betaine administration corrects ethanol-induced defective VLDL secretion. *Molecular and Cellular Biochemistry*. 327: 75–78.

Kim, S. J., Jung, Y. S., Kwon do, Y. and Kim, Y. C., 2008. Alleviation of acute ethanol-induced liver injury and impaired metabolomics of S-containing substances by betaine supplementation. *Biochemical and Biophysical Research Communications*. 368: 893–898.

Kim, S. J., Lee, J. W., Jung, Y. S., Kwon do, Y., Park, H. K., Ryu, C. S., Kim, S. K., Oh, G. T. and Kim, Y. C., 2009. Ethanol-induced liver injury and changes in sulfur amino acid metabolomics in glutathione peroxidase and catalase double knockout mice. *Journal of Hepatology*. 50: 1184–1191.

Lanthier, J. and Desrosiers, R. R., 2004. Protein L-isoaspartyl methyltransferase repairs abnormal aspartyl residues accumulated in vivo in type-I collagen and restores cell migration. *Experimental Cell Research*. 293: 96–105.

Leyendecker, M., Korsten, P., Reinehr, R., Speckmann, B., Schmoll, D., Scherbaum, W. A., Bornstein, S. R., Barthel, A. and Klotz, L. O., 2011. Ceruloplasmin expression in rat liver cells is attenuated by insulin: role of FoxO transcription factors. *Hormone and Metabolic Research*. 43: 268–274.

Li, J., Li, X. M., Caudill, M., Malysheva, O., Bardag-Gorce, F., Oliva, J., French, B. A., Gorce, E., Morgan, K., Kathirvel, E., Morgan, T. and French, S. W., 2011. Betaine feeding prevents the blood alcohol cycle in rats fed alcohol continuously for 1 month using the rat intragastric tube feeding model. *Experimental and Molecular Pathology*. 91: 540–547.

Lieber, C. S., Casini, A., DeCarli, L. M., Kim, C. I., Lowe, N., Sasaki, R. and Leo, M. A., 1990. S-adenosyl-L-methionine attenuates alcohol-induced liver injury in the baboon. *Hepatology*. 11: 165–172.

Lu, S. C., Huang, Z. Z., Yang, J. M. and Tsukamoto, H., 1999. Effect of ethanol and high-fat feeding on hepatic gamma-glutamylcysteine synthetase subunit expression in the rat. *Hepatology*. 30: 209–214.

Monetti, M., Levin, M. C., Watt, M. J., Sajan, M. P., Marmor, S., Hubbard, B. K., Stevens, R. D., Bain, J. R., Newgard, C. B., Farese, R. V., Sr., Hevener, A. L. and Farese, R. V., Jr., 2007. Dissociation of hepatic steatosis and insulin resistance in mice overexpressing DGAT in the liver. *Cell Metabolism*. 6: 69–78.

Mukherjee, S., 2011. Betaine and nonalcoholic steatohepatitis: back to the future? *World Journal of Gastroenterology*. 17: 3663–3664.

Nanji, A. A., 1998. Apoptosis and alcoholic liver disease. *Seminars in Liver Disease*. 18: 187–190.

Nishimaki-Mogami, T., Yao, Z. and Fujimori, K., 2002. Inhibition of phosphatidylcholine synthesis via the phosphatidylethanolamine methylation pathway impairs incorporation of bulk lipids into VLDL in cultured rat hepatocytes. *Journal of Lipid Research*. 43: 1035–1045.

Noga, A. A., Zhao, Y. and Vance, D. E., 2002. An unexpected requirement for phosphatidylethanolamine N-methyltransferase in the secretion of very low density lipoproteins. *Journal of Biological Chemistry*. 277: 42358–42365.

Oliva, J., Bardag-Gorce, F., Li, J., French, B. A., Nguyen, S. K., Lu, S. C. and French, S. W., 2009. Betaine prevents Mallory-Denk body formation in drug-primed mice by epigenetic mechanisms. *Experimental and Molecular Pathology*. 86: 77–86.

Oliva, J., Bardag-Gorce, F., Tillman, B. and French, S. W., 2011. Protective effect of quercetin, EGCG, catechin and betaine against oxidative stress induced by ethanol in vitro. *Experimental and Molecular Pathology*. 90: 295–299.

Orman, E. S., Odena, G. and Bataller, R., 2013. Alcoholic liver disease: pathogenesis, management, and novel targets for therapy. *Journal of Gastroenterology and Hepatology*. 28. Suppl 1: 77–84.

Osna, N. A., White, R. L., Donohue, T. M., Jr., Beard, M. R., Tuma, D. J. and Kharbanda, K. K., 2010. Impaired methylation as a novel mechanism for proteasome suppression in liver cells. *Biochemical and Biophysical Research Communications*. 391: 1291–1296.

Schiller, J. S., Lucas, J. W. and Peregoy, J. A., 2012. Summary Health Statistics for U.S. Adults: National Health Interview Survey, 2011. National Center for Health Statistics. *Vital Health Statistics*. 10: 1–218.

Schubert, H. L., Blumenthal, R. M. and Cheng, X., 2003. Many paths to methyltransfer: a chronicle of convergence. *Trends in Biochemical Sciences*. 28: 329–335.

Shi, Q. Z., Wang, L. W., Zhang, W. and Gong, Z. J., 2010. Betaine inhibits toll-like receptor 4 expression in rats with ethanol-induced liver injury. *World Journal of Gastroenterology*. 16: 897–903.

Shimizu, T., Matsuoka, Y. and Shirasawa, T., 2005. Biological significance of isoaspartate and its repair system. *Biological and Pharmaceutical Bulletin*. 28: 1590–1596.

Sidransky, H. and Farber, E., 1958. Sex difference in induction of periportal fatty liver by methionine deficiency in the rat. *Proceedings of the Society for Experimental Biology and Medicine*. 98: 293–297.

Song, Z., Zhou, Z., Deaciuc, I., Chen, T. and McClain, C. J., 2008. Inhibition of adiponectin production by homocysteine: a potential mechanism for alcoholic liver disease. *Hepatology*. 47: 867–879.

Song, Z., Zhou, Z., Song, M., Uriarte, S., Chen, T., Deaciuc, I. and McClain, C. J., 2007. Alcohol-induced S-adenosylhomocysteine accumulation in the liver sensitizes to TNF hepatotoxicity: possible involvement of mitochondrial S-adenosylmethionine transport. *Biochemistry & Pharmacology*. 74: 521–531.

Song, Z., Zhou, Z., Uriarte, S., Wang, L., Kang, Y. J., Chen, T., Barve, S. and McClain, C. J., 2004. S-adenosylhomocysteine sensitizes to TNF-alpha hepatotoxicity in mice and liver cells: a possible etiological factor in alcoholic liver disease. *Hepatology*. 40: 989–997.

Tang, H., Sebastian, B. M., Axhemi, A., Chen, X., Hillian, A. D., Jacobsen, D. W. and Nagy, L. E., 2012. Ethanol-induced oxidative stress via the CYP2E1 pathway disrupts adiponectin secretion from adipocytes. *Alcoholism: Clinical and Experimental Research*. 36: 214–222.

Tao, R., Wei, D., Gao, H., Liu, Y., DePinho, R. A. and Dong, X. C., 2012. Hepatic FoxOs regulate lipid metabolism via modulation of expression of the nicotinamide phosphoribosyltransferase gene. *Journal of Biological Chemistry*. 286: 14681–14690.

Tikhanovich, I., Kuravi, S., Campbell, R. V., Kharbanda, K. K., Artigues, A., Villar, M. T. and Weinman, S. A., 2013. Regulation of FOXO3 by phosphorylation and methylation in hepatitis C virus infection and alcohol exposure. *Hepatology*. 59: 58–70.

Torres, L., Garcia-Trevijano, E. R., Rodriguez, J. A., Carretero, M. V., Bustos, M., Fernandez, E., Eguinoa, E., Mato, J. M. and Avila, M. A., 1999. Induction of TIMP-1 expression in rat hepatic stellate cells and hepatocytes: a new role for homocysteine in liver fibrosis. *Biochimica et Biophysica Acta*. 1455: 12–22.

Tsukamoto, H., French, S. W., Reidelberger, R. D. and Largman, C., 1985. Cyclical pattern of blood alcohol levels during continuous intragastric ethanol infusion in rats. *Alcoholism: Clinical and Experimental Research*. 9: 31–37.

Varatharajalu, R., Garige, M., Leckey, L. C., Gong, M. and Lakshman, M. R., 2010. Betaine protects chronic alcohol and omega-3 PUFA-mediated down-regulations of PON1 gene, serum PON1 and homocysteine thiolactonase activities with restoration of liver GSH. *Alcoholism: Clinical and Experimental Research*. 34: 424–431.

Venkatesan, S., Ward, R. J. and Peters, T. J., 1988. Effect of chronic ethanol feeding on the hepatic secretion of very-low-density lipoproteins. *Biochimica et Biophysica Acta*. 960: 61–66.

Wang, L., Chen, L., Tan, Y., Wei, J., Chang, Y., Jin, T. and Zhu, H., 2013. Betaine supplement alleviates hepatic triglyceride accumulation of apolipoprotein E deficient mice via reducing methylation of peroxisomal proliferator-activated receptor alpha promoter. *Lipids in Health and Disease*. 12: 34.

Wang, Z., Yao, T. and Song, Z., 2010. Involvement and mechanism of DGAT2 upregulation in the pathogenesis of alcoholic fatty liver disease. *Journal of Lipid Research*. 51: 3158–3165.

World Health Organization, 2011. Global status report on alcohol and health, Geneva.

Wu, D. and Cederbaum, A. I., 2009. Oxidative stress and alcoholic liver disease. *Seminars in Liver Disease*. 29: 141–154.

Xu, A., Wang, Y., Keshaw, H., Xu, L. Y., Lam, K. S. and Cooper, G. J., 2003. The fat-derived hormone adiponectin alleviates alcoholic and nonalcoholic fatty liver diseases in mice. *Journal of Clinical Investigation*. 112: 91–100.

Yamagata, K., Daitoku, H., Takahashi, Y., Namiki, K., Hisatake, K., Kako, K., Mukai, H., Kasuya, Y. and Fukamizu, A., 2008. Arginine methylation of

FOXO transcription factors inhibits their phosphorylation by Akt. *Molecular Cell*. 32: 221–231.

You, M., Considine, R. V., Leone, T. C., Kelly, D. P. and Crabb, D. W., 2005. Role of adiponectin in the protective action of dietary saturated fat against alcoholic fatty liver in mice. *Hepatology*. 42: 568–577.

You, M., Fischer, M., Deeg, M. A. and Crabb, D. W., 2002. Ethanol induces fatty acid synthesis pathways by activation of sterol regulatory element-binding protein (SREBP). *Journal of Biological Chemistry*. 277: 29342–29347.

You, M., Matsumoto, M., Pacold, C. M., Cho, W. K. and Crabb, D. W., 2004. The role of AMP-activated protein kinase in the action of ethanol in the liver. *Gastroenterology*. 127: 1798–1808.

Young, A. L., Carter, W. G., Doyle, H. A., Mamula, M. J. and Aswad, D. W., 2001. Structural integrity of histone H2B in vivo requires the activity of protein L-isoaspartate O-methyltransferase, a putative protein repair enzyme. *Journal of Biological Chemistry*. 276: 37161–37165.

Zakhari, S., 2006. Overview: how is alcohol metabolized by the body? *Alcohol Research & Health*. 29: 245–254.

Zhong, W., McClain, C. J., Cave, M., Kang, Y. J. and Zhou, Z., 2010. The role of zinc deficiency in alcohol-induced intestinal barrier dysfunction. *American Journal of Physiology–Gastrointestinal and Liver Physiology*. 298: G625–G633.

Zhu, X., Song, J., Mar, M. H., Edwards, L. J. and Zeisel, S. H., 2003. Phosphatidylethanolamine-N-methyltransferase (PEMT) knockout mice have hepatic steatosis and abnormal hepatic choline metabolite concentrations despite ingesting a recommended dietary intake of choline. *Biochemical Journal*. 370: 987–993.

CHAPTER 18

# *Betaine, DNA Methylation and Nonalcoholic Fatty Liver Disease*

ZHU HUILIAN*[a] AND WANG LIJUN[a]

[a]Faculty of Nutrition, School of Public Health, Sun Yat-sen University (North Campus), 74 Zhongshan Road II, Guangzhou 510080, PR China
*E-mail: zhuhl@mail.sysu.edu.cn

## 18.1 Epidemiologic Character and Nutritional Risk Factors of NAFLD

Nonalcoholic fatty liver disease (NAFLD), which is histologically the same as alcoholic fatty liver disease (AFLD), is characterized by triglyceride accumulation exceeding 5% of hepatocytes in the absence of excessive alcohol consumption (<20 g per day for men and <10 g per day for women) and other chronic liver diseases. NAFLD encompasses a spectrum of progressive liver diseases from nonalcoholic fatty liver (NAFL), nonalcoholic steatohepatitis (NASH), and fibrosis to cirrhosis. NAFL is the initial, mild stage of NAFLD and NASH is a progressive stage exhibiting liver injury and inflammation. Ten to twenty-nine percent of NASH patients may develop cirrhosis within ten years and 4–27% of cirrhosis patients may develop hepatocellular carcinoma (Hsu and Kao, 2012).

During the past three decades, NAFLD has been a prevalent liver disease in Western countries, affecting 20–31% of the population in Europe (Blachier *et al.*, 2013). Data from the Third National Health and Nutrition Examination

Food and Nutritional Components in Focus No. 7
Betaine: Chemistry, Analysis, Function and Effects
Edited by Victor R Preedy

Published by the Royal Society of Chemistry, www.rsc.org

Survey (1988–1994) in the United States indicated that 19% of adults, an estimated 28.8 million people, were suffering from NAFLD nationwide (Lazo *et al.*, 2013). Recently, this trend has spread across the world. The prevalence of NAFLD in the Asia-Pacific region is 5–30%, which is comparable to that of Western countries (Amarapurkar *et al.*, 2007). Determining the true prevalence of NAFLD is challenging because existing studies have a wide variation in genders, ethnicities, exclusion criterion of alcohol consumption, study design and diagnosis method. NAFLD can be noninvasively detected by imaging techniques, such as ultrasonography, computed tomography and magnetic resonance imaging, which are both highly sensitive (>77%) and specific (>89%) but cannot distinguish steatohepatitis and fibrosis. A histological analysis by liver biopsy remains the gold diagnostic criterion for NAFLD, however, it is invasive, costly and affected by sampling error, the biopsy size and the conditions under which samples are stored. Many individuals with NALFD are asymptomatic and have normal biochemistries, and are therefore not aware of their status without routine screening. Of particular note is the serious prevalence of NAFLD among preadolescents and adolescents. In Europe, 36–44% of patients aged 6–20 years in the general population have been diagnosed with NAFLD by ultrasonography (Blachier *et al.*, 2013). A dramatic rise in NAFLD prevalence has also been found in the obese pediatric population over the past thirty years, jeopardizing children as young as two years old (Patton *et al.*, 2006). The prevalence of NALFD across the globe is therefore likely to be much larger than we know. NAFLD is rapidly emerging as a critical global public health problem.

The presence of dyslipidemia, obesity and Type II diabetes increases the risk of NAFLD. Similarly, NAFLD is an emerging threat to diabetic and obese individuals, affecting 50% of diabetic subjects, 57–74% of obese people and 90% of morbidly obese people (Patrick, 2002). NAFLD is considered the hepatic manifestation of metabolic syndrome. Imbalanced nutrition is considered the main pathogenic factor in chronic diseases such as dyslipidemia, obesity, Type II diabetes and NAFLD. In rodents, high fat, carbohydrate diets and methyl-donor deficiency (lacking in methionine, choline and folic acid) diets are well-proven diets to induce NAFLD. In NAFLD patients, food rich in saturated fats, cholesterol and fructose result in an accumulation of intrahepatic lipids, whereas a diet rich in monounsaturated fatty acids, polyunsaturated fatty acids, fiber and antioxidants are protective factors (Caporaso *et al.*, 2012). Dietary choline deficiency and the following impaired release of hepatic very low density lipoprotein (VLDL)-triglyceride into plasma can promote NAFLD. Seventy-seven percent of men and eighty percent of postmenopausal women who are deprived of dietary choline develop NAFLD (Fischer *et al.*, 2007). Therefore, the management of nutritional risk factors is extremely important in halting the progression of NAFLD.

Betaine is an oxidized product of choline. Although the relationship between plasma/serum betaine and NAFLD is not well understood, encouraging conclusions can be drawn from existing studies on the association between betaine levels and metabolic syndrome components. The concentration of plasma betaine is inversely associated with serum non-HDL cholesterol, triglyceride, body mass index and waist circumference and is positively associated with HDL and physical activity (Konstantinova *et al.*, 2008). In our unpublished study of 1070 participants, Zhu *et al.* observed that the concentration of serum betaine was higher in normal individuals than in NAFLD patients and that serum betaine was inversely associated with serum triglyceride, homocysteine and the CHDH rs9001 polymorphism. An elevated homocysteine concentration in the circulation is frequently seen in patients with NAFLD. Betaine supplies as a methyl donor that effectively converts homocysteine to methionine. Therefore, it is reasonable to assume that a low circulatory betaine concentration may be a factor in the development of NAFLD. Betaine supplementation for the treatment of NAFLD is thus worthy of attention.

## 18.2 Experimental and Clinical Evidence for Betaine Treatment of NAFLD

Although NAFLD is histologically identical to AFLD, it presents a more complex etiology and pathogenesis. The generally accepted theory of NAFLD pathogenesis is the "two hits" hypothesis. The "first hit" is the accumulation of excessive triglyceride in the hepatocytes due to insulin resistance, which is mild and potentially reversible. As the disease progresses, the liver is more susceptible to the "second hit" which include oxidative stress, the release of cytokines, endoplasmic stress and apoptosis, leading to hepatic inflammation, fibrosis and ultimately cirrhosis. Therefore, in order to reverse NAFLD as early as possible, treatments that can prevent and alleviate the accumulation of hepatic triglyceride receive much attention.

A growing body of evidence has shown that betaine is an important nutrient for the prevention of chronic diseases. Betaine has been demonstrated to be a useful lipotropic agent in the fatty liver induced by ethanol, toxins such as chloroform, lipopolysaccharides, methotrexate and carbon tetrachloride. Recent animal studies have been conducted in animals fed with a high-fat diet (Kwon *et al.*, 2009; Kathirvel *et al.*, 2010; Wang *et al.*, 2010; Pekkinen *et al.*, 2013; Zhang *et al.*, 2013) or a high-sucrose diet (Song *et al.*, 2007), to investigate the hepato-protective effect of betaine in NAFLD. As expected, betaine supplementation effectively attenuated hepatic steatosis and reduced liver injury in these studies. Table 18.1 summarizes the mechanisms by which betaine has been found to attenuate NAFLD, which mostly focus on the "two hits" hypothesis.

The hallmark of NAFLD is an aberrant accumulation of triglyceride within the hepatocytes. A dynamic triglyceride metabolism in the liver depend on a balance of four pathways, including the uptake of free fatty acids, the *de novo* lipogenesis, fatty acid oxidation and the synthesis/secretion of apolipoproteins (as shown in Figure 18.1). Insulin resistance contributes to triglyceride deposits in the liver by increasing the influx of free

**Table 18.1** A summary of the available evidence for the mechanisms by which betaine alleviates hepatic triglycerides, based on studies in NAFLD animal models and human patients.

| Type of study | Object profile | Dietary intervention | Main mechanisms | References |
|---|---|---|---|---|
| Experimental studies | Male mouse | High fat diet and 1% or 1.5% w/v betaine in the drinking water | 1. Improves signal transduction and alleviates insulin resistance<br>2. Improves adipokine production<br>3. Prevents ER stress<br>4. Increases the betaine and dimyristoyl glycerol concentration and decreases the homocysteine concentration in the liver | Kathirvel *et al.* (2010), Wang *et al.* (2010) |
| | Male rat | High fat diet and 1% w/v betaine in the drinking water for 3 weeks | 1. Increases the methionine, SAM and glutathione concentration<br>2. Decreases hypotaurine and taurine levels in the liver<br>3. Inhibits inflammation and improves the antioxidant capacity | Kwon *et al.* (2009) |
| | Female rat | High fat diet for 12 weeks and intragastric 400 mg $kg^{-1}$ betaine per day for 4 weeks | 1. Inhibits the HMGB1/TLR4 signaling pathways | Zhang *et al.* (2013) |
| | Male mouse | High fat diet and 1% w/v betaine in the drinking water for 4 weeks | 1. Increases the betaine level in the plasma, muscle and liver and increases the dimyristoyl glycerol and methionine levels in the liver<br>2. Enhances carnitine metabolism in the muscle and liver | Pekkinen *et al.* (2013) |
| | Male mouse | High sucrose diet and 1% w/v betaine in the drinking water for 16 weeks | 1. Increases the 5′AMP-activated protein kinase pathway<br>2. Inhibits the expression of the genes involved in *de novo* lipogenesis | Song *et al.* (2007) |

| | | | | |
|---|---|---|---|---|
| | Male goose | Overfeeding diet and 1 g betaine per day for 15 weeks | 1. Negative results for a betaine effect on the methylation of the Spot14α promoter for the alteration of its mRNA expression | Su *et al.* (2009) |
| | Male mouse | High fat diet and 1% or 2% w/v betaine in the drinking water for 12 weeks | 1. Reverses the dysregulated expression of genes in triglyceride metabolism<br>2. Downregulates the DNA methylation level of the *MTTP* promoter and subsequently upregulates its mRNA expression<br>3. Restores the methylation capacity and genomic methylation level | Wang *et al.* (2014) |
| | Male and female broilers | Basal diet without betaine for 56 days, or basal diet supplemented with 0.1% (1000 mg $kg^{-1}$) betaine for 66 days | 1. Downregulates gene expression in *de novo* lipogenesis in adipose tissue<br>2. Decreases the lipoprotein lipase mRNA expression, possibly by altering the methylation pattern of the lipoprotein lipase promoter | Xing *et al.* (2011) |
| Clinical trials | 191 patients with NASH diagnosed by ultrasound scans (male/female = 134/57) | 150 mg betaine twice a day for 8 weeks, combined with diethanolamine glucuronate and nicotinamide ascorbate | 1. Decreases hepatic steatosis, hepatomegaly, discomfort and pain in the upper right quadrant, and effectively normalizes the levels of alanine and aspartate transaminases and γ-glutamyl transpeptidase | Miglio *et al.* (2000) |
| | 10 patients with biopsy-proven NASH (male/female = 4/6) | 20 g betaine daily for 12 months | 1. Improves the serum alanine transaminase and aspartate transaminase and shows a trend of improvement in steatosis, necroinflammation and fibrosis | Abdelmalek *et al.* (2001) |
| | 55 patients with biopsy-proven NASH (male/female = 20/35) | 20 g betaine daily for 12 months | 1. May protect against worsening steatosis | Abdelmalek *et al.* (2009) |

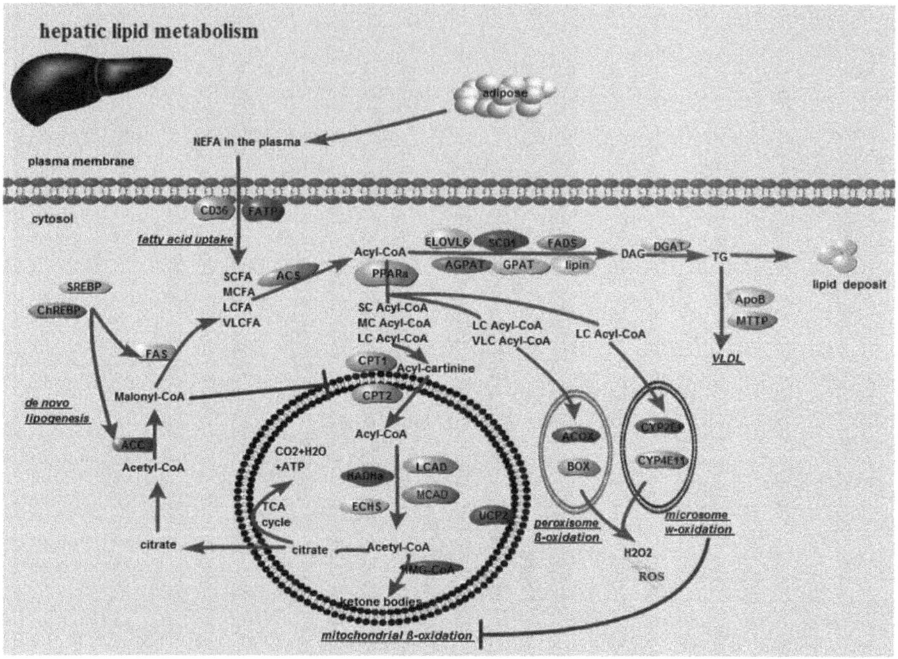


**Figure 18.1** Molecular mechanisms of hepatocellular fatty acid metabolism under normal conditions. Lipid metabolism in the liver relies mostly on the balance of four pathways: fatty acid uptake, *de novo* lipogenesis, fatty acid oxidation and VLDL transport. Dysregulation in one or more of these pathways can lead to NAFLD. Fatty acid uptake depends on the involvement of *FATP* and *CD36* to transport NEFA from plasma into hepatocyte across the plasma membrane. Acetyl-CoA is an essential substrate for *de novo* lipogenesis and is converted into malonyl-CoA by *ACC*. *FAS* then catalyzes both Acetyl-CoA and Malonyl-CoA to form fatty acid. Transcription factors such as *SREBP* and *ChREBP* promote the transcription of lipogenetic genes. In order to further synthesis and oxidation, fatty acids demand activation by *ACS* into Acyl-CoA. Acyl-CoA in the hepatocytes can be metabolized by two ways, one of which is oxidation, the other is esterified into triglyceride. Mitochondria is the major site for fatty acid oxidation. *CPT1*, expressed in the outer mitochondrial membrane, converts Acyl-CoA into Acyl-cartinine for the convenience of fatty acid transferring into mitochondria. *CPT2*, located in the inner mitochondrial membrane, converts Acyl-carnitine back into Acyl-CoA for further β-oxidation, in which *LCAD*, *MCAD*, *HADHα*, *ECHS* enzymes are involved. *UCP2*, an inner mitochondrial membrane protein, negatively impact cellular ATP production. Once the impairment of mitochondrial oxidation due to fatty acid overload occurs, peroxisomal β-oxidation and microsomal ω-oxidation oxidation would be activated to complement fatty acid oxidation, however, reactive oxygen species will be overproduced in the process. *ACOX* and *BOX* are the initial and rate-limiting enzymes of β-oxidation in peroxisomal. *CYP2E1* and *CYP4A11* are the major enzymes in microsomal ω-oxidation, their overexpression indicates the peroxidation of fatty acid. *PPARα*, an important transcriptional

fatty acids into the hepatocytes, enhancing synthesis of hepatic lipids and the decreasing synthesis of apolipoprotein B-100. *In vivo* and *in vitro* studies have revealed that betaine supplementation can improve insulin signal transduction in adipose tissue by activating protein kinases and attenuate insulin resistance, thereby enhancing the insulin-mediated suppression of lipolysis and decreasing the uptake of free fatty acids by the liver. Betaine supplementation has also been shown to activate the hepatic 5′ AMP-activated protein kinase pathway and subsequently inhibit the expression of genes involved in *de novo* lipogenesis (Song *et al.*, 2007; Kathirvel *et al.*, 2010; Wang *et al.*, 2010). Betaine also exerts a hepato-protective effect in the "second hit." The oxidation of fatty acids occurs mainly in the mitochondria. When mitochondrial β-oxidation is impaired under the stress of fatty acid overload, the alternative peroxisomal β-oxidation and microsomal ω-oxidation pathways are activated, in this case, the excessive generation of reactive oxygen species during the course of peroxisomal β-oxidation and microsomal ω-oxidation triggers oxidative stress in the liver. Reactive oxygen species also contribute to the release of proinflammatory cytokines. Betaine supplementation may prevent the reduction of antioxidant capacity against hydroxyl and peroxyl radicals in NAFLD mice by elevating the concentration of *S*-adenosyl methionine (SAM) and glutathione in the liver (Kwon *et al.*, 2009). Inflammation in NAFLD progression may also be inhibited by betaine supplementation through the suppression of TNF-α and the HMGB1/TLR4 signaling pathways (Kwon *et al.*, 2009; Zhang *et al.*, 2013).

After the promising results in animal studies, betaine has been recommended for the treatment of NAFLD patients. To date, only three clinical trials have been conducted on betaine treatment in NAFLD patients, as shown in Table 18.1. Miglio *et al.* (2000) reported a 25% decrease in hepatic steatosis after eight weeks of a combined therapy of betaine, diethanolamine glucuronate and nicotinamide ascorbate. Abdelmalek *et al.* studied the effect of betaine alone on NASH patients. Both their pilot study (Abdelmalek *et al.*, 2001) and a subsequent study with a larger sample (Abdelmalek *et al.*, 2009) showed an improvement in hepatic steatosis with betaine treatment. Although these results are encouraging, the safety and efficacy of betaine treatment require further study in a larger sample with different nationalities, ethnicities and genders.

factor, upregulates the expression of a suite of genes in mitochondrial, peroxisomal and microsomal oxidation. *ELOVL6*, *SCD1*, *FADS*, *AGPAT*, *GPAT* and lipin are involved in the esterification of Acyl-CoA to produce triglyceride. *MTTP* and *ApoB* are responsible for the VLDL assembly that facilitate to transport triglyceride from liver to peripheral tissues. If overproduced, triglyceride will deposit within the hepatocytes.

## 18.3 The Underlying Genetic Role of Betaine in the Treatment of NAFLD

Although the "two hits" theory in the pathogenesis of NAFLD has been widely accepted, there is growing evidence that NAFLD occurs as a consequence of multifactorial events that involve metabolomics, proteomics, genomics and epigenetics. Recent studies using microarrays and sequencing have provided new sights into the genomic and epigenetic mechanisms responsible for the pathogenesis of NAFLD and have enhanced our understanding of whether gene dysregulation in lipid metabolism contributes to triglyceride accumulation and the progression of NAFLD.

Microarray analyses of the differential gene expression in human NAFLD patients have yielded mixed results with respect to lipid metabolism. Extensive overexpression of the genes responsible for free fatty acid uptake, *de novo* lipid synthesis, mitochondrial β-oxidation and VLDL secretion was reported in 27 adolescent NASH patients. However, free fatty acid uptake and *de novo* synthesis were both upregulated more than the fatty acid oxidation and VLDL secretion pathways (Zhu *et al.* 2011). Kohjima *et al.* (2007) studied 26 NAFLD adult patients and found increased gene expression in *de novo* lipogenesis and free fatty acid uptake, impaired or fully activated gene expression in mitochondrial β-oxidation and overexpression of peroxisomal β-oxidation and microsomal ω-oxidation oxidation. The antioxidant pathways were then upregulated to neutralize the reactive oxygen species derived from peroxisomal and microsomal oxidation and the expression of lipid droplet formation-related genes was increased to aggravate triglyceride accumulation in the hepatocytes. Greco *et al.* (2008) studied the livers of 24 subjects who had elevated liver function tests and found that, compared with those with a low liver fat content, patients with a high liver fat content presented dysregulated gene expression, which contributed to decreased triglyceride lipolysis and increased free fatty acid uptake and fatty acid oxidation. Although the mechanisms for triglyceride accumulation are incompletely characterized, we can conclude that a defect in one or more of the lipid metabolism pathways may result in triglyceride deposits. Thus, correcting this defect is a promising strategy for the treatment of hepatic steatosis.

Although a comprehensive, systemic review of the effect of betaine on genetic regulation in NAFLD has not yet been performed, existing studies using microarrays to determine the effect of betaine treatment on rats with alcohol have shown that the most profound genetic effect of betaine is on the lipid and glucose metabolic pathways (Li *et al.*, 2010). Elevated homocysteine has a close relationship with disturbed gene expression in lipid metabolism (Sharma *et al.*, 2006). As betaine supplementation lowers homocysteine levels, betaine supplementation may reverse the detrimental effect on gene expression. Several studies linking betaine and genetic changes in metabolically disordered animals provide additional clues. In animals with a predisposition to hepatic steatosis resulting from dyslipidemia (Wang *et al.*, 2013) or a high-energy diet (Song *et al.*, 2007), betaine supplementation has been shown to effectively downregulate the expression of genes involved in hepatic *de novo* lipogenesis and upregulate the expression of genes involved in mitochondrial oxidation and VLDL transport.

## 18.4 Regulation of DNA Methylation by Betaine in the Treatment of NAFLD

### 18.4.1 Epigenetic Profile of NAFLD

Epigenetics, an emerging field, refers to the heritable, reversible modification of a phenotype without a change in the DNA sequence, and it mainly includes micro-RNAs, DNA methylation, histone modification and long noncoding RNA. In the past decade, the important role of epigenetic alterations in many chronic diseases has been increasingly recognized. Epigenetic regulation has shed light on the pathogenesis of NAFLD. The difference in the epigenetic status of liver cells even predetermines the individual susceptibility to hepatic steatosis in different mouse strains fed a lipogenic methyl-deficient diet (Pogribny *et al.*, 2009). Of these epigenetic regulation mechanisms, DNA methylation is the most extensively studied.

Normally, the majority of CpG islands within the promoter region are methylation free, whereas repetitive sequences and interspread CpG dinucleotides in the global DNA are heavily methylated. Aberrant DNA methylation can lead to changes in gene expression: the hypermethylation of specific gene promoter directly leads to gene silencing and the hypomethylation of the genome affects genomic stability and integrity.

A case-control study of 63 NAFLD patients observed the hypermethylation of the liver *PPARGC1A* promoter, an inverse correlation between the methylation level of the *PPARGC1A* promoter and its mRNA level and significant positive correlations between the *PPARGC1A* promoter methylation level and plasma fasting insulin levels and HOMA-insulin resistance, indicating that DNA methylation of the *PPARGC1A* promoter leads to repression of transcription and ultimately influences the insulin resistance phenotype (Sookoian *et al.*, 2010). Recently, Ahrens *et al.* (2013) systematically studied DNA methylation and mRNA expression profiles in the liver of patients with all stages of NAFLD before and after bariatric surgery. They found different methylation levels at CpG sites in the liver at different NAFLD stages and DNA methylation remodeling after postsurgery weight loss. Thus, DNA methylation must play an important role in the pathogenesis of NAFLD and may act as an interface for the causal relationship between the environment and genetics.

### 18.4.2 DNA Methylation Connects Methyl Donors and NAFLD

The effect of diet has been a focal point in DNA methylation studies. The DNA methylation reaction depends on the availability of methyl groups from SAM, and SAM levels thus indicate the cellular methylation capacity.

SAM synthesis requires abundant methyl groups derived from food, such as methionine, folate, choline and betaine. Dietary supplementation of these methyl donors may increase the hepatocellular SAM concentration, whereas dietary depletion will decrease the concentration. Therefore, DNA methylation is susceptible to nutritional status, especially methyl donors. Although the available human data are limited, massive datasets on the methylation changes in diet-induced NAFLD can be obtained in animal models.

In mouse models with chronic liver disease, a methyl-deficient diet has been generally proved to cause a profound loss of genomic methylation by the depletion of intracellular pools of SAM, and to result in the specific hyper- or hypomethylation of gene promoters in the liver (Tryndyak *et al.*, 2011). A methyl deficiency (lacking in folate and vitamin B12) during pregnancy and lactation disturbs the DNA methylation of gene promoters in their offspring and subsequently impairs mitochondrial fatty acid oxidation (Pooya *et al.*, 2012). If adequate dietary methyl is supplemented in the early stage of methyl-deficient feeding, the liver SAM levels and SAM/*S*-adenosyl homocysteine ratios are restored and genomic DNA methylation increases to normal levels (Pogribny *et al.*, 2006). In view of the depletion of methyl donors like SAM, betaine and choline and the elevation of homocysteine, a high-fat diet promotes the hypomethylation of genomic DNA and leads to the hypermethylation of the gene promoters involved in lipid metabolism in the tissue, such as the hypermethylation of the *MTTP* gene, which causes the downregulation of its expression in the liver,

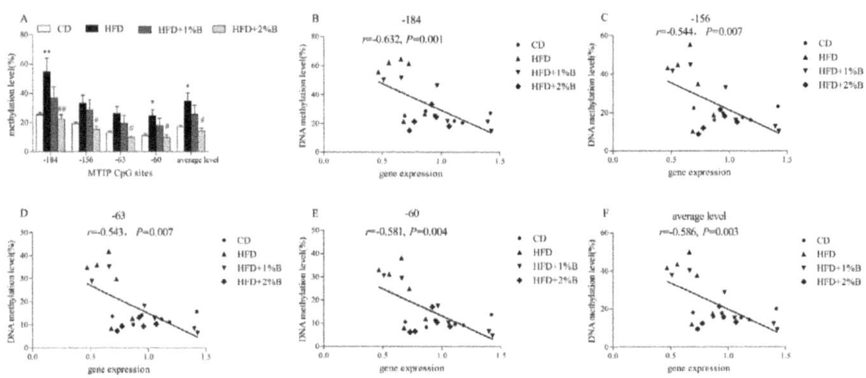


**Figure 18.2** Betaine supplementation reversed the hypermethylation of the *MTTP* promoter in mice fed with a high-fat diet. C57BL/6 mice were fed four different diets ($n = 6$) for 12 weeks: a control diet (CD), a high fat diet (HFD) and a high fat diet supplemented with either 1% or 2% betaine (HFD+1%B and HFD+2%B) in the drinking water (wt/vol). A: DNA methylation levels at four CpG sites and the average DNA methylation level of the *MTTP* promoter in the liver. B–F: The linear association between the *MTTP* mRNA expression level and the DNA methylation level at the −184, −156, −63 and −60 CpG sites. Data are presented as the means ± SEM. $^{*}P < 0.05$, $^{**}P < 0.01$ *vs.* CD group, $^{\#}P < 0.05$, $^{\#\#}P < 0.01$ *vs.* HFD group. *MTTP*, microsomal triglyceride transfer protein. Data are from our group (Wang *et al.*, 2014), with permission from the Publishers.

interfering with assembly lipoproteins and lipid export from the liver (Wang *et al.*, 2014). Another example is the hypermethylation of the PPARγ gene by a high-fat diet, which causes the downregulation of its expression in the visceral adipose, disturbing cytokine secretion and lipid metabolism (Fujiki *et al.*, 2009).

Both betaine and folate are required for the synthesis of methionine, which is the precursor to form SAM. Betaine has even been introduced into the diet of poultry, such as chicken and geese, to reduce fat deposits (Su *et al.*, 2009; Xing *et al.*, 2011). However, the DNA methylation profile in the regulation of gene expression in lipid metabolism (lipoprotein lipase and Spot14α genes) by betaine is not well understood, as the results obtained vary with species, disease model, selected genes and betaine concentration, among others. Further exploration of the role of DNA methylation by betaine in the regulation of lipid metabolism is therefore needed. A previous study showed the effect of

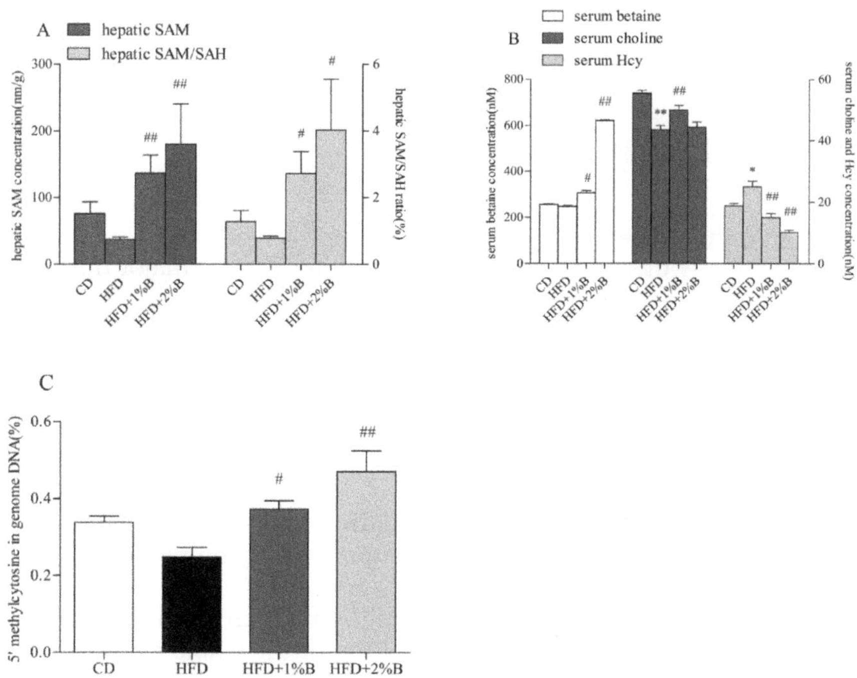


**Figure 18.3** Betaine supplementation improved the methylation capacity of mice fed with HFD. C57BL/6 mice were divided into four different groups ($n$ = 6) and fed a control diet (CD), a high-fat diet (HFD) and a high-fat diet supplemented with either 1% or 2% betaine (HFD+1%B and HFD+2%B) in the drinking water (wt/vol) for 12 weeks. A: Hepatic SAM concentrations and SAM/SAH ratios. B: Serum betaine, choline and Hcy levels. C: Genomic methylation levels in the liver. Data are presented as the means ± SEM. $^{*}P < 0.05$, $^{**}P < 0.01$ *vs.* CD group, $^{\#}P < 0.05$, $^{\#\#}P < 0.01$ *vs.* HFD group. SAM, *S*-adenosylmethionine; SAH, *S*-adenosylhomocysteine; Hcy, homocysteine.Data are from our group (Wang *et al.*, 2014), with permission from the Publishers.

betaine supplementation on the methylation of the PPARα promoter in dyslipidemia mice (Wang *et al.*, 2013). A recent study conducted in mice fed with a high-fat diet showed that 1% and 2% betaine supplementation in the drinking water dose-dependently alleviated triglyceride accumulation and reversed the dysregulated expression of six genes in triglyceride metabolism (*FAS*, *ACC*, *PPARα*, *ACOX*, *ApoB* and *MTTP*). Only the *MTTP* expression was directly influenced by the modification of the DNA methylation of its promoter, betaine effectively decreased the DNA methylation of the *MTTP* promoter and restored the methylation capacity and genomic methylation level in the high-fat diet mice (Wang *et al.*, 2014) (as shown in Figures 18.2 and 18.3).

In conclusion, this chapter focuses on what is currently known about betaine's protective effect on hepatic triglyceride metabolism from the point of DNA methylation. Although limited studies have been conducted and the underlying mechanisms are incompletely understood, the existing results prompt us to further investigate and enhance our understanding of a nutri-epigenetics strategy for evaluating the preventative and therapeutic role of betaine in NAFLD.

## Summary Points

- Hepatic triglyceride deposits are the prerequisite for the initiation and progression of nonalcoholic fatty liver disease (NAFLD).
- High-energy diets and methyl-deficient diets are the major nutritional risk factors for NAFLD.
- Betaine supplementation has been shown to relieve hepatic triglyceride accumulation in the treatment of NAFLD, by both experimental and clinical studies.
- Betaine supplementation has a hepato-protective effect, reversing the dysregulated expression of the genes involved in lipid metabolism in NAFLD animal models.
- Aberrant DNA methylation, including the hypermethylation of specific gene promoters and the hypomethylation of the genome, contributes to the pathogenesis of NAFLD.
- The hypermethylation of specific gene promoters generally leads to gene silencing and the hypomethylation of the whole genome affects genomic stability and integrity.
- As a key source of methyl donors, betaine may exert a lipotropic effect on the downregulation of DNA methylation in the promoter region of specific genes and restoring the genomic DNA in NAFLD animal models.

## Key Facts of Nutriepigenomics

1. Nutri-epigenomics is a discipline that links dietary nutrients and their biochemical effects on health or disease from epigenetic perspectives.

2. Epigenetics is labile and reversible in response to the environment, which makes epigenetics in chronic diseases a particularly attractive target for nutritional intervention.
3. No matter whether in prenatal and perinatal periods or adulthood, nutritional intake can affect epigenome, and these epigenetic alterations can lead to a permanent remodeling in tissue and organ.
4. DNA methylation is the most extensively studied epigenetic mechanism linking nutrition and disease.
5. Many nutrients have a significant effect on DNA methylation and disease phenotypes. Of these nutrients, methyl donors are the primary focus of attention.
6. As a methyl donor plays an important role in DNA methylation, optimal "EpiG diets" should be investigated as an prevention and treatment of DNA methylation-dependent diseases.

## Definitions of Words and Terms

**CpG island**: A genomic region approximately 500 bp long with a GC content of >50% and an observed to expected ratio of >60%. CpG islands within the promoter region are unmethylated and an elevated methylation level under abnormal circumstances directly represses gene expression.

**DNA methylation**: The transfer of a methyl group from *S*-adenosylmethionine to DNA by DNA methyltransferases, converting cytosine to 5-methylcytosine.

**Epigenetics**: A systematic study of heritable and reversible modifications of phenotypes independent of DNA sequence changes, due to microRNAs, DNA methylation, histones and long noncoding RNA modifications, among others.

**Hepatic lipid metabolism**: Dynamic balance of fatty acid metabolism in the hepatocytes is achieved by four pathways: free fatty acid uptake, *de novo* lipogenesis, fatty acid oxidation and very low density lipoprotein-triglyceride transfer.

**Metabolic syndrome**: A disorder of several metabolic components, diagnosed by central obesity and ethnicity-specific waist circumference, plus two or more of the following: raised serum triglycerides, reduced high-density lipoprotein-cholesterol level, elevated blood pressure, elevated fasting plasma glucose or previously diagnosed as diabetes.

**Methyl donors**: Metabolites that can supply methyl groups, which include *S*-adenosyl methionine, methionine, choline, betaine and folate. Choline, betaine and folate are mostly derived from diet.

**Microarray**: A high-throughput method that hybridizes biological materials and a solid substrate chip. It is used to detect the qualitative and quantitative properties of nucleic acids, proteins, peptides and cells, among others. RNA and DNA microarrays are the most widely used.

**Nonalcoholic fatty liver disease (NAFLD)**: A spectrum of liver diseases ranging from nonalcoholic fatty liver, nonalcoholic steatohepatitis and fibrosis to cirrhosis in the absence of excessive alcohol consumption (in men < 20 g per day and in women < 10 g per day) or other chronic liver diseases.

**Nonalcoholic fatty liver (NAFL)**: Histologically characterized by triglyceride accumulation exceeding 5% of hepatocytes, without evident inflammation or hepatocellular injury. NAFL is the initial, mild, reversible stage of NAFLD.

**Nonalcoholic steatohepatitis (NASH)**: Histologically characterized by steatosis and evident inflammation or ballooning of the hepatocytes. A serious stage that can progress to cirrhosis, fibrosis and even liver cancer.

## List of Abbreviations

| | |
|---|---|
| ACC | Acetyl-CoA carboxylase |
| ACOX | Straight-chain acyl-CoA oxidase |
| ACS | Acyl-CoA synthetases |
| AFLD | Alcoholic liver disease |
| AGPAT | 1-Acylglycerol-3- phosphate acyltransferase |
| ALT | Alanine amino transferase |
| Apo B100 | Apolipoprotein B100 |
| AST | Aspartate amino transferase |
| BOX | Branched-chain acyl-CoA oxidase |
| CD36(FAT) | Fatty acid translocase |
| CHDH | Choline dehydrogenase |
| ChREBP | Carbohydrate responsive element binding protein |
| DAG | Diacylglycerol |
| DGAT | Acyl-CoA:diacylglycerol acyltransferase |
| ECHS | Enoyl-CoA hydratase |
| ELOVL 6 | Long-chain fatty acid elongase 6 |
| ER | Endoplasmic reticulum |
| FADS | Fatty acid desaturase |
| FAS | Fatty acid synthase |
| FATP | Fatty acid transport protein |
| GPAT | Glycerol-3-phosphate acyltransferase |
| HADH$\alpha$ | 3-Hydroxyacyl-CoA dehydrogenase alpha |
| HDL | High-density lipoprotein |
| HMGB1 | High-mobility group box 1 |
| HMG-CoA synthase | 3-Hydroxy-3-methylglutaryl-CoA synthase |
| HOMA-IR | Homeostasis model assessment of insulin resistance |
| LCAD | Long-chain acyl-CoA dehydrogenase |
| LCFA | Long-chain fatty acid |
| MCAD | Medium-chain acyl-CoA dehydrogenase |
| MCFA | Medium-chain fatty acid |
| MTTP | Microsomal triglyceride transfer protein |
| NAFLD | Nonalcoholic fatty liver disease |

| | |
|---|---|
| NASH | Nonalcoholic steatohepatitis |
| NEFA | Nonesterified fatty acid |
| PPARα | Peroxisomal proliferator-activated receptor alpha |
| PPARγ | Peroxisomal proliferator-activated receptor gamma |
| PPARGC1α | Peroxisome proliferator-activated receptor gamma Coactivator 1 alpha |
| ROS | Reactive oxygen species |
| SAH | *S*-Adenosylhomocysteine |
| SAM | *S*-Adenosylmethionine |
| SCD1 | Stearoyl-CoA desaturase 1 |
| SCFA | Short-chain fatty acid |
| SREBP | Sterol regulatory element binding protein |
| SS | Simple steatosis |
| TLR4 | Toll-like receptors 4 |
| TNF-α | Tumor necrosis factor-alpha |
| VLCFA | Very long chain fatty acid |
| VLDL-TG | Very low density lipoprotein-triglyceride |
| γ-GT | Gamma-glutamyltransferase. |

## References

Abdelmalek, M. F., Angulo, P., Jorgensen, R. A., Sylvestre, P. B. and Lindor, K. D., 2001. Betaine, a promising new agent for patients with nonalcoholic steatohepatitis: results of a pilot study. *American Journal of Gastroenterology*. 96: 2711–2717.

Abdelmalek, M. F., Sanderson, S. O., Angulo, P., Soldevila-Pico, C., Liu, C., Peter, J., Keach, J., Cave, M., Chen, T., McClain, C. J. and Lindor, K. D., 2009. Betaine for nonalcoholic fatty liver disease: results of a randomized placebo-controlled trial. *Hepatology*. 50: 1818–1826.

Ahrens, M., Ammerpohl, O., von Schonfels, W., Kolarova, J., Bens, S., Itzel, T., Teufel, A., Herrmann, A., Brosch, M., Hinrichsen, H., Erhart, W., Egberts, J., Sipos, B., Schreiber, S., Hasler, R., Stickel, F., Becker, T., Krawczak, M., Rocken, C., Siebert, R., Schafmayer, C. and Hampe, J., 2013. DNA methylation analysis in nonalcoholic fatty liver disease suggests distinct disease-specific and remodeling signatures after bariatric surgery. *Cell Metabolism*. 18: 296–302.

Amarapurkar, D. N., Hashimoto, E., Lesmana, L. A., Sollano, J. D., Chen, P. J. and Goh, K. L., 2007. Asia-Pacific Working Party for NAFLD. How common is non-alcoholic fatty liver disease in the Asia-Pacific region and are there local differences? *Journal of Gastroenterology and Hepatology*. 22: 788–793.

Blachier, M., Leleu, H., Peck-Radosavljevic, M. and Valla, D. C. and Roudot-Thoraval, F., 2013. The burden of liver disease in Europe: a review of available epidemiological data. *Journal of Hepatology*. 58: 593–608.

Caporaso, N., Morisco, F., Camera, S., Graziani, G., Donnarumm, L. and Ritieni, A., 2012. Dietary approach in the prevention and treatment of NAFLD. *Frontiers in Bioscience*. 7: 2259–2268.

Fischer, L. M., DaCosta, K. A., Kwock, L., Stewart, P. W., Lu, T., Stabler, S. P., Allen, R. H. and Zeisel, S. H., 2007. Sex and menopausal status influence human dietary requirements for the nutrient choline. *American Journal of Clinical Nutrition*. 85: 1275–1285.

Fujiki, K., Kano, F., Shiota, K. and Murata, M., 2009. Expression of the peroxisome proliferator activated receptor gamma gene is repressed by DNA methylation in visceral adipose tissue of mouse models of diabetes. *BMC Biology*. 7: 38.

Greco, D., Kotronen, A., Westerbacka, J., Puig, O., Arkkila, P., Kiviluoto, T., Laitinen, S., Kolak, M., Fisher, R. M., Hamsten, A., Auvinen, P. and Yki-Järvinen, H., 2008. Gene expression in human NAFLD. *American Journal of Physiology: Gastrointestinal and Liver Physiology*. 294: G1281–G1287.

Hsu, C. S. and Kao, J. H., 2012. Non-alcoholic fatty liver disease: an emerging liver disease in Taiwan. *Journal of the Formosan Medical Association*. 111: 527–535.

Kathirvel, E., Morgan, K., Nandgiri, G., Sandoval, B. C., Caudill, M. A., Bottiglieri, T. and French, S. W., Morgan, T. R., 2010. Betaine improves nonalcoholic fatty liver and associated hepatic insulin resistance: a potential mechanism for hepatoprotection by betaine. *American Journal of Physiology: Gastrointestinal and Liver Physiology*. 299: G1068–G1077.

Kohjima, M., Enjoji, M., Higuchi, N., Kato, M., Kotoh, K., Yoshimoto, T., Fujino, T., Yada, M., Yada, R., Harada, N., Takayanagi, R. and Nakamuta, M., 2007. Re-evaluation of fatty acid metabolism-related gene expression in nonalcoholic fatty liver disease. *International Journal of Molecular Medicine*. 20: 351–358.

Konstantinova, S. V., Tell, G. S., Vollset, S. E., Nygard, O., Bleie, O. and Ueland, P. M., 2008. Divergent associations of plasma choline and betaine with components of metabolic syndrome in middle age and elderly men and women. *Journal of Nutrition*. 138: 914–920.

Kwon, D. Y., Jung, Y. S., Kim, S. J., Park, H. K., Park, J. H. and Kim, Y. C., 2009. Impaired sulfur-amino acid metabolism and oxidative stress in nonalcoholic fatty liver are alleviated by betaine supplementation in rats. *Journal of Nutrition*. 139: 63–68.

Lazo, M., Hernaez, R., Eberhardt, M. S., Bonekamp, S., Kamel, I., Guallar, E., Koteish, A., Brancati, F. L. and Clark, J. M., 2013. Prevalence of nonalcoholic fatty liver disease in the United States: the Third National Health and Nutrition Examination Survey, 1988-1994. *American Journal of Epidemiology*. 178: 38–45.

Li, J., Bardag-Gorce, F., Oliva, J., French, B. A., Dedes, J. and French, S. W., 2010. The effects of betaine treatment on rats fed an acute bolus of ethanol at 3 and 12 h post bolus: a microarray analysis. *Genes and Nutrition*. 5: 321–329.

Miglio, F., Rovati, L. C., Santoro, A. and Setnikar, I., 2000. Efficacy and safety of oral betaine glucuronate in non-alcoholic steatohepatitis. A double-blind, randomized, parallel-group, placebo-controlled prospective clinical study. *Arzneimittelforschung*. 50: 722–727.

Patrick, L., 2002. Nonalcoholic fatty liver disease: relationship to insulin sensitivity and oxidative stress. Treatment approaches using vitamin E, magnesium, and betaine. *Alternative Medicine Review*. 7: 276–291.

Patton, H. M., Sirlin, C., Behling, C., Middleton, M., Schwimmer, J. B. and Lavine, J. E., 2006. Pediatric nonalcoholic fatty liver disease: a critical appraisal of current data and implications for future research. *Journal of Pediatric Gastroenterology and Nutrition*. 43: 413–427.

Pekkinen, J., Olli, K., Huotari, A., Tiihonen, K., Keski-Rahkonen, P., Lehtonen, M., Auriola, S., Kolehmainen, M., Mykkanen, H., Poutanen, K. and Hanhineva, K., 2013. Betaine supplementation causes increase in carnitine metabolites in the muscle and liver of mice fed a high-fat diet as studied by nontargeted LC-MS metabolomics approach. *Molecular Nutrition & Food Research*. 57: 1959–1968.

Pogribny, I. P., Ross, S. A., Wise, C., Pogribna, M., Jones, E. A., Tryndyak, V. P., James, S. J., Dragan, Y. P. and Poirier, L. A., 2006. Irreversible global DNA hypomethylation as a key step in hepatocarcinogenesis induced by dietary methyl deficiency. *Mutation Research*. 593: 80–87.

Pogribny, I. P., Tryndyak, V. P., Bagnyukova, T. V., Melnyk, S., Montgomery, B., Ross, S. A., Latendresse, J. R., Rusyn, I. and Beland, F. A., 2009. Hepatic epigenetic phenotype predetermines individual susceptibility to hepatic steatosis in mice fed a lipogenic methyl-deficient diet. *Journal of Hepatology*. 51: 176–186.

Pooya, S., Blaise, S., Moreno, G. M., Giudicelli, J., Alberto, J. M., Gueant-Rodriguez, R. M., Jeannesson, E., Gueguen, N., Bressenot, A., Nicolas, B., Malthiery, Y., Daval, J. L., Peyrin-Biroulet, L., Bronowicki, J. P. and Guéant, J. L., 2012. Methyl donor deficiency impairs fatty acid oxidation through PGC-1alpha hypomethylation and decreased ER-alpha, ERR-alpha, and HNF-4alpha in the rat liver. *Journal of Hepatology*. 57: 344–351.

Sharma, P., Senthilkumar, R. D., Brahmachari, V., Sundaramoorthy, E., Mahajan, A., Sharma, A. and Sengupta, S., 2006. Mining literature for a comprehensive pathway analysis: a case study for retrieval of homocysteine related genes for genetic and epigenetic studies. *Lipids in Health and Disease*. 5: 1.

Song, Z., Deaciuc, I., Zhou, Z., Song, M., Chen, T., Hill, D. and McClain, C. J., 2007. Involvement of AMP-activated protein kinase in beneficial effects of betaine on high-sucrose diet-induced hepatic steatosis. *American Journal of Physiology: Gastrointestinal and Liver Physiology*. 293:G894–G902.

Sookoian, S., Rosselli, M. S., Gemma, C., Burgueno, A. L., Fernandez, G. T., Castano, G. O. and Pirola, C. J., 2010. Epigenetic regulation of insulin resistance in nonalcoholic fatty liver disease: impact of liver methylation of the peroxisome proliferator-activated receptor gamma coactivator 1alpha promoter. *Hepatology*. 52: 1992–2000.

Su, S. Y., Dodson, M. V., Li, X. B., Li, Q. F., Wang, H. W. and Xie, Z., 2009. The effects of dietary betaine supplementation on fatty liver performance, serum parameters, histological changes, methylation status and the mRNA expression level of Spot14α in Landes goose fatty liver. *Comparative Biochemistry and Physiology, Part A: Molecular & Integrative*. 154: 308–314.

Tryndyak, V. P., Han, T., Muskhelishvili, L., Fuscoe, J. C., Ross, S. A., Beland, F. A. and Pogribny, I. P., 2011. Coupling global methylation and gene expression profiles reveal key pathophysiological events in liver injury induced by a methyl-deficient diet. *Molecular Nutrition & Food Research*. 55:411–418.

Wang, L., Chen, L., Tan, Y., Wei, J., Chang, Y., Jin, T. and Zhu, H., 2013. Betaine supplement alleviates hepatic triglyceride accumulation of apolipoprotein E deficient mice via reducing methylation of peroxisomal proliferator-activated receptor alpha promoter. *Lipids in Health and Disease*. 12: 34.

Wang, L. J., Zhang, H. W., Zhou, J. Y., Liu, Yan., Yang, Y., Chen, X. L., Zhu, C. H., Zheng, R. D., Ling, W. H. and Zhu, H. L., 2014. Betaine attenuates hepatic steatosis by reducing methylation of the MTTP promoter and elevating genomic methylation in mice fed a high fat diet. *Journal of Nutritional Biochemistry*. 25: 329–336.

Wang, Z., Yao, T., Pini, M., Zhou, Z., Fantuzzi, G. and Song, Z., 2010. Betaine improved adipose tissue function in mice fed a high-fat diet: a mechanism for hepatoprotective effect of betaine in nonalcoholic fatty liver disease. *American Journal of Physiology: Gastrointestinal and Liver Physiology*. 298: G634–G642.

Xing, J., Kang, L. and Jiang, Y., 2011. Effect of dietary betaine supplementation on lipogenesis gene expression and CpG methylation of lipoprotein lipase gene in broilers. *Molecular Biology Reports*. 38: 1975–1981.

Zhang, W., Wang, L. W., Wang, L. K., Li, X., Zhang, H., Luo, L. P., Song, J. C. and Gong, Z. J., 2013. Betaine protects against high-fat-diet-induced liver injury by inhibition of high-mobility group box 1 and toll-like receptor 4 expression in rats. *Digestive Diseases and Sciences*. 58: 3198–3206.

Zhu, L., Baker, S. S., Liu, W., Tao, M. H., Patel, R., Nowak, N. J. and Baker, R. D., 2011. Lipid in the livers of adolescents with nonalcoholic steatohepatitis: combined effects of pathways on steatosis. *Metabolism*. 60: 1001–1011.

CHAPTER 19

# *The Beneficial Role of Dietary Betaine in the Suppression of Age-Related Inflammatory NF-κB*

HAE YOUNG CHUNG*[a], EUN KYEONG LEE[a], AND BYUNG PAL YU[b]

[a]Molecular Inflammation Research Center for Aging Intervention (MRCA), College of Pharmacy, Pusan National University, Gumjung-gu, Busan 609-735, South Korea; [b]Department of Physiology, The University of Texas Health Science Center at San Antonio, TX 78229-3900, USA
*E-mail: hyjung@pusan.ac.kr

## 19.1 Introduction

The important role of dietary betaine has been known for many years in many biological systems; in osmolytes for example, it protects against various cellular process associated with osmosis (Hoffmann *et al.*, 2013). Recent studies of the functional basis for betaine's nutritional benefit established it as a major source of a methyl donor that participates in the methionine-homocysteine cycle (Lawson-Yuen and Levy, 2006) and in interventions against the development of hepatic insulin resistance (Ji and Kaplowitz, 2003) and hepatic steatosis (Kathirvel *et al.*, 2010). However, the proposed protective roles of betaine against oxidative stress in the reduction of proinflammatory

Food and Nutritional Components in Focus No. 7
Betaine: Chemistry, Analysis, Function and Effects
Edited by Victor R Preedy

Published by the Royal Society of Chemistry, www.rsc.org

activities and the aging process have not been explored to date, and the experimental data are sparse.

In this chapter, we review the relevant literature, including our own data, focusing specifically on betaine's less-recognized anti-inflammatory and antioxidant roles. Our group has previously reported our findings from studies on betaine's antiaging effects in its ability to attenuate age-related oxidative stress and proinflammatory nuclear factor (NF)-κB activation (Go *et al.*, 2005, 2007). Based on these findings, we proposed that betaine restores the reduction–oxidation (redox) balance by maintaining thiol homeostasis during the aging process (Go *et al.*, 2005).

The relation between redox status and the aging process is highlighted by the significant role of NF-κB in aging-related inflammation states. Recent studies show that not only is NF-κB activation highly sensitive to the cellular redox status but that it also plays a key role in the pathogenesis of major age-related inflammatory diseases such as atherosclerosis, arthritis, and diabetes (Ginaldi *et al.*, 2005; Li and Fukagawa, 2010; Nguyen *et al.*, 2013). Discoveries on the close association between NF-κB and these redox-mediated, uncontrolled, chronic inflammatory diseases provide crucial support for the molecular inflammatory hypothesis of aging (Chung *et al.*, 2011). Additional data supporting this hypothesis were generated from studies on the antioxidative/anti-inflammatory effects of thiol molecules and flavonoids, all of which have been shown to modulate the activation of NF-κB and its target gene expression during aging (Lim *et al.*, 2012; Thakurta *et al.*, 2012).

Based on the results of these published studies, the present chapter examines data on betaine's protective mechanisms in aging models. Furthermore, from these data we propose betaine as a potential therapeutic agent targeting NF-κB in the treatment of age-related inflammatory diseases.

## 19.2 Modulation of NF-κB by Betaine in the Aging Process

### 19.2.1 Betaine as a Redox Mediator

Betaine is a well-established methyl donor that controls levels of homocysteine, which is known to have deleterious effects contributing to the development of cardiovascular disease, cancer, and neural tube defects (Lawson-Yuen and Levy, 2006). Studies demonstrated that betaine's potent oxidative stress reducing power attenuates free-radical production in rats exposed to adverse conditions, as betaine administration was shown to be effective in reducing free-radical and lipid peroxidation production in ethanol-induced hepatic steatosis (Balkan *et al.*, 2004) and isoprenaline-induced myocardial infarction (Ganesan *et al.*, 2010).

Our previous studies also revealed that betaine suppresses reactive oxygen species (ROS) production and NF-κB activation mediated by IκB kinase (IKK)/mitogen-activated protein kinases (MAPKs) in the aging kidney and aorta (Go *et al.*, 2005; Lee *et al.*, 2013). Although the precise mechanism underlying betaine's protective effect against oxidative damage-induced diseases has not been fully explored, betaine's efficacy is likely related to its influence on *S*-adenosylmethionine (SAM) levels.

SAM, synthesized from methionine, plays a central role in the metabolism of all cells as a precursor molecule involved in three main pathways: methylation, transsulfuration, and aminoprophylation as shown in Figure 19.1 (Bottiglieri, 2002). In methionine metabolism, SAM regulates the balance of concentrations between homocysteine and methionine. When methionine is needed, homocysteine is remethylated by methyltetrahydrofolate, but when methionine is in excess, SAM accelerates the catabolism of homocysteine *via* cystathionine β-synthase (CBS). In this process, CBS is an important link to the synthesis of reduced glutathione (GSH, a major cellular antioxidant) *via* cysteine and is a major degradation route for methionine. Many studies have reported betaine's protective effect against oxidative damage-induced liver disease. This effect may be associated with the restoration of SAM, which increases the supply of the substrate needed for GSH synthesis and ultimately protects cells from reactive metabolites and ROS (Kwon *et al.*, 2009). Data from our laboratory revealed

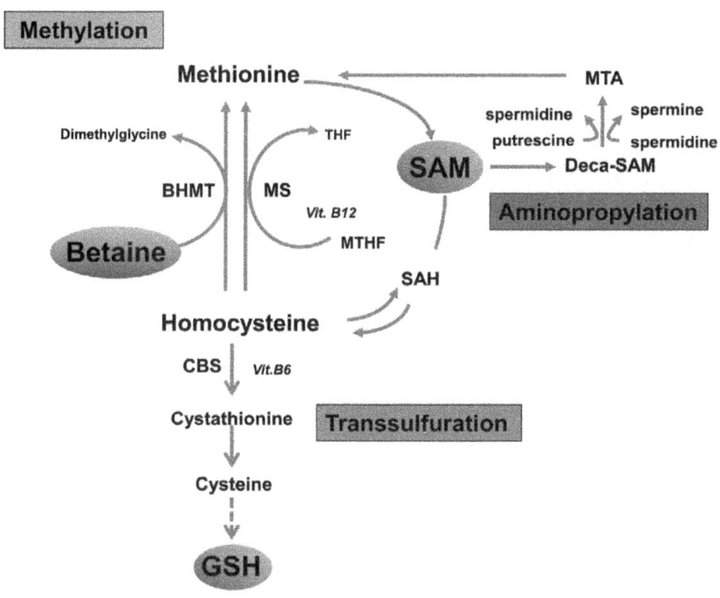


**Figure 19.1** *S*-adenosylmethionine metabolism. *S*-adenosylmethionine (SAM) is a metabolite of methionine metabolism and is metabolized by three pathways: methylation, transsulfuration, and aminopropylation. BHMT, betaine-homocysteine methyltransferase; CBS, cystathione β-synthase; GSH, reduced glutathione; MS, methionine synthase; MTA, 5′-methylthioadenosine; MTHF, methylenetetrahydrofolate; SAH, *S*-adenosyl homocysteine; SAM, *S*-adenosylmethionine; deca-SAM, decarboxylated *S*-adenosyl methionine; THF, tetrahydrofolate; vit, vitamin.

the possibility that the inhibitory effect of betaine on NF-κB activation and the expression of its target genes is due to its ability to maintain thiol homeostasis in aged rat kidney and kidney cells (Go *et al.*, 2007).

Glutathione is the most abundant low molecular weight thiol and plays major roles in both maintaining cellular redox status and protecting cells from oxidative attacks. Glutathione exists in both reduced (GSH) and oxidized (GSSG) forms, and the GSH/GSSG ratio is a reliable indicator of the cellular redox state, because glutathione concentrations are higher than those of other cellular redox systems, including the nicotinamide adenine dinucleotide phosphate system (NADPH/$NADP^+$) and the thioredoxin couplet (Trx$(SH)_2$/TrxSS). GSH is one of the most important ROS scavengers, and the GSH/GSSG ratio may be used as a biomarker of oxidative stress (Zitka *et al.*, 2012). Disturbance of the GSH redox status affects various biological processes, including the activation of redox-sensitive transcriptional factors, redox-regulated signaling pathways, inflammation, apoptosis, and cell proliferation (Rahman and MacNee, 2000). Among these processes, the activation of the redox-sensitive transcription factor NF-κB is linked to aging and age-related inflammatory diseases (Kumar *et al.*, 2004; Chung *et al.*, 2011).

### 19.2.2 Sensitivity of NF-κB to Redox Changes

NF-κB is a ubiquitous transcription factor that plays a crucial role in regulating inflammation, immunity, cell survival, and apoptosis (Kumar *et al.*, 2004). Dysregulation and chronic activation of NF-κB have been implicated in aging and age-related inflammatory diseases, such as atherosclerosis, asthma, arthritis, cancer, diabetes, inflammatory bowel disease, muscular dystrophy, and Alzheimer's disease (Kumar *et al.*, 2004; Chung *et al.*, 2011).

NF-κB is formed by the hetero- or homodimerization of NF-κB family proteins: p50, p52, p65 (RelA), c-Rel, and RelB. It exists in an inactive form in the cytoplasm and requires activation signaling that can be triggered by a variety of extracellular stimuli, including cytokines, growth factors, stress-inducing agents, and bacterial and viral pathogens. These activation signals lead to the phosphorylation and subsequent proteasome-mediated degradation of inhibitory molecules, that is, inhibitor of NF-κB (IκB) proteins (Bonizzi and Karin, 2004). This process then leads to NF-κB translocation into the nucleus, which initiates the expression of multiple proinflammatory target genes, including cytokines, chemokines, adhesion molecules, antiapoptotic factors, and enzymes such as inducible nitric oxide synthase (iNOS) and cyclo-oxygenase (COX)-2 (Bonizzi and Karin, 2004).

A unique characteristic of NF-κB is its extreme sensitivity to oxidative stress (Bowie and O'Neill, 2000). Several studies report that hydrogen peroxide, ultraviolet light, ionizing radiation, and oxidized lipids activate the cytoplasmic NF-κB complex and increase the nuclear binding activity of NF-κB during aging (Chung *et al.*, 2011). Several studies show that many sites involved in NF-κB signaling are redox-regulated, such as the following (Figure 19.2) (Janssen-Heininger,*et al.*, 2000; Jung *et al.*, 2009): (a) Two serine residues of IκBα are phosphorylated by IκB kinases β (IKKβ), leading to ubiquitination and degradation by the 26S proteasome. Degradation of IκBα enables the nuclear localization of NF-κB, where it binds to DNA. (b) Phosphorylation of IκBα at tyrosine 42 causes dissociation of IκBα without degradation *via* the proteasome pathway. NF-κB is released from phosphorylated IκBα and translocates into the nucleus. Phosphorylated IκBα interacts with phosphoinositide-3-kinase (PI3K). (c) The p65 subunit of NF-κB is phosphorylated by kinases in the cytoplasm (*e.g.*, protein kinase A [PKA], casein kinase II [CKII], and IKKα/β) or in the nucleus (*e.g.*, mitogen- and stress-activated kinase-1 [MSK1] and IKKα). Phosphorylation of p65 enhances the recruitment of the p300/CREB-binding protein (CBP) and p300/CBP-associated factor (PCAF) transcriptional coactivators, which, in turn, acetylate both p65 and the surrounding histones. These post-translational events contribute to the activation of NF-κB-regulated genes (d) ROS increase protein tyrosine kinase (PTK) activity but decrease protein tyrosine phosphatases (PTP) activity, resulting in a PTK/PTP imbalance. The PTK/PTP imbalance increases phosphorylation of NIK/IKK and MAPKs, leading to NF-kB activation. (e) NF-κB contains thiol groups (*e.g.*, cysteine62 on p50) that are critical for binding to DNA. Upon stimulation, NF-κB containing an oxidized cysteine62 in the p50 subunit translocates into the nucleus. TRX and Ref-1 directly or synergistically reduce the oxidized cysteine and enhance the DNA-binding activity of NF-κB (Figure 19.2).

### 19.2.3 Changes in NF-κB by Thiol Status

NF-κB activation is modulated by the intracellular thiol redox status. Several reports have shown that NF-κB is enhanced by GSH-depleting agents but is inhibited by antioxidants.

Treatment with α-lipoic acid, an antioxidant, has been found to suppress NF-κB activation induced by tumor necrosis factor α (TNFα) or by phorbol ester in human T cells (Paker *et al.*, 1995). This study demonstrated that pretreatment of human Jurkat cells with α-lipoic acid resulted in an increase in intracellular glutathione levels. Other reducing thiol agents also influence NF-κB activation. For example, pyrrolidine dithiocarbamate attenuates hippocampal matrix metallopeptidase-9 (MMP-9) activation through NF-κB inactivation and improves cognitive deficits in streptozotocin-induced diabetic rats (Zhao *et al.*, 2013). *N*-acetylcysteine, a precursor in the formation of GSH, has been found to inhibit the activation of NF-κB induced by various stimuli *in vitro* and *in vivo* (Yi *et al.*, 2013). Dithiolethione, a thiol antioxidant with chemopreventive and proglutathione properties, is also reported to inhibit the DNA-binding activity of NF-κB through a covalent reaction in breast cancer cells (Switzer *et al.*, 2012).

Our observations of betaine treatment are consistent with these reports (Go *et al.*, 2007). Betaine suppressed NF-κB activation through the enhancement of the thiol status in HEK 293T cells. Our data also show that betaine treatment restored the loss of intracellular GSH levels when challenged by the GSH depletion agents, dimethyl maleate and buthionine sulfoximine. In addition, our *in vitro* study using HEK293T cells to investigate redox maintenance reflects betaine's suppression of increased ROS production.

Therefore, the exogenous dietary betaine and other thiol antioxidants may be useful investigative tools in the regulation of endogenous glutathione systems to elucidate redox-sensitive gene expression.

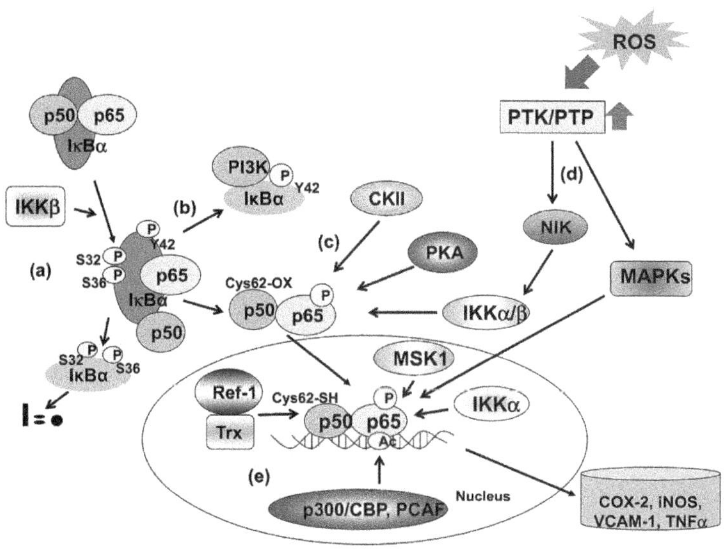


**Figure 19.2** Schematic pathways in NF-κB signaling that are redox regulated. Many sites in NF-κB signaling are regulated by the redox state: (a) Phosphorylation of IκBα (serine 32 and 36) by IKKβ leads to the ubiquitination and proteasomal degradation of IκBα. (b) Phosphorylation of IκBα at tyrosine 42 causes dissociation of the NF-κB and IκBα complex without degradation *via* the proteasome pathway. (c) The p65 subunit of NF-κB is modified by phosphorylation of PKA, CKII, and IKKα/β in the cytosol or MSK1 and IKKα in the nucleus. Phosphorylation of p65 enhances transcriptional activity by acetylation of the transcriptional coactivators p300/CBP and PCAF. (d) ROS increases PTK activity but decreases PTP activity, thereby causing PTK/PTP imbalance. Subsequently, the PTK/PTP imbalance increases phosphorylation of NIK/IKK and MAPKs, leading to NF-kB activation. (e) TRX and Ref-1-mediated reduction of p50 in the nucleus enhances the DNA-binding activity of NF-κB. CKII, caseine kinase II; COX-2, cyclo-oxygenase-2; p300/CBP, p300/CREB-binding protein Cys, cysteine; IKKα/β, IκB kinase α/β; iNOS, inducible nitric oxide synthase; MSK1, mitogen and stress-activated kinase 1; NF-κB, nuclear factor-κB; NIK, NF-κB inducing kinase; P, phosphorylation; PCAF, p300/CBP-associated factor; PKA, protein kinase A; PTK, protein tyrosine kinase; PTP, protein tyrosine phosphatase; Ref-1, redox effector factor-1; S, serine; TNFα, tumor necrosis factor α; TRX, thioredoxin; VCAM-1, vascular cell adhesion molecule-1.

### 19.2.4 Activation of NF-κB during Aging

The evidence is strong that increased oxidative stress is one of the most important contributors in accelerating the aging process. Numerous studies show clear evidence of age-related alterations in structure and function at the molecular, cellular, tissue, and organ levels as a result of oxidative stress (Rattan, 2006). Molecular studies of oxidative stress have identified alterations in gene regulation during aging, thus providing strong evidence for the effects of oxidative stress at the genetic level (Kim *et al.*, 2002). One of the major gene-modulation systems is triggered by the NF-κB signaling pathway. Previous studies report that the upregulation of NF-κB activity is accompanied by increased ROS production during aging and that a ROS-mediated imbalance in PTK/PTP leads to phosphorylation and activation of NF-κB signaling, including NF-κB-inducing kinase (NIK)/IKKα/β, and MAPKs (Kim *et al.*, 2002).

Moreover, activation of NF-κB leads to the expression of proinflammatory genes, such as TNFα, interleukin (IL)-6, iNOS, vascular cell adhesion molecule (VCAM)-1, and COX-2 during aging, and thus, the upregulation of inflammatory genes leads to a chronic proinflammatory state (Chung *et al.*, 2011) (Figure 19.3). Conversely, various antioxidants (*e.g.*, betaine, kaempferol, and baicalin) are shown to modulate the proinflammatory state involving activation of NF-κB by maintaining redox balance during aging and oxidative stress-related inflammatory diseases, suggesting that antioxidant therapies may be beneficial (Park *et al.*, 2009; Lim *et al.*, 2012; Lee *et al.*, 2013).

## 19.3 Protective Roles of Betaine Against Age-Related Changes

Betaine may exert its anti-inflammatory effects through its ability to modulate NF-κB signaling during aging. The possible regulatory mechanisms of betaine are described in the following sections.

### 19.3.1 Attenuation of Oxidative Stress-Induced NF-κB Pathway in the Aged Kidney

Due to its high level of metabolic activity, the kidney is vulnerable to ROS damage and is sensitive to inflammatory responses during aging (Ozbek, 2012). For illustrative purposes, we detail our own data on betaine's anti-inflammatory effect on NF-κB in the aged kidney (Go *et al.*, 2005). We investigated betaine's effect on oxidative stress-induced NF-κB signaling at the molecular level using two models. First, we fed betaine (3, 6, or 12 mg $kg^{-1}$) to aged Fischer 344 rats (21 months old) for 10 days; secondly, we applied betaine to YPEN-1 cells (a rat endothelial cell line) *in vitro*.

The results show that in aged kidney, betaine treatment reduced NF-κB activation and the expression of its target genes (COX-2, iNOS, and VCAM-1) compared to the non-fed betaine group. Furthermore, in YPEN-1 cells,

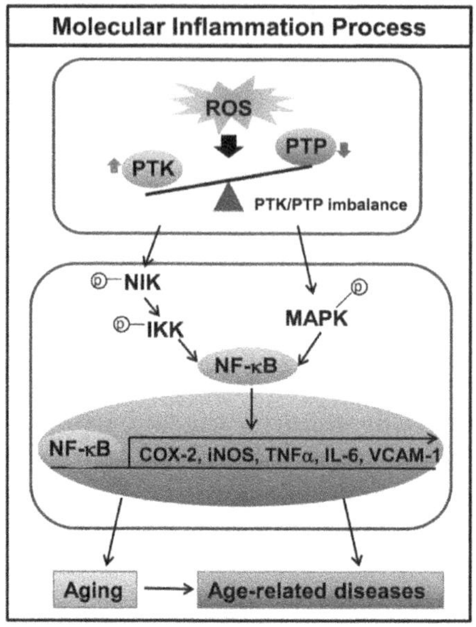


**Figure 19.3** A schematic illustration of the activation of NF-κB during aging. Age-related ROS can increase PTK activity but decrease PTP activity, indicating a PTK/PTP imbalance. Subsequently, the PTK/PTP imbalance stimulates phosphorylation of MAPK and NIK/IKK α/β, leading to NF-κB activation. Upregulation of NF-κB leads to the expression of inflammatory genes, COX-2, iNOS, adhesion molecules (VCAM-1) and cytokines (TNFα, IL-6) and thus, aggravating aging and age-related inflammatory diseases. COX-2, cyclooxygenase-2; IKKα/β, IκB kinase α/β; IL-6, Interleukin-6; iNOS, inducible nitric oxide synthase; MAPK, mitogen-activated protein kinase; NF-κB, nuclear factor-κB; NIK, NF-κB inducing kinase; P, phosphorylation; PTK, protein tyrosine kinase; PTP, protein tyrosine phosphatase; ROS, reactive oxygen species; TNFα, tumor necrosis factor α; VCAM-1, vascular cell adhesion molecule-1.

betaine attenuated oxidative stress-induced NF-κB *via* NIK/IKK and MAPKs pathways. Based on these results, we concluded that betaine suppressed age-related NF-κB activation through the downregulation of oxidative stress-induced NIK/IKK and MAPKs.

### 19.3.2 Suppression of Adhesion Molecule Expression

Lysophosphatidylcholine (LPC) is a major phospholipid component of oxidized low-density lipoprotein and has been implicated as a critical factor in the development of vascular diseases (Zou *et al.*, 2009). Recently, our group showed that LPC is a potent mediator of endothelial dysfunction by regulating the expression of adhesion molecules (AMs) during aging (Zou *et al.*, 2009). More broadly, we demonstrated betaine's protective effects against LPC-induced inflammation in endothelial cells by reducing AM expression and suppressing the NF-κB signaling pathway during aging (Lee *et al.*, 2013).

These experiments were performed using young (7 month-old) and old (21 month-old) rats; two groups of old rats were fed betaine (3 or 6 mg $kg^{-1}$ per day for 10 days). To investigate betaine's anti-inflammatory effects in aged vasculature, we used aged rat aorta and YPEN-1 cells. Betaine inhibited the expression of LPC-related AMs, VCAM-1, and P-selectin in both the serum and tissue of aged rats without affecting the levels of LPC in the serum. Importantly, betaine also prevented ROS generation, thereby maintaining the redox balance by enhancing the thiol status during aging. Furthermore, betaine attenuated NF-κB activation *via* the dephosphorylation of IKK and MAPKs in the aged aorta and LPC-treated YPEN-1 cells. These data strongly suggest that dietary betaine supplementation may protect against chronic vascular inflammatory disease and aging.

### 19.3.3 Enhancement of Thiol Status in Attenuating Age-Related NF-κB

Redox homeostasis plays a critical role in protecting cells from both internal and external oxidative stress and other forms of cellular stress, while maintaining a regulatory role for redox-sensitive transcription factors, such as NF-κB. Involvement of the redox status in aging process highlights the importance of NF-κB in the development of aging-related inflammatory diseases (Kumar, 2004). Based on our previous data demonstrating betaine's protective effects in the kidney and aorta, we investigated the mechanism by which betaine suppresses NF-κB during aging (Go *et al.*, 2007). To assess betaine's ability to restore the redox status during aging, we used betaine-fed aged rats (age: 21 months; betaine treatment: 3, 6, or 12 mg $kg^{-1}$ per day for 10 days) and human embryonic kidney 293T (HEK293T) cells.

In rats, betaine treatment decreased ROS generation and NF-κB phosphorylation, but increased thiol levels in the aged kidney. In addition, betaine treatment increased GSH levels and reduced ROS formation under conditions of GSH depletion; thus, betaine prevented the nuclear translocation of NF-κB and upregulation of the inflammatory genes, TNFα and COX-2. It is possible that betaine increases cellular GSH levels by serving as a methyl-group donor in the conversion of homocysteine to methionine.

Based on the betaine's thiol-enhancing action, we expect that betaine supplementation could provide much broader protective effects against various age-related oxidative insults by replenishing the thiol pool, results that warrant further indepth investigation.

## 19.4 Summary

In this review, we summarized the literature supporting the benefits of dietary betaine in its protection against the uncontrolled, chronic inflammation associated with aging. During aging, betaine suppresses NF-κB activation by modulating the cellular redox status, specifically by increasing thiol levels.

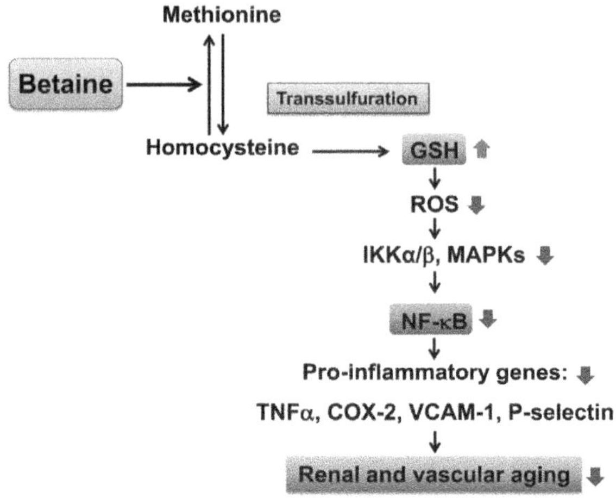


**Figure 19.4** Proposed mechanisms for the role of betaine in aging. Betaine increases GSH levels through methionine metabolism, which upregulates the transsulfuration pathway. Increased GSH inhibits ROS and suppresses NF-κB activation through IKKα/β and MAPKs. These conditions suppress NF-κB-dependent gene expression (TNFα, COX-2, VCAM-1, and P-selectin) and might attenuate renal and vascular aging. COX-2, cyclo-oxygenase-2; GSH, reduced glutathione; IKKα/β, IκB kinase α/β; MAPKs, mitogen-activated protein kinases; NF-κB, nuclear factor-κB; ROS, reactive oxygen species; TNFα, tumor necrosis factor α; VCAM-1, vascular cell adhesion molecule-1.

Betaine can transfer its one methyl group to homocysteine to form methionine. Methionine is subsequently converted to GSH, an antioxidant agent that upregulates the transsulfuration pathway. Increased GSH levels attenuate ROS, resulting in the downregulation of NF-κB activation through IKKα/β and MAPKs and subsequent attenuation of the expression of inflammatory genes, such as TNFα, COX-2, VCAM-1, and P-selectin (Figure 19.4). Data from animal and *in vitro* studies support the notion that betaine enhances the thiol status through methionine metabolism and stably maintains the cellular redox status. Further data suggest that betaine may be a useful therapeutic agent for attenuating renal and vascular aging and in the further progression of other age-related diseases.

# Summary Points

- Evidence is strong for increased oxidative stress as a leading contributor in accelerating the aging process.
- Activation of nuclear factor-κB (NF-κB) under oxidative stress conditions is implicated in aging and age-related inflammatory diseases.

- The focus of this chapter is the anti-inflammatory mechanism of betaine in experimental models of aging.
- Betaine is a well-established methyl donor that controls levels of homocysteine, which has deleterious effects that contribute to the development of cardiovascular disease, cancer, and neural tube defects
- Betaine enhances thiol levels to function as an antioxidative defense system.
- Betaine exerts its anti-inflammatory effects by modulating NF-κB signaling in aging kidney and aorta.
- Betaine treatment can regulate the age-associated reduced glutathione decrease and enhanced reactive oxygen species that in turn would prevent the translocation of nuclear NF-κB and upregulation of the inflammatory genes, tumor necrosis factor α and cyclooxygenase-2.

## Key Facts in the Development of Betaine as an Antiaging Agent

- Betaine was first discovered in the juice of sugar beets (*Beta vulgaris*) in the 19th century and is now found widely in various micro-organisms, plants, and animals.
- The biological function of betaine is proposed to protect cells under stress by supplying methyl groups needed for methionine metabolism.
- Betaine is found to protect cells by reducing oxidative stress in many tissues.
- Although the precise mechanism of the protective effect of betaine has not fully been explored, it is generally accepted that betaine's efficacy is likely related to the restoration of intracellular reduced glutathione (GSH) levels.
- GSH is an important antioxidant found in plants, animals, fungi and in some bacteria and archaea that can prevent damage to important cellular components caused by oxidative attacks.
- A deficit in GSH is associated with aging processes and age-related disease such as cardiovascular, diabetes, and neurodegenerative diseases.
- In this chapter, the importance of the antioxidative efficacy of betaine is highlighted by describing its contribution to curtail aging processes and age-related diseases.

## Definition of Words and Terms

**Aging**: A time-dependent functional decline that leads to an organism's incapacity to withstand external and internal challenges.
**Adhesion molecule (AM)**: Proinflammatory proteins that play a crucial role in cell–cell/cell–matrix interactions. It is implicitly involved in the immune response and inflammation process.

**Glutathione**: A tripeptide formed from cysteine, glutamate, and glycine. It is an intracellular thiol redox regulator that plays a major role in both the maintenance of redox status and in the protection of cells from electrophilic and oxidative attacks.

**NF-κB**: Transcription factors shown to be exquisitely sensitive to oxidative stress conditions. Most of the genes activated by NF-κB transcription factor are shown to be proinflammatory and involved in inflammation processes.

**Oxidant**: Oxidizes another chemical by taking electrons, hydrogen, or by adding oxygen ($\cdot O_2^-$, $H_2O_2$, $ONOO^-$ and $\cdot OH$, *etc.*).

**Oxidative stress**: A disturbance in the balance between the production of oxidants and antioxidant defenses that can potentially lead to damage.

**Reactive oxygen species (ROS)**: Chemically reactive molecules containing oxygen including superoxide anion ($\cdot O_2^-$), hydrogen peroxide ($H_2O_2$) and hydroxyl radical ($\cdot OH$).

**Reductant**: Reduces another chemical by supplying electrons, hydrogen, or by removing oxygen (GSH, TRX, catalase, superoxide dismutase, *etc.*).

**Redox (reduction–oxidation) status**: Balance between oxidant and reductant in the cell. In order to maintain a healthy status, oxidants and antioxidants should be in equilibrium. When this equilibrium between oxidant and antioxidant is disrupted, tilting the equilibrium toward an oxidized state, oxidative stress is produced.

**Thiol**: An organosulfur compound that contains a sulfhydryl (–SH) group. Among all the antioxidants that are available in the body, thiols constitute the major portion of the total body antioxidants and they play a significant role in defense against ROS.

## List of Abbreviations

| | |
|---|---|
| AM | Adhesion molecule |
| CBS | Cystathionine β-synthase |
| CBP | p300/CREB-binding protein |
| CK II | Casein kinase II |
| COX-2 | Cyclo-oxygenase-2 |
| GSH | Reduced glutathione |
| GSSG | Oxidized glutathione |
| IκB | Inhibitor of NF-κB |
| IKK α/β | IκB kinase α/β |
| IL-6 | Interleukin-6 |
| iNOS | Inducible nitric oxide synthase |
| LPC | Lysophosphatidylcholine |
| MAPKs | Mitogen-activated protein kinases |
| MSK1 | Mitogen- and stress-activated kinase 1 |
| MMP-9 | Matrix metalloproteinases-9 |
| NF-κB | Nuclear factor (NF)-κB |
| NIK | NF-κ B-Inducing kinase |
| PCAF | p300/CBP-Associated factor |

| | |
|---|---|
| PI3K | Phosphoinositide-3-kinase |
| PKA | Protein kinase A |
| PTK | Protein tyrosine kinase |
| PTP | Protein tyrosine phosphatases |
| Redox | Reduction–oxidation |
| ROS | Reactive oxygen species |
| SAM | *S*-Adenosylmethionine |
| TNFα | Tumor necrosis factor α |
| TRX | Thioredoxin |
| VCAM-1 | Vascular cell adhesion molecule-1 |

## Acknowledgements

This work was supported by the National Research Foundation of Korea (NRF) grant funded by the Korea government (MSIP) (No. 2009-0083538). We are grateful to the 'Aging Tissue Bank' for supplying research information.

## References

Balkan, J., Oztezcan, S., Küçük, M., Cevikbaş, U., Koçak-Toker, N. and Uysal, M., 2004. The effect of betaine treatment on triglyceride levels and oxidative stress in the liver of ethanol-treated guinea pigs. *Experimental and Toxicologic Pathology*. 55: 505–509.

Bonizzi, G. and Karin, M., 2004. The two NF-kappaB activation pathways and their role in innate and adaptive immunity. *Trends in Immunology*. 25: 280–288.

Bottiglieri, T., 2002. S-Adenosyl-L-methionine (SAMe): from the bench to the bedside–molecular basis of a pleiotrophic molecule. *American Journal of Clinical Nutrition*. 76: 1151S–1157S.

Bowie, A. and O'Neill, L. A., 2000. Oxidative stress and nuclear factor-kappaB activation: a reassessment of the evidence in the light of recent discoveries. *Biochemical Pharmacology*. 59: 13–23.

Chung, H. Y., Lee, E. K., Choi, Y. J., Kim, J. M., Kim, D. H., Zou, Y., Kim, C. H., Lee, J., Kim, H. S., Kim, N. D., Jung, J. H. and Yu, B. P., 2011. Molecular inflammation as an underlying mechanism of the aging process and age-related diseases. *Journal of Dental Research*. 90: 830–840.

Ganesan, B., Buddhan, S., Anandan, R., Sivakumar, R. and AnbinEzhilan, R., 2010. Antioxidant defense of betaine against isoprenaline-induced myocardial infarction in rats. *Molecular Biology Reports*. 37: 1319–1327.

Ginaldi, L., Di Benedetto, M. C. and De Martinis, M., 2005. Osteoporosis, inflammation and ageing. *Immunity and Ageing*. 2: 14.

Go, E. K., Jung, K. J., Kim, J. Y., Yu, B. P. and Chung, H. Y., 2005. Betaine suppresses proinflammatory signaling during aging: the involvement of nuclear factor-kappaB via nuclear factor-inducing kinase/IkappaB kinase and mitogen-activated protein kinases. *Journals of Gerontology. Series A, Biological Sciences and Medical Sciences*. 60: 1252–1264.

Go, E. K., Jung, K. J., Kim, J. M., Lim, H., Lim, H. K., Yu, B. P. and Chung, H. Y., 2007. Betaine modulates age-related NF-kappaB by thiol-enhancing action. *Biological and Pharmaceutical Bulletin*. 30: 2244–2249.

Hoffmann, L., Brauers, G., Gehrmann, T., Häussinger, D., Mayatepek, E., Schliess, F. and Schwahn, B. C., 2013. Osmotic regulation of hepatic betaine metabolism. *American Journal of Physiology. Gastrointestinal and Liver Physiology*. 304: G835–G846.

Janssen-Heininger, Y. M., Poynter, M. E. and Baeuerle, P. A., 2000. Recent advances towards understanding redox mechanisms in the activation of nuclear factor kappaB. *Free Radical Biology and Medicine*. 28: 1317–1327.

Ji, C., Kaplowitz, N., 2003. Betaine decreases hyperhomocysteinemia, endoplasmic reticulum stress, and liver injury in alcohol-fed mice. *Gastroenterology*. 124: 1488–1499.

Jung, K. J., Lee, E. K., Yu, B. P. and Chung, H. Y., 2009. Significance of protein tyrosine kinase/protein tyrosine phosphatase balance in the regulation of NF-kappaB signaling in the inflammatory process and aging. *Free Radical Biology and Medicine*. 47: 983–991.

Kathirvel, E., Morgan, K., Nandgiri, G., Sandoval, B. C., Caudill, M. A., Bottiglieri, T., French, S. W. and Morgan, T. R., 2010. Betaine improves nonalcoholic fatty liver and associated hepatic insulin resistance: a potential mechanism for hepatoprotection by betaine. *American Journal of Physiology: Gastrointestinal and Liver Physiology*. 299: G1068–G1077.

Kim, H. J., Yu, B. P. and Chung, H. Y., 2002. Molecular exploration of age-related NF-kappaB/IKK downregulation by calorie restriction in rat kidney. *Free Radical Biology and Medicine*. 32: 991–1005.

Kumar, A., Takada, Y., Boriek, A. M. and Aggarwal, B. B., 2004. Nuclear factor-kappaB: its role in health and disease. *Journal of Molecular Medicine*. 82: 434–448.

Kwon, D. Y., Jung, Y. S., Kim, S. J., Park, H. K., Park, J. H. and Kim, Y. C., 2009. Impaired sulfur-amino acid metabolism and oxidative stress in nonalcoholic fatty liver are alleviated by betaine supplementation in rats. *Journal of Nutrition*. 139: 63–68.

Lawson-Yuen, A. and Levy, H. L., 2006. The use of betaine in the treatment of elevated homocysteine. *Molecular Genetics and Metabolism*. 88: 201–207.

Lee, E. K., Jang, E. J., Jung, K. J., Kim, D. H., Yu, B. P. and Chung, H. Y., 2013. Betaine attenuates lysophosphatidylcholine-mediated adhesion molecules in aged rat aorta: modulation of the nuclear factor-κB pathway. *Experimental Gerontology*. 48: 517–524.

Li, M., Fukagawa, N. K., 2010. Age-related changes in redox signaling and VSMC function. *Antioxidants and Redox Signaling*. 12: 641–655.

Lim, H. A., Lee, E. K., Kim, J. M., Park, M. H., Kim, D. H., Choi, Y. J., Ha, Y. M., Yoon, J. H., Choi, J. S., Yu, B. P. and Chung, H. Y., 2012. PPARγ activation by baicalin suppresses NF-κB-mediated inflammation in aged rat kidney. *Biogerontology*. 13: 133–145.

Nguyen, D., Samson, S. L., Reddy, V. T., Gonzalez, E. V. and Sekhar, R. V., 2013. Impaired mitochondrial fatty acid oxidation and insulin resistance in aging: novel protective role of glutathione. *Aging Cell*. 12: 415–425.

Ozbek, E., 2012. Induction of oxidative stress in kidney. *International Journal of Nephrology*. 2012: 465897.

Packer, L., Witt, E. H. and Tritschler, H. J., 1995. alpha-Lipoic acid as a biological antioxidant. *Free Radical Biology and Medicine*. 19: 227–250.

Park, M. J., Lee, E. K., Heo, H. S., Kim, M. S., Sung, B., Kim, M. K., Lee, J., Kim, N. D., Anton, S., Choi, J. S., Yu, B. P. and Chung, H. Y., 2009. The anti-inflammatory effect of kaempferol in aged kidney tissues: the involvement of nuclear factor-kappaB via nuclear factor-inducing kinase/IkappaB kinase and mitogen-activated protein kinase pathways. *Journal of Medicinal Food*. 12: 351–358.

Rahman, I. and MacNee, W., 2000. Regulation of redox glutathione levels and gene transcription in lung inflammation: therapeutic approaches. *Free Radical Biology and Medicine*. 28: 1405–1420.

Rattan, S. I., 2006. Theories of biological aging: genes, proteins, and free radicals. *Free Radical Research*. 40: 1230–1238.

Switzer, C. H., Cheng, R. Y., Ridnour, L. A., Murray, M. C., Tazzari, V., Sparatore, A., Del Soldato, P., Hines, H. B., Glynn, S. A., Ambs, S. and Wink, D. A., 2012. Dithiolethiones inhibit NF-κB activity via covalent modification in human estrogen receptor-negative breast cancer. *Cancer Research*. 72: 2394–2404.

Thakurta, I. G., Chattopadhyay, M., Ghosh, A. and Chakrabarti, S., 2012. Dietary supplementation with N-acetyl cysteine, α-tocopherol and α-lipoic acid reduces the extent of oxidative stress and proinflammatory state in aged rat brain. *Biogerontology*. 13: 479–488.

Yi, D., Hou, Y., Wang, L., Ding, B., Yang, Z., Li, J., Long, M., Liu, Y. and Wu, G., 2014. Dietary N-acetylcysteine supplementation alleviates liver injury in lipopolysaccharide-challenged piglets. *British Journal of Nutrition*. 111: 46–54.

Zhao, Z., Huang, G., Wang, B. and Zhong, Y., 2013. Inhibition of NF-kappaB activation by Pyrrolidine dithiocarbamate partially attenuates hippocampal MMP-9 activation and improves cognitive deficits in streptozotocin-induced diabetic rats. *Behavioural Brain Research*. 238: 44–47.

Zitka, O., Skalickova, S., Gumulec, J., Masarik, M., Adam, V., Hubalek, J., Trnkova, L., Kruseova, J., Eckschlager, T. and Kizek, R., 2012. Redox status expressed as GSH: GSSG ratio as a marker for oxidative stress in paediatric tumour patients. *Oncology Letters*. 4: 1247–1253.

Zou, Y., Kim, D. H., Jung, K. J., Heo, H. S., Kim, C. H., Baik, H. S., Yu, B. P., Yokozawa, T. and Chung, H. Y., 2009. Lysophosphatidylcholine enhances oxidative stress via the 5-lipoxygenase pathway in rat aorta during aging. *Rejuvenation Research*. 12: 15–24.

CHAPTER 20

# Betaine in Athletic Performance and Body Composition

STUART A. S. CRAIG*[a]

[a]DuPont Nutrition & Health, 660 White Plains Rd, Suite 610, Tarrytown, NY 10589, USA
*E-mail: stuart.craig@dupont.com

## 20.1 Introduction

Betaine plays several important physiological roles as an organic osmolyte and methyl donor in micro-organisms, plants and animals (Craig, 2004). The physiological function of betaine is regulated by the enzyme betaine homocysteine methyl transferase (BHMT). Downregulation of BHMT conserves existing intracellular betaine to function as an osmolyte, whereas upregulation catalyzes methyl (one-carbon) group donation in the methionine cycle. In a landmark publication in "Science", Yancey and coworkers established that there are osmolyte systems in almost all water-stressed organisms and that osmolytes are compatible with macromolecular structure and function, even at high concentrations (Yancey *et al.*, 1982). Osmolytes maintain cellular function by reducing the need for protein modification when intracellular contents become concentrated, and by counteracting perturbing solutes (*e.g.* urea, salts, *etc.*) without interacting directly with substrates and cofactors (Somero and Yancey, 2010). Alternatively, methyl groups produced from betaine *via* BHMT contribute to over a hundred different reactions essential for normal cell function. These include synthesis of *S*-adenosyl methionine, creatine, carnitine, epinephrine, and DNA methylation (Craig, 2004).

Food and Nutritional Components in Focus No. 7
Betaine: Chemistry, Analysis, Function and Effects
Edited by Victor R Preedy

Published by the Royal Society of Chemistry, www.rsc.org

Betaine is a significant nutrient in the human diet with several foods being rich dietary sources including grains (*i.e.* wheat, rye, quinoa), spinach, shellfish, and beets. However, modern diets may be lacking in betaine and other osmolytes or methyl donors (Likes *et al.*, 2007). With that in mind, a series of studies were conducted to determine whether dietary supplementation of betaine could improve human physical performance. The first studies investigated the acute effect of betaine primarily as an osmolyte. Betaine can regulate the movement of water and electrolytes across the intestine and can influence intracellular and extracellular osmotic forces that may counteract stress during exercise. The protocols looked at exercise in high heat and humidity, which is relevant to those seeking to counteract the detrimental effects of dehydration and environmental heat stress on exercise performance, such as athletes, soldiers, construction workers, *etc*. This early evidence was sufficiently compelling to support a US Patent (Craig, 2010). Later studies focused more on chronic intake and measures of strength, power and endurance that may utilize both the osmolyte and methyl-donor properties of betaine. The following describes the current body of evidence on the ergogenic effects of betaine.

## 20.2 Human Performance Studies

### 20.2.1 Acute Dose Studies

The first ergogenic study conducted (Armstrong *et al.*, 2008) investigated betaine supplementation in rehydration fluids (with and without carbohydrate and electrolytes) for ergogenic effect during exercise in the heat. The design was a double-blind, placebo-controlled, randomized, crossover with 10 physically active college male runners. They were dehydrated to −2.7% of body weight, then partially rehydrated to −1.4% of body weight 90 min prior to a controlled exercise test using one of four different rehydration solutions (1L): (a) flavored noncaloric water (W); (b) flavored noncaloric water + betaine (5 g) (W + B); (c) carbohydrate–electrolyte drink (C); and (d) carbohydrate–electrolyte drink + betaine (5 g) (C + B). Subjects ran on a treadmill in an environmental chamber (31.1 °C, 88.0 °F) for 75 min at 65% $VO_2$ max followed by timed sprint until exhaustion at 84% $VO_2$ max. The results showed that rehydration with fluids containing betaine resulted in significant differences in plasma volume, oxygen consumption, plasma lactate concentration, and thermal sensation. Most other measures showed no differences, but there was a nonsignificant trend of longer sprint times to exhaustion for betaine trials when compared with nonbetaine trials (carbohydrate beverage 16% longer; water beverage 21% longer), see Figure 20.1.

Therefore, this study was supportive of an ergogenic benefit of betaine, and the authors suggested that the observed increases of both aerobic and anaerobic metabolism warranted further investigation of betaine as a nutritional supplement using other types of exercise.

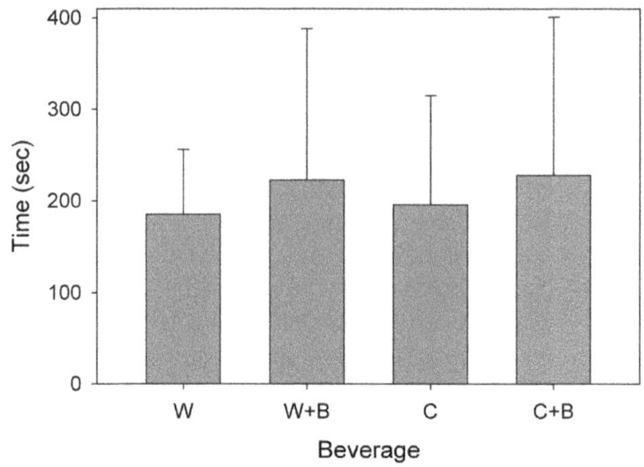


**Figure 20.1** Mean sprint duration to volitional exhaustion at 84% $VO_2$ max (following 75 min of continuous running (65% $VO_2$max) in the heat). Subjects who took rehydration beverages containing betaine (compared to water or carbohydrate–electrolyte beverages) tended to be able to sprint for a longer time until exhaustion, although the difference was not statistically significant.

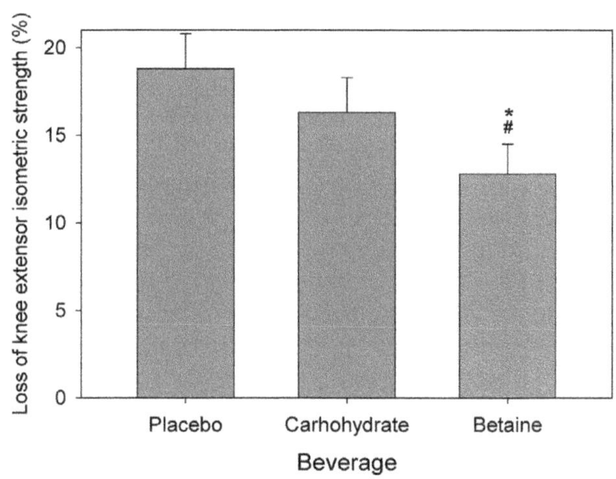


**Figure 20.2** Loss of knee extensor isometric strength following cycling in the heat. * = Significantly different ($p < 0.05$) between betaine and carbohydrate. # = Significantly different ($p < 0.05$) between betaine and placebo. Subjects who took beverages containing betaine (compared to placebo or carbohydrate beverages) had less muscle fatigue following exercise.

A subsequent study (Millard-Stafford *et al.*, 2005) also looked at the effects of betaine on fluid replacement in the heat. They conducted a double-blind, placebo-controlled, randomized, cross-over study with 8 trained cyclists. All subjects participated in three exercise tests using a different drink each time – placebo, carbohydrate–electrolyte drink and carbohydrate–electrolyte drink plus 6 g betaine (0.5% w/v). The drinks were consumed in several doses before and during the trial. Cycling in the heat for 120 min was followed by a 15 min time trial and muscle tests. Betaine did not improve performance or hydration status during cycling, but total work performed during the time trial tended to be higher for betaine (10%) and carbohydrate (14%) groups compared to placebo. Betaine significantly improved maintenance of muscular strength following exercise in the heat compared to both placebo and carbohydrate groups, see Figure 20.2.

Therefore, this study was the first to show an improvement in an aspect of muscle strength/fatigue due to betaine supplementation.

### 20.2.2 Chronic Dose Studies

The demonstration of improvement in muscular strength (Millard-Stafford *et al.*, 2005) led to a series of studies to confirm and expand on that finding. In addition, studies were conducted with lower levels of betaine (2.5 g) but ingested over a longer span of time (1–6 weeks) to ensure a homeostatic concentration in tissues. The next study (Lee *et al.*, 2010) was conducted on the ergogenic effects of betaine supplementation on strength and power performance. Twelve resistant-trained college males completed two 14 day experimental trials separated by a 14 day washout period, in a balanced, randomized, double-blind, repeated measures, crossover design. Subjects completed two consecutive days of a standardized high intensity strength/power resistance exercise challenge prior to (Days 1 and 2) and following 14 days (Days 14 and 15) of twice-daily placebo or betaine supplementation. This two-day exercise challenge was conducted to determine whether betaine might provide a performance benefit on consecutive days and improve recovery. Performance included bench, squat, and jump tests. For power tests, bench throw power was increased on D14 and D15 (betaine supplementation) compared to D1 and D2 (presupplementation) and corresponding placebo, see Figure 20.3. There were no differences in power output measures for the four vertical jumps combined. However, there were significant increases in power output for jumps 3 and 4 performed on D14 and D15 (betaine supplementation) compared to corresponding D1 and D2 (presupplementation). For strength tests, isometric bench press force was increased on D14 and D15 (betaine supplementation) compared to D1 and D2 (presupplementation) and corresponding placebo, see Figure 20.4. Isometric squat force was also greater on D14 and D15 (betaine supplementation) compared to presupplementation. There were no differences in jump squat power or the number of bench press or squat repetitions. Therefore, this study supported the previous finding of improved strength by demonstrating that betaine supplementation increased force, in addition to demonstrating improved power, maintenance and recovery.

The next study (Hoffman *et al.*, 2009) examined betaine supplementation (15 days) on muscle endurance, power performance and rate of fatigue in active college-aged men. Twenty-four male subjects were randomly assigned

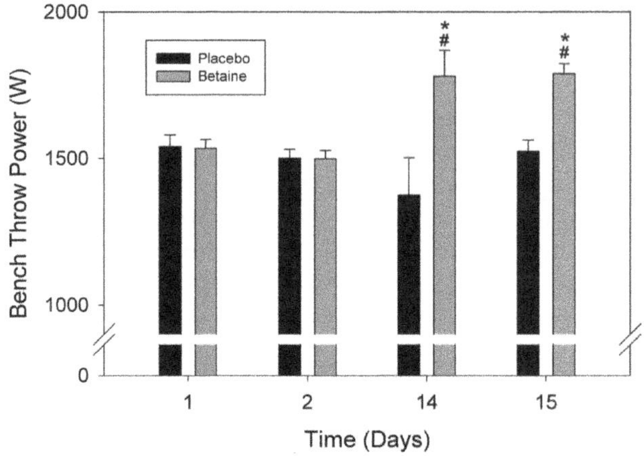


**Figure 20.3** Bench throw power. * = Significantly different ($p < 0.05$) between post- and presupplementation. # = Significantly different ($p < 0.05$) between betaine and placebo. Subjects who took beverages containing betaine (compared to placebo beverages) had more power in the bench throw exercise.

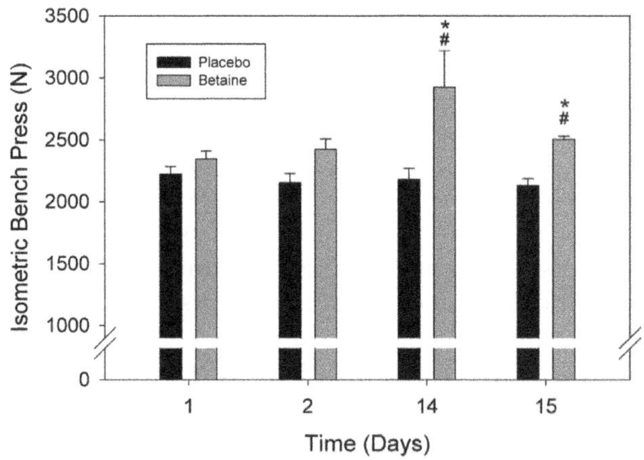


**Figure 20.4** Isometric bench press force. * = Significantly different ($p < 0.05$) between post- and presupplementation. # = Significantly different ($p < 0.05$) between betaine and placebo. Subjects who took beverages containing betaine (compared to placebo beverages) had more strength in the isometric bench press exercise.

into two groups where they consumed either placebo (240 mL of sports drink) or the supplement (1.25 g of betaine mixed in 240 mL of a sport drink) twice per day. Subjects were tested prior to the onset of supplementation (*i.e.* 0 days) and 7 and 14 days following supplementation. Tests included vertical jump power, bench press throw power, squat and bench press maximum repetitions at 75% of 1-RM and 30 s Wingate anaerobic power tests. No difference in any power assessment or bench press repetitions was seen between the groups. For squat, the number of repetitions performed for betaine *versus* placebo was 30% greater at 7 days ($p < 0.05$) and 47% greater at 14 days ($p = 0.06$), see Figure 20.5. Also, the number of repetitions performed at ≥90% of peak power in the squat exercise was significantly greater for betaine at both 7 days and 14 days compared to placebo, see Figure 20.6.

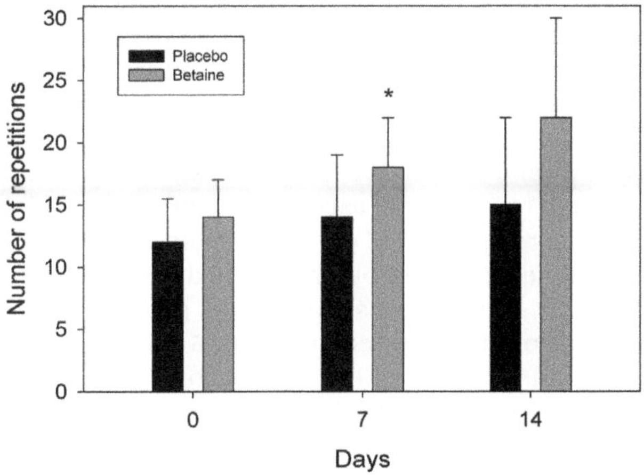


**Figure 20.5** Total number of squat repetitions. * = Significantly different ($p < 0.05$) between betaine and placebo. Subjects who took beverages containing betaine (compared to placebo beverages) performed more squat repetitions at 7 and 14 days, which is a measure of improved muscle endurance or strength.

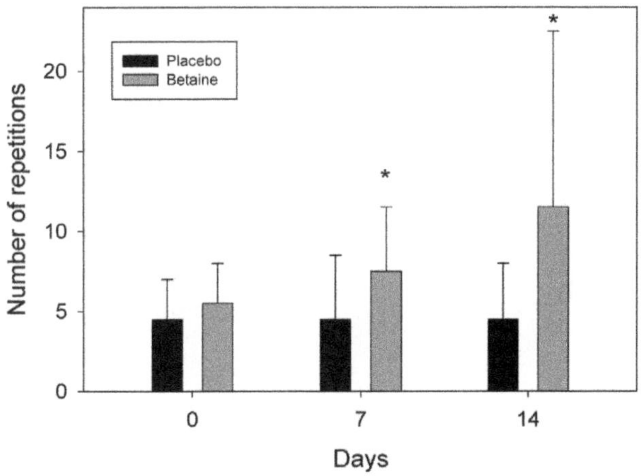


**Figure 20.6** Total number of squat repetitions performed at ≥90% of peak power. b: Total number of squat repetitions performed at ≥90% of mean power. * = Significantly different ($p < 0.05$) between betaine and placebo. Subjects who took beverages containing betaine (compared to placebo beverages) performed more squat repetitions at ≥90% of mean power at 7 and 14 days, which is a measure of workout quality.

Therefore, one and two weeks of betaine supplementation in active, college males improved muscle endurance of the squat exercise, and increased the quality of repetitions performed. This study also supported previous findings on strength, but did not find improvements in power. It was the first to show how betaine improves workout quality (number of repetitions at ≥90% peak power).

Further work by this group (Hoffman *et al.*, 2011) examined 15 days of betaine supplementation on peak concentric and eccentric force during isokinetic exercise. Eleven active college-aged men were randomly assigned into two groups where they consumed either placebo (500 mL of sports drink) or supplement (2.5 g of betaine in 500 mL of sports drink) each day for 15 days. Subjects reported for a testing and isokinetic chest press exercise on 5 occasions during this period, separated by 72 h. After each exercise protocol, subjects rated their fatigue and muscle soreness on a 15 cm visual analog scale. Subjects then consumed no daily betaine for 4 weeks but maintained familiarity with the exercise device once per week. After the washout period, subjects resumed the study using the opposite drink and repeated the same 15 day protocol. No differences were noted in maximum concentric or eccentric force output between pre- and postworkouts in both placebo and betaine. No differences in subjective measures of soreness and fatigue were seen, but a significant reduction in Δ fatigue was observed in betaine compared to placebo, see Figure 20.7.

Therefore, 15 days of betaine supplementation did not increase peak concentric or eccentric force outputs as measured using an isokinetic dynamometer. However, betaine did reduce subjective measures of fatigue to the exercise protocol and was the first to observe this improvement using a visual analog scale.

The next study (Trepanowski *et al.*, 2011) looked at the effects of chronic betaine supplementation on exercise performance, skeletal muscle oxygen

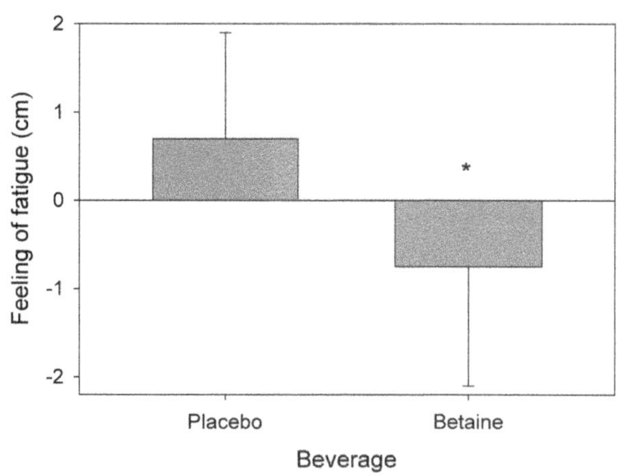


**Figure 20.7** Δ Subjective measures of fatigue. * = Significantly different ($p < 0.05$) between betaine and placebo. Subjects who took beverages containing betaine (compared to placebo beverages) had less fatigue from the exercise protocol.

saturation, and associated biochemical parameters. Resistance-trained men were randomly assigned in a double-blind manner using a crossover design to consume betaine (2.5 g of betaine mixed in 500 mL of Gatorade or a placebo (500 mL of Gatorade) for 14 days, with a 21 day washout period. Before and after each treatment period, tests of lower- and upper-body muscular power and isometric force were conducted, including a test of upper-body muscular endurance (10 sets of bench-press exercise to failure). Muscle tissue oxygen saturation during the bench-press protocol was measured by near-infrared

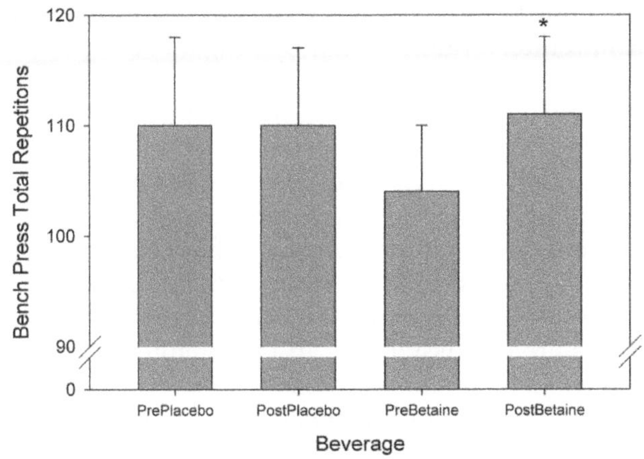


**Figure 20.8** Bench-press total repetitions. * = Significantly different ($p < 0.05$) between post- and prebetaine. Subjects who took beverages containing betaine (compared to preintervention) performed more bench-press repetitions.

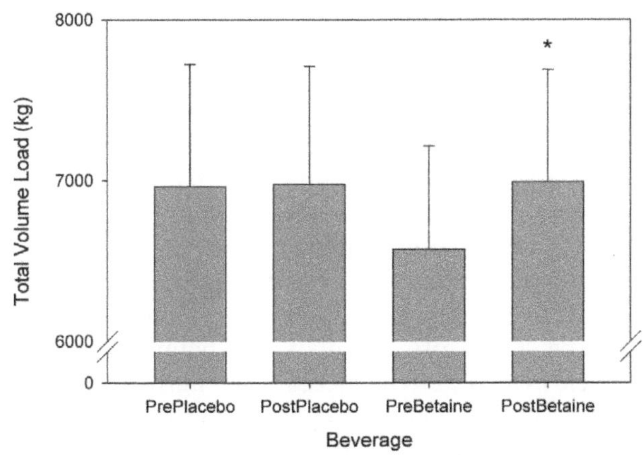


**Figure 20.9** Bench-press total volume load. * = Significantly different ($p < 0.05$) between post- and prebetaine. Subjects who took beverages containing betaine (compared to preintervention) performed more bench-press total volume load.

spectroscopy and blood samples were collected before and after exercise for analysis of lactate, nitrate/nitrite, and malondialdehyde. No significant differences were noted between conditions for exercise performance variables. However, betaine supplementation led to an increase in total repetitions, see Figure 20.8 and total volume load, see Figure 20.9, in the 10-set bench press protocol of about 6.5% comparing preintervention to postintervention.

Malondialdehyde was relatively unchanged, postexercise blood lactate increased to a lesser extent with betaine supplementation (210%) compared with placebo administration (270%), although not of statistical significance ($p = 0.14$), and there was a trend ($p = 0.06$) for lower $NO_x$ postintervention compared with preintervention. When pooling all sets, there was a slightly higher $StO_2$ value after 2 weeks of betaine supplementation and lower values at the end of exercise during betaine supplementation. This translated into a greater $StO_2$ difference, and suggests enhanced muscle oxygen consumption. This was supported by the increased repetitions/volume load and a slight (but nonsignificant) decrease in blood lactate accumulation during the bench-press exercise for betaine treatment. The osmoprotectant properties of betaine help cells continue to perform under stress, which may include maintenance of muscle oxygen consumption and energy production. Possible mechanisms are discussed later in this chapter. Therefore, this study found that betaine supplementation results in a moderate increase in total repetitions and volume load in the bench press with enhanced muscle oxygen consumption as a contributing factor.

A study investigated the role of betaine supplementation on muscle phosphorylcreatine (PCr) content and strength performance in untrained subjects (Del Favero *et al.*, 2012). They compared the ergogenic and physiological responses to betaine *versus* creatine supplementation and tested the possible additive effects of creatine and betaine supplementation in a double-blind, randomized, placebo-controlled study. Subjects were assigned to receive betaine (2 g per day), creatine (20 g per day), betaine plus creatine (2 + 20 g per day, respectively) or placebo. At baseline and after 10 days of supplementation, they assessed muscle strength and power, muscle PCr content, and body composition. The creatine and betaine + creatine groups had a greater increase in muscle PCr content than placebo. PCr content was not increased for betaine or betaine + creatine. Creatine and betaine + creatine data showed greater squat and bench-press performance, whereas betaine did not show significant differences for strength and power *versus* placebo or for creatine *versus* betaine + creatine. Body composition did not differ between the groups. Therefore, in this study betaine supplementation did not increase muscle PCr content or increase strength and power performance in untrained subjects. Two factors may have contributed to the null results for betaine. This study compared 20 g of creatine to 2 g of betaine, so the 10- fold greater level of creatine may have overwhelmed any additional effect of betaine. Also, strength and power tests were not conducted until 3 and 6 days (respectively) after the 10 days of betaine supplementation had ceased. This delay may have diluted the physiological benefits (Cholewa *et al.*, 2013).

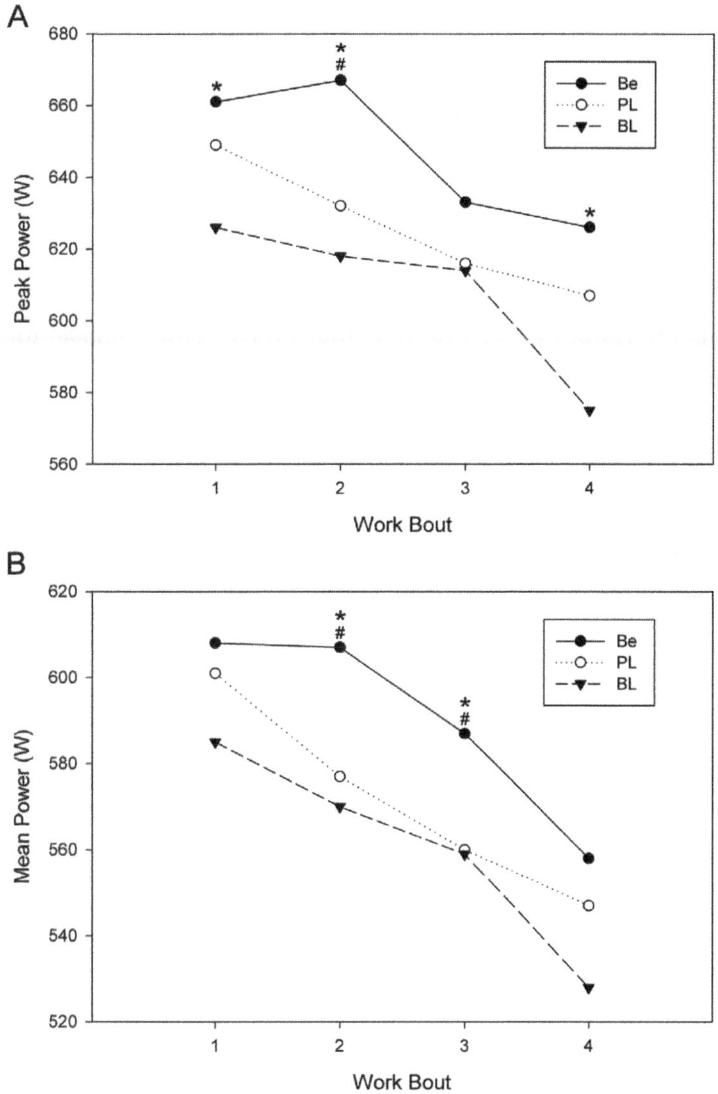


**Figure 20.10** Individual cycle runs power comparison. A: peak power; B: mean power. * = Significantly different ($p < 0.05$) between betaine and baseline. # = Significantly different ($p < 0.05$) between betaine and placebo. Subjects who took beverages containing betaine (compared to placebo) exhibited greater power when conducting repeated high-intensity cycling bouts.

Another study (Pryor *et al.*, 2012) examined the effect of betaine supplementation on cycling sprint performance. Sixteen recreationally active subjects (7 females and 9 males) completed three sprint tests (baseline, placebo and betaine) in a double-blind crossover design with a 3 week washout between trials. The placebo test was preceded by seven days of daily consumption of carbohydrate–electrolyte beverage (591 mL) and the betaine test was preceded by consumption of the same beverage which contained 0.42% betaine (approximately 2.5 g of betaine a day). Half the beverage was consumed in the morning and the other half in the afternoon. Each sprint test consisted of four 12 s efforts against a resistance (equal to 5.5% of body weight) separated by 2.5 min of cycling at zero resistance. Compared to baseline, betaine ingestion increased average peak power (6.4%), maximum peak power (5.7%), average mean power (5.4%), and maximum mean power (4.4%) for all subjects combined. Compared to placebo, betaine ingestion significantly increased average peak power (3.4%), maximum peak power max (3.8%), average mean power (3.3%), and maximum mean power (3.5%) for all subjects combined. There were no differences between the placebo and baseline trials. Figure 20.10 shows an individual cycle runs power comparison for all subjects. In addition, analysis was conducted on the data from just the males and just the females. Both groups had significantly increased mean and peak power for some individual cycle runs when comparing betaine *versus* placebo and baseline. This is particularly noteworthy as the first data showing an ergogenic benefit of betaine in females. Therefore, one week of betaine ingestion improved cycling sprint power in recreationally active males and females.

A recent study (Cholewa *et al.*, 2013) investigated the effects of long-term betaine supplementation on body composition, performance, and homocysteine thiolactone (HCTL) in experienced strength-trained men. Twenty-three subjects were matched for training experience and body-fat percentage, randomly assigned to either a placebo or betaine group and then conducted a 6 week periodized training program consisting of 3 two-week microcycles. Body composition was determined by arm and thigh cross-sectional area (girth and skin fold). Performance was measured *via* bench press, back squat and vertical jump. Fasting urine was collected at baseline, weeks 2, 4 and 6, and assayed for HCTL. Body composition (body fat %, fat mass, lean body mass) improved significantly in betaine but not placebo, see Figures 20.11 and 20.12. Arm (but not thigh) cross sectional area increased significantly for betaine but not placebo, see Figure 20.13. Bench-press training volume was significantly improved for betaine compared to placebo at microcycles one and three. Back squat training volume increased significantly for both groups throughout training. No differences were found in other performance variables between groups, except there was a trend ($p = 0.07$) for increased vertical jump power in betaine *versus* placebo. A significant interaction existed for HCTL, with increases from baseline to week 2 in placebo, but not betaine, and HCTL remained elevated at week 4 in placebo, but not betaine. This study is most significant for being the first to show improvement in body composition in humans (several studies have shown this benefit in animals and are outlined later in this chapter). The results show how the addition of betaine supplementation to resistance training can provide differentiating benefits. In conclusion, six weeks of betaine supplementation improved body composition, arm size, bench-press work capacity, attenuated the rise in urinary HCTL, and tended to improve power but not strength.

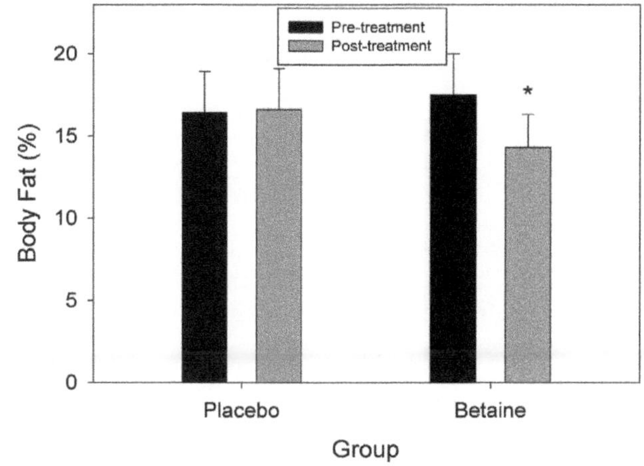


**Figure 20.11** Body fat percentage. * = Significantly different ($p < 0.05$) between post- and prebetaine. Subjects who took supplements containing betaine (compared to placebo) lost more body fat as measured by skin-fold calipers.

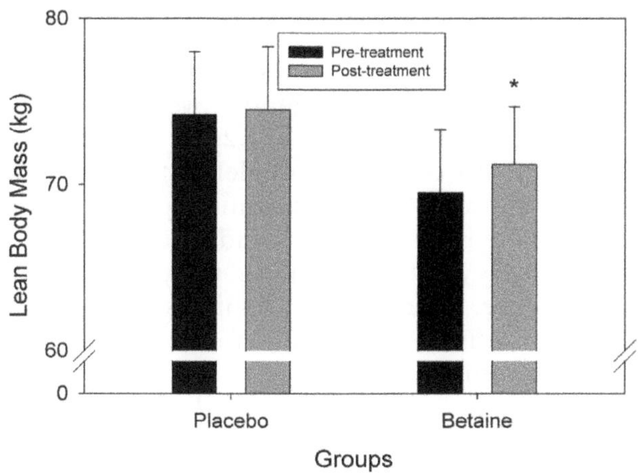


**Figure 20.12** Lean body mass. * = Significantly different ($p < 0.05$) between post- and prebetaine. Subjects who took supplements containing betaine (compared to placebo) gained more arm lean body mass as calculated from body fat percentage and body weight.

## 20.3 Mechanism Studies

### 20.3.1 In Humans

One study investigated whether the performance benefits of betaine supplementation might be linked to increased plasma nitric oxide (Bloomer *et al.*, 2011). There were three different study designs (acute intake of betaine at

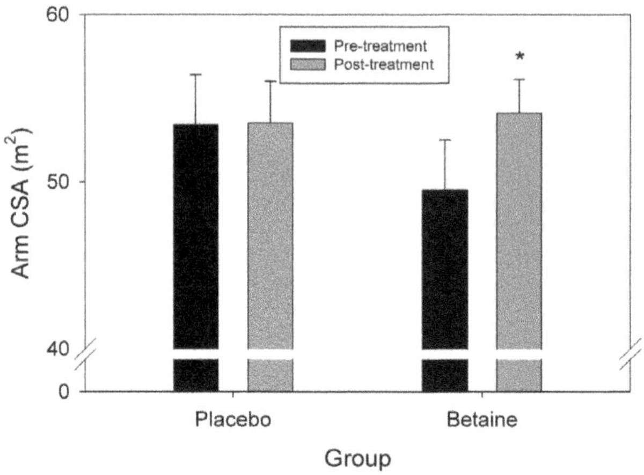


**Figure 20.13** Arm cross-sectional area. * = Significantly different ($p < 0.05$) between post- and prebetaine. Subjects who took supplements containing betaine (compared to placebo) gained more arm muscle size as measured by cross-sectional area.

1.25 and 5.00 g, chronic intake of betaine at 2.5 g per day for 14 days, and chronic [6 g of betaine per day for 7 days] followed by acute intake [6 g]), all using exercise-trained men. Blood samples were collected before and at 30, 60, 90, and 120 min after ingestion of 1.25 and 5.00 g of betaine (Study 1); before and after 14 days of betaine supplementation at a dosage of 2.5 g (Study 2); and before and after 7 days of betaine supplementation at a dosage of 6 g, followed by acute ingestion of 6 g and blood measures at 30 and 60 min postingestion (Study 3). In all 3 studies, there was no significant effect on plasma nitrate/nitrite. Therefore, mechanisms other than increased circulating nitric oxide are likely responsible for the performance-enhancing effects of betaine supplementation.

A human muscle biopsy study (Apicella *et al.*, 2013) examined the effect of betaine supplementation on selected circulating hormonal measures and Akt muscle signaling proteins after an acute exercise session. Twelve trained college age men underwent 2 weeks of supplementation with either betaine (1.25 g twice per day) or placebo. Following a 2 week washout period, subjects underwent supplementation with the other treatment. Before and after each 2 week period, subjects performed an acute exercise session. Circulating growth hormone (GH), insulin growth factor (IGF-1), cortisol, and insulin were measured. *Vastus lateralis* biopsy samples were analyzed for signaling proteins (Akt, p70 S6k, AMPK). For the circulating blood measures, betaine (*vs.* placebo) supplementation significantly increased IGF-1 (11.7%), approached a significant increase in GH (6.4%; $p = 0.06$), and significantly decreased cortisol (12.2%). There was no difference in insulin. For the muscle biopsy measures, betaine (*vs.* placebo) increased resting total Akt and potentiated phosphorylation of Akt (Ser(473)) and p70 S6 k (Thr(389)).

Phosphorylation of AMPK (Thr(172)) decreased during both treatments. Therefore, betaine (*vs.* placebo) supplementation enhanced both the anabolic endocrine profile and the corresponding anabolic signaling environment, suggesting increased protein synthesis.

Another study looked at the betaine content of sweat from exercising females (Craig *et al.*, 2010). Sweat patches were placed on eight trained adolescent Highland dancers, who then participated in a dance class for 2 h. Patches were removed, sweat was recovered *via* centrifugation, and subsequently analyzed for betaine, choline, sodium, potassium, chloride, lactate, glucose, urea and ammonia. Betaine was present in the sweat of all subjects ($232 \pm 84\ \mu mol\ L^{-1}$) and therefore may be lost from the body in significant amounts if there is a large sweat volume.

## 20.3.2 In Animals

Several studies have investigated the effect of betaine supplementation on increasing muscle and decreasing fat mass (Eklund *et al.*, 2005). For example, a study looked at replacement of choline by betaine on broiler performance and carcass characteristics (Jahanian and Rahmani,2008). Three betaine replacement levels (0, 50 and 100% in substitution for choline) were used in two different basal diets in a $2 \times 3$ factorial arrangement with four replicates of 10 birds. Two hundred-forty day-old broiler chicks were fed with the experimental diets from 1 to 49 days of age and at 49 days of age, two birds from each replicate were selected randomly for blood sampling and comparison of carcass characteristics. The results showed that betaine inclusion had no effect on feed intake, but significantly increased body weight gain (at 1–3 and 3–5 weeks of age) and feed conversion ratio (at 3–5 weeks of age). Replacing choline with betaine increased dressing and breast meat percentages and reduced abdominal fat percent, but had no significant effect on thigh and liver weight percentages. Plasma levels of cholesterol and LDL were not affected by dietary substitution of betaine for choline. Dietary betaine replacement caused a significant decrease in plasma triglycerides and VLDL and significant increase in HDL. These findings indicate that dietary betaine replacement for choline had little benefit in terms of performance parameters, but resulted in favorable changes in body composition (abdominal fat and breast meat percentages).

Recently, some animal studies have also investigated mechanisms. For example, one group from Zhejiang University investigated the effect of dietary betaine (0%, 0.125%) on growth performance, growth hormone (GH) pulsatile secretion and serum metabolites in crossbred finishing pigs (Huang *et al.*, 2007). Three replications of eight pigs (four barrows and four gilts) were used for each treatment, and blood samples for the determination of GH pulsatile secretion were collected every 15 min for 3 h. The results showed that betaine supplementation resulted in a 5.5% increase in average daily gain (average daily feed intake and feed to gain ratio were not affected). Betaine supplementation increased serum basal GH, mean GH and GH pulse amplitude by 42%, 49% and 35%, respectively (GH pulse frequency and pulse duration were unchanged). Betaine supplementation reduced serum urea nitrogen concentration by 22% and increased total protein level. The group further reported (Huang *et al.*,2008) that betaine supplementation increased carcass lean proportion and loin muscle area and decreased carcass fat proportion and average backfat thickness (betaine did not influence the dressing proportion, proportion of skin and bone of the carcass or leaf fat weight). Betaine supplementation decreased the activities of acetyl-CoA carboxylase, FAS and malic enzyme in subcutaneous adipose tissue by 18.0%, 18.8% and 14.5%, respectively (the activities of glucose-6-phosphate dehydrogenase and isocitrate dehydrogenase were not affected). Semiquantitative reverse transcription polymerase chain reaction analysis revealed a 25% decrease in FAS mRNA expression in pigs fed betaine. A significant correlation between mRNA expression level and potential enzyme activity for FAS was found ($r = 0.93$). The studies suggest that betaine supplementation in finishing pigs could promote growth by enhancing GH secretion. The reduction in adipose tissue might result from diminished rates of lipogenesis (due to decreased activities and gene expression of lipogenic enzymes).

Another study (Choe *et al.*, 2010) investigated the effects of dietary betaine on insulin growth factor (IGF-I, IGFBP-3 and IGFBP-1) secretion and IGF-I mRNA gene expression in the serum and liver of laying hens. Seventy two ISA-Brown laying hens were fed with four different levels of betaine (0, 300, 600, 1200 ppm) based on a corn–soybean meal diet containing 2800 kcal $kg^{-1}$ of metabolizable energy (ME) and 16% crude protein (CP) for four weeks. The results showed significantly higher serum and liver IGF-I concentrations, and a correlated increase in liver IGF-1 gene expression, in the laying hens fed with 600 and 1200 ppm betaine compared to controls. Serum IGFBP-3 concentrations were elevated significantly in the groups fed 600 ppm of betaine, but IGFBP-1 secretion in the liver of laying hens fed on 600 and 1200 ppm of betaine was significantly lower than in the controls. Therefore, this study suggests that dietary betaine supplementation plays a pivotal role in changes of the IGFs system in laying hens.

Finally, several studies have looked at the osmolyte function of betaine and support a protective role in stress conditions. For example, the effect of betaine on the absorption of nutrients in duodenal loop tissue of *Eimeria acervulina*-infected chicks was determined (Augustine and Danforth, 1999). Methionine (but not glucose) absorption was higher and intestinal cell ultrastructure improved (duodenal tissue heavier, epithelial cells less electron dense) in chicks fed a basal diet supplemented with betaine (0.15%) as compared to unsupplemented basal diet. Therefore, betaine affects the movement of nutrients and water across the intestinal epithelium and has a role in the osmoregulation of small intestine parasite developmental stages.

Another study (Kettunen *et al.*, 2001) looked at the water-holding capacity of broiler chick intestinal tissue when fed diets with or without betaine (0.2%) supplementation in the drinking water. Slices from duodenum and jejunum were incubated in iso-osmotic (300 mM) or hyperosmotic (600 mM) saline with or without betaine (10 mM). The ileal and duodenal tissues incubated in the hyperosmotic saline accumulated less water than those incubated in iso-osmotic saline. Duodenal slices incubated in hyperosmotic saline with the presence of betaine showed a water content similar to slices incubated in iso-osmotic saline. The data therefore suggest that betaine helped the duodenal, but not jejunal, epithelium to maintain water balance in hyperosmotic conditions.

Therefore, animal studies show how betaine helps increase lean body mass, decrease fat mass, and increase muscle size. Also, betaine helped diminish rates of lipogenesis, due to decreased gene expression and activity of lipogenic enzymes, and enhanced growth hormone and IGF secretion. Finally, betaine improved nutrient absorption and hydration of the digestive epithelium.

### 20.3.3 *In Vitro*

Betaine can act as an osmolyte to maintain cell hydration and function under stress by helping maintain optimal osmotic pressure and allowing macromolecules to maintain native folded conformation and stability without perturbing other cellular processes. Betaine is one of the most effective organic osmolytes, as demonstrated when comparing several osmoprotectants by vapor pressure osmometry of *in vitro* osmolyte–protein systems (Courtenay *et al.*, 2000). They reported iso-osmolal preferential interaction coefficients ($\Gamma_{\mu1}$) and the corresponding dialysis preferential interaction coefficient ($\Gamma_{\mu1,\mu3}$) over a wide range of concentrations for interactions between native bovine serum albumin (BSA) and six small solutes. All the solutes gave negative $\Gamma_{\mu1}$ and $\Gamma_{\mu1,\mu3}$ values, indicating preferential exclusion from a BSA solution. Of the solutes investigated, betaine was the most excluded ($\Gamma_{\mu1,\mu3}/m_3^{bulk} = -49 \pm 1$ $m^{-1}$) in the relative order of betaine >> proline > TMAO > trehalose ≈ $K^+Glu^-$ > glycerol. This represents their effectiveness as osmoprotectants. For betaine, the value provides a minimum estimate of the hydration of native BSA of approximately $2.8 \times 10^3$ $H_2O$/BSA which corresponds to slightly less than a monolayer (estimated to be approximately $3.2 \times 10^3$ $H_2O$). The authors concluded that of the solutes investigated, only betaine might be suitable for use in osmotic-stress experiments *in vitro* as a direct probe to quantify changes in hydration of protein surface in biopolymer processes.

A cell-free protein synthesis system was examined to mimic what might happen in cells exposed to hypertonicity due to increased osmolarity and ionic strength (Brigotti *et al.*, 2003). The *in vitro* system measures translation of globin mRNA by rabbit reticulocyte lysate, which decreased by 30–60% when osmolality was increased from 0.35 to 0.53 osmol/kg of water by the addition of inorganic salts. In contrast, equivalent additions of compatible osmolytes (betaine or myo-inositol) caused a 40–50% increase in the rate of translation. Amino acids (50–135 mM) that are transported *via* system A had no significant effect. Addition of KCl (75 mM) dramatically decreased the amount of the 43 S preinitiation complex, whereas it was completely preserved when osmolarity was similarly increased by the addition of betaine (150 mM). The formation of a nonenzymic initiation complex (between rabbit [(3)H]Phe-tRNA, poly(U) and the 80 S ribosomes) was unaffected by the addition of 75 mM NaCl or KCl. However, subsequent translation of the complex decreased by 70%. Addition of 150 mM betaine had no effect on reticulocyte extracts translating endogenous mRNA, whereas addition of 75 mM KCl caused a marked decrease in the polysome peak and an increase in the proportion of 80 S ribosomes and ribosomal subunits. In total, these results show that initiation and elongation are inhibited by high concentrations of inorganic ions, but not by the compatible osmolytes such as betaine. Therefore, betaine can protect protein synthesis under osmotic-stress conditions.

This research group conducted several other *in vitro* studies on betaine, including exposure of muscle (C2C12) cells to hypertonic stress (Alfieri *et al.*, 2006). The cells were seeded and cultured for 48 h in isotonic (0.3 osmol (kg $H_2O)^{-1}$) medium and then incubated in hypertonic medium (0.53 osmol (kg $H_2O)^{-1}$) with or without the addition of 20 mM osmolyte (betaine, myo-inositol, taurine or creatine). Culture was continued for 24 h and cell growth was estimated by cell counting. Compared to isotonic, hypertonic medium led to severe reduction of muscle growth and survival (~15% of control). Betaine was the most effective osmolyte at maintaining muscle cell survival (~90% of control).

Myosin belongs to the family of motor proteins that transform the chemical energy of ATP into mechanical work and works together with actin for muscle contraction. Urea can inhibit this activity by altering myosin tertiary structure, and a study (Ortiz-Costa *et al.*, 2008) examined the effects of urea and betaine on the catalytic activity and structure of myosin subfragment-1 (the catalytic globular portion of myosin). They used intrinsic and extrinsic fluorescence to show that urea inactivates subfragment-1 in parallel with its ability to induce exposure of the enzyme hydrophobic domains. Both effects were counteracted by betaine, which alone did not significantly affect subfragment-1. Urea also enhanced the accessibility of thiol groups, promoted aggregation and decreased the α-helix content of S1. These effects were also counteracted by betaine. Therefore, urea-induced inactivation of myosin ATPase was restored when betaine was present.

One group (Bounedjah *et al.*, 2012) investigated how hypertonic stress leads to a rapid increase in cellular ionic strength, macromolecular crowding and assembly of stress granules (SGs; mRNA-protein aggregates) that contributes to reduced cellular function. Slow accumulation of betaine in the cytoplasm reduces macromolecular crowding and ionic strength, leading to dissociation of SGs. Also, when cells have preaccumulated a large amount of betaine, SG formation is severely impaired and cells increase their chances of survival to another hypertonic episode. These results indicate the impact of betaine on mRNA-associated machineries that may play an important role in cell resistance and adaption to hyperosmolarity in many tissues.

Finally, a recent study (Senesi *et al.*, 2013) looked at how betaine affected skeletal muscle differentiation and hypertrophy using C2C12 cells and examining neo myotubes maturation and differentiation. $RT^2$-PCR array,

Western blot and immunofluorescence analysis revealed the effects of betaine on morphological features of C2C12 and on signaling pathways involved in muscle differentiation and hypertrophy. The authors performed a dose-response study, establishing that betaine (10 mM) was the dose able to stimulate morphological changes and hypertrophic process in neo myotubes. $RT^2$-PCR array methodology identified the expression profile of genes encoding proteins involved in the IGF-1 pathway and betaine (10 mM) was found to promote IGF-1 receptor expression. Western blot and immunofluorescence analysis of neo myotubes showed that betaine (10 mM) improved IGF-1 signaling, synthesis of myosin heavy chain and neo myotubes length. They also investigated myoblast proliferation and differentiation. Betaine did not modify C2C12 proliferative rate, but promoted myogenic induction, enhancing MyoD protein content and cellular elongation. During differentiation, betaine increased muscle-specific markers and IGF-1 R protein levels. This study is the first to show that betaine can promote muscle fiber differentiation and increase myotube size by IGF-1 pathway activation. The authors concluded that betaine may be useful not only in sport performance but also in clinical conditions characterized by muscle-function impairment.

Therefore, *in vitro* studies show how betaine helps cells under stress by facilitating a protective monolayer of water around biopolymers, protecting mRNA translation and protein synthesis, maintaining myosin ATPase activity, and enhancing muscle-cell survival, function and hypertrophy. This is well aligned with the human and animal studies discussed earlier.

## 20.4 Conclusions

Nine studies have examined the ergogenic effects of betaine supplementation in humans. All but one found an improvement in athletic performance, including power, strength, endurance, stamina, fatigue, and sprint time. One of these studies also showed improvements in body composition (lean body mass, fat mass, % body fat, arm cross-sectional area) supported by previous animal studies. Human, animal and *in vitro* studies have pointed to possible mechanisms. Increased protein synthesis (lean body mass) may be due to improved circulating anabolic endocrine profile (growth hormone, insulin growth factor, cortisol, and insulin) and corresponding muscle anabolic signaling (Akt, p70 S6k, AMPK). Reduced adipose tissue (fat mass) may result from diminished rates of lipogenesis that is a consequence of decreased lipogenic enzyme gene expression and activities. Other findings include improved muscle-cell survival, preserved myosin ATPase activity and enhanced muscle oxygen consumption ($StO_2$). The dual physiological roles of betaine as an organic osmolyte and methyl donor may be the underlying factor in these benefits. As an osmolyte, betaine accumulates in stressed cells to help maintain optimal cellular function. This protective effect involves redistribution of water in cells, leading to more effective biopolymer hydration and increased cytoplasmic osmolality. This may allow cells to extend metabolism further when betaine is present. As a methyl donor, betaine contributes to over a hundred different reactions essential for normal cell function, such as synthesis of s-adenosyl methionine, creatine, carnitine, epinephrine, and DNA methylation. Therefore, all of the above mechanisms may manifest as improved athletic performance and body composition from betaine ingestion.

## Summary Points

- This chapter focuses on the effects of betaine on athletic performance and body composition.
- Several human studies have shown that acute or chronic betaine supplementation can improve strength, power, endurance and body composition.
- Performance benefits have manifested in prolonged sprint time to exhaustion, increased (and greater retention of) strength, reduced fatigue, increased power, and improved quality of workout.
- Body composition benefits have manifested in decreased body fat % and fat mass, increased lean body mass, and increased muscle size (cross-sectional area).
- The likely mechanism of action is from both the osmolyte and methyl-donor properties of betaine.
- Osmolytes accumulate in cells to help maintain optimal cellular function in the presence of stressors such as dehydration and accumulating electrolytes.
- The osmolyte effect includes redistribution of cellular water, increased cytoplasmic osmolality, more effective biopolymer hydration, protection of cellular machinery and function (*e.g.* protein synthesis, enzyme activity, oxygen consumption) and cell survival.
- Measured biomarkers include increase in growth hormone, IGF-1, reduced cortisol.
- Increased protein synthesis (*i.e.* lean body mass, increased muscle size, improved recovery) may be due to improved circulating anabolic endocrine profile (growth hormone, insulin growth factor, cortisol, and insulin) and corresponding muscle anabolic signaling (Akt, p70 S6k, AMPK).
- Reduced adipose tissue (fat mass) may result from diminished rates of lipogenesis that is a consequence of decreased lipogenic enzyme gene expression and activities.
- The human evidence is supported by animal and *in vitro* data.

## Key Facts

### Key Facts on Betaine as a Nutrient

1. Betaine is a naturally occurring component of most foods.
2. Particularly good sources of betaine are whole wheat, spinach, beets and shellfish.
3. Some dietary choline can be oxidized to betaine in the body.

4. Epidemiological studies suggest that the average daily intake of betaine is ~200 mg $day^{-1}$.
5. These studies also show that higher levels of betaine intake or higher serum concentrations of betaine are correlated with lower risk of many chronic diseases, such as heart, cancer, diabetes, cataracts, brain health, *etc.*
6. A recommended daily intake (RDI) of betaine has not yet been determined.
7. Refined food ingredients (*e.g.* grains and sugars) have less betaine than unrefined.
8. Betaine is extracted from sugar beet commercially to produce a crystalline powder that can be added to foods and supplements.

## Definitions of Words and Terms

**Betaine**: A quaternary ammonium compound also known as trimethylglycine, glycine betaine, lycine, and oxyneurine.

**Osmolyte**: Osmolytes are soluble compounds that affect osmosis and help maintain cell volume and hydration. Inorganic osmolytes (*i.e.* electrolytes) stabilize cell-water balance and organic osmolytes (*e.g.* betaine) stabilize cell biopolymers and machinery (*i.e.* mRNA, proteins, *etc.*). They work together to help cells cope with the osmotic stress.

**Ergogenic**: Enhancing physical performance, stamina or recovery, and reducing fatigue.

**Strength**: The ability to exert force (Newtons) that can be measured in muscles by a variety of techniques.

**Power**: The rate at which work is performed (J $s^{-1}$ or W) that can be measured in muscles by a variety of techniques.

**Hypertrophy**: Growth in the size or mass of skeletal muscle tissue due to an increase in the size of component cells. The most common stimulus is exercise.

**mRNA**: Messenger ribonucleic acid carries information from DNA in the cell nucleus as a template for protein synthesis.

**Enzymes**: A protein produced by living cells that catalyze a specific biochemical reaction.

**Hormones**: A regulatory substance (usually a peptide or steroid) produced in tissues and transported in the bloodstream to stimulate specific cells or tissues into activity, such as growth or metabolism.

**Osmotic pressure**: The pressure exerted by the flow of water between two solutions with different concentrations of solute when separated by a semipermeable membrane. The higher the osmotic pressure of a solution, the more water tends to move into it. Cells actively regulate osmotic pressure to maintain homeostasis of the water content so as to keep the organism's fluids from becoming too diluted or too concentrated.

**Anabolic**: Metabolism in living tissues where simple substances are converted into the more complex materials.

**Dehydration**: A significant loss of body water with disruption of metabolic processes such as enzyme activity and protein synthesis. Water loss, heat and high electrolyte accumulation causes cellular proteins to unfold from their native state.

## List of Abbreviations

| | |
|---|---|
| 1-RM | One repetition maximum |
| Akt | Protein kinase B |
| AMPK | 5′ AMP-Activated protein kinase |
| ATP | Adenosine triphosphate |
| BHMT | Betaine homocysteine methyl transferase |
| BSA | Bovine serum albumin |
| C2C12 | Mouse muscle myoblast cells |
| DNA | Deoxyribonucleic acid |
| FAS | Fatty acid synthase |
| GH | Growth hormone |
| HCTL | Homocysteine thiolactone |
| HDL | High-density lipoprotein |
| IGF-1 | Insulin-like growth factor |
| IGFBP | Insulin-like growth factor binding protein |
| KCl | Potassium chloride |
| $K^+Glu^-$ | Potassium glutamate |
| LDL | Low-density lipoprotein |
| mRNA | Messenger ribonucleic acid |
| MyoD | Myogenic differentiation protein |
| NaCl | Sodium chloride |
| NOx | Nitrate/nitrite |
| PCr | Phosphorylcreatine |
| PCR | Polymerase chain reaction |
| SG | Stress granules |
| $StO_2$ | Muscle-tissue oxygen saturation |
| TMAO | Trimethylamine N-oxide |
| VLDL | Very low density lipoprotein |
| $VO_2$ max | Maximal aerobic power |

## References

Alfieri, R. R., Bonelli, M. A., Cavazzoni, A., Brigotti, M., Fumarola, C., Sestili, P., Mozzoni, P., De Palma, G., Mutti, A., Carnicelli, D., Vacondio, F., Silva, C., Borghetti, A. F., Wheeler, K. P. and Petronini, P. G., 2006. Creatine as a compatible osmolyte in muscle cells exposed to hypertonic stress. *Journal of Physiology.* 576(Pt 2): 391–401.

Apicella, J. M., Lee, E. C., Bailey, B. L., Saenz, C., Anderson, J. M., Craig, S. A., Kraemer, W. J., Volek, J. S. and Maresh, C. M., 2013. Betaine supplementation enhances anabolic endocrine and Akt signaling in response to acute bouts of exercise. *European Journal of Applied Physiology*. 113(3): 793–802.

Armstrong, L. E., Casa, D. J., Roti, M. W., Lee, E. C., Craig, S. A. S., Sutherland, J. W., Fiala, K. A. and Maresh, C. M., 2008. Influence of betaine consumption on strenuous running and sprinting in a hot environment. *Journal of Strength & Conditioning Research*. 22(3): 851–860.

Augustine, P. C. and Danforth, H. D., 1999. Influence of betaine and salinomycin on intestinal absorption of methionine and glucose and on the ultrastructure of intestinal cells and parasite developmental stages in chicks infected with Eimeria acervulina. *Avian Diseases*. 43(1): 89–97.

Bloomer, R., Farney, T., Trepanowski, J., McCarthy, C. and Canale, R., 2011. Effect of betaine supplementation on plasma nitrate/nitrite in exercise-trained men. *Journal of the International Society of Sports Nutrition*. 8(1): 5.

Bounedjah, O., Hamon, L., Savarin, P., Desforges, B., Curmi, P. A. and Pastre, D., 2012. Macromolecular crowding regulates assembly of mRNA stress granules after osmotic stress: new role for compatible osmolytes. *Journal of Biological Chemistry*. 287(4): 2446–2458.

Brigotti, M., Petronini, P. G., Carnicelli, D., Alfieri, R. R., Bonelli, M. A., Borghetti, A. F. and Wheeler, K. P., 2003. Effects of osmolarity, ions and compatible osmolytes on cell-free protein synthesis. *Biochemical Journal*. 369(2): 369–374.

Choe, H. S., Li, H. L., Park, J. H., Kang, C. W. and Ryu, K. S., 2010. Effects of dietary betaine on the secretion of insulin-like growth factor-I and insulin-like growth factor binding protein-1 and -3 in laying hens. *Asian-Australasion Journal of Animal Sciences*. 23(3): 379–384.

Cholewa, J. M., Wyszczelska-Rokiel, M., Glowacki, R., Jakubowski, H., Matthews, T., Wood, R., Craig, S. A. and Paolone, V., 2013. Effects of betaine on body composition, performance, and homocysteine thiolactone. *Journal of the International Society of Sports Nutrition*. 10(1): 39.

Courtenay, E. S., Capp, M. W., Anderson, C. F. and Record, M. T., Jr., 2000. Vapor pressure osmometry studies of osmolyte-protein interactions: implications for the action of osmoprotectants in vivo and for the interpretation of "osmotic stress" experiments in vitro. *Biochemistry*. 39(15): 4455–4471.

Craig, S. A. S., 2004. Betaine in human nutrition. *American Journal of Clinical Nutrition*. 80(3): 539–549.

Craig, S. A. S., 2010. Use of betaine to enhance exercise performance. US774087882. US.

Craig, S. S., Craig, S. A. S., Ganio, M. S., Maresh, C. M., Horrace, G., Da Costa, K. A. and Zeisel, S. H., 2010. The betaine content of sweat from adolescent females. *Journal of the International Society of Sports Nutrition*. 7(1): 3–8.

Del Favero, S., Roschel, H., Artioli, G., Ugrinowitsch, C., Tricoli, V., Costa, A., Barroso, R., Negrelli, A. L., Otaduy, M. C., da Costa Leite, C., Lancha-Junior, A. H. and Gualano, B., 2012. Creatine but not betaine supplementation increases muscle phosphorylcreatine content and strength performance. *Amino Acids*. 42(6): 2299–2305.

Eklund, M., Bauer, E., Wamatu, J. and Mosenthin, R., 2005. Potential nutritional and physiological functions of betaine in livestock. *Nutrition Research Reviews*. 18: 31–48.

Hoffman, J. R., Ratamess, N. A., Kang, J., Gonzalez, A. M., Beller, N. A. and Craig, S. A. S., 2011. Effect of 15 days of betaine ingestion on concentric and eccentric force outputs during isokinetic exercise. *Journal of Strength & Conditioning Research*. 25(8): 2235–2241.

Hoffman, J. R., Ratamess, N. A., Kang, J., Rashti, S. L. and Faigenbaum, A. D., 2009. Effect of betaine supplementation on power performance and fatigue. *Journal of the International Society of Sports Nutrition*. 6(1): 7–17.

Huang, Q. C., Xu, Z. R., Han, X. Y. and Li, W. F., 2007. Effect of betaine on growth hormone pulsatile secretion and serum metabolites in finishing pigs. *Journal of Animal Physiology and Animal Nutrition (Berlin)*. 91(3–4): 85–90.

Huang, Q. C., Xu, Z. R., Han, X. Y. and Li, W. F., 2008. Effect of dietary betaine supplementation on lipogenic enzyme activities and fatty acid synthase mRNA expression in finishing pigs. *Animal Feed Science and Technology*. 140: 365–375.

Jahanian, R. and Rahmani, H. R., 2008. The effect of dietary fat level on the response of broiler chicks to betaine and choline supplements. *Journal of Biological Sciences*. 8(2): 362–367.

Kettunen, H., Peuranen, S. and Tiihonen, K., 2001. Betaine aids in the osmoregulation of duodenal epithelium of broiler chicks, and affects the movement of water across the small intestinal epithelium in vitro. *Comparative Biochemistry and Physiology Part A: Molecular & Integrative Physiology*. 129(2–3): 595–603.

Lee, E. L., Maresh, C. M., Kraemer, W. J., Yamamoto, L. M., Hatfield, D. L., Bailey, B. L., Armstrong, L. E., Volek, J. S., McDermott, B. P. and Craig, S. A. S., 2010. Ergogenic effects of betaine supplementation on strength and power performance. *Journal of the International Society of Sports Nutrition*. 7(1): 27.

Likes, R., Madl, R. L., Zeisel, S. H. and Craig, S. A. S., 2007. The betaine and choline content of a whole wheat flour compared to other mill streams. *Journal of Cereal Science*. 46(1): 93–95.

Millard-Stafford, M., Warren, G. L., Hitchcock, K. M., Welling, R. L., Rosskopf, L. B. and Snow, T. K., 2005. Fluid replacement in the heat-effects of betaine. *Medicine & Science in Sports Exercise*. 37(5): S28.

Ortiz-Costa, S., Sorenson, M. M. and Sola-Penna, M., 2008. Betaine protects urea-induced denaturation of myosin subfragment-1. *FEBS Journal*. 13: 3388–3396.

Pryor, J. L., Craig, S. A. and Swensen, T., 2012. Effect of betaine supplementation on cycling sprint performance. *Journal of the International Society of Sports Nutrition*. 9(1): 12–18.

Senesi, P., Luzi, L., Montesano, A., Mazzocchi, N. and Terruzzi, I., 2013. Betaine supplement enhances skeletal muscle differentiation in murine myoblasts via IGF-1 signaling activation. *Journal of Translational Medicine*. 11(1): 174.

Somero, G. N. and Yancey, P. H., 2010. *Osmolytes and Cell-volume Regulation: Physiological and Evolutionary Principles. Comprehensive Physiology*. John Wiley & Sons, Inc., pp. 441–484.

Trepanowski, J. F., Farney, T. M., McCarthy, C. G., Schilling, B. K., Craig, S. A. and Bloomer, R. J., 2011. The effects of chronic betaine supplementation on exercise performance, skeletal muscle oxygen saturation and associated biochemical parameters in resistance trained men. *Journal of Strength & Conditional Research*. 25(12): 3461–3471.

Yancey, P. H., Clark, M. E., Hand, S. C., Bowlus, R. D. and Somero, G. N., 1982. Living with water stress: evolution of osmolyte systems. *Science*. 217(4566): 1214–1222.

CHAPTER 21

# Betaine, Osmoprotection, and Health of the Ocular Surface in Dry-Eye Disease

QIAN GARRETT*[a] AND MARK D. P. WILLCOX[a]

[a]School of Optometry and Vision Science, University of New South Wales, Sydney, Australia
*E-mail: q.garrett@unsw.edu.au

## 21.1 The Ocular Surface and Tear Film

The ocular (eye) surface is the interface between the eye and outer environment. It comprises the cornea and conjunctiva, bordered by upper and lower lids and covered by a thin layer of tear film. Preserving a stable tear film is fundamental for maintaining ocular surface health, for clear vision and comfort, and provides protection from infection and desiccation. The lacrimal glands, tear film, ocular surface epithelium (cornea, conjunctiva and meibomian glands), eyelids, and interconnecting sensory and motor nerves maintain the homeostasis of the ocular surface and its health (Stern *et al.*, 2004). This highly dynamic integrated function unit provides compositional and hydrodynamic support of a stable tear film. The compositional factors include mucins from ocular surface epithelia; lipids from meibomian glands, and aqueous fluids from lacrimal glands. The hydrodynamic factors involve eyelid blinking to express meibum lipids and help tear/lipid spreading and clearance, and eyelid closure to avoid exposure of the ocular surface to environmental assault.

Food and Nutritional Components in Focus No. 7
Betaine: Chemistry, Analysis, Function and Effects
Edited by Victor R Preedy

Published by the Royal Society of Chemistry, www.rsc.org

The tear film has a lipid layer, anterior to both an aqueous and mucin layer, which reduces evaporation and surface tension to allow spreading of the tears over the ocular surface. The tear film contains a mixture of the components derived primarily from goblet cells, meibomian glands, lacrimal glands, and vascular leakage, which acts as a physical buffer, a cleaning fluid, and source of nutrients for the underlying cornea and conjunctiva. It helps to maintain a highly refractive optical surface, lubricate eyelids, and protect the cornea from infection and desiccation (Rolando and Zierhut, 2001).

## 21.2 Dry-Eye Disease and Tear-Film Hyperosmolarity

Disturbance (disease or damage) to any components of the ocular surface functional unit can result in abnormality in either the quantity or quality of any primary tear component with decreased tear production, altered tear composition and/or altered tear-film distribution. Damage to the lacrimal glands, for example, can lead to aqueous tear-deficient dry eye due to failure of lacrimal tear secretion. Meibomian gland abnormalities such as gland obstruction or gland dropout can significantly affect lipid production, leading to evaporative of dry eye.

Dry eye is one of the most prevalent ocular surface diseases. Increasing with age, the prevalence of dry eye is estimated at 5–30% of the people over fifty years (Moss *et al.*, 2000; Epidemiology Subcommittee of the International Dry Eye Workshop, 2007b). It is estimated that 25% of the US population suffers from dry eyes or some abnormality of the ocular surface (Schaumberg *et al.*, 2003; Nichols *et al.*, 2005). The actual number affected is probably much greater if we include undiagnosed cases such as those who complain of dry eyes due to contact lens use, environmental factors (such as air conditioning and pollution), and surgery (such as laser-assisted stromal *in situ* keratomileusis (LASIK)), consumption of medications, and use of computer or other portable electronic devices. Dry eye can significantly affect the quality of life causing ocular discomfort, ocular irritation, redness and itching, fatigue, and impaired vision that interferes with daily-life activities such as reading and driving, and in severe cases, can lead to an increased risk of blindness due to the increased risk of ocular infection. The economic burden of dry-eye disease has been estimated to be approximately US$1000 per year for the sufferer and US$3.5 billion per year for the healthcare system (Yu *et al.*, 2011).

Dry-eye disease is defined by The 2007 International Dry Eye Workshop (DEWS) as a multifactorial disease of the tears and ocular surface that can result in ocular discomfort, visual impairment and tear-film instability, with potential damage to the ocular surface (Definition and Classification Subcommittee of the International Dry Eye Workshop, 2007a). The common feature of all forms of dry-eye syndrome is the increased osmolarity of the tear film. Tear hyperosmolarity results from reduced aqueous tear flow (aqueous-deficient dry eye) and/or increased evaporation of the aqueous tear from the exposed ocular surface (evaporative dry eye) due to lipid deficiency or compromised lipid quality (Bron *et al.*, 2009). For dry-eye patients, osmolarity measured from the lower meniscus of tears can reach values of up to 360 mOsm, while the concomitant values in normal subjects range between 296 and 302 mOsm (Tomlinson *et al.*, 2006; Lemp *et al.*, 2011). It has been predicted that tear osmolarity might rise to levels as high as 600 mOsm over the cornea (Liu *et al.*, 2009; Gaffney *et al.*, 2010). Tear-film hyperosmolarity is the primary cause of ocular discomfort and has been associated with apoptosis of ocular surface epithelial cells and inflammation in dry eye (Huang and Tunnacliffe, 2004; Luo *et al.*, 2004; Luo *et al.*, 2005; Li *et al.*, 2006; Definition and Classification Subcommittee of the International Dry Eye Workshop, 2007a; Corrales *et al.*, 2008; Pflugfelder *et al.*, 2008; Gaffney *et al.*, 2010; Wang *et al.*, 2011). Hyperosmolarity of the tear film correlates with the severity of dry-eye disease (Sullivan *et al.*, 2010).

### 21.2.1 Hyperosmolarity, Inflammation and Apoptosis

Tear-film hyperosmolarity is an inflammatory stimulus. In models of dry eye, hyperosmolarity induces ocular-surface damage and initiates an inflammatory cascade involving mitogen-activated protein kinases (MAPKs) (Luo *et al.*, 2004) and the nuclear factor-κB (NF-κB) (Lan *et al.*, 2012) signaling pathways from the ocular surface epithelial cells that stimulates production and release of inflammatory cytokines (such as Interleukin (IL)-1, IL-6, IL-8, IL-17, and tumor necrosis factor (TNF)-α) (Li *et al.*, 2006) and tissue modulating factor matrix metalloproteinases (MMP-9) (Li *et al.*, 2004). These inflammatory responses lead to further ocular-surface damage and the development of a self-perpetuating inflammatory cycle with recruitment and activation of inflammatory cells ($CD4^+$ T cells) to the ocular surface and lacrimal glands (Research Subcommittee of the International Dry Eye Workshop, 2007c). Dry-eye patients receiving topical treatment with immunosuppressants, corticosteroids (Pflugfelder *et al.*, 2004) or cyclosporin A (Brignole *et al.*, 2001) have reported improved clinical signs and symptoms.

Ocular-surface damage, which can be visualized as patches of fluorescein staining on the ocular surfaces, is another cellular response to the tear hyperosmolarity. It is believed that epithelial damage involves cell death by apoptosis, a loss of goblet cells, which results in a reduction in mucus secretion, and subsequent instability. Reduced cell volume and elevated intracellular ionic strength resulting from rapid loss of water from cells (Burg *et al.*, 2007; Lang, 2007) can inhibit DNA replication, transcription and translation, and depolarize mitochondria, causing DNA damage and cell-cycle arrest, and ultimately leading to apoptotic cell death (McManus *et al.*, 1995; Orlov and Novikov, 1996; Waldegger *et al.*, 1998; Lang, 2007; Luo *et al.*, 2007). Hyperosmolarity-induced ocular epithelial apoptosis has been seen in *in vitro* studies using cultured human ocular-surface epithelial cells (Luo *et al.*, 2007; Chen *et al.*, 2008; Png *et al.*, 2011; Clouzeau *et al.*, 2012; Garrett *et al.*, 2013) and animal dry-eye models (Yeh *et al.*, 2003; Chen *et al.*, 2009; Chen *et al.*, 2013). Hyperosmolarity-induced apoptosis of human corneal epithelial cells is through cytochrome c death pathways (Luo *et al.*, 2007; Chen *et al.*, 2008; Png *et al.*, 2011; Clouzeau *et al.*, 2012; Garrett *et al.*, 2013) mediated by c-Jun *N*-terminal kinase (JNK)1 MAPK signaling pathways (Chen *et al.*, 2008). MAPKs are the cell-signaling mediators that play a vital role in cellular response to stress. MAPKs include the extracellular signal regulated kinases (ERK), JNK and P38 MAPK. The activation and functional role of each MAPK cascade depends on the cell types and the stimuli to which cells are exposed. JNK is known to be involved in osmotic-stress responses and is a key mediator in osmoregulation (Galcheva-Gargova *et al.*, 1994). A direct link between hyperosmotic exposure of cells and activation of JNK signaling pathway is seen in human corneal epithelial cells (Chen *et al.*, 2008).

### 21.2.2 Osmoprotection – Organic Osmolytes

The maintenance of osmotic balance between the cytoplasmic and extracellular compartments of cells is vital for the control of cell volume, and thereby health and functioning of the cell. When such a balance is disturbed by hyperosmotic stress, as occurs in dry-eye disease, an immediate response of the cell to the challenge is a rapid volume correction (within minutes) by regulatory volume increase in which ions (potassium, sodium and chloride) move across the cell membrane along with water (Lang *et al.*, 1998). This early process is mediated by the ion-transport systems controlled by p38 MAPK activation including the $Na^+$-$K^+$-$2Cl^-$ cotransporter (NKCC), the $Na^+/H^+$ and the $Cl^-/HCO_3^-$ exchangers present on the human corneal epithelial cell plasma membrane (Bildin *et al.*, 2000; Capó-Aponte *et al.*, 2007; Lee *et al.*, 2008). Accumulation of inorganic ions in cells can interfere with normal cellular processes. Cells counteract this increased ionic strength by substituting inorganic ions with small organic osmolytes that do not interfere with cell metabolism or destabilize proteins, even at relatively high intracellular concentrations (Burg *et al.*, 2007), thus maintaining intracellular water homeostasis without impairing normal biochemical function. In mammalian cells, the most frequently used organic osmolytes are neutral amino acids or their derivatives such as taurine, polyols such as sorbitol and myo-inositol, and methylamines such as betaine and carnitine. This adaptation process, osmoprotection, is characterized by increased synthesis of either membrane transporters essential for uptake of these osmolytes (also termed compatible solutes) (Kempson and Montrose, 2004), or enzymes for their synthesis (also termed metabolic production) (Yancey *et al.*, 1990).

Osmoprotective therapies for dry eye presently in the market include artificial tears that contain compatible solutes, erythritol and L-carnitine. Use of such therapies might lead to improved ocular-surface health, improvement in the common signs and symptoms of dry eye and reduced dry eye-associated conjunctival fluorescein or lissamine green staining (Guillon *et al.*, 2010).

## 21.3 Betaine and Ocular Osmoregulation

Betaine is often accumulated in tissues to assist cell-volume regulation (Schliess and Haussinger, 2002; Wehner *et al.*, 2003; Lang, 2007; Verbalis, 2010; Zhou *et al.*, 2012). Accumulation in kidney, liver, brain, intestine and skin has been well documented (Warskulat *et al.*, 2004; Lim *et al.*, 2007; Venkatesu *et al.*, 2009; Zhou *et al.*, 2012). The protective role of betaine, mediated by its transporter BGT-1(betaine/γ-aminobutyric acid (GABA) transporter) has been well studied in cells derived from the renal medulla that is constantly exposed to high extracellular osmolarity during urinary concentration *in vivo* (Venkatesu *et al.*, 2009). Although there is a very limited number of studies in betaine ocular osmoregulation, several studies have reported that betaine might possess a similar function in response to hyperosmolarity in human corneal epithelial cells (Corrales *et al.*, 2008; Chen *et al.*, 2013; Garrett *et al.*, 2013).

### 21.3.1 Betaine Presence in Human Tears

Humans obtain betaine from foods or through betaine synthesis from choline oxidation in the liver and kidney. In healthy humans, the betaine concentration in blood plasma ranges from 20 to 144 μM (Holm *et al.*, 2003; Kempson and Montrose, 2004). This provides a source of betaine for uptake into various cells. Ocular accumulation and transport of betaine has not been reported and whether betaine is present in the ocular system to act to balance extracellular hyperosmolarity is not known. We conducted a pilot clinical study on tears (basal nonreflex) to determine whether betaine is present in the human ocular system. Tears of ten healthy subjects were collected using a sterile microcapillary tube from the lower eyelid margin of eyes.

The collected tears were analyzed for the presence of betaine by liquid chromatography-mass spectrometry using a Thermo Scientific TSQ Quantum XLS (Thermo Scientific) and a Synergi™ Fusion-RP LC Column (4 μm, 80 Å, 250 × 1 mm, Phenomenex). One μl of the diluted tear sample (1 : 20) was loaded on the column. The betaine detection was monitored by measuring signals at 118.2 *m*/*z* (betaine) and 58 *m*/*z* (dimethylglycine). The betaine concentration in tears was calculated against a standard calibration curve ranging from 0–500 fmol $\mu l^{-1}$. A betaine-d9 internal standard (Medical Isotopes Inc., Pelham, NH) was used.

Betaine was detectable in the tears of all subjects. Concentration ranged from 0.555 to 6.337 μM with an average of 2.506 ± 1.878 μM (mean ± SD). This level is one to two magnitudes lower than betaine concentration in blood plasma (20–144 μM), but at the same magnitude level as that of L-carnitine (another methylamine osmolyte) in tears of healthy individuals (8.94 ± 0.50 μM) (Pescosolido *et al.*, 2009). The subjects enrolled did not fast before their tears were collected, and so this may affect the betaine level in tears, which might lead to the high variation in concentration.

### 21.3.2 Ocular Betaine Transport

Cellular accumulation of betaine is mediated *via* the osmoregulated betaine/γ-aminobutyric acid (GABA) transporter, designated BGT-1 (Yamauchi *et al.*, 1992). Expression of BGT1 in response to osmolar flux is controlled

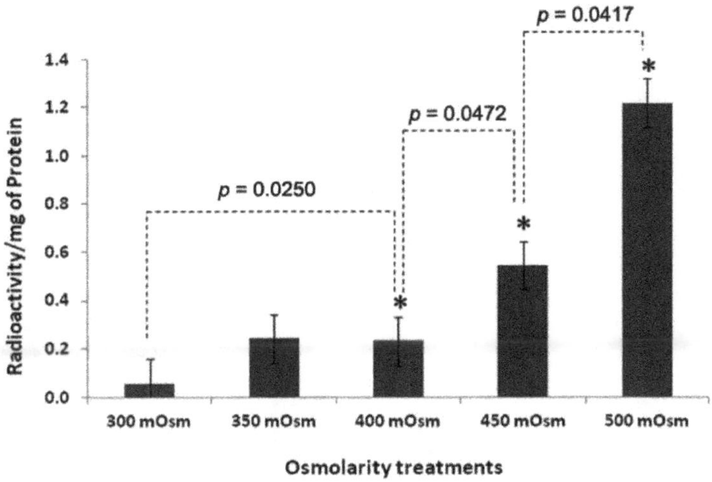


**Figure 21.1** Osmolarity dependence of $^{14}C$- betaine uptake in HCLE cells (mean ± SD of eight samples; $n = 8$). * Statistically significant difference compared to cells treated with 300 mOsm buffer. Reprinted from Experimental Eye Research. 2013; 108 : 33–41. Garrett *et al.* Betaine stabilizes cell volume and protects against apoptosis in human corneal epithelial cells under hyperosmotic stress, with permission from Elsevier.

by the transcription factor tonicity-responsive enhancer binding protein (TonEBP) also known as Nuclear Factor of Activated T cells (NFAT5). TonEBP/NFAT5 is a master transcriptional regulator that promotes cellular accumulation of the major organic osmolytes in the renal medulla by transcriptional stimulation of plasma membrane transporters, such as BGT-1 (Miyakawa *et al.*, 1999; Lopez-Rodriguez *et al.*, 2001; Woo *et al.*, 2002; Kempson and Montrose, 2004; ). The $Na^+$- and $Cl^-$-dependent GABA/betaine acid transporter was initially cloned from Madin–Darby canine kidney (MDCK) cells and is expressed in many other mammalian tissues (Olsen *et al.*, 2005; Warskulat *et al.*, 2008). BGT1 is predominantly expressed in the liver and kidneys, and at lower levels at the brain surface (Zhou *et al.*, 2012).

BGT1 is expressed in cultured human corneal limbal epithelial (HCLE) cells (Garrett *et al.*, 2013). Betaine increases in cultured HCLE cells when the extracellular environment becomes hyperosmotic (Figure 21.1), similar to that in the renal medulla (Kempson and Montrose, 2004). The master regulator NFAT5 mentioned above is induced and translocated to the nucleus of HCLE under hyperosmotic stress (Lee *et al.*, 2008). We speculate this mechanism is also operating in our observed increased expression of BGT-1 with the concomitant influx of betaine.

When corneal epithelial cells, which were exposed to 16 h of hyperosotic medium (500 mOsm), were returned to isotonic conditions (300 mOsm), there was a progressive increase in the amount of betaine released into

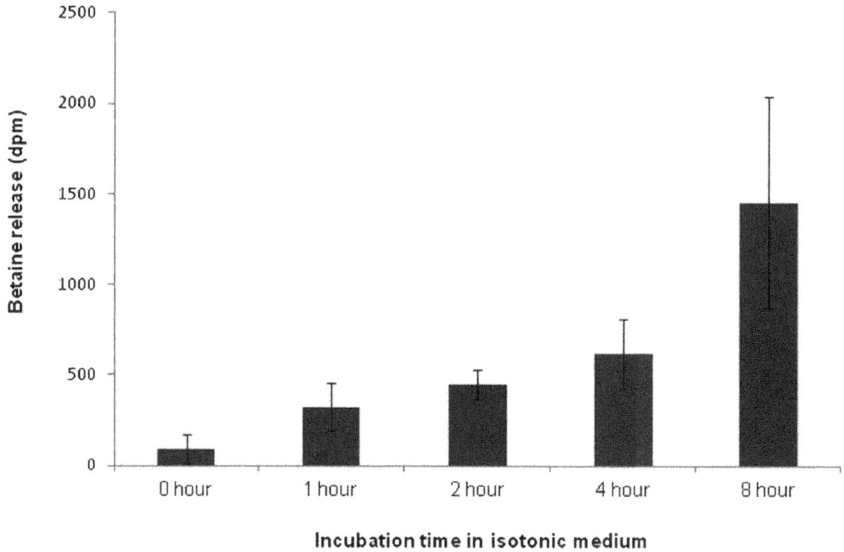


**Figure 21.2** Release of betaine into an isotonic medium (300 mOsm) by human corneal limbal epithelial cells after 16 h exposure to a hyperosmotic culture medium (500 mOsm).

the isotonic medium (Figure 21.2). This process might involve an efflux of betaine through specific release pathways and possible downregulation of BGT-1.

### 21.3.3 Betaine and Hyperosmotic Stress-Induced Cell Shrinkage

Ocular epithelial cells can adapt to changes in osmotic strength through the influx of the osmolyte betaine and restore cell volume reduced by hyperosmotic stress (Garrett *et al.*, 2013). Osmoregulation of betaine can be measured using flow cytometry to sort subpopulations of different sizes as a result of exposure to iso-, or hyperosmolarity with or without betaine (Figure 21.3). The analysis is based on the principle that forward scatter (FSC) of the laser light produced by the flow cytometer is proportional to the volume of the normal cell, while the side scatter (SSC) represents cells with greater structural complexity and granularity, hence shrunken cells (Edwards *et al.*, 2004). Exposure of human corneal epithelial cells to a hyperosmolar environment causes a reduced population of normal cells with intact cell volume and an increased population of shrunken cells (Figure 21.3) compared to exposure to isotonic conditions. When betaine is present in the hyperosmolar extracellular fluid (culture medium), the percentage of shrunken cells is reduced by more than 2.5-fold compared to cells under the same conditions in the absence of betaine (Figure 21.3).

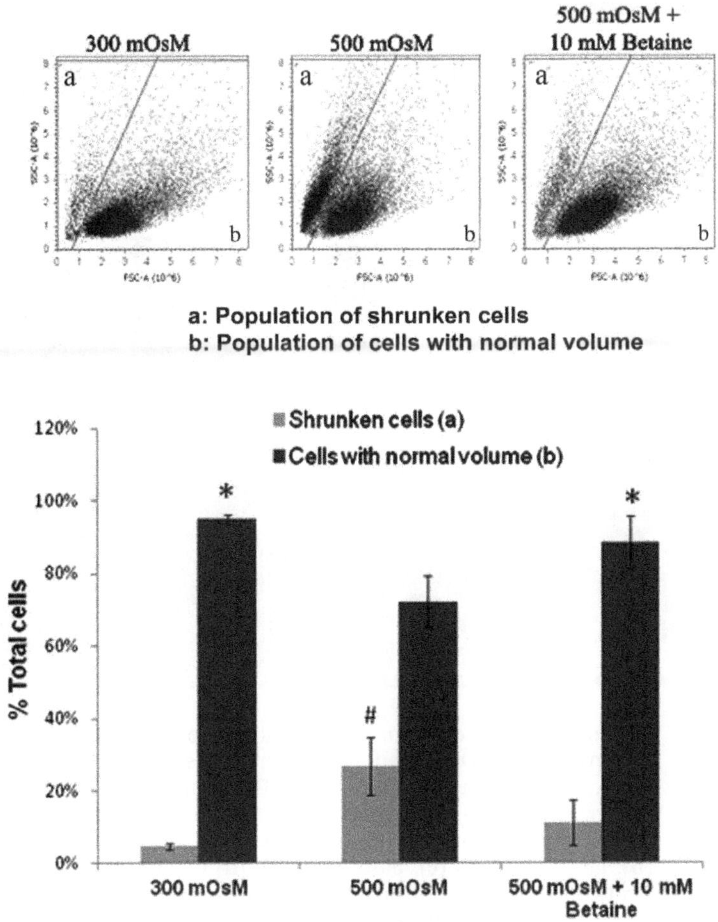


**Figure 21.3** Flow cytometry analysis of the effect of 300 mOsm and 500 mOsm media in the presence or absence of betaine (10 mM) on the cell volume/size. (a) population of cells with normal cell volume; (b) population of shrunken cells. * Statistically significant difference ($p < 0.05$) compared to exposure to hyperosmotic stress (500 mOsm) without betaine. [#] Statistically significant difference between exposure to hyperosmotic stress with and without betaine ($p = 0.003$). Adapted from Experimental Eye Research. 2013; 108 : 33–41. Garrett *et al.* Betaine stabilizes cell volume and protects against apoptosis in human corneal epithelial cells under hyperosmotic stress, with permission from Elsevier.

### 21.3.4 Betaine and Hyperosmotic Stress-Induced Cell Apoptosis

Cell shrinkage is an early event in conditions of hyperosmotic stress that trigger apoptotic cell death. (Lang *et al.*, 1998) Corneal epithelial apoptosis as a consequence of chronic hyperosmotic stress is detected in cultured human corneal epithelial cells under acute and prolonged exposure to hyperosmotic

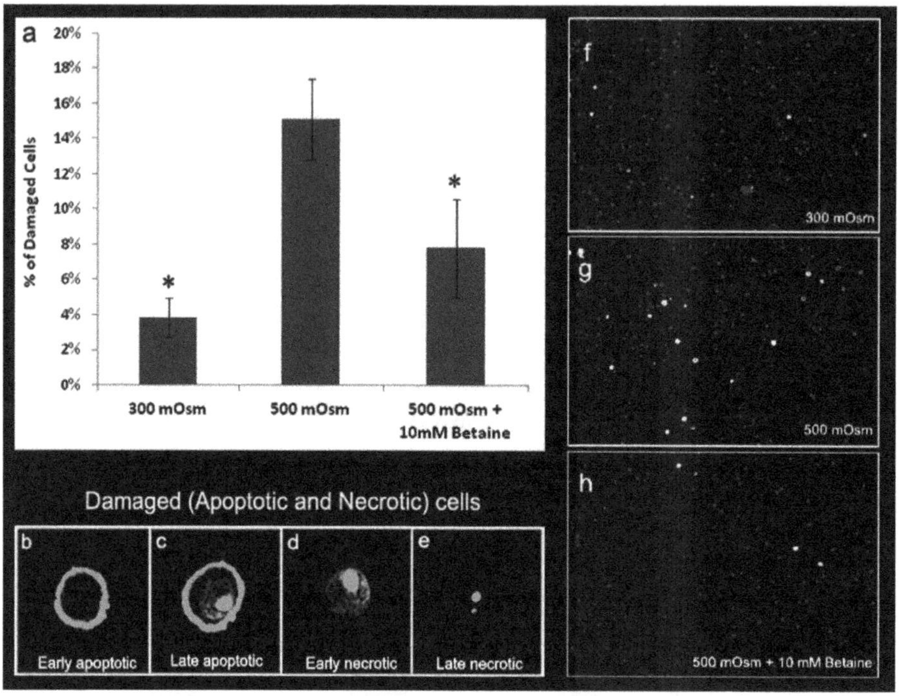


**Figure 21.4** Confocal microscopic analysis of cellular apoptosis in response to 300 mOsm or 500 mOsm media in the presence or absence of betaine (10 mM). (a) Percentage of damaged cells. * Statistically significant difference with 500 mOsm media without betaine compared to with betaine ($p < 0.001$) or 300 mOsm ($p < 0.001$) ($n = 12$). Cells are in early apoptosis (b), late apoptosis (c), early necrosis (d) or late necrosis (e). Magnification 60 × using confocal microscopy. Figures (f), (g) and (h) are representative of cells in 300 mOsm, 500 mOsm and 500 mOsm + 10 mM betaine respectively (magnification 10× ). Reprinted from Experimental Eye Research. 2013; 108 : 33–41. Garrett *et al.* Betaine stabilizes cell volume and protects against apoptosis in human corneal epithelial cells under hyperosmotic stress, with permission from Elsevier.

culture medium (Luo *et al.*, 2007; Chen *et al.*, 2008; Png *et al.*, 2011; Clouzeau *et al.*, 2012; Garrett *et al.*, 2013).

The antiapoptotic effect of betaine, possibly as a result of reduced cell shrinkage, on HCLE cells is further illustrated through confocal microscopy (Figure 21.4). HCLE cells exposed to the hyperosmotic culture medium (500 mOsm) in the presence of betaine show significantly reduced numbers of apoptotic and necrotic cells compared to hyperosmotic exposure without the presence of betaine. This is further confirmed by flow-cytometry analysis of full cell populations contained in samples (Figure 21.5). These data further support the osmoprotection of betaine on HCLE against hyperosmolar induced biochemical disequilibrium that leads to an irreversible apoptotic cascade (Yu and Choi, 2000; Alfieri *et al.*, 2002; Schliess and Haussinger, 2002).

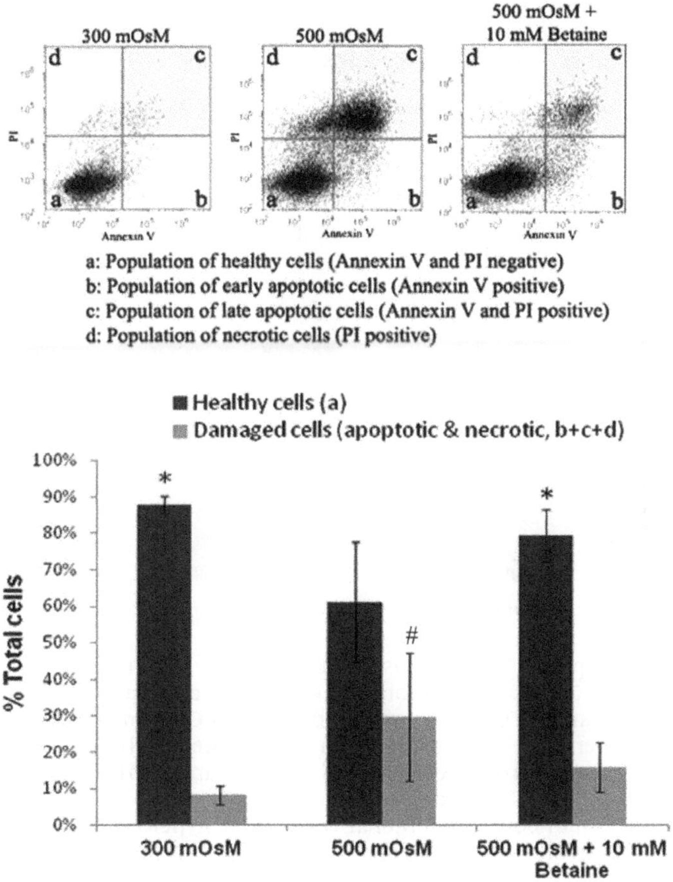


**Figure 21.5** Flow cytometry analysis of the effect of 300 mOsm and 500 mOsm media in the presence or absence of betaine (10 mM) on HCLE cells. Quadrant (a) viable cells that did not stain with PI (inecrotic) or Annexin V (apoptotic). Quadrant (b) cells stained with Annexin V only (early apoptosis). Quandrant (c) cells positively stained with Annexin V and PI (late apoptosis). Quadrant (d) cells stained with PI only (necrosis). * Statistically significant difference ($p < 0.05$) compared to exposure to hyperosmotic stress without betaine ($n = 6$). # Statistically significant difference between 300 mOsm and 500 mOsm media ($p = 0.015$). Adapted from Experimental Eye Research. 2013; 108:33–41. Garrett *et al.* Betaine stabilizes cell volume and protects against apoptosis in human corneal epithelial cells under hyperosmotic stress, with permission from Elsevier.

Two pathways are known to be involved in the induction of apoptosis (Fan *et al.*, 2005; Elmore, 2007). The first involves interaction of death ligands such as TNF-α with their respective cell-surface receptors. This ligation subsequently activates caspase-8 (Xia *et al.*, 1995; Chen and Goeddel, 2002) which can form a death-signaling complex with TRADD (tumor necrosis

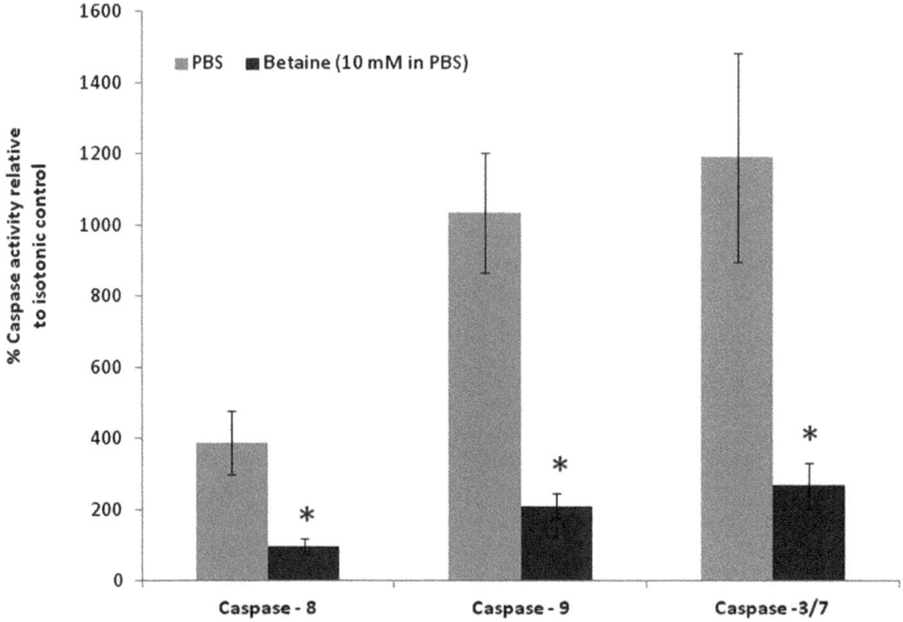


**Figure 21.6** Caspase activities (caspase-8, -9 and -3/7) in response to incubation in 500 mOsm solution for 16 h and treatment with betaine (10 mM; mean ± SD, $n = 6$) compared to 300 mOsm media. * Significant difference ($p < 0.001$) between cells treated with and without betaine. Adapted from Experimental Eye Research. 2013; 108:33–41. Garrett *et al.* Betaine stabilizes cell volume and protects against apoptosis in human corneal epithelial cells under hyperosmotic stress, with permission from Elsevier.

factor receptor type 1-associated death domain protein) and FADD (Fas-associated protein with death domain) that further induces effector caspases such as caspase-3 (Schneider-Brachert *et al.*, 2004; Li and Lin, 2008). The second pathway involves the mitochondrial release of cytochrome c to activate caspase-9 and subsequent activation of the downstream effector caspase-3. The active downstream caspases cleave a broad spectrum of cellular proteins to ultimately result in apoptotic cell death. Both mitochondrial-mediated and death receptor-mediated apoptotic pathways are implicated in hyperosmolarity-induced human corneal epithelial apoptosis, as suggested by the observed increase in activity of caspase-8, -9, -3/7 (Figure 21.6) and elevation in TNF-α production (Figure 21.7) in 500 mOsm solutions.

Although it is not fully ascertained how the cell shrinkage induced by hyperosmolarity signals the cascade event of apoptosis, MAPKs have been reported to play an important role in signaling (Xia *et al.*, 1995; Chuang *et al.*, 2000). Cytochrome c released from mitochondria of cultured human corneal epithelial cells under hyperosmotic stress triggers activation of JNK, ERK MAPK, and p38, and downstream activation of caspase-3 (Luo *et al.*, 2007).

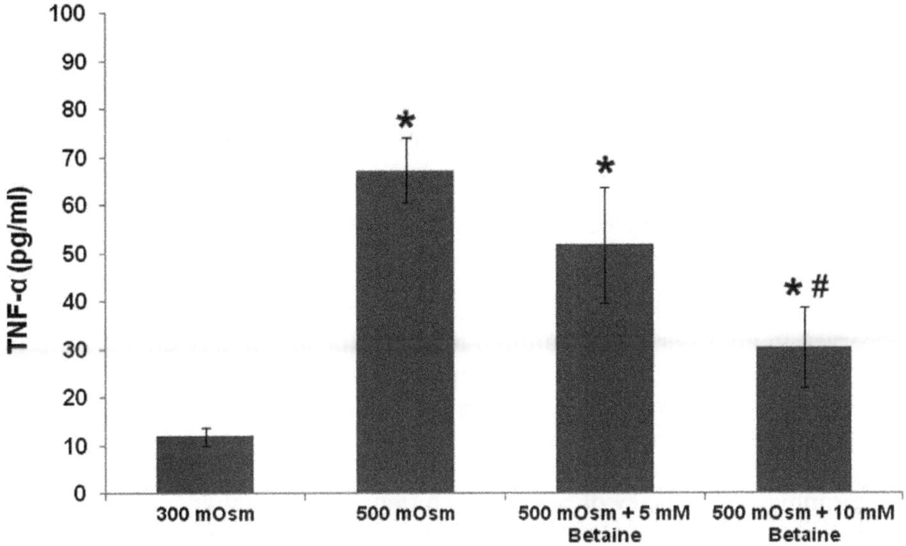


**Figure 21.7** Production of TNF-α by HCLE cells after exposure to 300 mOsm or 500 mOsm media with or without betaine (5 mM or 10 mM) for 16 h (mean ±SD, $n$ = 5). * Statistically significant difference compared to 300 mOms; all with $p < 0.01$). # Statistically significant difference with 500 mOsm with betaine (10 mM) or without betaine ($p$ =0.0005). Reprinted from Experimental Eye Research. 2013; 108:33–41. Garrett *et al.* Betaine stabilizes cell volume and protects against apoptosis in human corneal epithelial cells under hyperosmotic stress, with permission from Elsevier.

In addition, Polo-like kinase 3 (Plk3) activated by hypersmotic stress can induce activation of ATF-2 (activating transcription factor) to regulate the formation of AP-1 (activator protein 1) transcription factor complex that acts as a central switch to convert extracellular signals into genetic responses to determine cell apoptosis (Wang *et al.*, 2011). Another separate signaling pathway involves translocation of AIF (apoptosis-inducing factor) from the mitochondria to the nucleus where AIF binds to DNA, leading to chromatin condensation and ocular epithelial cell death (Clouzeau *et al.*, 2012).

Betaine stabilizes human corneal epithelial cell shrinkage induced by hyperosmolality and prevents activation of caspase-3/7, -8, -9 and TNF-α (Figures 21.6 and 21.7), as well as reducing apoptosis in cells. Addition of betaine reduces phosphorylation ratios of JNK and p38, and hyperosmolarity-induced activation of MAP kinases (Corrales *et al.*, 2008). Betaine enhances human corneal epithelial cell survival and recovery during hyperosmotic exposure, as shown by a significant increase in mitochondrial activity (indicating cell viability) and cell numbers compared to hyperosmotic treatment without betaine (Garrett *et al.*, 2013). In addition to functioning as an osmolyte and preserving cell volume, betaine, a zwitterionic methylamine osmolyte, also acts as a chemical chaperone to limit damage to proteins and thereby stabilizes protein structure and preserves enzyme function (Yancey *et al.*, 1982). The precise mechanism by which betaine protects human corneal epithelial cells from hypertonicity-induced apoptosis remains unknown. The osmoregulation and protein stabilizing effects of betaine might both contribute to protection from hyperosmolarity-induced apoptosis.

### 21.3.5 Betaine in Prevention and Treatment of Animal Dry Eye

Osmoprotection of the ocular surface by betaine from dry-eye insult is also observed in an experimental animal model. Exposure to a relative humidity of 13%, airflow of 2 m $s^{-1}$ and a temperature of 22 °C (a so-called "intelligently controlled environmental system" (ICES)) for 21 days induces dry eye similar to that observed in humans (Chen *et al.*, 2013). Two systematic topical administration schedules were used in the animal study; one to explore the prophylactic efficacy of betaine with delivery of betaine contemporaneously upon admission to ICES, and the other for the efficacy of betaine in the treatment of established dry eye. Betaine as a prophylaxis for dry eye limits progression of dry eye during desiccation and associated ocular surface inflammatory responses. Once dry eye has developed, treatment with betaine shows little effect in reducing the severity of such dry-eye conditions (Chen *et al.*, 2013). Although betaine has been reported to also possess anti-inflammatory properties, (Detopoulou *et al.*, 2008) the animal study further supports the predominant role of betaine as an osmolyte unlike L-carnitine that shows both prophylaxis and a therapeutic effect on desiccation-induced animal dry eye (Chen *et al.*, 2013).

## 21.4 Summary

Osmoprotection of the ocular environment is essential for ocular clarity, ocular comfort, and protection against ocular-surface damage and infection. Dry eye resulting from increased tear hyperosmolarity and instability of tear film is the most prevalent ocular-surface disease that affects millions with an associated reduction in quality of life. Hyperosmolarity is linked directly to ocular surface epithelial damage (apoptosis) and inflammation as a result of cell shrinkage.

Betaine, a known and well-studied osmolyte in the renal system, is also present in human tear fluid. Expression of the betaine transporter BGT1 in human corneal epithelial cells, the role of BGT1 in facilitating betaine accumulation in the human corneal epithelial cells under hyperosmotic stress, and the efflux of betaine from the cells upon returning to physiological isotonic conditions, highlight a potential role of exogenous betaine in tears in osmoprotection of ocular epithelial cells. Betaine attenuates hyperosmolarity-induced human corneal epithelial cell shrinkage and subsequent apoptosis *in vitro* and helps limit progression of environmentally induced animal dry eye.

Betaine offers potential as a prophylaxis solution to provide osmoprotection of eyes in desiccating environments from developing dry-eye disease.

## Summary Points

- This chapter focuses on a potential role of betaine as an organic osmolyte in restoring the health of the ocular surface in dry-eye disease.
- Betaine is a well-studied organic osmolyte in renal medullary cells that protects the cells from harmful effects of the high interstitial sodium chloride and urea concentrations. However, little is known about the osmoprotective role of betaine in ocular applications, in particular the dry-eye diseases that are accompanied by increased osmolarity of the tear film.
- Betaine is present in human tear fluid.
- The potential of exogenous betaine to play a role in the osmoprotection of ocular epithelial cells is evidenced by expression of the betaine transporter, betaine/γ-aminobutyric acid transporter-1 (BGT-1) in human corneal epithelial cells; the role of this transporter in facilitating betaine accumulation in the human corneal epithelial cells under hyperosmotic stress; and the efflux of betaine from the cells upon returning to physiological isotonic conditions.
- Betaine attenuates hyperosmolarity-induced human corneal epithelial cell shrinkage and subsequent apoptosis *in vitro* and helps limit progression of environmentally induced animal dry eye.
- Betaine offers potential as a prophylaxis solution to provide osmoprotection of eyes in desiccating environments from developing dry-eye disease.

## Key Facts

1. **The ocular surface and tear film:** The ocular surface is the interface between the eye and outer environment and comprises the cornea and conjunctiva, upper and lower lids and the tear film. The tear film is a vital component of the eye that controls vision, transparency, comfort and protection against infection. Efficient tear-film production, turnover and stability are fundamental for maintaining ocular-surface health.
2. **Dry-eye diseases:** Dry-eye disease can be classified as either evaporative or aqueous deficient dry eye. Both result in symptoms of discomfort, visual disturbance, and tear-film instability with potential damage to the surface of the eye. Dry-eye disease is accompanied by increased osmolarity (a measure of solute concentration of particles; hyperosmolarity) of the tear film and inflammation of the ocular surface. Evaporative dry eye is most common. The prevalence of all dry-eye disease is 5–30%, and increases with age. Women are mostly affected.

3. **Evaporative dry eye**: Evaporative dry eye results from dysfunction of meibomian glands, which are located within the upper and lower eyelids. These glands produce oily secretions to form the outermost lipid layer of the tear film that is involved in preventing tear-film evaporation.
4. **Aqueous-deficient dry eye**: Aqueous tear-deficient dry eye is an eye disorder in the lacrimal glands, which are situated within the orbit on the outer portion of the upper eye. These glands produce the water layer of tears (known as the aqueous layer), which contains proteins and other solutes. Failure to produce enough of the watery component of tears results in dry eyes and an unhealthy ocular surface.
5. **Tear-film hyperosmolarity**: Abnormal tear osmolarity is a failure of homeostatic regulation. Tear hyperosmolarity arises as a result of either abnormal water evaporation from the exposed ocular surface (evaporative dry-eye disease), low aqueous tear production (aqueous tear-deficient dry-eye disease), or a combination of both. Tear hyperosmolarity is linked directly to ocular surface epithelial cell damage and inflammation as a result of cell shrinkage, and also results in tear-film instability. This instability exacerbates ocular-surface hyperosmolarity, establishing a vicious circle in dry eye.
6. **Osmoregulation**: The maintenance of osmotic balance between the cell and its outside environment is vital for the control of cell volume, cell function and health. The body is normally able to regulate osmolarity through various mechanisms. An immediate response of the cell to the hyperosmotic challenge is rapid volume correction in which inorganic ions (potassium, sodium and chloride) move across the cell membrane along with water. Accumulation of inorganic ions in cells interferes with normal cellular processes. Cells can counteract this increased ionic strength by substituting inorganic ions with small organic osmolytes (also known as compatible solutes) that do not interfere with cell metabolism.
7. **Organic osmolytes/compatible solutes**: Organic osmolytes (also known as compatible solutes) are small solutes used by cells and tissues to maintain cell volume. These solutes are amino acids and derivatives, polyols and sugars, methylamines, methylsulfonium compounds, and are usually neutral at physiological pH. Compatible solutes do not interact with macromolecules in detrimental ways and can be increased or decreased in concentration with minimal impact on cellular functions.
8. **Betaine and osmoprotection of the ocular surface**: Betaine is present in human tears. It can be transported into ocular-surface cells *via* specific transporters, the transported being increased in times of osmotic stress. Betaine attenuates hyperosmolarity-induced human corneal epithelial cell shrinkage and subsequent apoptosis *in vitro*. *In vivo*, it helps to limit the progression of environmentally induced dry eye in an animal model. Betaine offers potential as an osmoprotectant for eyes in desiccating environments and for dry-eye disease.

## Definitions of Words and Terms

**Aqueous-deficient dry eye:** Aqueous tear-deficient dry eye – lacrimal glands that produce the water and certain proteins in tears fail to produce enough to keep eyes moist and maintain a healthy eye.

**Apoptosis**: Apoptosis is a form of cell death where a programmed sequence of events leads to the elimination of cells without releasing harmful substances. Characteristic apoptotic cell morphological changes include blebbing, cell shrinkage, nuclear fragmentation, chromatin condensation and chromosmal DNA fragmentation.

**Caspase**: Caspase – intracellular cysteine proteases that are involved in cell apoptosis. Present as inactive proenzymes that are activated by proteolytic cleavage. Initiator caspases (*e.g.* caspase-8 and -9) cleave inactive proforms of effector caspases (*e.g.* caspase-3 and -7), thereby activating them and leading to destruction of other cellular proteins.

**Compatible solutes**: Compatible solutes – small molecules compatible with the cellular metabolism that accumulate in the cytoplasm of the cell to maintain cell volume in times of osmotic stress. They include sugars, polyols, and amino acids.

**Cytokines:** Cytokines – small proteins released by cells for cell-to-cell communication in immune responses and to stimulate cell movement. Examples include interleukins, lymphokines and interferons.

**Dry-eye disease**: Dry eye is an eye disease in which tear-film evaporation is high or tear production is low. It causes symptoms of discomfort, visual disturbance, and tear-film instability with damage to the surface of the eye. It is accompanied by increased osmolarity and inflammation of the ocular surface.

**Evaporative dry eye**: Evaporative dry eye results from dysfunction of the lipid or oily part of tears that normally retards evaporation of tears.

**Lacrimal gland**: The lacrimal gland – situated within the orbit on the outer portion of the upper eye. It secretes the "water" layer of tears that moisten, lubricate and protect the surface of the eye.

**Mebomian gland**: Meibomian glands – sebaceous glands located within the eyelids near the lashes, responsible for producing oily secretions, known as meibum. Meibum consists of polar and nonpolar lipids and proteins that make up the film lipid layer.

**Ocular Surface:** The ocular surface – the surface of the eye that is exposed to the outside world, including the cornea and the conjunctiva and the tear film.

**Tear-film hyperosmolarity**: Abnormal increase in salt concentration of tears. Tear hyperosmolarity arises as a result of either (1) abnormal water evaporation from the exposed ocular surface, (2) low aqueous tear production or (3) a combination of both. Tear hyperosmolarity is regarded as the central mechanism causing ocular surface inflammation, damage, and symptoms of dry eye.

## List of Abbreviations

| | |
|---|---|
| AIF | Apoptosis-Inducing Factor |
| AP | Activator Protein |
| ATF-2 | Activating Transcription Factor |
| BGT-1 | Betaine/γ-aminobutyric acid transporter-1 |
| $CD4^+$ T | Cluster of Differentiation 4 T cell |
| DNA | Deoxyribonucleic acid |
| DEWS | Dry Eye Workshop |
| ERK | Extracellular signal Regulated Kinases |
| FADD | Fas-Associated protein with Death Domain |
| FSC | Forward Scatter |
| GABA | γ-Aminobutyric Acid |
| HCLE | Human Corneal Limbal Epithelial |
| ICES | Intelligently Controlled Environmental System |
| IL | Interleukin |
| JNK | Jun N-terminal Kinase |
| LASIK | Laser Assisted Stromal *In Situ* Keratomileusis |
| MAPK | Mitogen-Activated Protein Kinase |
| MDCK | Madin–Darby Canine Kidney |
| MMP | Matrix Metalloproteinases |
| NFAT | Nuclear Factor of Activated T cells |
| NF-κB | Nuclear Factor-κB |
| NKCC | $Na^+$-$K^+$-$2Cl^-$ cotransporter |
| Plk3 | Polo-like kinase 3 |
| SSC | Side Scatter |
| TNF-α | Tumor Necrosis Factor-alpha |
| TonEBP | Tonicity-Responsive Enhancer Binding Protein |
| TRADD | Tumor Necrosis Factor Receptor-Associated Death Domain |

## Acknowledgements

The authors wish to thank Jennie Diec, Thomas Naduvilath, and Percy Lazon from Brien Holden Vision Institute, and Sonia Bustamante from the Bioanalytical Mass Spectrometry Facility of the University of New South Wales for their contributions to betaine detection in tears, and Judith Flanagan from Brien Holden Vision Institute for her critical review.

## References

Alfieri, R. R., Cavazzoni, A., Petronini, P. G., Bonelli, M. A., Caccamo, A. E., Borghetti, A. F. and Wheeler, K. P., 2002. Compatible osmolytes modulate the response of porcine endothelial cells to hypertonicity and protect them from apoptosis. *Journal of Physiology*. 540(2): 499–508.

Bildin, V. N., Yang, H., Crook, R. B., Fischbarg, J. and Reinach, P. S., 2000. Adaptation by corneal epithelial cells to chronic hypertonic stress depends on upregulation of Na:K:2Cl cotransporter gene and protein expression and ion transport activity. *Journal of Membrane Biology*. 177(1): 41–50.

Brignole, F., *et al.*, 2001. Flow cytometric analysis of inflammatory markers in KCS: 6-month treatment with topical cyclosporin A. *Investigative Ophthalmology & Visual Science*. 42(1): 90–95.

Bron, A. J., Yokoi, N., Gaffney, E. and Tiffany, J. M., 2009. Predicted phenotypes of dry eye: proposed consequences of its natural history. *Ocular Surface*. 7(2): 78–92.

Burg, M. B., Ferraris, J. D. and Dmitrieva, N. I., 2007. Cellular response to hyperosmotic stresses. *Physiological Reviews*. 87(4): 1441–1474.

Capó-Aponte, J. E., Wang, Z., Bildin, V. N., Pokorny, K. S. and Reinach, P. S., 2007. Fate of hypertonicity-stressed corneal epithelial cells depends on differential MAPK activation and p38MAPK/Na–K–2Cl cotransporter1 interaction. *Experimental Eye Research*. 84(2): 361–372.

Chen, G. and Goeddel, D. V., 2002. TNF-R1 signaling: a beautiful pathway. Science. 296(5573): 1634–1635.

Chen, W., *et al.*, 2009. Trehalose protects against ocular surface disorders in experimental murine dry eye through suppression of apoptosis. *Experimental Eye Research*. 89(3): 311–318.

Chen, W., *et al.*, 2013. Efficacy of osmoprotectants on prevention and treatment of murine dry eye. *Investigative Ophthalmology & Visual Science*. 54(9): 6287–6297.

Chen, Z., *et al.*, 2008. Hyperosmolarity-Induced cornification of human corneal epithelial cells is regulated by JNK MAPK. *Investigative Ophthalmology & Visual Science*. 49(2): 539–549.

Chuang, S. M., Wang, I. C. and Yang, J. L., 2000. Roles of JNK, p38 and ERK mitogen-activated protein kinases in the growth inhibition and apoptosis induced by cadmium. *Carcinogenesis*. 21(7): 1423–1432.

Clouzeau, C., *et al.*, 2012. Hyperosmolarity potentiates toxic effects of benzalkonium chloride on conjunctival epithelial cells in vitro. *Molecular Vision*. 18: 851–863.

Corrales, R. M., Luo, L., Chang, E. Y. and Pflugfelder, S. C., 2008. Effects of osmoprotectants on hyperosmolar stress in cultured human corneal epithelial cells. *Cornea*. 27(5): 574–579.

Definition and Classification Subcommittee of the International Dry Eye Workshop, 2007a. The definition and classification of dry eye disease: report of the Definition and Classification Subcommittee of the International Dry Eye Workshop. *Ocular Surface*. 5(2): 75–92.

Detopoulou, P., Panagiotakos, D. B., Antonopoulou, S., Pitsavos, C. and Stefanadis, C., 2008. Dietary choline and betaine intakes in relation to concentrations of inflammatory markers in healthy adults: the ATTICA study. *American Journal of Clinical Nutrition*. 87(2): 424–430.

Edwards, B. S., Oprea, T., Prossnitz, E. R. and Sklar, L. A., 2004. Flow cytometry for high-throughput, high-content screening. *Current Opinion in Chemical Biology*. 8(4): 392–398.

Elmore, S., 2007. Apoptosis: a review of programmed cell death. *Toxicologic Pathology*. 35(4): 495–516.

Epidemiology Subcommittee of the International Dry Eye Workshop, 2007b. The Epidemiology of dry eye disease: report of the Epidemiology Subcommittee of the International Dry Eye WorkShop. *Ocular Surface*. 5(2): 93–107.

Fan, T. J., Han, L. H., Cong, R. S. and Liang, J., 2005. Caspase family proteases and apoptosis. *Acta Biochimica et Biophysica Sinica (Shanghai)*. 37(11): 719–727.

Gaffney, E. A., Tiffany, J. M., Yokoi, N. and Bron, A. J., 2010. A mass and solute balance model for tear volume and osmolarity in the normal and the dry eye. *Progress in Retinal and Eye Research*. 29(1): 59–78.

Galcheva-Gargova, Z., Derijard, B., Wu, I. and Davis, R., 1994. An osmosensing signal transduction pathway in mammalian cells. *Science*. 265(5173): 806–808.

Garrett, Q., *et al.*, 2013. Betaine stabilizes cell volume and protects against apoptosis in human corneal epithelial cells under hyperosmotic stress. *Experimental Eye Research*. 108(0): 33–41.

Guillon, M., Maissa, C. and Ho, S., 2010. Evaluation of the effects on conjunctival tissues of Optive eyedrops over one month usage. *Contact Lens and Anterior Eye*. 33(2): 93–99.

Holm, P. I., Ueland, P. M., Kvalheim, G. and Lien, E. A., 2003. Determination of choline, betaine, and dimethylglycine in plasma by a high-throughput method based on normal-phase chromatography–tandem mass spectrometry. *Clinical Chemistry*. 49(2): 286–294.

Huang, Z. and Tunnacliffe, A., 2004. Response of human cells to desiccation: comparison with hyperosmotic stress response. *Journal of Physiology*. 558(1): 181–191.

Kempson, S. A. and Montrose, M. H., 2004. Osmotic regulation of renal betaine transport: transcription and beyond. *Pflügers Archiv–European Journal of Physiology*. 449(3): 227–234.

Lan, W., Petznick, A., Heryati, S., Rifada, M. and Tong, L., 2012. Nuclear factor-κB: central regulator in ocular surface inflammation and diseases. *Ocular Surface*. 10(3): 137–148.

Lang, F., 2007. Mechanisms and significance of cell volume regulation. *Journal of the American College of Nutrition*. 26(suppl. 5): 613S–623S.

Lang, F., *et al.*, 1998. Functional significance of cell volume regulatory mechanisms. *Physiological Reviews*. 78(1): 247–306.

Lee, J. H., *et al.*, 2008. NFAT5 induction and its role in hyperosmolar stressed human limbal epithelial cells. *Investigative Ophthalmology & Visual Science*. 49(5): 1827–1835.

Lemp, M. A., *et al.*, 2011. Tear osmolarity in the diagnosis and management of dry eye disease. *American Journal of Ophthalmology*. 151(5): 792–798.e1.

Li, D.-Q., Chen, Z., Song, X. J., Luo, L. and Pflugfelder, S. C., 2004. Stimulation of matrix metalloproteinases by hyperosmolarity via a JNK pathway in human corneal epithelial cells. *Investigative Ophthalmology & Visual Science*. 45(12): 4302–4311.

Li, D.-Q., *et al.*, 2006. JNK and ERK MAP kinases mediate induction of IL-1β, TNF-α and IL-8 following hyperosmolar stress in human limbal epithelial cells. *Experimental Eye Research*. 82(4): 588–596.

Li, H. and Lin, X., 2008. Positive and negative signaling components involved in TNFα-induced NF-κB activation. *Cytokine*. 41(1): 1–8.

Lim, C. H., Bot, A. G. M., de Jonge, H. R. and Tilly, B. C., 2007. Osmosignaling and volume regulation in intestinal epithelial cells. In: H. Dieter and S. Helmut (ed.), *Methods in Enzymology*. Academic Press, pp. 325–342.

Liu, H., *et al.*, 2009. A link between tear instability and hyperosmolarity in dry eye. *Investigative Ophthalmology & Visual Science*. 50(8): 3671–3679.

Lopez-Rodriguez, C. *et al.*, 2001. Bridging the NFAT and NF-kappaB families: NFAT5 dimerization regulates cytokine gene transcription in response to osmotic stress. *Immunity*. 15(1): 47–58.

Luo, L., *et al.*, 2004. Experimental dry eye stimulates production of inflammatory cytokines and MMP-9 and activates MAPK signaling pathways on the ocular surface. *Investigative Ophthalmology & Visual Science*. 45(12): 4293–4301.

Luo, L., Li, D. Q., Corrales, R. M. and Pflugfelder, S. C., 2005. Hyperosmolar saline is a proinflammatory stress on the mouse ocular surface. *Eye and Contact Lens*. 31(5): 186–193.

Luo, L., Li, D. Q. and Pflugfelder, S. C., 2007. Hyperosmolarity-induced apoptosis in human corneal epithelial cells is mediated by cytochrome c and MAPK pathways. *Cornea*. 26(4): 452–460.

McManus, M. L., Churchwell, K. B. and Strange, K., 1995. Regulation of cell volume in health and disease. *New England Journal of Medicine*. 333(19): 1260–1267.

Miyakawa, H., Woo, S. K., Dahl, S. C., Handler, J. S. and Kwon, H. M., 1999. Tonicity-responsive enhancer binding protein, a rel-like protein that stimulates transcription in response to hypertonicity. *Proceedings of the National Academy of Sciences of the United States of America*. 96(5): 2538–2542.

Moss, S. E., Klein, R. and Klein, B. K., 2000. Prevalence of and risk factors for dry eye syndrome. *Archives of Ophthalmology*. 118(9): 1264–1268.

Nichols, J. J., Ziegler, C., Mitchell, G. L. and Nichols, K. K., 2005. Self-reported dry eye disease across refractive modalities. *Investigative Ophthalmology & Visual Science*. 46(6): 1911–1914.

Olsen, M., Sarup, A., Larsson, O. M. and Schousboe, A., 2005. Effect of hyperosmotic conditions on the expression of the betaine-GABA-transporter (BGT-1) in cultured mouse astrocytes. *Neurochemical Research*. 30(6–7): 855–865.

Orlov, S. N. and Novikov, K. N., 1996. The regulation of cell volume: the mechanisms, coupled cellular reactions and pathophysiological significance. *Ross Fiziol Zh Im I M Sechenova*. 82(8–9): 1–15.

Pescosolido, N., Imperatrice, B., Koverech, A. and Messano, M., 2009. L-carnitine and short chain ester in tears from patients with dry eye. *Optometry and Vision Science*. 86(2): E132–E138.

Pflugfelder, S. C., de Paiva, C. S., Li, D.-Q. and Stern, M. E., 2008. Epithelial-immune cell interaction in dry eye. *Cornea*. 27: S9–S11 10.1097.

Pflugfelder, S. C., *et al.*, 2004. A randomized, double-masked, placebo-controlled, multicenter comparison of loteprednol etabonate ophthalmic suspension, 0.5%, and placebo for treatment of keratoconjunctivitis sicca in patients with delayed tear clearance. *American Journal of Ophthalmology*. 138(3): 444–457.

Png, E., *et al.*, 2011. Hyperosmolarity-mediated mitochondrial dysfunction requires transglutaminase-2 in human corneal epithelial cells. *Journal of Cellular Physiology*. 226(3): 693–699.

Research Subcommittee of the International Dry Eye Workshop, 2007c. Research in dry eye: report of the Research Subcommittee of the International Dry Eye WorkShop. *Ocular Surface*. 5(2): 179–193.

Rolando, M. and Zierhut, M., 2001. The ocular surface and tear film and their dysfunction in dry eye disease. *Survey of Ophthalmology*. 45(suppl. 2): S203–S210.

Schaumberg, D. A., Sullivan, D. A., Buring, J. E. and Dana, M. R., 2003. Prevalence of dry eye syndrome among US women. *American Journal of Ophthalmology*. 136(2): 318–326.

Schliess, F. and Haussinger, D., 2002. The cellular hydration state: a critical determinant for cell death and survival. *Biological Chemistry*. 383(3–4): 577–583.

Schneider-Brachert, W., *et al.*, 2004. Compartmentalization of TNF receptor 1 signaling: internalized TNF receptosomes as death signaling vesicles. *Immunity*. 21(3): 415–428.

Stern, M. E., Gao, J., Siemasko, K. F., Beuerman, R. W. and Pflugfelder, S. C., 2004. The role of the lacrimal functional unit in the pathophysiology of dry eye. *Experimental Eye Research*. 78(3): 409–416.

Sullivan, B. D. *et al.*, 2010. An objective approach to dry eye disease severity. *Investigative Ophthalmology & Visual Science*. 51(12): 6125–6130.

Tomlinson, A., Khanal, S., Ramaesh, K., Diaper, C. and McFadyen, A., 2006. Tear film osmolarity: determination of a referent for dry eye diagnosis. *Investigative Ophthalmology & Visual Science*. 47(10): 4309–4315.

Venkatesu, P., Lee, M. J. and Lin, H. M., 2009. Osmolyte counteracts urea-induced denaturation of alpha-chymotrypsin. *Journal of Physical Chemistry B*. 113(15): 5327–5338.

Verbalis, J. G., 2010. Brain volume regulation in response to changes in osmolality. *Neuroscience*. 168(4): 862–870.

Waldegger, S., *et al.*, 1998. Mechanisms and clinical significance of cell volume regulation. *Nephrology Dialysis Transplantation*. 13(4): 867–874.

Wang, L., Dai, W. and Lu, L., 2011. Hyperosmotic stress-induced corneal epithelial cell death through activation of Polo-like kinase 3 and c-Jun. *Investigative Ophthalmology & Visual Science*. 52(6): 3200–3206.

Warskulat, U., *et al.*, 2008. Ultraviolet A induces transport of compatible organic osmolytes in human dermal fibroblasts. *Experimental Dermatology*, 17(12): 1031–1036.

Warskulat, U., Reinen, A., Grether-Beck, S., Krutmann, J. and Haussinger, D., 2004. The osmolyte strategy of normal human keratinocytes in maintaining cell homeostasis. *Journal of Investigative Dermatology*. 123(3): 516–521.

Wehner, F., Olsen, H., Tinel, H., Kinne-Saffran, E. and Kinne, R. K. H., 2003. *Cell Volume Regulation: Osmolytes, Osmolyte Transport, and Signal Transduction, Reviews of Physiology, Biochemistry and Pharmacology.* Springer Berlin Heidelberg, pp. 1–80.

Woo, S. K., Lee, S. D. and Kwon, H. M., 2002. TonEBP transcriptional activator in the cellular response to increased osmolality. *Pflügers Archiv*. 444(5): 579–585.

Xia, Z., Dickens, M., Raingeaud, J., Davis, R. J. and Greenberg, M. E., 1995. Opposing effects of ERK and JNK-p38 MAP kinases on apoptosis. *Science*. 270(5240): 1326–1331.

Yamauchi, A., *et al.*, 1992. Cloning of a Na(+)- and Cl(−)-dependent betaine transporter that is regulated by hypertonicity. *Journal of Biological Chemistry*. 267(1): 649–652.

Yancey, P., Clark, M., Hand, S., Bowlus, R. and Somero, G., 1982. Living with water stress: evolution of osmolyte systems. *Science*. 217(4566): 1214–1222.

Yancey, P. H., Burg, M. B. and Bagnasco, S. M., 1990. Effects of NaCl, glucose, and aldose reductase inhibitors on cloning efficiency of renal medullary cells. *American Journal of Physiology: Cell Physiology*. 258(1): C156–C163.

Yeh, S., *et al.*, 2003. Apoptosis of ocular surface cells in experimentally induced dry eye. *Investigative Ophthalmology & Visual Science*. 44(1): 124–129.

Yu, J., Asche, C. V. and Fairchild, C. J., 2011. The economic burden of dry eye disease in the United States: a decision tree analysis. *Cornea*. 30(4): 379–387.

Yu, S. P. and Choi, D. W., 2000. Ions, cell volume, and apoptosis. *Proceedings of the National Academy of Sciences of the United States of America*. 97(17): 9360–9362.

Zhou, Y., *et al.*, 2012. The betaine-GABA transporter (BGT1, slc6a12) is predominantly expressed in the liver and at lower levels in the kidneys and at the brain surface. *American Journal of Physiology: Renal Physiology*. 302(3): F316–F328.

CHAPTER 22

# *Betaine and Choline and Cancer*

JUNG EUN LEE[a] AND EUNYOUNG CHO*[b,c]

[a]Department of Food and Nutrition, Sookmyung Women's University, Cheongpa-ro 47-gil 100, Yongsan-gu, Seoul, South Korea; [b]The Warren Alpert Medical School of Brown University, 339 Eddy Street, Providence, RI, USA; [c]Channing Division of Network Medicine, Department of Medicine Brigham and Women's Hospital, Harvard Medical School, 181 Longwood Avenue, Boston, MA, USA
*E-mail: eunyoung.cho@channing.harvard.edu

## 22.1 Introduction

Betaine and choline are important nutrients with several biological functions. Along with folate and other B vitamins, betaine and choline contribute to one-carbon metabolism that constitutes a network of integrated biochemical pathways transferring one-carbon (methyl) groups from one compound to another (Mason, 2003). Choline is oxidized to betaine. Both folate and betaine donate methyl groups to homocysteine and produce methionine, which transfer a methyl-group to *S*-adenosylmethionine, a universal methyl-group donor for methylation of DNA and RNA. Through this process, one-carbon metabolism affects genomic stability and expression and mediates nucleotide synthesis.

Methylation of homocysteine by betaine is confined to the liver and kidney, but that by folate exists in all body cells (Olthof and Verhoef, 2005). Methylation pathways mediated by choline and betaine and folate are considered

Food and Nutritional Components in Focus No. 7
Betaine: Chemistry, Analysis, Function and Effects
Edited by Victor R Preedy

Published by the Royal Society of Chemistry, www.rsc.org

to be interrelated; disruption of one pathway may affect the others. Animal and human studies support the possibility. A choline-deficient diet reduces hepatic folate levels in animals (Selhub *et al.*, 1991). Similarly, a folate-deficient diet lowered hepatic choline levels in animals (Kim *et al.*, 1994). A double-blind clinical trial of folate supplementation increased plasma betaine levels in a dose-dependent manner following folate supplementation (Melse-Boonstra *et al.*, 2005). Plasma choline levels were responsive to depletion and subsequent repletion of folate intake (Jacob *et al.*, 1999). The inverse association between plasma betaine and homocysteine levels was most noticeable when serum folate levels were low (Holm *et al.*, 2005).

Betaine and choline have other biological functions not shared by each other. Betaine is an osmolyte and protects cells, proteins, and enzymes from environmental stress (Craig, 2004). Choline is a precursor for phospholipids such as phosphatidylcholine and sphingomyelin, which are incorporated into cellular membrane and are involved in signal transduction (Zeisel and Blusztajn, 1994). Choline is also a precursor for the neurotransmitter acetylcholine and is involved in brain development and normal memory function (Meck *et al.*, 1988; Pyapali *et al.*, 1998; Albright *et al.*, 1999; Meck and Williams, 1999; Meck and Williams, 2003).

Extensive epidemiologic research has identified low folate intake as a risk factor for several cancer sites (Kennedy *et al.*, 2011; Chen *et al.*, 2014; Tio *et al.* 2014). However, epidemiologic studies on dietary betaine and choline are few because a food-composition database has been available since 2003 (Zeisel *et al.*, 2003a,b; U.S. Department of Agriculture, 2004). Prolonged deficiency of dietary choline promotes development of hepatic tumor even in the absence of any additional carcinogens in animals, suggesting its potential biological importance in cancer development (Locker *et al.*, 1986; Henning and Swendseid, 1996; Zeisel, 2000).

## 22.2 Assessment of Betaine and Choline Intake by Food Frequency Questionnaire (FFQ) in Epidemiological Studies

To assess the role of betaine and choline intake in relation to disease endpoints in epidemiological studies, it is necessary to validate betaine and choline intake measured by dietary assessment tool. A few studies have published data on the association between dietary betaine and choline intake measured by a FFQ and plasma total homocysteine level (Cho *et al.*, 2006; Chiuve *et al.*, 2007; Lee *et al.*, 2010). Because betaine intake would influence homocysteine levels as a methyl-group donor, the plasma total homocysteine level served as a biological marker potentially responsive to betaine and choline intake. Studies have found that intakes of folate and vitamins B6 and B12, other nutrients involved in methyl-group metabolism, predict homocysteine concentrations (Brattstrom, L., *et al.*, 1998; Nygard *et al.*, 1998; Jacques *et al.*, 2001; Poirier *et al.*, 2001; De Bree *et al.*, 2002; Schnyder *et al.*, 2002; Mason, 2003; Nurk *et al.*, 2004; Shai *et al.*, 2004). Higher betaine and choline intake measured by FFQ was associated with lower homocysteine levels, supporting the validity of betaine and choline intake assessment using FFQ within the range consumed by a general population (Cho *et al.*, 2006; Chiuve *et al.*, 2007; Lee *et al.*, 2010b). Furthermore, choline and betaine intake had a stronger association with homocysteine levels, especially when folate intake was low (Cho *et al.*, 2006), which further supports the interrelationship between betaine and choline and folate in methyl-group metabolism.

In epidemiologic studies, intakes of betaine and choline were almost always examined together. Intakes of these nutrients were assessed by FFQ and calculated using USDA food composition database (US Department of Agriculture, 2004), regardless of whether the studies were conducted in the US or in other countries. Total choline intake was calculated as the sum of intake from free choline, phosphocholine, glycerophosphocholine, phosphatidylcholine (lecithin), and sphingomyelin.

## 22.3 Assessment of Circulating Levels of Betaine and Choline in Epidemiological Studies

Circulating levels of betaine and choline can serve as an indicator of choline and betaine status in epidemiological studies and are often measured using mass spectrometry (Holm *et al.*, 2003; Midttun *et al.*, 2005). The laboratory method of measurement of choline is regarded as accurate and assay values and has good agreement between different methods or different laboratories (Food and Nutrition Board, 2000). Plasma or serum levels of betaine and choline have been examined in relation to cancers of prostate (Johansson *et al.*, 2009) and liver (Butler *et al.*, 2013) and colorectal adenoma (de Vogel *et al.*, 2011) in epidemiologic studies.

## 22.4 Intake Distribution and Food Sources of Betaine and Choline

Although choline is endogenously synthesized, humans still need to obtain choline from diet. The recommended daily intake was set in 1998 at 550 mg per day for men and 425 mg per day for women in the US (Yates *et al.*, 1998). Although choline is widely available in food, studies show that most choline intake in the general population comes from limited food sources. Animal-based foods, including red meat, poultry, milk, and eggs are the major sources of choline (Cho *et al.*, 2006; Cho *et al.*, 2007; Zeng *et al.*, 2013). Humans can obtain betaine either from diet or from endogenous synthesis from choline. The recommended daily intake is not established for betaine. Most betaine intake in the general population also come from limited food sources. Grain products and vegetables such as spinach and beets are major sources of betaine (Cho *et al.*, 2007; Xu *et al.*, 2009; Zeng *et al.*, 2013).

## 22.5 Case-Control Studies of Dietary Betaine and Choline Intake and Cancer Risk

There have been several case-control studies of betaine and choline intake and cancer risk from the US, China, and Australia (Table 22.1). These studies evaluated cancers of breast, esophagus, lung, nasopharynx, and ovary. Two of them investigated breast cancer. A study of breast cancer was based on a population of the Long Island Breast Cancer Study Project and found no association between betaine and choline intake and breast cancer risk (Xu *et al.*, 2008; Xu and Chen, 2009). The study also examined the groups of choline-containing compounds and found that high free (unesterified) choline intake was inversely associated with breast cancer (OR = 0.74; 95% CI = 0.53–1.02; P for trend = 0.04) (Xu *et al.*, 2009). However, dietary intake of other types of choline-containing compounds including sphingomyelin, glycerophosphocholine, phosphatidylcholine, and phosphocholine were similar between case and control groups. On the other hand, a study conducted in China found that choline and betaine intake were each associated with a reduced risk of breast cancer (OR for top *vs.* bottom quartile of intake 0.40 (95% CI 0.28–0.57) for choline and 0.58 (95% CI 0.42–0.80) for betaine) (Zhang *et al.*, 2013). When each choline-containing compound was examined, free choline and choline from glycerophosphocholine, phosphocholine, phosphatidylcholine, and sphingomyelin were inversely associated with breast cancer risk. A study from Australia investigated choline and betaine intakes in relation to risk of esophageal cancer (Ibiebele *et al.*, 2011). Betaine intake was associated with a reduced risk of Barret's esophagus but was not associated with either esophageal adenocarcinoma or squamous cell carcinoma. A study on lung cancer in the US found that higher betaine intake was associated with about 30% lower risk of lung cancer (OR = 0.71, 95% CI = 0.61–0.84) (Ying *et al.*, 2013) and the inverse association was mainly driven by former and current smokers; ORs (95% CIs) were 1.03 (0.70–1.52; *P* for trend = 0.87) among never smokers, 0.75 (0.59–0.96; *P* for trend = 0.02) among former smokers, and 0.55 (0.42–0.73; *P* for trend = 0.0001) among current smokers. In contrast, in this case-control study of lung cancer, the association for choline intake was more pronounced among never smokers than those among former or current smokers (OR = 0.68, 95% CI = 0.47–0.98 among never smokers). A case-control study on nasopharyngeal cancer in China found that higher intakes of both betaine and choline were associated with a lower risk of nasopharyngeal cancer (Zeng *et al.*, 2013). In this Chinese study, eggs contributed 26.9% of total choline intake and spinach, amaranth and water spinach contributed 79.0% of betaine intake. Also, intakes of free choline and choline from phosphatidylcholine, glycerophosphocholine, and phosphocholine were associated with lower risk of nasopharyngeal cancer. A large case-control study of Australia found that neither choline nor betaine intake were associated with ovarian cancer risk (Webb *et al.*, 2011).

**Table 22.1** Choline and betaine intake and cancer risk in case-control studies.[a]

| Author (year) | Country | Study design | Sex | No. of cases/ controls | Total Choline category | Total choline odds ratio (95% CI) | Betaine category | Betaine odds ratio (95% CI) | Variables adjusted |
|---|---|---|---|---|---|---|---|---|---|
| **Breast cancer** | | | | | | | | | |
| Xu *et al.* (2009) | USA | Case-control | W | 1479/1522 | (mg per day) | | (mg per day) | | Age, and total energy intake |
| | | | | | <122.7 | 1.00 | <61.11 | 1.00 | |
| | | | | | 122.7–160.0 | 0.91 (0.72–1.15) | 61.11–89.95 | 0.94 (0.75–1.19) | |
| | | | | | 160.0–196.9 | 0.89 (0.69–1.15) | 89.95–123.95 | 0.95 (0.75–1.21) | |
| | | | | | 196.9–247.2 | 0.79 (0.59–1.06) | 123.95–179.71 | 1.04 (0.81–1.33) | |
| | | | | | >247.2 | 0.85 (0.61–1.18) | >179.71 | 0.96 (0.73–1.25) | |
| | | | | | | *P* for trend = 0.25 | | *P* for trend = 0.93 | |
| Zhang *et al.* (2013) | China | Case-control | W | 807/807 | (mg per day) | | (mg per day) | | Age, residence, occupation, BMI, age at menarche, live births, age at first live birth, mother/sister/daughter with breast cancer, passive smoke, alcohol intake, physical activity, total energy intake, and study stage |
| | | | | | <117.65 | 1.00 | <140.14 | 1.00 | |
| | | | | | 117.65–163.62 | 0.70 (0.53–0.93) | 140.14–246.87 | 1.13 (0.84–1.50) | |
| | | | | | 163.62–217.47 | 0.47 (0.34–0.64) | 246.87–408.04 | 0.83 (0.61–1.12) | |
| | | | | | >217.47 | 0.40 (0.28–0.57) | >408.04 | 0.58 (0.42–0.80) | |
| | | | | | | *P* for trend<0.001 | | *P* for trend<0.001 | |

| | | | | | | | | | |
|---|---|---|---|---|---|---|---|---|---|
| **Esophageal cancer** | | | | | | | | | |
| Ibiebele *et al.* (2011) | Australia | Case-control | C | Esophageal adenocarcinoma 613/1490 | Esophageal adenocarcinoma Median(range) (mg per day) | | Esophageal adenocarcinoma Median(range) (mg per day) | | Age, sex, education, BMI, frequency of heartburn, acid reflux 10 y prior to diagnosis, smoking status, alcohol intake, NSAID use, and total energy intake |
| | | | | | Q1: 380 (124–435) | 1.00 | Q1: 85 (36–100) | 1.00 | |
| | | | | | Q2: 510 (435–618) | 1.34 (0.97–1.85) | Q2: 112 (100–124) | 0.88 (0.65–1.19) | |
| | | | | | Q3: 754 (618–1003) | 1.25 (0.91–1.73) | Q3: 138 (124–154) | 0.75 (0.56–1.03) | |
| | | | | | Q4: 1171 (1003–2984) | 1.25 (0.91–1.72) | Q4: 185 (154–725) | 0.79 (0.58–1.07) | |
| | | | | | | *P* for trend = 0.48 | | *P* for trend = 0.12 | |
| | | | | Esophageal squamous cell carcinoma 225/1490 | Esophageal squamous cell carcinoma Median(range) (mg per day) | | Esophageal squamous cell carcinoma Median(range) (mg per day) | | |
| | | | | | Q1: 380 (124–435) | 1.00 | Q1: 85 (36–100) | 1.00 | |
| | | | | | Q2: 510 (435–618) | 1.67 (1.03–2.69) | Q2: 112 (100–124) | 0.95 (0.60–1.52) | |
| | | | | | Q3: 754 (618–1003) | 1.48 (0.91–2.40) | Q3: 138 (124–154) | 0.72 (0.44–1.18) | |
| | | | | | Q4: 1171 (1003–2984) | 1.39 (0.88–2.30) | Q4: 185 (154–725) | 1.30 (0.83–1.99) | |
| | | | | | | | | *P* for trend = 0.21 | |

(*continued*)

**Table 22.1** (*continued*)

| Author (year) | Country | Study design | Sex | No. of cases/controls | Total Choline category | Total choline odds ratio (95% CI) | Betaine category | Betaine odds ratio (95% CI) | Variables adjusted |
|---|---|---|---|---|---|---|---|---|---|
| **Lung cancer** | | | | | | | | | |
| Ying *et al.* (2013) | USA | Case-control | C | 2821/2923 | Q1 | 1.00 | Q1 | 1.00 | Age, sex, race/ethnicity, emphysema, hay fever, asthma, addiction index, family history of cancer, dust exposure, smoking status, pack-years, second-hand smoke, time since cessation, alcohol intake, and total energy intake |
| | | | | | Q2 | 0.79 (0.68–0.93) | Q2 | 0.85 (0.73–1.00) | |
| | | | | | Q3 | 0.82 (0.70–0.96) | Q3 | 0.81 (0.69–0.95) | |
| | | | | | Q4 | 0.92 (0.78–1.07) | Q4 | 0.71 (0.61–0.84) | |
| | | | | | | *P* for trend = 0.334 | | *P* for trend <0.0001 | |
| **Nasopharyngeal carcinoma** | | | | | | | | | |
| Zeng *et al.* (2013) | China | Case-control | C | 600/600 | Median (mg per day) | | Median (mg per day) | | Age, sex, household type, education, occupation, marital status, household income, BMI, chronic rhinitis history, smoking status, alcohol intake, physical activity, exposure to potential toxic substances, multivitamin supplement, total energy intake, and folate intake |
| | | | | | Q1: 121 | 1.00 | Q1: 101 | 1.00 | |
| | | | | | Q2: 173 | 0.80 (0.57–1.11) | Q2: 206 | 0.69 (0.50–0.96) | |
| | | | | | Q3: 216 | 0.45 (0.31–0.66) | Q3: 329 | 0.56 (0.40–0.80) | |
| | | | | | Q4: 286 | 0.42 (0.29–0.61) | Q4: 548 | 0.50 (0.35–0.72) | |
| | | | | | | *P* for trend <0.001 | | *P* for trend <0.001 | |

| | | | | | | | | | |
|---|---|---|---|---|---|---|---|---|---|
| **Ovarian cancer** | | | | | | | | | |
| Webb *et al.* (2011) | Australia | Case-control | W | 1363/1414 | (mg per day) | | (mg per day) | | Age, state of residence, education, parity, alcohol intake, oral contraceptive use, and total energy intake |
| | | | | | <433 | 1.00 | <92.9 | 1.00 | |
| | | | | | 433–622 | 0.87 (0.70–1.09) | 92.9–114 | 1.08 (0.87–1.34) | |
| | | | | | 623–1022 | 0.92 (0.74–1.14) | 115–143 >143 | 0.85 (0.69–1.06) | |
| | | | | | >1022 | 0.89 (0.71–1.10) | | 1.09 (0.88–1.36) | |
| | | | | | | *P* for trend = 0.5 | | *P* for trend = 0.7 | |

[a]Abbreviations: C, men and women; M, men; W, women; Q, quantile; FFQ, food frequency questionnaire; BMI, body mass index.

It is of note that two Chinese case-control studies showed a significantly lower risk of either breast cancer (Zhang *et al.*, 2013) or nasopharyngeal cancer (Zeng *et al.*, 2013) with dietary choline and betaine intake, whereas case-control studies conducted in the US or Australia found a lack of association or less clear association. It is possible that betaine and choline intake may be more important in undernourished populations rather than well-nourished populations. This merits further studies on interaction with nutritional status or other nutrients such as folate status.

## 22.6 Cohort Studies of Dietary Betaine and Choline Intake and Cancer Risk

There have been 10 cohort studies of betaine and choline intake and cancer from the US, China, Sweden and Norway (Table 22.2). All of the US studies were based on cohort studies of health professionals that used the same FFQ. Cohort studies of both pre- (Cho *et al.*, 2007) and postmenopausal women (Cho *et al.*, 2010) did not observe any significant associations for betaine or choline intake in relation to breast-cancer risk. Among postmenopausal women, none of the individual choline-containing compounds were related to breast-cancer risk (Cho *et al.*, 2010). The same group of investigators examined the association between betaine and choline intake and colorectal cancer risk in the male health professionals and found no associations (Lee *et al.*, 2010). Also, in the latency analysis of this study, betaine or choline intake 0–4, 4–8, 8–12, or 12–16 years prior to diagnosis was not associated with colorectal cancer risk. However, for colorectal adenoma, a precursor of colorectal cancer, positive association for total choline intake and non-linear inverse association for betaine intake were observed in a study of women (Cho *et al.*, 2007). Further studies are needed to elucidate the role of betaine and choline in the colorectal carcinogenesis. A combined analysis of two cohort studies of female nurses found no association between betaine and choline intake and ovarian-cancer risk (Kotsopoulos *et al.*, 2010). Total choline intake was not associated with endometrial cancer risk in a 26 year follow-up study of female nurses (the Nurses' Health Study) (Liu *et al.*, 2013). For prostate cancer, one cohort study that examined dietary betaine and choline intake found increased risk of lethal prostate cancer with high choline intake (Richman *et al.*, 2012); RR comparing top with bottom quintiles was 1.70 (95% CI = 1.18–2.45; *P* for trend = 0.005). In this study, betaine intake was not associated with the incidence of lethal prostate cancer. A pooled analysis of the Nurses' Health Study and Health Professionals Follow-up Study found little evidence that betaine and choline intake was associated with renal cell cancer risk (Cho *et al.*, 2013).

**Table 22.2** Choline and betaine intake and cancer risk in prospective cohort studies.[a]

| Author (year) | Country | Study design | Follow-up year | Sex | No. of cases/total | Total choline category | Total choline relative risk (95% CI) | Betaine category | Betaine relative risk (95% CI) | Variables adjusted |
|---|---|---|---|---|---|---|---|---|---|---|
| **Breast cancer** | | | | | | | | | | |
| Cho *et al.* (2007a) | USA | Cohort | 1991–2003 | W | All 1032/90 663 | All median (mg per day) | | All median (mg per day) | | Age, calendar year, BMI, height, age at menarche, parity, age at first birth, family history of breast cancer, history of benign breast disease, smoking status, alcohol intake, oral contraceptive use, total energy intake, and animal-fat intake |
| | | | | | | Q1: 263 | 1.00 | Q1: 114 | 1.00 | |
| | | | | | | Q2: 301 | 0.80 (0.66–0.97) | Q2: 155 | 1.07 (0.87–1.32) | |
| | | | | | | Q3: 327 | 0.91 (0.75–1.10) | Q3: 190 | 1.01 (0.82–1.25) | |
| | | | | | | Q4: 354 | 0.79 (0.64–0.96) | Q4: 230 | 1.19 (0.97–1.46) | |
| | | | | | | Q5: 397 | 0.88 (0.72–1.07) | Q5: 305 | 0.99 (0.79–1.22) | |
| | | | | | | | *P* for trend = 0.26 | | *P* for trend = 0.88 | |
| | | | | | ER(−) 221/90 663 | ER(−) | | ER(−) | | |
| | | | | | | Q1 | 1.00 | Q1 | 1.00 | |
| | | | | | | Q2 | 0.65 (0.42–1.00) | Q2 | 0.93 (0.61–1.43) | |
| | | | | | | Q3 | 0.84 (0.56–1.25) | Q3 | 0.90 (0.59–1.38) | |
| | | | | | | Q4 | 0.74 (0.49–1.13) | Q4 | 0.94 (0.61–1.45) | |
| | | | | | | Q5 | 0.86 (0.57–1.30) | Q5 | 0.85 (0.54–1.33) | |
| | | | | | | | *P* for trend = 0.67 | | *P* for trend = 0.55 | |

(*continued*)

**Table 22.2** (*continued*)

| Author (year) | Country | Study design | Follow-up year | Sex | No. of cases/ total | Total choline category | Total choline relative risk (95% CI) | Betaine category | Betaine relative risk (95% CI) | Variables adjusted |
|---|---|---|---|---|---|---|---|---|---|---|
| Cho *et al.* (2010) | USA | Cohort | 1984–2004 | W | 3990/74 584 | Median (range) (mg per day)<br>Q1: 260 (57–280)<br>Q2: 294 (280–307)<br>Q3: 319 (307–332)<br>Q4: 346 (332–364)<br>Q5: 396 (364–1427) | 1.00<br>1.09 (0.99–1.21)<br>1.10 (1.00–1.22)<br>1.09 (0.98–1.20)<br>1.10 (0.99–1.22)<br>*P* for trend = 0.14 | Median- (range) (mg per day)<br>Q1: 71 (10–79)<br>Q2: 86 (79–93)<br>Q3: 99 (93–<br>Q4: 106)<br>Q4: 114 (106–125)<br>Q5: 144 (125–1595) | 1.00<br>1.01 (0.91–1.11)<br>1.01 (0.91–1.12)<br>1.07 (0.97–1.19)<br>0.98 (0.89–1.09)<br>*P* for trend = 0.96 | Age, calendar year, BMI at age 18, height, weight change between age 18 and current, age at menarche, parity, age at first birth, family history of breast cancer, history of benign breast disease, smoking status, alcohol intake, physical activity, postmenopausal hormone use, total energy intake, and folate intake |
| **Colorectal cancer** | | | | | | | | | | |
| Lee *et al.* (2010a) | USA | Cohort | 1986–2004 | M | 987/47 302 | Q1<br>Q2<br>Q3<br>Q4<br>Q5 | 1.00<br>1.03 (0.84–1.27)<br>1.01 (0.82–1.24)<br>1.08 (0.88–1.32)<br>0.97 (0.79–1.20)<br>*P* for trend = 0.87 | Q1<br>Q2<br>Q3<br>Q4<br>Q5 | 1.00<br>0.99 (0.81–1.21)<br>0.93 (0.75–1.13)<br>1.07 (0.88–1.30)<br>0.94 (0.77–1.16)<br>*P* for trend = 0.79 | Age, BMI, family history of colorectal cancer, history of endoscopy, smoking status before age 30, alcohol intake, aspirin dose, total energy intake, and total folate |

| | | | | | | | | | | |
|---|---|---|---|---|---|---|---|---|---|---|
| **Distal colorectal adenoma** | | | | | | | | | | |
| Cho *et al.* (2007b) | USA | Cohort | 1984–2002 | W | 2408/39 246 | Median (mg per day) | | Median (mg per day) | | Age, BMI, family history of colon cancer, history of endoscopy, year of endoscopy, smoking status, alcohol intake, physical activity, aspirin use, menopausal status, postmenopausal hormone use, total energy intake, and intakes of folate, total fiber, and calcium |
| | | | | | | Q1: 261 | 1.00 | Q1: 122 | 1.00 | |
| | | | | | | Q2: 293 | 1.03 (0.90–1.18) | Q2:158 | 0.94 (0.83–1.07) | |
| | | | | | | Q3: 315 | 1.01 (0.88–1.16) | Q3:186 | 0.85 (0.75–0.97) | |
| | | | | | | Q4: 340 | 1.23 (1.07–1.41) | Q4: 218 | 0.86 (0.75–0.98) | |
| | | | | | | Q5: 383 | 1.45 (1.27–1.67) | Q5: 276 | 0.90 (0.78–1.04) | |
| | | | | | | | *P* for trend <0.001 | | *P* for trend = 0.09 | |
| **Epithelial ovarian cancer** | | | | | | | | | | |
| Kotsopoulos *et al.* (2010) | USA | Cohort | 1984–2006 (NHS I) 1991–2005 (NHS II) | W | 526/159 957 | Q1 | 1.00 | Q1 | 1.00 | Age, calendar year, BMI, height, age at menarche, parity, family history of breast or ovarian cancer, alcohol intake, oral contraceptive use, tubal ligation, hysterectomy, menopausal status, postmenopausal hormone use, total energy intake, and caffeine intake |
| | | | | | | Q2 | 1.05 (0.79–1.41) | Q2 | 1.06 (0.80–1.41) | |
| | | | | | | Q3 | 1.27 (0.96–1.68) | Q3 | 1.19 (0.90–1.56) | |
| | | | | | | Q4 | 1.10 (0.82–1.46) | Q4 | 1.05 (0.79–1.39) | |
| | | | | | | Q5 | 0.98 (0.73–1.31) | Q5 | 0.98 (0.74–1.31) | |
| | | | | | | | *P* for trend = 0.79 | | *P* for trend = 0.51 | |

(*continued*)

**Table 22.2** (*continued*)

| Author (year) | Country | Study design | Follow-up year | Sex | No. of cases/total | Total choline category | Total choline relative risk (95% CI) | Betaine category | Betaine relative risk (95% CI) | Variables adjusted |
|---|---|---|---|---|---|---|---|---|---|---|
| **Endometrial cancer** | | | | | | | | | | |
| Liu *et al.* (2013) | USA | Cohort | 1980–2006 | W | 788/NA | Median (mg per day)<br>Q1: 227.3<br>Q2: 281.5<br>Q3: 323.6<br>Q4: 370.6<br>Q5: 444.8 | <br>1.00<br>1.09 (0.86–1.37)<br>1.12 (0.89–1.41)<br>1.17 (0.94–1.47)<br>1.02 (0.81–1.29)<br>*P* for trend = 0.66 | | | Age, calendar year, smoking status, BMI, race, age at menarche, oral contraceptive use, menopausal status, post-menopausal hormone use, and parity |
| **Lethal prostate cancer** | | | | | | | | | | |
| Richman *et al.* (2012) | USA | Cohort | 1986–2008 | M | 695/47 896 | Median (mg per day)<br>Q1: 305<br>Q2: 351<br>Q3: 385<br>Q4: 425<br>Q5: 509 | <br>1.00<br>1.24 (0.93–1.65)<br>1.38 (1.01–1.87)<br>1.55 (1.11–2.16)<br>1.70 (1.18–2.45)<br>*P* for trend = 0.005 | <br>Q1<br>Q2<br>Q3<br>Q4<br>Q5 | <br>1.00<br>1.07 (0.83–1.37)<br>1.08 (0.84–1.39)<br>1.22 (0.95–1.57)<br>1.02 (0.78–1.33)<br>*P* for trend = 0.78 | Age, time period, BMI, smoking status, physical activity, total energy intake, and intakes of coffee, calcium, cholesterol, zinc, saturated fat, lycopene, phosphorus and protein |

| | | | | | | | | | | |
|---|---|---|---|---|---|---|---|---|---|---|
| **Renal cell cancer** | | | | | | | | | | |
| Cho *et al.* (2013) | USA | Cohort | 1984–2008 (NHS) | W | W: 225/77 208 | NHS median (mg per day) | | NHS median (mg per day) | | Age, calendar year, BMI, parity in NHS, history of hypertension, history of diabetes, smoking status, alcohol intake, physical activity, total energy intake, and intakes of fruit and vegetable |
| | | | 1986–2008 (HPFS) | M | M: 211/47 886 | Q1: 267.1 | 1.00 | Q1: 70.5 | 1.00 | |
| | | | | | | Q2: 300.8 | 1.33 (0.88–1.99) | Q2: 85.6 | 0.84 (0.55–1.29) | |
| | | | | | | Q3: 325.1 | 1.04 (0.67–1.60) | Q3: 97.7 | 0.98 (0.65–1.49) | |
| | | | | | | Q4: 351.9 | 1.07 (0.70–1.66) | Q4: 112 | 1.09 (0.72–1.65) | |
| | | | | | | Q5: 399.8 | 0.78 (0.48–1.27) | Q5: 138.4 | 0.78 (0.49–1.24) | |
| | | | | | | | *P* for trend = 0.17 | | *P* for trend = 0.57 | |
| | | | | | | HPFS median (mg per day) | | HPFS median (mg per day) | | |
| | | | | | | Q1: 308.3 | 1.00 | Q1: 84.9 | 1.00 | |
| | | | | | | Q2: 348.0 | 0.89 (0.58–1.38) | Q2: 106.2 | 1.09 (0.70–1.70) | |
| | | | | | | Q3: 376.9 | 1.22 (0.81–1.84) | Q3: 124 | 1.35 (0.88–2.08) | |
| | | | | | | Q4: 409.5 | 0.62 (0.38–1.01) | Q4: 145.5 | 0.95 (0.59–1.52) | |
| | | | | | | Q5: 471.7 | 0.96 (0.62–1.48) | Q5: 186.4 | 1.3 (0.83–2.02) | |
| | | | | | | | *P* for trend = 0.55 | | *P* for trend = 0.4 | |

(*continued*)

**Table 22.2** (*continued*)

| Author (year) | Country | Study design | Follow-up year | Sex | No. of cases/ total | Total choline category | Total choline relative risk (95% CI) | Betaine category | Betaine relative risk (95% CI) | Variables adjusted |
|---|---|---|---|---|---|---|---|---|---|---|
| | | | | C | | Pooled | | Pooled | | |
| | | | | | | Q1 | 1.00 | Q1 | 1.00 | |
| | | | | | | Q2 | 1.10 (0.75–1.62) | Q2 | 0.95 | |
| | | | | | | | 1.13 (0.84–1.52) | | (0.70–1.30) | |
| | | | | | | Q3 | 0.83 (0.49–1.41) | Q3 | 1.15 | |
| | | | | | | | 0.87 (0.63–1.21) | | (0.84–1.57) | |
| | | | | | | Q4 | *P* for trend = 0.19 | Q4 | 1.03 | |
| | | | | | | | | | (0.75–1.40) | |
| | | | | | | Q5 | | Q5 | 1.01 | |
| | | | | | | | | | (0.62–1.65) | |
| | | | | | | | | | *P* for trend = 0.69 | |

[a]Abbreviations: C, men and women; M, men; W, women; Q, quantile; FFQ, food frequency questionnaire; BMI, body mass index; ER, estrogen receptor-negative; NHS, Nurses' Health Study; HPFS, Health Professionals Follow-up Study; NA, not available.

## 22.7 Circulating Levels of Betaine and Choline and Cancer Risk

A cross-sectional study in Norway examined whether plasma levels of betaine and choline were associated with distal colorectal adenomas and found that high plasma levels of betaine was related to reduced prevalence of high-risk colorectal adenomas (adenomas ≥ 10 mm in diameter and/or villous components and/or severe dysplasia) (de Vogel *et al.*, 2011); OR comparing top with bottom quartiles was 0.74 (95% CI 0.54–1.02; *P* for trend = 0.03). However, plasma choline levels were not associated with colorectal adenomas. In this study, mean plasma levels of betaine adjusted for age, sex, study center, smoking status and alcohol intake were 37.30 μmol $L^{-1}$ among non-cases, 36.76 μmol $L^{-1}$ among cases with low-risk colorectal adenomas, and 35.80 μmol $L^{-1}$ among those with high-risk colorectal adenomas. In a nested case-control study of Chinese men living in a hepatocellular cancer high-risk region, an inverse association of serum choline levels with hepatocellular cancer risk was found (OR = 0.35; 95% CI = 0.16–0.78; *P* for trend = 0.01) (Butler *et al.*, 2013) (Table 22.3). This inverse association remained significant in those with serum hepatitis B surface antigen negative (HBsAg(−)) and whom did not have cirrhosis (OR = 0.38; 95% CI = 0.13–1.06; *P* for trend = 0.06). No association was found with plasma betaine levels. Two nested case-control studies of prostate cancer evaluated plasma choline and betaine levels. A Norwegian study including approximately 3000 cases and matched controls found no significant association for serum betaine levels (de Vogel *et al.*, 2014). However, another study in Sweden showed a positive association of prostate cancer risk with plasma choline levels (Johansson *et al.*, 2009).

## 22.8 Mortality Studies of Dietary Betaine and Choline Intake among Cancer Patients

The survival studies typically evaluated betaine and choline intake prior to diagnosis of cancer and followed the cancer patients until death (Table 22.4). Epidemiologic studies began from a case-control study design could ascertain all-cause or cancer-specific mortality through linkage to the national death statistics or active follow-up among case patients. Betaine and choline levels were asked when investigators recruited case and control groups. Because a primary purpose of the case-control studies was to determine risk factors for cancer development, participants were generally asked about their diet before the cancer diagnosis such as the year before the diagnosis. Survival analysis was performed only among case group in the study population (Xu *et al.*, 2009; Galván-Portillo *et al.*, 2010; Dixon *et al.*, 2013). Case-only survival analysis in the Long Island Breast Cancer Study found that higher intakes of betaine, phosphocholine and free choline were associated with lower risk of all-cause and breast cancer-specific deaths (Xu *et al.*, 2009). However, two other studies did not find any

**Table 22.3** Circulating choline and betaine levels and cancer risk in prospective cohort studies.[a]

| Author (year) | Country | Study design | Follow-up year | Sex | No. of cases/controls | Total choline category | Total choline relative risk (95% CI) | Betaine category | Betaine relative risk (95% CI) | Choline betain assessment | Variables adjusted |
|---|---|---|---|---|---|---|---|---|---|---|---|
| **Hepatocellular carcinoma** | | | | | | | | | | | |
| Butler *et al.* (2013) | China | Nested case-control | 1986–2012 | M | 294/631 | (μmol $L^{-1}$)<br>Q1: <16.8<br>Q2: 16.8–19.8<br>Q3: 19.9–24.0<br>Q4: 24.1–30.6<br>Q5: >30.6 | 1.00<br>0.76 (0.45–1.30)<br>0.83 (0.49–1.40)<br>0.59 (0.32–1.08)<br>0.35 (0.16–0.78)<br>*P* for trend = 0.01 | (μmol $L^{-1}$)<br>Q1: <48.5<br>Q2: 48.5–57.8<br>Q3: 57.9–65.3<br>Q4: 65.4–75.8<br>Q5: >75.8 | 1.00<br>1.06 (0.62–1.81)<br>0.96 (0.55–1.67)<br>0.79 (0.45–1.40)<br>1.37 (0.78–2.41)<br>*P* for trend = 0.41 | Serum levels | Date of birth, date of blood draw, residence, liver cirrhosis, smoking status, alcohol intake, seropositivity for hepatitis B surface antigen, and one-carbon metabolites |

| **Prostate cancer** | | | | | | | | | | | |
|---|---|---|---|---|---|---|---|---|---|---|---|
| Johansson *et al.* (2009) | Sweden | Nested case-control | 1985–2006 | M | 522/959 | Q1 | 1.00 | Q1 | 1.00 | Plasma levels | Age and date of blood draw |
| | | | | | | Q2 | 0.91 (0.65–1.28) | Q2 | 1.16 (0.84–1.61) | | |
| | | | | | | Q3 | 1.05 (0.74–1.47) | Q3 | 1.31 (0.94–1.81) | | |
| | | | | | | Q4 | 1.48 (1.07–2.04) | Q4 | 1.36 (0.98–1.88) | | |
| | | | | | | | *P* for trend = 0.03 | | *P* for trend = 0.08 | | |
| de Vogel *et al.* (2014) | Norway | Nested case-control | 1973–2007 | M | 2941/2951 | — | — | (μmol $L^{-1}$) | | Serum levels | Education, BMI, smoking status, physical activity, and serum creatinine concentration |
| | | | | | | | | Q1: <36.1 | 1.00 | | |
| | | | | | | | | Q2: 36.1–<41.1 | 1.04 (0.88–1.22) | | |
| | | | | | | | | Q3: 41.1–<45.6 | 1.02 (0.87–1.20) | | |
| | | | | | | | | Q4: 45.6–<52.0 | 1.05 (0.89–1.24) | | |
| | | | | | | | | Q5: ≥52.0 | 0.98 (0.83–1.16) | | |
| | | | | | | | | | *P* for trend = 0.81 | | |

[a]Abbreviations: M, men; Q, quantile; BMI, body mass index.

**Table 22.4** Choline and betaine intake and mortality among cancer patients.[a]

| Author (year) | Country | Study design | Follow-up years | Sex | No. of deaths/total | Total choline category | Total choline hazard ratio (95% CI) | Betaine category | Betaine hazard ratio (95% CI) | Variables adjusted |
|---|---|---|---|---|---|---|---|---|---|---|
| ***All-cause deaths*** | | | | | | | | | | |
| **Breast cancer** | | | | | | | | | | |
| Xu *et al.* (2009)[b] | USA | Cohort | (1996–1997)–2005 (mean = 8.0 yr) | W | 308/1508 | (mg per day) | | (mg per day) | | Age at diagnosis, and total energy intake |
| | | | | | | <142 | 1.00 (ref.) | <79 | 1.00 (ref.) | |
| | | | | | | 142–205 | 0.79 (0.58–1.08) | 79–138 | 0.76 (0.58–1.01) | |
| | | | | | | >205 | 1.04 (0.71–1.52) | >138 | 0.64 (0.46–0.89) | |
| | | | | | | | *P* for trend = 0.92 | | *P* for trend = 0.01 | |
| **Ovarian cancer** | | | | | | | | | | |
| Dixon *et al.* (2013)[b] | Australia | Cohort | (2002–2006)–2011 | W | 792/1132 | (mg per day) | | (mg per day) | | Age, cancer stage, residual tumor, grade, and total energy intake |
| | | | | | | ≤428 | 1.00 | ≤93 | 1.00 | |
| | | | | | | 429–609 | 0.92 (0.75–1.14) | 94–114 | 1.12 (0.91–1.38) | |
| | | | | | | 610–984 | 0.92 (0.75–1.14) | 115–145 | 1.02 (0.82–1.26) | |
| | | | | | | >984 | 0.94 (0.76–1.16) | >145 | 1.09 (0.88–1.36) | |
| | | | | | | | *P* for trend = 0.7 | | *P* for trend = 0.6 | |
| | | | | | | Per 100 µg per day | 0.99 (0.97–1.02) | Per 100 µg per day | 1.05 (0.94–1.17) | |

| | | | | | | | | | | |
|---|---|---|---|---|---|---|---|---|---|---|
| ***Disease-specific deaths*** | | | | | | | | | | |
| **Breast cancer** | | | | | | | | | | |
| Xu *et al.* (2009)[b] | USA | Cohort | (1996–1997)–2005 (mean = 8.0 yr) | W | 164/1508 | (mg per day) | | (mg per day) | | Age at diagnosis, and total energy intake |
| | | | | | | <142 | 1.00 | <79 | 1.00 | |
| | | | | | | 142–205 | 0.79 (0.52–1.19) | 79–138 | 0.65 (0.43–0.97) | |
| | | | | | | >205 | 0.72 (0.43–1.22) | >138 | 0.65 (0.41–1.00) | |
| | | | | | | | *P* for trend = 0.22 | | *P* for trend = 0.05 | |
| **Gastric cancer** | | | | | | | | | | |
| Galván-Portillo *et al.* (2010)[b] | Mexico | Cohort[b] | (2004–2005)–2006 | C | 128/248 | (mg per day) | | (mg per day) | | Age, sex, tumor stage, smoking status, alcohol intake, total energy intake, Helicobacter pylori CagA infection seropositivity, MTHFR C677 T polymorphism |
| | | | | | | 170.0–582.0 | 1.00 | 74.9–498.6 | 1.00 | |
| | | | | | | 582.1–776.6 | 1.06 (0.64–1.73) | 498.7–716.9 | 0.78 (0.46–1.30) | |
| | | | | | | 776.7–1542.5 | 1.07 (0.61–1.91) | 717.0–1493.5 | 0.92 (0.52–1.64) | |
| | | | | | | | *P* for trend = 0.80 | | *P* for trend = 0.81 | |

(*continued*)

Table 22.4 (*continued*)

| Author (year) | Country | Study design | Follow-up years | Sex | No. of deaths/total | Total choline category | Total choline hazard ratio (95% CI) | Betaine category | Betaine hazard ratio (95% CI) | Variables adjusted |
|---|---|---|---|---|---|---|---|---|---|---|
| **Lethal prostate cancer** | | | | | | | | | | |
| Richman *et al.* (2012) | USA | Cohort | 1986–2008 | M | 271/4282 | (Median mg per day) | | | | Age, time period, time since diagnosis, total energy intake, treatment, Gleason sum, clinical T-stage, BMI, smoking, vigorous activity, and intakes of calcium, coffee, phosphorus, saturated fat, cholesterol, and polyunsaturated fat |
| | | | | | | Q1: (297) | 1.0 | Q1 | 1.0 | |
| | | | | | | Q2: (341) | 1.64 (1.06, 2.54) | Q2 | 1.16 (0.78, 1.73) | |
| | | | | | | Q3: (371) | 1.30 (0.80, 2.13) | Q3 | 1.07 (0.70, 1.62) | |
| | | | | | | Q4: (405) | 1.40 (0.82, 2.41) | Q4 | 1.12 (0.74, 1.69) | |
| | | | | | | Q5: (471) | 1.69 (0.93, 3.09) | Q5 | 1.08 (0.70, 1.67) | |
| | | | | | | | *P* for trend = 0.20 | | *P* for trend = 0.92 | |

[a]Abbreviations: C, men and women; M, men; W, women; Q, quantile; FFQ, food frequency questionnaire; BMI, body mass index; MTHFR, methylenetetrahydrofolate reductase.

[b]Study participants were obtained from case group diagnosed with cancer in a case-control study.

significant association for either betaine or choline in relation to survival among patients of ovarian cancer (Dixon *et al.*, 2013) or gastric cancer (Galván-Portillo *et al.*, 2010).

One cohort study, the Health Professionals Follow-up Study (Richman *et al.*, 2012), examined whether postdiagnostic intake of betaine and choline was associated with lethal prostate cancer, defined as distant organ metastases from prostate cancer or prostate cancer death. In this study, because the FFQs have been administered almost every 4 years, investigators were able to assess dietary betaine and choline intake after prostate-cancer diagnosis and participants with prostate cancer were followed until the distant organ metastases or deaths. No significant associations were observed for either betaine or choline intake. The small numbers of studies or inconsistent findings among studies warrants further prospective studies examining both pre- and postdiagnostic status of dietary betaine and choline in relation to survival.

## 22.9 Summary and Conclusion

In summary, choline and betaine are important nutrients to consider evaluating the association in relation to cancer risk. However, due to lack of a food-composition database for choline and betaine intake until relatively recently, epidemiologic studies have just started to emerge on this topic. Even fewer studies exist evaluating blood levels of these nutrients in relation to cancer. These nutrients would be especially important in relation to cancer risk in undernourished populations where intakes of certain nutrients such as folate intake are low. There should be more epidemiologic studies in diverse populations with various cancer sites to better understand the role of betaine and choline in cancer development and progression.

## Summary Points

- This chapter has summarized the epidemiologic evidence on the roles of betaine and choline in the cancer development and progression.
- Betaine and choline are important nutrients in humans with a variety of functions in the body.
- Epidemiologic studies of dietary betaine and choline and chronic diseases are sparse because a food-composition database for the nutrients only became available in 2003.
- Several case-control and prospective cohort studies have recently evaluated the association between betaine and choline intake in relation to risk of cancer.
- A few studies have evaluated betaine and choline intake in relation to mortality among cancer patients.

# Key Facts

## Key Facts of Betaine and Choline

- Choline is a precursor of betaine, a methyl-group donor.
- Provision of a methyl group is important for DNA methylation and synthesis.
- Betaine is also an osmolyte.
- Choline is essential for the structural integrity of cell membranes.
- Choline is involved in the synthesis and release of acetylcholine, an important neurotransmitter in memory storage, muscle control, and many other functions.
- Major food sources of choline are eggs, milk, and meat.
- Major food sources of betaine are spinach and grain products.

## Key Facts of Measures in Epidemiologic Studies

- Odds ratio (OR) is a measure of association that compares the odds of disease in exposed to the odds of disease in nonexposed and used in case-control studies.
- Hazard ratio (HR) or relative risk (RR) is a measure of how often a particular event happens in one group compared to how often it happens in another group, over time and used in prospective cohort studies.
- A 95% confidence interval (95% CI) is an estimated range of values where the true value would be included in 95%.

# Definitions of Words and Terms

**Betaine**: Betaine is a methyl donor for the betaine-homocysteine methyltransferase (BHMT)-mediated reaction. BHMT is an enzyme that catalyzes the conversion of betaine and homosyteine to dimethylglycine and methionine.

**Choline**: Choline is a precursor for acetylcholine, phospholipids, and betaine. Choline is available as free choline or is bound as esters including phosphocholine, glycerophosphocholine, sphingomyelin, and phosphatidylcholine.

**Food frequency questionnaire (FFQ)**: FFQ measures long-term intake of foods by assessing one's usual frequency (and portion) of individual food intake. The basic FFQ consists of two components: a food list and a frequency response section for subjects to report how often each food was eaten. Often, the FFQ also includes further details of quantity of each food consumed. Nutrients are calculated from the food intake measured by FFQ.

## Abbreviations

| | |
|---|---|
| BHMT | Betaine-homocysteine methyltransferase |
| CI | Confidence interval |
| FFQ | Food frequency questionnaire |
| HR | Hazard ratio |
| OR | Odds ratio |
| RR | Relative risk |

## Acknowledgements

None of the authors had any personal or financial conflicts of interest.

## References

Albright, C. D., Tsai, A. Y., *et al.*, 1999. Choline availability alters embryonic development of the hippocampus and septum in the rat. *Brain Research. Developmental Brain Research*. 113(1–2): 13–20.

Brattstrom, L., Landgren, F., Israelsson, *et al.*, 1998. Lowering blood homocysteine with folic acid based supplements: meta-analysis of randomised trials. *British Medical Journal*. 316(7135): 894–898.

Butler, L. M., Arning, E., *et al.*, 2013. Prediagnostic levels of serum one-carbon metabolites and risk of hepatocellular carcinoma. *Cancer Epidemiology, Biomarkers & Prevention*. 22(10): 1884–1893.

Chen, P., Li, C., *et al.*, 2014. Higher dietary folate intake reduces the breast cancer risk: a systematic review and meta-analysis. *British Journal of Cancer*. 110(9): 2327–2338.

Chiuve, S. E., Giovannucci, E. L., *et al.* (2007). The association between betaine and choline intakes and the plasma concentrations of homocysteine in women. *American Journal of Clinical Nutrition*. 86(4): 1073–1081.

Cho, E., Giovannucci, E. L., *et al.*, 2013. Nutrients related to one-carbon metabolism and risk of renal cell cancer. *Cancer Causes & Control*. 24(2): 373–382.

Cho, E., Holmes, M. D., *et al.*, 2010. Choline and betaine intake and risk of breast cancer among post-menopausal women. *British Journal of Cancer*. 102(3): 489–494.

Cho, E., Holmes, M., *et al.*, 2007a. Nutrients Involved in One-Carbon Metabolism and Risk of Breast Cancer among Premenopausal Women. *Cancer Epidemiology, Biomarkers & Prevention*. 16(12): 2787–2790.

Cho, E., Willett, W. C., *et al.*, 2007b. Dietary choline and betaine and the risk of distal colorectal adenoma in women. *Journal of National Cancer Institute*. 99(16): 1224–1231.

Cho, E., Zeisel, S. H., *et al.*, 2006. Dietary choline and betaine assessed by food-frequency questionnaire in relation to plasma total homocysteine concentration in the Framingham Offspring Study. *American Journal of Clinical Nutrition*. 83(4): 905–911.

Craig, S. A. 2004. Betaine in human nutrition. *American Journal of Clinical Nutrition.* 80(3): 539–549.

De Bree, A., Verschuren, W. M., *et al.*, 2002. Homocysteine determinants and the evidence to what extent homocysteine determines the risk of coronary heart disease. *Pharmacological Reviews.* 54(4): 599–618.

de Vogel, S., Schneede, J., *et al.*, 2011. Biomarkers related to one-carbon metabolism as potential risk factors for distal colorectal adenomas. *Cancer Epidemiology, Biomarkers & Prevention.* 20(8): 1726–1735.

de Vogel, S., Ulvik, A. *et al.*, 2014. Sarcosine and other metabolites along the choline oxidation pathway in relation to prostate cancer—a large nested case-control study within the JANUS cohort in Norway. *International Journal of Cancer.* 134(1): 197–206.

Dixon, S. C., Ibiebele, T. I., *et al.*, 2013. Dietary folate and related micronutrients, folate-metabolising genes, and ovarian cancer survival. *Gynecologic Oncology.* 132(3): 566–572.

Food and Nutrition Board, 2000. *Dietary Reference Intakes for Thiamin, Riboflavin, Niacin, Vitamin $B_6$, Folate, Vitamin $B_{12}$, Pantothenic Acid, Biotin, Choline.* National Academy of Sciences, Washington, DC.

Galván-Portillo, M. V., Oñate-Ocaña, L. F., *et al.*, 2010. Dietary folate and vitamin B12 intake before diagnosis decreases gastric cancer mortality risk among susceptible MTHFR 677TT carriers. *Nutrition.* 26(2): 201–208.

Henning, S. M. and Swendseid, M. E., 1996. The role of folate, choline, and methionine in carcinogenesis induced by methyl-deficient diets. *Advances in Experimental Medicine and Biology.* 399: 143–155.

Holm, P. I., Ueland, P. M., *et al.*, 2003. Determination of choline, betaine, and dimethylglycine in plasma by a high-throughput method based on normal-phase chromatography-tandem mass spectrometry. *Clinical Chemistry.* 49(2): 286–294.

Holm, P. I., Ueland, P. M., *et al.*, 2005. Betaine and folate status as cooperative determinants of plasma homocysteine in humans. Arteriosclerosis, Thrombosis, and Vascular Biology. 25(2): 379–385.

Ibiebele, T. I., Hughes, M. C., *et al.*, 2011. High intake of folate from food sources is associated with reduced risk of esophageal cancer in an Australian population. *Journal of Nutrition.* 141(2): 274–283.

Jacob, R. A., Jenden, D. J., *et al.*, 1999. Folate nutriture alters choline status of women and men fed low choline diets. *Journal of Nutrition.* 129(3): 712–717.

Jacques, P. F., Bostom, A. G., *et al.*, 2001. Determinants of plasma total homocysteine concentration in the Framingham Offspring cohort. *American Journal of Clinical Nutrition.* 73(3): 613–621.

Johansson, M., Van Guelpen, B., *et al.*, 2009. One-carbon metabolism and prostate cancer risk: prospective investigation of seven circulating B vitamins and metabolites. *Cancer Epidemiology, Biomarkers & Prevention.* 18(5): 1538–1543.

Kennedy, D. A., Stern, S. J., *et al.*, 2011. Folate intake and the risk of colorectal cancer: a systematic review and meta-analysis. *Cancer Epidemiology.* 35(1): 2–10.

Kim, Y. I., Miller, J. W., *et al.*, 1994. Severe folate deficiency causes secondary depletion of choline and phosphocholine in rat liver. *Journal of Nutrition*. 124(11): 2197–2203.

Kotsopoulos, J., Hankinson, S. E., *et al.*, 2010. Dietary betaine and choline intake are not associated with risk of epithelial ovarian cancer. *European Journal of Clinical Nutrition*. 64(1): 111–114.

Lee, J. E., Giovannucci, E., *et al.*, 2010a. Choline and betaine intake and the risk of colorectal cancer in men. *Cancer Epidemiology, Biomarkers & Prevention*. 19(3): 884–887.

Lee, J. E., Jacques, P. F., *et al.*, 2010b. Are dietary choline and betaine intakes determinants of total homocysteine concentration? *American Journal of Clinical Nutrition*. 91(5): 1303–1310.

Liu, J. J., Hazra, A., *et al.*, 2013. One-carbon metabolism factors and endometrial cancer risk. *British Journal of Cancer*. 108(1): 183–187.

Locker, J., Reddy, T. V., *et al.*, 1986. DNA methylation and hepatocarcinogenesis in rats fed a choline-devoid diet. *Carcinogenesis*. 7(8): 1309–1312.

Mason, J. B., 2003. Biomarkers of nutrient exposure and status in one-carbon (methyl) metabolism. *Journal of Nutrition*. 133(suppl. 3): 941S–947S.

Meck, W. H., Smith, R. A., *et al.*, 1988. Pre- and postnatal choline supplementation produces long-term facilitation of spatial memory. *Developmental Psychobiology*. 21(4): 339–353.

Meck, W. H. and Williams, C. L., 1999. Choline supplementation during prenatal development reduces proactive interference in spatial memory. *Brain Research. Developmental Brain Research*. 118(1–2): 51–59.

Meck, W. H. and Williams, C. L., 2003. Metabolic imprinting of choline by its availability during gestation: implications for memory and attentional processing across the lifespan. *Neuroscience & Biobehavioral Reviews*. 27(4): 385–399.

Melse-Boonstra, A., Holm, P. I., *et al.*, 2005. Betaine concentration as a determinant of fasting total homocysteine concentrations and the effect of folic acid supplementation on betaine concentrations. *American Journal of Clinical Nutrition*. 81(6): 1378–1382.

Midttun, O., Hustad, S., *et al.*, 2005. Multianalyte quantification of vitamin $B_6$ and $B_2$ species in the nanomolar range in human plasma by liquid chromatography-tandem mass spectrometry. *Clinical Chemistry*. 51(7): 1206–1216.

Nurk, E., Tell, G. S., *et al.*, 2004. Changes in lifestyle and plasma total homocysteine: the Hordaland Homocysteine Study. *American Journal of Clinical Nutrition*. 79(5): 812–819.

Nygard, O., Refsum, H., *et al.*, 1998. Major lifestyle determinants of plasma total homocysteine distribution: the Hordaland Homocysteine Study. *American Journal of Clinical Nutrition*. 67(2): 263–270.

Olthof, M. R. and Verhoef, P., 2005. Effects of betaine intake on plasma homocysteine concentrations and consequences for health. *Current Drug Metabolism*. 6(1): 15–22.

Poirier, L. A., Wise, C. K., *et al.*, 2001. Blood determinations of S-adenosylmethionine, S-adenosylhomocysteine, and homocysteine: correlations with diet. *Cancer Epidemiology, Biomarkers & Prevention*. 10(6): 649–655.

Pyapali, G. K., Turner, D. A., *et al.*, 1998. Prenatal dietary choline supplementation decreases the threshold for induction of long-term potentiation in young adult rats. *Journal of Neurophysiology*. 79(4): 1790–1796.

Richman, E. L., Kenfield, S. A., *et al.*, 2012. Choline intake and risk of lethal prostate cancer: incidence and survival. *American Journal of Clinical Nutrition*. 96(4): 855–863.

Schnyder, G., Roffi, M., *et al.*, 2002. Effect of homocysteine-lowering therapy with folic acid, vitamin $B_{12}$, and vitamin $B_6$ on clinical outcome after percutaneous coronary intervention: the Swiss Heart study: a randomized controlled trial. *Journal of American Medical Association*. 288(8): 973–979.

Selhub, J., Seyoum, E., *et al.*, 1991. Effects of choline deficiency and methotrexate treatment upon liver folate content and distribution. *Cancer Research*. 51(1): 16–21.

Shai, I., Stampfer, M. J., *et al.*, 2004. Homocysteine as a risk factor for coronary heart diseases and its association with inflammatory biomarkers, lipids and dietary factors. *Atherosclerosis*. 177(2): 375–381.

Tio, M., Andrici, J., *et al.*, 2014. Folate intake and the risk of upper gastrointestinal cancers: a systematic review and meta-analysis. *Journal of Gastroenterology and Hepatology*. 29(2): 250–258.

U.S. Department of Agriculture, 2004. *USDA Database for the Choline Content of Common Foods*. From http://www.nal.usda.gov/fnic/foodcomp/Data/Choline/Choline.html.

Webb, P. M., Ibiebele, T. I., *et al.*, 2011. Folate and related micronutrients, folate-metabolising genes and risk of ovarian cancer. *European Journal of Clinical Nutrition*. 65(10): 1133–1140.

Xu, X. and Chen, J., 2009. One-carbon metabolism and breast cancer: an epidemiological perspective. *Journal of Genetics and Genomics = Yi chuan xue bao*. 36(4): 203–214.

Xu, X., Gammon, M. D., *et al.*, 2008. B-vitamin intake, one-carbon metabolism, and survival in a population-based study of women with breast cancer. *Cancer Epidemiology, Biomarkers & Prevention*. 17(8): 2109–2116.

Xu, X., Gammon, M. D., *et al.*, 2009. High intakes of choline and betaine reduce breast cancer mortality in a population-based study. *FASEB Journal*. 23(11): 4022–4028.

Yates, A. A., Schlicker, S. A., *et al.*, 1998. Dietary Reference Intakes: the new basis for recommendations for calcium and related nutrients, B vitamins, and choline. *Journal of American Dietetic Association*. 98(6): 699–706.

Ying, J., Rahbar, M. H., *et al.*, 2013. Associations between dietary intake of choline and betaine and lung cancer risk. *PLoS One*. 8(2): e54561.

Zeisel, S. H., 2000. Choline: an essential nutrient for humans. *Nutrition*. 16(7–8): 669–671.

Zeisel, S. H. and Blusztajn, J. K., 1994. Choline and human nutrition. *Annual Review of Nutrition*. 14: 269–296.

Zeisel, S. H., Mar, M.-H., *et al.*, 2003a. Erratum: Concentrations of choine-containing compounds and betaine in common foods. *Journal of Nutrition.* 133: 1302–1307, 2918–2919.
Zeisel, S. H., Mar, M. H., *et al.*, 2003b. Concentrations of choline-containing compounds and betaine in common foods. *Journal of Nutrition.* 133(5): 1302–1307.
Zeng, F. F., Xu, C. H., *et al.*, 2013. Choline and betaine intakes are associated with reduced risk of nasopharyngeal carcinoma in adults: a case-control study. *British Journal of Cancer.* 110(3): 808–816.
Zhang, C. X., Pan, M. X., *et al.*, 2013. Choline and betaine intake is inversely associated with breast cancer risk: a two-stage case-control study in China. *Cancer Science.* 104(2): 250–258.

# Subject Index

Illustrations and figures are in **bold**. Tables are in *italics*.


abiotic stress *see also* drought stress; freezing stress; osmotic stress; salt stress (plants)
  betaine accumulation 32, 222–3
  betaine function 18, 30, 203
  definition 30, 178–9, 195
absorption 3, 76, 204–5, **210**, 215, 250, 254
abundance 56
accumulation (plants)
  abiotic stress 32, 203, 222–3
  biosynthesis 9
  dietary sources *31*
  evolutionary development 33–5
  genetic engineering 21, 23, 30–1
  osmoprotection 10, 20, 30–1
  sugar beet 11–14, 17
accuracy
  capillary electrophoresis/mass spectrometry (CE/MS) 129
  definition 134
acetonitrile (ACN) 142, 144
acetyl-carnitine 123
acetyl CoA **316**
acetyl-CoA carboxylase 358
*N*-acetyl-*S*-farnesyl-L-cysteine 294
acetylcholine 250, 391, 412
*N*-acetylcysteine 333
*Achillea* genus 192
*Achyranthes aspera* Linn *160*, 162–7, 172
actin 360
activating transcription factor (ATF-2) 379
activator protein 1 (AP-1) 379
acyl-carnitines 120, 123, 212
acyl-CoA dehydrogenases **316**
acyl-CoA-diglyceride acyltransferase (DGAT) **292**, 294–5
acyl-CoA oxidases **316**
adenosine 16, 288
adenosine-5'-triphosphate (ATP) 79, 288, 360
adenosine kinase 16
adenosine monophosphate 16
*S*-adenosylhomocysteine hydrolase (SAHH) **287**, 288, 295
*S*-adenosylhomocysteine (SAH) *see also* methionine cycle
  alcohol effects 288, **289**, **290**
  apoptosis 295
  biosynthesis 206, 301
  hydrolysis 288
  plants 16, 181, **182**
*S*-adenosylmethionine (SAM/AdoMet) *see also* methionine cycle
  alcohol effects 288, **289**, 299
  betaine deficiency effect 4
  betaine supplementation 317, 331
  biosynthesis 90, 268
  carnitine biosynthesis 256

DNA methylation 319–20
function 4, 196, 206, 247, 261, 301, 331
methionine adenosyltransferases (MAT) 288
plants 15, 16, 32, 181, **182**
pregnancy 277
salt stress (plants) 22
VLDL synthesis 292
adhesion molecules (AM) 332, 335, 336–7, **338**, 339
adipocytes
definition 216
hypoxia 111
supplementation 213, 214
adipokines 213, 216 *see also* adiponectin; interleukin-6; leptin; resistin; tumor necrosis factor alpha
adiponectin 214, 291, **292**, 293–4
adipose tissue
alcohol effects 291, **292**, 293–4
betaine supplementation 208, 213, 214, 215, 257, 291, 293–4, 317, 358
DNA methylation 320–1
endocrine function 212–13
hypoxia 213
metabolic stress 212–13
adipose tissue-liver crosstalk 286
adiposity 257
AdoMet *see* *S*-adenosylmethionine
adultaration of foods 131–3
aging 330–9
agricultural uses 18–21
Akt *see* protein kinases
alanine aminotransferase (AST) 299
alanine betaine **52**, *188*, 192
albumin 4
alcohol consumption 285, 286, 300
alcohol dehydrogenase 300
alcoholic liver disease (ALD)
animal models 4–5
betaine supplementation 290–9
definition 301
future research requirements 299–300
pathogenesis 5, 286, 289, **290**
polymorphisms 299–300
risk factors 286
aldehyde dehydrogenases 33, **34**, 71
aldose reductase 79
alfalfa 192, 193
algae 160
alkaloids 84
Alzheimer's disease 4, 224, 236, 240, 332
Amaranthaceae 14, *160*
*Amaranthus* 18, 36
amides, betaine osmolyte function 65
amino acid homology
betaine-homocysteine methyltransferase (BHMT) encoding 93
definition 99
amino acid identity, definition 99
amino acid sequence analysis, definition 40
amino acid transport system A 80, 82
amino acids
osmolyte function 62
sulfur dietary intake 92
aminoaldehydes 33
aminopropylation **331**
amniotic fluid 149
AMP-activated protein kinase (AMPK) system 208, **292**, 294, 317, 356–7
anabolic, definition 363
anabolic signaling 212
anencephaly 270
angiogenesis 213
animal feed *31*, 161, 255
animal models *see also* water-immersion restraint stress
adhesion molecules (AM) 337
alcoholic liver disease 4–5, 291, 293
betaine-GABA transporter (BGT1) 78, 80, 83

animal models (*continued*)
body composition 357–8
DNA methylation 320
dry-eye diseases 380
embyrogenesis 270
fat metabolism 207–8
high-carbohydrate diets 312, 313, *314*
high-fat diets 208, 212, 214, 257, 312, 313, *314–15*
hyperhomocysteinemia 295
nonalcoholic fatty liver disease (NAFLD) 249, *314–15*
obesity 256
oxidative stress 330
redox homeostasis 337
storage 250
supplementation 4–5, 81, 207–8, 212, 214, 293, *314–15*
antiaging 330, 339
antifolates 67, 71, 72
antioxidants
aging 335
betaine 339
catechin 224, 236, 239
glutathione (GSH) 339
thiol agents 333–4
*Aphanothece halophytica* 21, 22, 25
aphid attractant 21
apolipoproteins 313, 317
apoptosis
alcohol effects 295
betaine supplementation 375–80
definition 301, 383
dry-eye diseases 370–1, 375–80
pathways 377–9
apoptosis-inducing factor (AIF) 379
aqueous-deficient dry eye 382, 383
*Arabidopsis* 16, 17, 21, 22–3, 37
aracyl trifluoromethanesulfonates 184
arginine 297
aromatic carbon interactions 65
arsenobetaine 56
arthritis 332
artificial tears 371
aspartate aminotransferase (ALT) 299
asthma 332
atherosclerosis 249, 332
athletic performance *see* exercise
ATP *see* adenosine-5'-triphosphate
*Atriplex* 18
autism 252

bacteria *see also Escherichia coli*
antibiotic resistance 69
betaine analogs 56
betaine biosynthesis 21, 25
choline catabolism 29
dihydrofolate reductase (DHFR) function 69–70
3-dimethylsulfoniopropionate (DMSP) 179
BADH *see* betaine aldehyde dehydrogenase
bariatric surgery *see also* obesity
benefits 261
betaine deficiency 248–9
betaine levels *253*, 258, 259–60
definition 260
DNA methylation 319
future research requirements 259–60
hyperhomocysteinemia 248, 258, 259, 260
side effects 248, 259
barley (*Hordeum vulgare*) 9, 17–18, 19, 36, 151
basolateral, definition 216
beetroot 13, 179
benzene ring interactions 65
*Beta vulgaris see* sugar beet
betaine
classes 120–1
discovery 3, 9, 47, 76
formula 49, 159
functional chemistry overview 49–54

functional overview 5, 24, 29, 39, 47–8, 77, 106, 223, **248**, 329, 390, 391 *see also specific functional main entries*
nomenclature 3, 9, 48–9, 105, 159
properties 49, *50*, 105–6, 204
structure 3, **4**, **30**, 49, **50**, 55, **91**, **107**, **167**, 203, **204**
betaine aldehyde 14, 31–2, 161, **246**
betaine aldehyde dehydrogenase (BADH)
aldehyde substrates 33, **34**
betaine binding 71
betaine biosynthesis 14, 32, 33–5, **34**, 181, 204, **246**
docking 33
plant gene expression 16
betaine-GABA transporter (BGT1)
brain 82–3, 84
central nervous system (CNS) 237–8
difference to GAT2 78
discovery 77–8
expression **210**
eyes 372, 373, 380
keratinocytes 211
kidneys 78–80, 84, 205–6, 210, 237, 239
liver 80–2, 84, 205, **207**, 208
oxidative stress 211
regulation 205–6, 372–3
betaine-homocysteine methyltransferase (BHMT) *see also* liver
active site **96**
alcohol effects 288, **289**, 297
betaine plasma levels 250, 252
betaine supplementation 291, 293
cirrhosis 249
deficiency 257
definition 261
dimethylglycine (DMG) 92, 97, 99, 252
distribution 90, 98
Down's syndrome 254
embryogenesis 270
expression 207
fetal development 270, 276, 279
folate 270, 273–4
function 3–4, 48, 57, 81, **82**, 90–4, 111, 245–7, **246**, **259**, 268
hepatocytes 80, **207**
kidneys 210–11
mechanism 95–8, 99
methionine synthase pathway regulation 269, 277, 278, 279
mutations 98
NMR spectroscopy 111
plasma homocysteine 206–7
polymorphism 270–1, 279
regulation 344
sodium 92
structure 94–8
sulfur amino acids 92
zinc 93, 94, 95, 97, 98, 99
betaine hydrochloride 56
betaine nicotinate *see* trigonelline
betaine nonaccumulators
aldehyde dehydrogenases 33–5
definition 25
genetic engineering 30, 37, 38, 39
betonicine (hydroxyproline betaine) 114, 184, 186, 188, 192
BGT1 *see* betaine-GABA transporter
BHMT *see* betaine-homocysteine methyltransferase
BHMT-2 92
binary complex, definition 99
bioavailability 203, 215
biosynthesis
acetylcholine 391
adenosine in plants 16
*S*-adenosylhomocysteine (SAH) 206, 301
*S*-adenosylmethionine (SAM/AdoMet) 90, 268
betaine 3, 76, 78, 148, 204, **207**, **246**, 268

biosynthesis (*continued*)
betaine in plants **11**, 13–17, **23**, 24, 31–8, 181–2
carnitine 193, 256
carnitine in plants 33
choline 92, **246**
choline in plants 14–16, 37–8, 181
cyanobacteria 21, 25
3-dimethylsulfoniopropionate (DMSP) 179
ethanolamine in plants 16
evolutionary development 33–5, 38
folate 67
GABA (gamma-aminobutyric acid) *83*
glutathione (GSH) **287**, 288, 331
homocysteine 57, 84
homocysteine in plants 16, 181, **182**
methionine 90, **91**, 258, **259**
*see also* methionine cycle
NF-κB 332
phosphatidylcholine 257, 391
QACs (quaternary ammonium compounds) 181–4
serine in plants **22**, 23
sphingomyelin 245, 391
triglycerides 294, **316**
biotic stress
definition 178, 195
QACs (quaternary ammonium compounds) 179–80
birth weight 271
blood-alcohol cycling 298
blood–brain barrier 250
blood plasma *see* plasma levels
blood pressure 257
BMI (body mass index) 257, 258–9, 313
body composition 256–7, 352, 354, **355**, 357–8, 361
*Boletus edulis* 194
bovine serum albumin (BSA) 63, 64–5
brain
alcohol 236
betaine concentration 83
betaine-GABA transporter (BGT1) 82–3, 84, 206
carnitine 111
choline 391
myo-inositol 83
seizures 82, 85
stress 223, 236
taurine 83
transport 250
breast cancer 249, 333, 393, *394*, 398, *399–400*, 405, *408*, *409*
Brix degree 186, 195
bulk density 49, *50*
butanol 122, 130, 133
γ-butyrobetaine *188*, 192

c-Jun *N*-terminal kinase (JNK) 371, 378, 379
calcium, betaine-GABA transporter (BGT1) regulation 79
cancer
antifolates 67
breast 249, 333, 393, *394*, 398, *399–400*, 405, *408*, *409*
choline intake 391
colorectal 249, *253*, 254, 398, *400–1*, 405
endometrial 398, *402*
esophageal 393, *395*
folate intake 391
gastric 112–13, 115, *409*
liver 80, 84, 391, *404*
lung 113, 393, *396*
mortality studies 405–11
nasopharyngeal 393, 398
NF-κB 332
ovarian 393, *397*, 398, *401*, *408*, 411
prostate 398, *402*, 405, *407*, *410*, 411
renal cell 398, *403–4*
*Cannabis sativa* 170–1, 172

canola (*Brassica napus*) 16
capillary electrophoresis/mass spectrometry (CE/MS)
 food analysis 121–3, 129–33, 143–4
 overview 134
 sample preparation 130
 sensitivity 123
 separation methods 124–6, 134
 validation 126–9
capillary zone electrophoresis (CZE) 124
carbohydrate responsive element binding protein (ChREBP) **316**
*S*-(δ-carboxybutyl)-L-homocysteine (CBHcy)
 function 91–2
 structure **91**
carcass modifier 255, 257, 261
carcinogenesis 4 *see also* cancer
cardiovascular disorders
 betaine deficiency effect 4
 homocysteine 206
 hyperhomocysteinemia 249
 plasma levels 110
 supplementation 254
carnitine
 betaine supplementation 208, 212, 255
 betaine transport inhibition 82
 biosynthesis 193, 256
 biosynthesis in plants 33
 classification 120
 eyes 371, 372
 function 255–6, 261
 mass spectrometry 121, 123, **125**, 126, **127**, *128*
 mass-to-charge ratio (m/z) 126
 muscle cells 212
 NMR spectroscopy 111, **112**, 113
 serum levels *106*, 111
 structure **107**
 urine levels *106*
 vegetable oil levels *131*, **132**
carnitine palmitoyltransferase (CPT1) 255–6, 261
casein kinase II (CKII) 333, **334**
caspases 295, 377, 378, 379, 383
catalase 19, 64, 298
catalytic activity, definition 40
catechin
 antioxidant function 224, 236, 239
 memory function 231, 236
 sources 239
catecholamines 261
cation–π interactions 70–1, 72
cell diffusion 70
cell-volume homeostasis 17, 53, 208, 209, 370, 371–2, 374, 382
cellular pressure 17, 20
central nervous system (CNS) 223, 237
cereals 150, 151, 191
cerebrospinal biofluid (CSF)
 function 114
 hydrophilic interaction liquid chromatography (HILIC) *145*, 149
chaperone function 17, 29, 63–4, 71, 379–80
CHDH *see* choline dehydrogenase
chemical chaperone, definition 72
chemokines 332
Chenopodiaceae 9–10, 14, 16, 170
*Chenopodium botrys* L. (Jerusalem oak) 159
chickens 204–5, 211, 357, 358
children
 congenital heart defects 252, *253*
 nonalcoholic fatty liver disease (NAFLD) 312
chloroplasts 11, 19, 21, 32
cholesterol 255, 257, 291, 294, 313, 357
choline
 betaine accumulation 21
 betaine biosynthesis 3, 14, 31, **81**, 181, 204, **207**, 268

choline (*continued*)
  betaine-homocysteine methyltransferase (BHMT) expression 92
  betaine supplementation 207
  biosynthesis 92, **246**
  biosynthesis in plants 37–8, 181
  brain 250
  deficiency 80–1, 286, 391
  dietary sources **246**
  distribution 205
  formula 57
  function 57, 80, 390, 412
  high-pressure liquid chromatography (HPLC) 161
  hydrophilic interaction liquid chromatography (HILIC) 144, 148–9, 151–2
  ischemic strokes 111, 113
  liver 245, **246**
  mass spectrometry *188*
  methionine synthase pathway 268
  NMR spectroscopy 111
  oxidation pathway 279
  phosphatidylethanolamine *N*-methyltransferase (*PEMT*) 277
  plasma levels 257
  pregnancy 272, *273*, 275–7, 278
  recommended daily intake (RDI) 392
  salt stress (plants) 17
  serum levels *106*
  structure **107**
  total intake calculation 392
  transport in plants 18, 21, 37–8
  ultraperformance liquid chromatography-mass spectrometry (UPLC-MS) 149
  urine levels *106*
  VLDL synthesis 312
choline-betaine-dependent pathway 81, **82**, 84
choline binding protein 71
choline dehydrogenase (CHDH) 76, 78, **82**, 84, 204, **246**, 250, 313
choline mono-oxygenase (CMO)
  betaine biosynthesis 14, 31–2, 36–7, 181
  function 25
  genetic engineering 21
  leaf maturity 17
  peroxisomal 36
  plant gene expression 16
  structure 14, 36
  substrate specificity 37
α-chymotrypsin 64
CID *see* collision-induced dissociation
cirrhosis
  betaine-homocysteine methyltransferase (BHMT) 249
  epidemiology 311, 313
citrus fruit 112, 113, 182, 184, 185, 186, 190, 192–3
cluster of differentiation 14 (CD14) 299
CMO *see* choline mono-oxygenase
cobalamin *see* vitamin $B_{12}$
coenzyme, definition 40
coffee 112, 192
cold resistance (plants) 179
collagen 211, 289, 296
collision-induced dissociation (CID) 187–8, 190, 195
colon 211
colorectal cancer 249, *253*, 254, 398, *400–1*, 405
compatible solutes *see also* osmoprotectants
  betaine 10, 13, 48
  definition 25, 195, 382, 383
confidence intervals
  capillary electrophoresis/mass spectrometry (CE/MS) 129
  definition 412
congenital heart defects 252, *253*
corn flour 151
corn oil 131
corticosteroids 370

corticosterone 226–7, 230, 231, **233**
cortisol 356
cosmetics industry 49
cotton plants 20–1
COX-2 *see* cyclo-oxygenase-2
CpG islands 268, 319, **320**, 323
CPT1 *see* carnitine palmitoyltransferase
creatine 110, 352
CREB-binding protein (CBP) 333, **334**
crop production uses 18–21
crowding agents 70, 71
cryoprotection 179
crystal structure 55
cyanobacteria 21, 25
cycling 353–4
cyclo-oxygenase-2 (COX-2) 209, 223, 235, 299, 332, 335, **336**, 337, **338**
cyclosporin A 370
cystathionine 254
cystathionine β-synthase (CBS) 254, 255, **287**, 299, 331
cysteine 288, 331
cystic fibrosis 252–4
cytochrome c 378
cytokines 209, 230, 232, 317, 332, 370, 383
cytoplasm
  betaine 11, 13–14
  compatible solutes 10

daily intake
  average levels 254
  requirements 3, 392
deficiency *see also* insufficiency
  bariatric surgery 248–9, 258, 259–60
  betaine 4, 5, 150, 248, 249–50, 258
  betaine-homocysteine methyltransferase (BHMT) 257
  choline 80–1, 286, 312, 391
  folate 67, 258, 269, 312, 391, 411
  methionine 286, 312
  pregnancy 271, 272
  vitamin $B_{12}$ 258, 269
dehydration 10, 345, 364
density functional theory (DFT), definition 58
DEPT NMR *see* distortionless enhancement by polarization transfer
derivatization
  capillary electrophoresis/mass spectrometry (CE/MS) 123, **124**, **127**, 130, 133
  definition 134
  gas chromatography (GC) 143
  high-pressure liquid chromatography (HPLC) 184
dermal fibroblast 216
dermal fibroblasts 211
developmental defects 109, 267, 270–1, 279, 320
DGAT (acyl-CoA-diglyceride acyltransferase) **292**, 294–5
DHFR *see* dihydrofolate reductase
diabetes mellitus
  betaine insufficiency 48
  excretion 80
  microangiopathy 252, *253*
  nephropathy 252
  NF-κB 332
  nonalcoholic fatty liver disease (NAFLD) 312
  retinopathy 252
diet
  fasting 252
  high-fat 257, 320–2
  high fiber 252
  methyl donors 312, 319–20
dietary assessment 391–2, 412
dietary sources
  betaine 3, *31*, 48, 76, 121, 150, 159, **246**, 345, 392, 412
  choline 190, **246**, 392, 412
  cooking methods 190
  folate 67
  methyl donors 206
  proline betaine 112
  QACs (quaternary ammonium compounds) 190
diethanolamine glucuronate 317

diets
- energy-limited 211
- high-carbohydrate 312, 313, *314*
- high-fat 208, 212, 214, 215, 312, 313, *314–15*
- high-fiber 258

diffusion, cellular 70
digestion 204–5 *see also* absorption
dihydrofolate reductase (DHFR)
- betaine effects 65–70
- crowding agents 70
- function 72
- methotrexate 67
- R67 63, 66, **67**, 69–70, 73
- structure **66**

3,4-dihydroxyphenylalanine (L-DOPA) 180
*N*,*N*-dimethylglycine 3, **91**, 161, 261
dimethylglycine (DMG)
- betaine-homocysteine methyltransferase (BHMT) 92, 97, 99, **246**, **287**
- betaine supplementation 207
- biofluid analysis 148–9
- biosynthesis 268
- chromatography 143
- Down's syndrome 254
- fasting plasma total homocysteine (tHcy) 269
- fetal development 276
- plasma levels 252
- pregnancy 272, *273*, 274, 275, *276*, 277
- transmethylation cycle 207

dimethylsulfonioacetate 92, 98
dimethylsulfoniopropionate 92, 98
3-dimethylsulfoniopropionate (DMSP) 179
dipole moment vector 52, **53**
dissociation constant ($pK_a$) 49, *50*
distortionless enhancement by polarization transfer (DEPT NMR) 168, 173
distribution
- betaine-homocysteine methyltransferase (BHMT) 90, 98
- chickens 205
- mammals 48, 76
- sugar beet (*Beta vulgaris*) 11–12

dithiolethione 333
DMG *see* dimethylglycine
DNA binding
- alcohol effects 294
- lac repressor 63

DNA methylation
- *S*-adenosylmethionine (SAM) 4
- *S*-adenosylmethionine (SAM/AdoMet) 319–20
- bariatric surgery 319
- betaine supplementation 4–5
- definition 5, 323
- future research requirements 321
- high-fat diets 320–2
- nonalcoholic fatty liver disease (NAFLD) 319–22
- one-carbon metabolism (1-C) 267

docking
- betaine aldehyde dehydrogenase (BADH) 33
- definition 40

dopamine 180
Down's syndrome 254, 270
Dragendorff's reagent 170, 173
drought stress (water stress) 12, 18, 19, 20, 179
dry-eye diseases 369–71, 381, 383
dry mouth syndrome 211
duodenum 205
dyslipidemia 312, 318

EIEs *see* extracted ion electropherograms
electron density **53**
Electronic Reference To access *In vivo* Concentrations (ERETIC) 108–9

electrospray ionization (ESI)
betaine 54, 168, 169, 185–6, 194
hydrophilic interaction liquid chromatography (HILIC) 142, 144, 153
MS/MS 124, 194
overview 134, 154, 173, 195
QACs (quaternary ammonium compounds) 186, 194
elicitors 179–80
embyrogenesis 270
endocrine system, adipose tissue 212–13
endometrial cancer 398, *402*
endoplasmic reticulum (ER) stress 214, 289, **292**, 294, 295, 313
endotoxins
alcohol effects 299
tumor necrosis factor alpha (TNF-a) 209
enoyl-CoA hydratase **316**
enzyme catalysis 39–40
enzymes, definition 363
epidemiological studies, cancer 391–2
epidermis 211, 216
epigenetics
definition 319, 323
nonalcoholic fatty liver disease (NAFLD) 319
epinephrine 261, 298
epithelial cells 211
ERETIC (Electronic Reference To access *In vivo* Concentrations) 108–9
ergogenic function 345, 361, 363
ergothioneine *188*, 193–4
ERK *see* extracellular signal regulated kinase (ERK) pathway
erythritol 371
*Escherichia coli*
antibiotic resistance 69
betaine function 64, 68, 70
esophageal cancer 393, *395*
ethanol *see* alcohol consumption
ethanolamine 14–15, 32, 181
ethanolamine kinase 16
European Food Safety Authority (EFSA) 121
evaporative dry eye 381, 382, 383
evaporative light-scattering detection (ELSD) 147
excretion 77, 80, 207, 252
exercise
acute supplementation 345–7
betaine function 345
chronic supplementation 347–54, **355**
levels 257, 313
mechanism studies 355–61
resistance training 354
sweat 357
exons, definition 99
extra virgin olive oils *see* olive oils
extracellular signal regulated kinase (ERK) pathway 293, 294, 371, 378
extracted ion electropherograms (EIEs) **125**, 126, 129
eyes
betaine 372–80
diseases 369–71, 381, 383
structure 368–9, 381, 383

FADD (Fas-associated protein with death domain) 378
fast atom bombardment (FAB) mass spectrometry 185
fasting 252
fat metabolism
alcohol effects 291, 293–4
animal models 207–8
betaine supplementation 212, 214, 318
nonalcoholic fatty liver disease (NAFLD) 313, **316**, 317, 318
fatigue 347, 350
fatty acid ß-oxidation
alcohol effects **292**, 294
betaine function 214, 256
definition 261
nonalcoholic fatty liver disease (NAFLD) 313, **316**, 317, 318, 320

fatty acid synthase (FAS) 256, **316**, 358
fatty liver 285–6
ferrodoxin 25, 36
fertility 84
fetal development 109, 268, 270–1, 275–6, 279, 320
FIA *see* flow injection analysis
fibroblasts, dermal 211
flavonoids 330
flow cytometry 374, **375**, 376, **377**
flow injection analysis (FIA) 129, 186, 196
fluorescein 370, 371
fodder beets 10, 11–12
folate *see also* 5-methyltetrahydrofolate
bariatric surgery 248, 258, 259
betaine-homocysteine methyltransferase (BHMT) 270, 273–4
betaine interactions 65–70
biosynthesis 67
deficiency 67, 269, 391, 411
dietary sources 67
distribution 390
fasting plasma total homocysteine (tHcy) 269, 271
methionine cycle 287, 321
one-carbon metabolism 267–8, 271
pregnancy 272, 273–5, 277, 278
structure **68**
supplementation 250, 258, 260, 270, 271, 272, *278*, 391
folate cycle 84, 280
food analysis
adultaration 131–3
betaine levels 191
capillary electrophoresis/mass spectrometry (CE/MS) 129–33, 143–4
cooking methods 190
high-pressure liquid chromatography (HPLC) 184, 185–6
hydrophilic interaction liquid chromatography (HILIC) *145–6*, 150–2, 184–5
method validation 126–9
NMR spectroscopy 107
QACs levels *191*
QACs (quaternary ammonium compounds) 190, 192–4
regulations 121
sample preparation 186, **187**
separation methods 124–6
food fortification, folate 67, 268, 269, 272, *278*
food frequency questionnaire (FFQ) 391–2, 412
food sources *see* dietary sources
forkhead box O3 (FOXO3) 297
formula
betaine 49, 159
choline 57
Fourier transform infrared (FT-IR) spectrometry
betaine 54
overview 57–8
fragmentation pattern, definition 196
freezing stress 10, 20, 179
*Fructus Lycii* *146*, 147, 148, 162, 185
fungi, ergothioneine 193–4
fungi growth substrate 21

GABA 84
GABA (gamma-aminobutyric acid) *83*, 230, 231, 237
GABA transporter (GAT) 78, 239 *see also* betaine-GABA transporter (BGT1); GAT2
gamma-aminobutyric acid *see* GABA
gas chromatography (GC) 143
gastric cancer 112–13, 115, *409*
gastric ulcers 224
GAT2 78, 81, 224, 234–5, 237–8
gender differences 3, 250, *251*

gene expression *see also* DNA methylation
    alcohol effects 294, 297
    DNA methylation definition 5
    fetal development 270
    NF-κB 330
    nonalcoholic fatty liver disease (NAFLD) 318–19
    plants 16–17
    triglycerides 318
gene transcription
    alcohol effects **292**, 297
    CpG methylation 268, 319, **320**
genetic engineering
    definition 25
    plants 21–3, 30–1, 37, 38
glomerular filtration rate 272
glucocorticoid 223
glucose metabolism, betaine supplementation 318
glucose serum levels 257
glutamate *83*, 230, 231
glutamic acid 223
glutathione (GSH)
    *S*-adenosylmethionine (SAM/AdoMet) **289**, 317, 331
    betaine function 333–4, 337, **338**, 339
    biosynthesis **287**, 288, 331
    function 298–9, 332, 339, 340
    structure 340
glutathione peroxidase 298, 299
glycerophosphorylcholine 210
glycine 207, **246**
glycine betaine *see* betaine
glycine decarboxylase 21
glycine methylation 21–3
goblet cells 369, 370
goji berry *see Lycium chinense*
grains 345
green tea 224
growth hormone (GH) 257, 356, 357–8
GTPases **290**
guanidinoacetate 249
Guayule shrub 159
gut epithelium 211
gut-liver crosstalk 286

$^{1}$H correlation spectroscopy (COSY) NMR spectroscopy 168, 172
haemodilution 272
haemoglobin solvation 4, 5
haloacid dehalogenase (HAD) family 16
halophytes
    definition 25
    evolutionary development 35
hazard ratio (HR), definition 412
heat denaturation 10
heme oxygenase-1 232, 237
hemp 170
hepatic lipid metabolism 323
hepatic steatosis 208
hepatic tumor 391
hepatitis B surface antigen negative (HBsAg(−)) 405
hepatitis C virus (HCV) 297
hepatocellular carcinoma 80, 84, *404*, 405, *406*
hepatocytes *see also* liver
    alcohol effects 293, 295
    betaine-GABA transporter (BGT1) 205
    betaine-homocysteine methyltransferase (BHMT) 80, 207
    definition 216, 301
    triglycerides 294, 313, **316**, 317, 318
hepatotoxicity 208
hercynine *188*, 193
high density lipoproteins (HDL) 357
high-pressure liquid chromatography (HPLC)
    betaine 161–2, 168, 184, 185–6, 194
    choline 161
    derivatization 184
    overview 115, 174
    QACs (quaternary ammonium compounds) 194

high-pressure liquid chromatography (HPLC) (*continued*)
radioenzymatic assay 161
sample preparation 186, **187**
hippocampus 223, 232, 237, 333
histones 296, 297, 333
Hofmann degradation 174
homocysteine
alcohol effects **292**
betaine deficiency effect 4
betaine-homocysteine methyltransferase (BHMT) 92–3, 95–7, 99, 206–7, 223, 245–7
betaine intake 392
betaine plasma levels 250
betaine supplementation 254
biosynthesis 57, 84
biosynthesis in plants 16, 181, **182**
choline intake 392
dietary assessment 391
disease risks 330
Down's syndrome 254
fasting plasma total (tHcy) 269, 271, 272, *273*, 274, 280, 391
folate levels 269, 391
formula 84
function 57, 280
kidneys 390
liver 80, 289, 390
metabolism defects 206
methionine cycle 3–4, 48, 81, **82**, 92, 268, 286–7, 331
obesity 258
plasma levels 91–2
post methionine load (PML) test 269, 280
pre-eclampsia 252
pregnancy 272, *273*, 274, 277
structure **91**
testis 84
transulfuration pathway **287**, 288
vitamin $B_6$ 391
vitamin $B_{12}$ (cobalamin) 391
homocysteine thiolactone (HCTL) 299, 354
homocysteinemia
definition 5
dietary effects 4
homocystinuria
betaine supplementation 81, 254–5, 258
definition 5, 174
dietary effects 4
homocystinurics, definition 84
homologs, definition 40
homostachydrine *see* pipecolic acid betaine
*Hordeum vulgare see* barley
hormones 363
Huggins constant 204
hydration **51**, 52, **54**
hydration number 204
hydrogen peroxide, plants 179
hydrophilic interaction liquid chromatography (HILIC)
advantages 140, 144
apparatus **140**
betaine analysis 144–52, 184–5
development 139
mobile phases 142, 144
overview 153, 174
separation mechanisms 140–2, 152
stationary phases 142–3, 152
validation 147, 148
hydrophilicity 204
hydroxyproline betaine *see* betonicine
5-hydroxytryptamine (5-HT, serotonin) 180, **183**, *188*, **189**
hyperhomocysteinemia
alcohol consumption 289, 293–4
bariatric surgery 248, 258, 259, 260
betaine-homocysteine methyltransferase (BHMT) 98
betaine supplementation 249, 293–4

cardiovascular disorders 249
definition 260
hyperhomocystinuria 57
hyperosmolarity, tear-film 369–71, 372, 378, 381, 382, 383
hyperosmosis 208–9, 211
hypertonicity 64, 79, 206, 209, 359–60, 360
hypertrophy 360–1, 363
hypomethylation
betaine deficiency effect 4
definition 5
hypothalamic–anterior pituitary gland–adrenal cortex (HPA) system 223
hypoxia 111, 213, 216

IκB kinases (IKK) 330, 332–3, **334**, 335, **336**, 337, **338**
imino acid transporter (SLC36A1) 205, **207**
immunoglobulin A nephropathy (IAN) 110, 114–15
immunosuppressants 370
*in vivo* studies *see also* animal models
betaine chaperone function in *E. coli* 64
dihydrofolate reductase (DHFR) function 69–70
inducible nitric oxide synthase (iNOS) 223, 235, 236, 298, 299, 332, 335, **336**
industrial uses 49
infants, congenital heart defects 252, *253*
infertility 271
inflammation *see also* proinflammatory molecules
adipocytes 111, 213–14, 215
betaine effects 249, 254
NF-κB 332, 335, **336**
nonalcoholic fatty liver disease (NAFLD) 317
oxidative stress 223, 236
pathogenesis 330
redox homeostasis 337
tear-film
hyperosmolarity 370–1
tumor necrosis factor alpha (TNF-a) 114
inflammatory bowel disease 332
infrared (Fourier transform, FT) (FT-IR) spectrometry
betaine 54
overview 57–8
inhibitor of NF-κB (IκB) proteins 332, **334**
iNOS *see* inducible nitric oxide synthase
inositol 209–10
insufficiency 48 *see also* deficiency
insulin
definition 216
exercise 356
insulin-like growth factor 1 (IGF-1) 257, 356, 358, 361
insulin resistance 216, 313–17
insulin sensitivity 208, 214
interferon (IFN) 299
interleukin-1β (IL-1β) 232, **234**, 236, 237, 238, 370
interleukin-6 (IL-6) 213, 335, **336**, 370
interleukin-18 299
International Conference on Harmonization (ICH) guidelines 126
intestines
absorption 76, 80, 205, **207**, 254
betaine function 4, 211
betaine supplementation 358–9
distribution 205
hypertonicity 209
osmolality 64
pathophysiology 250
stress 358–9
transport 4, 345
intrauterine growth retardation 271
introns, definition 99
ion chromatography (IC) 143

ischemic strokes 111, 113
isoaspartyl residues 296
isoleucine betaine 170–1
isoprenylcysteine carboxyl methyltransferase (ICMT) **290**, **292**, 294, 295

JAK-signal transducers 297
Janus kinase 3 299
Jikoppi 161
JNK (c-Jun *N*-terminal kinase) 371, 378, 379

Kancolla 168–70, 172
keratinocytes 206, 209, 211, 215, 216
kidneys
 accumulation 76, 205
 betaine concentration 48, 78
 betaine-GABA transporter (BGT1) 78–80, 84, 205–6, 210, 239, 373
 betaine-homocysteine methyltransferase (BHMT) 90, 207, 210–11
 hypertonicity 209
 immunoglobulin A nephropathy (IAN) 110, 114–15
 nephropathy 252
 NF-κB 335–6
 osmolality 64, 77, 79, **210**, 223
 oxidative stress 223, 335–6
 redox homeostasis 337
 renal failure 252
 urea 77, 209–10
Kraut reagent 170, 174
Kupffer cells 80, 206, 208, 216, 301

lac repressor 63
lacrimal glands 368, 369, 370, 382, 383
lactate after exercise 345, 352
leaf accumulation 11–12, 13–14, 17 *see also* chloroplasts
lean mass gain 212, 214, 255, 354, **355**, **356**, 358, 359, 361
learning function 223 *see also* memory function
leptin 213, 214, 258
leptomeninges 83, 84
ligand binding 65, **67**, 68, 95–7
limit of detection (LOD)
 capillary electrophoresis/mass spectrometry (CE/MS) *128*, 129
 definition 134, 154, 174
 electrospray ionization (ESI) 147
 evaporative light-scattering detection (ELSD) 147–8
 high-pressure liquid chromatography (HPLC) 161, 162
 hydrophilic interaction liquid chromatography (HILIC) 144, 149, 152
 liquid chromatography (LC) 147
limit of quantification (LOQ)
 capillary electrophoresis/mass spectrometry (CE/MS) *128*, 129
 definition 134, 154, 174
 evaporative light-scattering detection (ELSD) 147–8
 high-pressure liquid chromatography (HPLC) 162
 hydrophilic interaction liquid chromatography (HILIC) 152
linearity
 capillary electrophoresis/mass spectrometry (CE/MS) 128–9
 definition 134
 hydrophilic interaction liquid chromatography (HILIC) 149
lipid disorders 48
lipid metabolism
 alcohol effects 291, 293–4

animal models 207–8
betaine supplementation 212, 214, 318
nonalcoholic fatty liver disease (NAFLD) 313, **316**, 317, 318
lipid peroxidation 228, 299
lipogenesis **292**, 294, 313, **316**, 317, 358
lipogenic enzymes 261
α-lipoic acid 333
lipopolysaccharide (LPS) 240
lipopolysaccharide (LPS)-induced memory impairment 224, 234–5, 236
lipoproteins
betaine supplementation 357
HDL 357
LDL 336, 357
paraoxonase 299
VLDL **290**, 291–3, 294, 312, 318, 357
liquid chromatography (LC) *see also* hydrophilic interaction liquid chromatography
biofluid analysis 148–9
food analysis 121, 129
mobile phase 153
normal-phase (NPLC) 141, 153
overview 134
reversed-phase (RPLC) 141, 143, 144, 153
sensitivity 130
separation mechanisms 140–2
stationary phase 154
ultraperformance (UPLC) 149–50
liquid chromatography-tandem mass spectrometry (LC-MS/MS) 154
liquid–liquid extraction (LLE) 130, 134, 135
liver *see also* alcoholic liver disease; cirrhosis; hepatocytes; nonalcoholic fatty liver disease
accumulation 76, 205
alcohol consumption effects 285–6, **289**, **290**
betaine biosynthesis 204
betaine concentration 48, 77, 80
betaine-GABA transporter (BGT1) 80–2, 84, 373
betaine-homocysteine methyltransferase (BHMT) 90, 270, 287
betaine supplementation 80, 207–8
cancer 80, 84, 391, *404*, 405, *406*
carnitine 111
crosstalk with other tissues 286
distribution 205
fat metabolism 207–8, 291, **292**, 313, **316**, 317
high-fat diets 257
homocysteine 80, 289
hypertonicity 209
Kupffer cells 80, 206, 208, 216
Mallory-Denk bodies 297, 301
nonparenchymal cells 80, 301
osmolality 208
phosphatidylcholine 245, **246**
structure 80, 301
liver steatosis
alcohol consumption 286
betaine deficiency effect 4
betaine supplementation 291, 294–5, 317
definition 5
one-carbon metabolism 80
pathogenesis **290**, **292**, 300
low density lipoproteins (LDL) 357
lung cancer 113, 393, *396*
*Lycium barbarum* Linn. 160
*Lycium chinense* 143, *145*, 147, 148, 161, 162

lysophosphatidylcholine (LPC) 336
lysozymes, osmolyte influence 51

α-2-macroglobulin 299
macrophages
definition 216
intestines 211
liver 208–9
malic enzyme 358
Mallory-Denk bodies 297, 301
malondialdehyde (MDA) 228, 234, 237, 240, 352
mangrove 17, 36
marijuana *see Cannabis sativa*
mass spectrometry *see also* capillary electrophoresis/mass spectrometry
derivatization 123, **124**, **127**, 133
fast atom bombardment 185
food analysis 121–2, 129–33, 185–6
MS/MS 122, 135, 188, **189**
overview 57, 115, 173
plasma levels 392
seed analysis 165, **167**
sensitivity 122
separation methods 124–6, 134, 143
tandem 129, **185**, 186, 187–90, 194, 196
mass-to-charge ratio (m/z) 123, 126, 135, 173
MAT *see* methionine adenosyltransferases
maternal plasma betaine levels 272, *273*, **274**, **275**
matrix metalloproteinases (MMP) 333, 370
*Medicago* genus 182, 192
medicinal plants 159, *160*, 161, 162, 170, 172
meibomian glands 368, 369, 382, 383
melting point 49, *50*
memory function
betaine supplementation 231, **233–4**
catechin 231, 236
choline 391
lipopolysaccharide (LPS)-induced memory impairment 224, 234–5, 236
stress 223, 229
mesophytes 35
metabolic syndrome 48, 214, 312–13, 323
metabolism of betaine 3–4, 114, **207**, 212, **246**
metabolomics 107, 122, 318
methanol 49
methionine
betaine supplementation 207
deficiency 286, 312
3-dimethylsulfoniopropionate (DMSP) biosynthese 179
Down's syndrome 254
fetal development 275
structure **91**
methionine adenosyltransferases (MAT) 288, **289**, 297
methionine cycle
acyl-CoA-diglyceride acyltransferase (DGAT) 294
*S*-adenosylmethionine (SAM/AdoMet) 331
alcohol effects 286, 288–9, **290**, **292**
betaine deficiency effect 4
betaine function 81, **82**, 321
betaine-homocysteine methyltransferase (BHMT) 90, **91**, 247, 268
folate 287, 321
plants 181, **182**
process steps 3–4, 268, 286–8
methionine synthase pathway
alcohol effects 288, **289**
betaine deficiency 258
betaine-homocysteine methyltransferase (BHMT) pathway regulation 269, 270, 277, 278
definition 280
folate 273–4

5-methyltetrahydrofolate (5-MTHF) 92
plant gene expression 16
process steps 81, **82**
vitamin $B_{12}$ 268, 287
methotrexate 67, 71
methyl donors
*S*-adenosylmethionine (SAM/AdoMet) **247**, 390
betaine 48, 80, 81, **82**, 148, 212, 223, 291, 344
choline *276*
definition 323
diet 312, 319–20
function 344
metabolic functions 206
5-methyltetrahydrofolate (5-MTHF) **82**, 84
one-carbon metabolism 267–8
plasma levels 250
methyl methanethiosulfonate 93, 94
methylamines 49, 62
methylation
*S*-adenosylmethionine (SAM/AdoMet) 206, 288, **331**
arginine 297
choline biosynthesis 15
definition 6, 216, 301
embyrogenesis 270
histone 297
neurodegenerative diseases 254
pathway relationships 390–1
protein-isoaspartate methyltransferases (PIMT) 296–7
methylation potential (SAM:SAH ratio) 289, 290–1, **292**, 293, 294, 295, 296–7, 301
methylenetetrahydrofolate reductase (MTHFR) 84, 255, 268, 271, **287**
N-methylglycine *see* sarcosine
5-methyltetrahydrofolate (5-MTHF) **82**, 84, 92, 258, **259**, 287, 331 *see also* folate
N-methyltransferases
definition 196
plant betaine biosynthesis 32, 181–2
methyltransferases 268, **289**, **292**, 301 *see also* betaine-homocysteine methyltransferase (BHMT)
microangiopathy 252, *253*
microarray, definition 323
microsomal triglyceride transfer protein (MTTP) **320**, 322
miscarriage 271
mitochondria
betaine function 247
betaine synthesis **207**
fatty acid transport 256, **316**, 318
lipid metabolism 212
oxidation pathways **316**, 317, 320
respiratory chain 298
mitogen-activated protein kinases (MAPKs)
apoptosis 371, 378
betaine function 379
inflammation **336**, 370
NF-κB 333, **334**, 335, 337, **338**
osmoprotection 371
oxidative stress 330
mitogen- and stress-activated kinase-1 (MSK1) 333, **334**
mobile phase, definition 153
molasses, betaine levels 10
molecular dynamics (MD)
betaine 52
definition 58
molecular weight *50*
mRNA, definition 216, 363
mucosa 211, 216
multiple reaction monitoring (MRM) 135, 149, 154, **185**
muscle tissue
anabolic signaling 357, 362
betaine levels 250
betaine metabolism 212
carnitine metabolism 212
distribution 205
hypertonicity 360

muscle tissue (*continued*)
hypertrophy 360–1, 363
lean mass gain 212, 214, 255, 354, **355**, **356**, 358, 359, 361
oxygen saturation 350–2
phosphorylcreatine (PCr) 352
power 347, 348–50, **353**, 354, 363
strength **346**, 347, **349**, 350, 352, 363
muscular dystrophy 332
mushrooms *see* fungi
mutagenesis
betaine-homocysteine methyl-transferase (BHMT) 93–4, 95
random 99
semirandom 99
site-directed 40, 99
myelomeningocele 270
myo-inositol 79, 83
myosin 4, 360, 361

NADPH *see* nicotinamide adenine dinucleotide phosphate
NAFLD *see* nonalcoholic fatty liver disease
NASH *see* nonalcoholic steatohepatitis
nasopharyngeal cancer 393, 398
neo myotubes 360–1
nephropathy 252
neural tube defects (NTDs) 268, 270, 271, 279
neurodegenerative diseases 224, 236, 237, 239, *253*, 254
neurotransmitters 83–4, 223, 391
NF-κB 223, 330–6, 337, 338, 339, 340, 370
nicotinamide adenine dinucleotide phosphate (NADPH/NADP$^+$) 63, 65–6, 332
nicotinamide ascorbate 317
nitric oxide
alcohol effects 298
brain 223
exercise 352, 355–6
plants 180
nitrogen fertilization 13
nitrogen interactions 65
NMR spectroscopy
betaine 54–5, 107–13, 165, **166**, *167*, 168–9
betaine-folate interactions 69
development 107, 113
distortionless enhancement by polarization transfer (DEPT) 168, 173
$^1$H correlation (COSY) 168, 172
limitations 184
proton (1H-NMR) 69, 107–14, 115, **166**
single quantum coherence filtering (SQCF) 111, **112**
nomenclature 3, 9, 48–9, 105, 159
nonaccumulators *see* betaine nonaccumulators
nonalcoholic fatty liver disease (NAFLD)
animal models 4–5, 208, 249
betaine supplementation 80, 208, 214, 215, 257, 313–22
definition 84, 216, 311, 323–4
DNA methylation 319–22
epidemiology 311–12
epigenetics 319
future research requirements 317
pathogenesis 313–17, 318, 319, 322
nonalcoholic steatohepatitis (NASH) 249, 311, 317, 318, 324
nonparenchymal cells 80, 301
nontargeted metabolite profiling 207, 212, 216
norepinephrine 261, 298
normal-phase liquid chromatography (NPLC) 141, 153
Nuclear Factor of Activated T cells (NFAT 5) 373

nuclear magnetic resonance (NMR) spectroscopy 57
nutriepigenomics 322–3 *see also* deficiency
nutrional status 398, 411

oats 151
obesity 213–14, *253*, 256, 257–8, 312 *see also* bariatric surgery
ocular surface 368, 381, 383
odds ratio (OR), definition 412
oestrogen 276
olive oils 130, 131–3
omphalocele 270
one-carbon metabolism (1-C) 80, 254, 267–9, 271, 275, 280, 390
optical rotation, definition 174
oral mucosa 211
organic osmolyte, definition 216
osmolality 64, 77, 79, 208, **210**, 223
osmolytes
  brain 83
  chemistry 51
  classes 62
  definition 10, 57, 62, 72, 260, 363, 382
  embryogenesis 270
  function 53, 62, 77, 344, 359, 362, 371
  intestines 358–9
  kidneys 209–10
  preferential exclusion model 63, 64–5, 66–7
  preferential interaction model 63
osmoprotectants
  bacteria 69–70
  betaine function 4, 10, 30–1, 47–8, 121, 208–9, 344, 359, 362, 374, 382, 391
  definition 6, 24, 25, 40, 72, 85, 196
  dry-eye diseases 371
  plants 33, 121, 179
osmoregulation 209, 211, 217, 223, 237, 371–2, 374, 382
osmotic pressure
  betaine synthesis regulation 13–14
  definition 363
  exogenous betaine application 20
  plant sucrose concentrations 13
  preferential interaction model 63
osmotic stress
  definition 40, 72, 196
  JNK (c-Jun *N*-terminal kinase) 371
  plants 13–14, 30–1
  *in vitro* experiments 359
ovarian cancer 393, *397*, 398, *401*, *408*, 411
oxidant, definition 340
oxidative phosphorylation 298
oxidative stress
  aging 335
  alcohol effects 298
  animal models 230
  betaine-GABA transporter (BGT1) 211
  betaine supplementation 208, 214, 235–6, 330–1, 335–6
  definition 240, 340
  GSH/GSSG ratio 332
  mental disorders 224
  NF-κB 223, 332
  nonalcoholic fatty liver disease (NAFLD) 313, 317
  skin 211
oxygen consumption, exercise 345
oxygen interactions 65
oxygen saturation, muscles 350–2

P-selectin **338**
p300/CBP-associated factor (PCAF) 333, **334**
palmytoil-carnitine 123
pancreas 90
paraoxonase 299
partition coefficients ($K_p$) 65

passive avoidance test 225, 226, 229, 231
passive diffusion 70
pathogen-related proteins (PRP) 180
PEAMT *see* phosphoethanolamine *N*-methyltransferase
peas 192
pentylentetrazol 82, 85
performance, exercise 345, 355–7
peroxisome proliferator activated receptor alpha (PPARα) 256, 261, **292**, 294, **316**, 321, 322
PGDH *see* 3-phosphoglycerate dehydrogenase
pH, uptake in plants 18
phagocytosis 208–9, 211, 217
2'-phenanthreneacyltriflate 184
phenylethanolamine methyltransferase 298
phosphate interactions 65
phosphatidylcholine
    betaine methyl donor function 257
    biosynthesis 391
    definition 261
    liver 245, **246**, 292, 293
    plant function 15
    pregnancy 277
    signal transduction 391
    VLDL synthesis 292
phosphatidylethanolamine (PE) 292, 293
phosphatidylethanolamine methyltransferase (PEMT) 276, 277, 279, **290**, 292–3
phosphocholine 15
phosphoethanolamine 15, 32
phosphoethanolamine *N*-methyltransferase (PEAMT) 15–16, 17, 21, 38
3-phosphoglycerate dehydrogenase (PGDH) 22
phosphoinositide-3-kinase (PI3K) 333
phospholipase D2 (PLD2) **246**
phospholipids
    biosynthesis 391
    brain 250
    hydrophilic interaction liquid chromatography (HILIC) 151–2
phosphoniobetaine 56
phosphorylcholine *106*, **107**, 111
phosphorylcreatine (PCr) 352
photorespiration 19
photosynthesis 11, 18–19, 23
photosystem II 19, 25
phylogenetic family, definition 40
physical activity
    acute supplementation 345–7
    betaine function 345
    chronic supplementation 347–54, **355**
    levels 257, 313
    mechanism studies 355–61
    resistance training 354
    sweat 357
phytoalexins 179
phytoplankton 179
pigs 357–8
pipecolic acid betaine (homostachydrine) 188, 192
plants *see also* food analysis; seed analysis
    accumulation *see* accumulation (plants)
    biosynthesis **11**, 13–17, **23**, 24, 31–8
    cellular pressure 17, 20
    compatible solutes 10, 179
    dimethylsulfoniopropionate 92
    diurnal rhythms 16–17
    elicitors 179
    exogenous application 18–21
    gene expression 16–17
    genetic engineering 21–3
    high-pressure liquid chromatography (HPLC) 147–8
    mass spectrometry 122

nitrogen fertilization 13
QACs (quaternary ammonium compounds) 181–4
stress 12–14, 178–80
transport 17–18
plasma biofluid, definition 114
plasma levels
average levels 76, 250–2, 372
bariatric surgery *253*
cardiovascular disorders 110
choline 257, 392
cystic fibrosis 252
diabetes mellitus 252, *253*
dimethylglycine (DMG) 252
distribution 205
Down's syndrome *253*, 254
fatty acids 291
fetal malfomation detection 109
folate supplementation 391
gender differences 3, 250, *251*
high-pressure liquid chromatography (HPLC) 161
homocysteine 91–2, 252
hydrophilic interaction liquid chromatography (HILIC) *145*, 148–9, 150
lactate after exercise 345, 352
methyl donors 250
neurodegenerative diseases *253*, 254
obesity *253*
peak concentration 204
pre-eclampsia 252
pregnancy 252, 272, *273*
renal failure 252, *253*
supplementation 254–5
triglycerides 357
vascular disease 252
*Pleurotus ostreatus* 194
Plumbaginaceae family 192
poliomyelitis 249
Polo-like kinase 3 (Plk3) 379
polyamines, catabolism 33
polyols, betaine osmolyte function 65
post methionine load (PML) test 268, 280
post-traumatic stress disorder (PTSD) 224, 239
potassium ions
betaine adduct formation 190
plants 12–13, 20
power, muscles 347, 348–50, **353**, 354, 363
PPARα *see* peroxisome proliferator activated receptor alpha
pre-eclampsia 252, 271
precision
capillary electrophoresis/mass spectrometry (CE/MS) 126–8, 129, 133
definition 135
preferential exclusion 63, 64–5, 66–7, 73, 359
preferential interaction 63, 64–5, 73, 359
pregnancy 109, 252, 268, 270–9, 320
preterm birth 271
ProAlgaZyme 110
proinflammatory molecules 223, 235, 236–7, 238, 240, 317, 335 *see also* inflammation; NF-κB
proline 20, 22, 182
proline betaine (stachydrine)
citrus fruit 112, 113, 184, 186, 192
function 112, 182
mass spectrometry **125**, 126, **127**, *128*, 188
NMR spectroscopy 112
serum levels *106*
structure **107**, 121
urine levels *106*
vegetable oil levels *131*, **132**
proline transporters 17–18
prostate cancer 398, *402*, 405, *407*, *410*, 411
proteasomes 297, 333
protein arginine methyltransferase (PRMT) **290**, 297

protein-isoaspartate methyltransferases (PIMT) **290**, 296–7
protein kinases (Akt/PKA) 79, 317, 333, **334**, 356
protein-protein interactions **290**, 297
protein tyrosine kinase (PTK) 333, **334**, 335, **336**
protein tyrosine phosphatases (PTP) 333, **334**, 335, **336**
proteins
  animal feed supplementation 255
  betaine function in plants 17
  binding 70–1
  ethanol-induced damage **290**, 295–7
  folding 63–4, 209, 359
  function 295
  isoaspartyl residues 296
  muscle cells 212
  osmolyte effects 62
  repair **290**
  synthesis 17, 212, 255, 359–60, 362
  unfolding 62
  urea 62, 209
proton nuclear magnetic resonance (1H-NMR) spectroscopy 69, 107–14, 115, **166**
ProX 70–1
pterins 71
pyridoxal 5'-phosphate-dependent L-serine decarboxylase 16
pyridoxal-5'-phosphate (PLP)-dependent enzymes 180
pyrrolidine dithiocarbamate 333

QACs (quaternary ammonium compounds)
  analysis 184–90, 194
  biosynthesis 181
  food analysis 190, 192–4
  plant function 178, 179–80
  structure 196
quantitative polymerase chain reaction (qPCR), definition 217
quinoa (*Chenopodium quinoa*) 162, 172, 345 *see also* Kancolla

raffinose 24
random mutagenesis, definition 99
reactive oxygen species (ROS)
  *S*-adenosylmethionine (SAM/AdoMet) 331
  aging 335, **336**
  alcohol effects 298
  betaine function 330, 335–6, 337, **338**, 339
  fatty acid overload 317, 318
  glutathione (GSH) 332
  plants 179
  stress 224, 234
  structure 340
real-time RT-PCR 227, 237, 240
receptor, definition 217
recombinant enzymes 93–4
recommended daily intake (RDI), choline 392
red-beet *see* beetroot
redox effector factor-1 (Ref-1) 333, **334**
redox homeostasis 337, 340
reductant, definition 340
rehydration fluids 345, 346–7
relative risk (RR), definition 412
relaxation rates **51**
remethylation
  definition 280
  homocysteine 206–7, 258, 268, 273–4, 277
  one-carbon metabolism 267–8
renal cell cancer 398, *403–4*
renal cells *see* kidneys
renal failure 252
repeatability, capillary electrophoresis/mass spectrometry (CE/MS) 128
reproduction 48, 267
resistance training 354

resistin 214
respiratory chain, mitochondria 298
retinopathy 252
reverse transcription *see* real-time RT-PCR
reversed-phase liquid chromatography (RPLC) 141, 143, 144, 153
rice 17, 19, 37
rice flour 151
roots (plants) 11, 12–13, 20, 23, 24
rye 345

SAH *see* *S*-adenosylhomocysteine
salicylic acid, plants 180
salinization *see* salt stress
salt stress (plants) 11–12, 16–17, 18–20, 22
SAM *see* *S*-adenosylmethionine
SAM:SAH ratio *see* methylation potential
sarcosine 207, 209–10, **246**, 261, 277
Scheibler, Carl 9, 47
sea beet 10
seed analysis 159–60, 162–71, 172
seizures 82, 85
selectivity
 capillary electrophoresis/mass spectrometry (CE/MS) 126, 133
 definition 135
 hydrophilic interaction liquid chromatography (HILIC) 142
 liquid–liquid extraction (LLE) 130
semirandom mutagenesis, definition 99
sensitivity
 capillary electrophoresis/mass spectrometry (CE/MS) 123, 129, 130, 133
 hydrophilic interaction liquid chromatography (HILIC) 144
 liquid chromatography (LC) 130
 mass spectrometry 122
separation methods 124–6, 129, 134, 153 *see also* hydrophilic interaction liquid chromatography
serine
 biosynthesis in plants **22**, 23
 choline biosynthesis 92
 ethanolamine biosynthesis 16
 salt stress (plants) 22
serine decarboxylase 16
serine hydroxymethyltransferase (SHMT) 21
serotonin *see* 5-hydroxytryptamine
serum biofluid
 betaine levels *106*, 109, 113, 257–8
 definition 114
 function 114
 glucose levels 257
 hydrophilic interaction liquid chromatography (HILIC) 148–9, 150
 mass spectrometry 122
 N-methylated compounds levels *106*, 111
 obesity 257–8
 peak concentration 254
shellfish 345
SHMT *see* serine hydroxymethyltransferase
signal-to-noise ratio (S/N) 126, 129, 154
signal transducers and activators of transcription (STAT) 297
signaling pathways inhibition 208, **290**
SIT1 80
site-directed mutagenesis, definition 40, 99
skin 211 *see also* keratinocytes

SMIT1 *see* sodium/myo-inositol cotransporter 1
smoking 257, 393
sodium chloride *see also* osmotic stress; salt stress (plants)
betaine-homocysteine methyltransferase (BHMT) 92
brain osmolytes 83
plants 10, 13–14
sodium ions
betaine adduct formation 190
liver betaine transport 81
plants 12–13
sodium/myo-inositol cotransporter 1 (SMIT1) 79, 83
soil salinity 24 *see also* salt stress (plants)
solubility 49, *50*
solute potential, definition 25
solvent-substitution 68
sorbitol 79
sorghum (*Sorghum bicolor*) 9
sources 3, *31*, 48, 76, 121, 150, 159, **246**, 345, 392, 412
soybean oil 131, **132**
spectral data 54–5
spermatogenesis 84
sphingomyelin 245, 391
spina bifida 270
spinach (*Spinacia oleracea*)
betaine 9, 10, 11, 179, 191
betaine aldehyde dehydrogenase (BADH) 33, 71
choline biosynthesis 38
choline mono-oxygenase (CMO) 36
phosphoethanolamine *N*-methyltransferase (PEAMT) 15–16, 21
spontaneous alternation behaviour 225–6, 229, 231
stachydrine *see* proline betaine
standardization (herbal drugs) 165, 174
stationary phase, definition 154
steatohepatitis, definition 85
steatosis
alcohol consumption 286, 291–2
betaine deficiency effect 4
betaine supplementation 291, 294–5, 317
definition 5, 85
lipid metabolism 318
one-carbon metabolism 80
pathogenesis **290**, **292**, 300
stellate cells 301
sterol regulatory element-binding proteins (SREBPs) 294, **316**
stopped-flow studies 68
strength, muscles **346**, 347, **349**, 352, 363
stress 223, 358, 360 *see also* oxidative stress
stress granules 360
stroke risk, betaine deficiency effect 4
structure
betaine 3, **4**, **30**, 49, **50**, 55, **91**, **107**, **167**, 203, **204**
betaine-homocysteine methyltransferase (BHMT) 94–8
*S*-(δ-carboxybutyl)-L-homocysteine (CBHcy) **91**
carnitine **107**
choline **107**
choline mono-oxygenase (CMO) 14
dihydrofolate reductase (DHFR) **66**
*N*,*N*-dimethylglycine **91**
folate **68**
homocysteine **91**
5-hydroxytryptamine (5-HT) **183**
isoleucine betaine **171**
methionine **91**
phosphorylcholine **107**
proline betaine **107**
ProX 70–1
QACs (quaternary ammonium compounds) 196
reactive oxygen species (ROS) 340

trigonelline **107**, 121, **169**
trimethylamine oxide (TMAO) **107**
trimethylamine (TMA) **107**
*N,N,N*-trimethyltryptamine **183**
tryptamine **183**
tryptophan **183**

*Suaeda maritima* 11
substrate specificity
betaine aldehyde dehydrogenase (BADH) 33, **34**
choline mono-oxygenase (CMO) 37
definition 40

sucrose, plants 12, 13, 24
sugar beet (*Beta vulgaris*)
ancestry 10
betaine accumulation 11–14, 17, 24, 181, 191
betaine distribution 11–12
betaine function 10
betaine isolation 9, 47
betaine transport 17–18
choline biosynthesis 15
choline mono-oxygenase (CMO) 36
drought stress 12
economic value 10
growing conditions 10
medical uses *160*
production volume 10

sugar (refined), analysis *145*, 150–1, 161
sulfur amino acid intake 92
sunflower oil 131
sunflowers 36
superoxide dismutase (SOD) 298
supplementation *see also* animal feed
adipocytes 208, 213, 214, 215, 257, 291, 293–4, 317, 358
alcoholic liver disease 290–9
animal models 4–5, 81, 207–8, 212, 214, 293
animal nutrition 31, 56
bariatric surgery 249, 258
body composition 257, 357–8, 361, 362
cardiovascular disorders 254
creatine 352
effect on homocysteine levels 254
exercise performance 345–55, 361–2
folate 250, 258, 260, 270, 271, 272, *278*, 391
future research requirements 259–60, 299–300, 317, 321
glucose metabolism 318
insulin resistance 317
lipid metabolism 212, 214, 318
liver function 80, 207–8
memory function 230–1
muscle cells 212
nonalcoholic fatty liver disease (NAFLD) 313–22
oxidative stress 214–15
pharmacokinetics 254–5
pretreatment 213, 214–15
testis function 84

sweat 77, 112, 357
Swiss chard 11–12
symmetry 49

tandem mass spectrometry 129, **185**, 186, 187–90, 194, 196
taurine 83
taurine transporter (TAUT) 79, 83
tear-film
function 368, 369, 381
hyperosmolarity 369–71, 372, 378, 382, 383

ternary complex, definition 99
testis
accumulation 76
betaine function 84

tetrahydrofolate (THF) 67, **259**, 261, 280
thermal sensation 345
thin-layer chromatography 164, **165**, 174–5

thiol
betaine
supplementation 333–4
definition 99, 340
redox homeostasis 330, 332, 333, 337
thiol agents 333
thiol-enzyme, definition 6
thiolate
betaine-homocysteine methyltransferase (BHMT) 93
definition 99
thioredoxin (TRX) 332, 333, **334**
thrombosis, definition 99
thyroid hormones 257
tight junctions 211, 217
tissue inhibitor of metalloproteinases 289
TMA *see* trimethylamine
TMAO *see* trimethylamine oxide
tobacco 16, 21, 37–8
toll-like receptors (TLR) 208, 299
tomatoes 17, 18–19
tonicity 4
tonicity-responsive enhancer binding protein 79
tonicity-responsive enhancer sequences 79
toothpaste 211
toxicity 56
TRADD (tumor necrosis factor receptor type 1-associated death domain protein) 377–8
trafficking defects 301
transcription factor tonicity-responsive enhancer binding protein (TonEBP) 373
transcription factors 297, 337 *see also* NF-κB
transgenic plants *see also* genetic engineering, definition 25, 40
transmethylation 4, 6, 16, 207, 208
transport
bacteria 70–1
blood–brain barrier 250
fatty acids 256
intestines 205, 254
kidneys 79, 80, 210
liver 81–2
plants **11**, 17–18
transulfuration pathway **287**, 288, 298–9, **331**, **338**
trehalose 24
triacylglycerol 255, 257
triglycerides
alcohol effects 291, 294
betaine 208, 291, 294, 313, 322, 357
biosynthesis 294, **316**
definition 301
gene expression 318
insulin resistance 313–17
*Trigonella corniculata 160*
trigonelline
food analysis 192–3
gastric adenocarcinoma marker 112–13
mass spectrometry 123, **124**, **125**, 126, **127**, *128*, *188*
seed analysis 168–9
serum levels *106*
structure **107**, 121, **169**
urine levels *106*, 112–13
vegetable oil levels *131*, **132**
trimethylamine oxide (TMAO) 62, 63–4, *106*, **107**, 112
trimethylamine (TMA) *106*, **107**, 112, 143
trimethylaminuria 112
trimethyllysine *188*, 190, 193
*N,N,N*-trimethyltryptamine 180, **183**, *188*, **189**
*N,N,N*-trimethyltyramine *188*, **189**
*Triticum aestivum see* wheat
tryptamine 180, **183**, 184
tryptophan 70, 180, **183**
tubercidin 295
tubulin 296
tumor necrosis factor alpha (TNF-α)
alcohol effects 295, 299
betaine function 110, 209, 223, 235, 337, **338**
catechin 236

definition 114
dry-eye diseases 370, 379
function 209, 377
NF-κB 333, 335, **336**
obesity 213
turgor 20, 30, 70
tyramine 180
tyrosine 180
tyrosine decarboxylase (TYDC) 180

ultraperformance liquid chromatography-mass spectrometry (UPLC-MS) 149–50
urea
accumulation 62
betaine supplementation 358, 360
kidneys 77, 209–10
myosin 360
protein function 62, 63–4, 65, 209–10
urine
betaine isolation 161
betaine levels *106*, 109, 110–11, 113
betaine recovery 210
composition 114
excretion 77
high-pressure liquid chromatography (HPLC) 161
hydrophilic interaction liquid chromatography (HILIC) *145*, 149
N-methylated compounds levels *106*, 112–13
UV absorption detection 122, 123, 129–30
UV radiation exposure 211

vacuole, definition 25
validation, definition 135
vapor pressure osmometry (VPO) 64–5, 69, 73, 359
vascular disease 48, 252, 336
vegetable oils analysis 129–33
very low density lipoproteins (VLDL) **290**, 291–3, 294, 312, 318, 357
visceral fat 213–14, 321
vitamin $B_6$ 391
vitamin $B_{12}$ (cobalamin)
bariatric surgery 248, 258, 259, 260
deficiency 269
fasting plasma total homocysteine (tHcy) 271
fasting plasma total (tHcy) 391
methionine synthase pathway 268, 287
pregnancy 277, *278*
VPO *see* vapor pressure osmometry

waist circumference 257, 258, 313
water, betaine hydration 52–3, **54**, 204, 209
water balance *see* osmoregulation
water-immersion restraint stress (WIRS)
betaine effects 236, 237
data analysis 228
definition 239
memory function 223–4
methodology 224–8
proinflammatory molecules 236–7, 238
results 229–35
water potential, definition 25
water retention 4 *see also* osmolality
water stress (plants) *see* drought stress
wheat (*Triticum aestivum*) 9, 20, 151, 160, 179, 345
wine, analysis *145*, 150, 151, 161

X-ray crystallography
betaine 55
definition 40, 58

Y-maze test 225–6, 229, 231

zinc 93, 94, 95, 97, 98, 99
zwitterion 6, 49